The Sun, the Solar Wind, and the Heliosphere

IAGA Special Sopron Book Series

Volume 4

Series Editor

Bengt Hultqvist
The Swedish Institute of Space Physics, Kiruna, Sweden

The International Association of Geomagnetism and Aeronomy is one of the eight Associations of the International Union of Geodesy and Geophysics (IUGG).

IAGA's Mission
The overall purpose of IAGA is set out in the first statute of the Association:

- to promote studies of magnetism and aeronomy of the Earth and other bodies of the solar system, and of the interplanetary medium and its interaction with these bodies, where such studies have international interest;
- to encourage research in these subjects by individual countries, institutions or persons and to facilitate its international coordination;
- to provide an opportunity on an international basis for discussion and publication of the results of the researches; and
- to promote appropriate standardizations of observational programs, data acquisition systems, data analysis and publication.

Volumes in this series:

The Earth's Magnetic Interior
Edited by E. Petrovský, D. Ivers, T. Harinarayana and E. Herrero-Bervera

Aeronomy of the Earth's Atmosphere and Ionosphere
Edited by M.A. Abdu, D. Pancheva and A. Bhattacharyya

The Dynamic Magnetosphere
Edited by W. Liu and M. Fujimoto

The Sun, the Solar Wind, and the Heliosphere
Edited by M.P. Miralles and J. Sánchez Almeida

Geomagnetic Observations and Models
Edited by M. Mandea and M. Korte

For titles published in this series, go to
http://www.springer.com/series/8636

The Sun, the Solar Wind, and the Heliosphere

Editors

Mari Paz Miralles
Harvard-Smithsonian Center for Astrophysics,
Cambridge, Massachusetts, USA

Jorge Sánchez Almeida
Instituto de Astrofísica de Canarias, La Laguna (Tenerife),
Spain

Springer

planetearth
Earth Sciences for Society

IAGA

Editors
Dr. Mari Paz Miralles
Harvard-Smithsonian Center
for Astrophysics
60 Garden Street, Cambridge
Massachusetts 02138
USA
mmiralles@cfa.harvard.edu

Dr. Jorge Sánchez Almeida
Instituto de Astrofísica de Canarias
E-38205 La Laguna, Tenerife
Spain
jos@iac.es

ISBN 978-90-481-9786-6 e-ISBN 978-90-481-9787-3
DOI 10.1007/978-90-481-9787-3
Springer Dordrecht Heidelberg London New York

Cover illustration: Collage of representative images from this book. Image courtesy of Mari Paz Miralles and Jorge Sánchez Almeida.

Printed on acid-free paper

Springer is part of Springer Science+Business Media (www.springer.com)

Foreword by the Series Editor

The IAGA Executive Committee decided in 2008, at the invitation of Springer, to publish a series of books, which should present the status of the IAGA sciences at the time of the IAGA 2009 Scientific Assembly in Sopron, Hungary, the "IAGA Special Sopron Series". It consists of five books, one for each of the IAGA Divisions, which together cover the IAGA sciences:

Division I – Internal Magnetic Field
Division II – Aeronomic Phenomena
Division III – Magnetospheric Phenomena
Division IV – Solar Wind and Interplanetary Field
Division V – Geomagnetic Observatories, Surveys and Analyses.

The teams of Editors of the books contain members of the IAGA Executive Committee and of the leadership of the respective Division, with, for some of the books, one or a few additional leading scientists in the respective fields.

The IAGA Special Sopron Series of books are the first ever (or at least in many decades) with the ambition to present a full coverage of the present status of all the IAGA fields of the geophysical sciences. In order to achieve this goal each book contains "overview papers", which together summarize the knowledge of all parts of the respective field. In book no. 5, on geomagnetic observations, all papers are of that kind. These major review papers are, in the other four books, complemented with invited reviews of special questions presented in Sopron. Finally, in some of the books a few short "contributed" papers of special interest are included. Thus, we hope the books will be of interest to both those who want a relatively concise presentation of the status of the sciences and to those who seek the most recent achievements.

I want to express my thanks to the editors and authors who have prepared the content of the books and to Petra van Steenbergen at Springer for good cooperation.

Kiruna, Sweden
November 2010

Bengt Hultqvist

Preface

The solar environment is one of the most studied plasma systems, yet it also presents numerous challenges due to its complexity. For example, the wind streams coming from the Sun, the *solar wind*, consist of charged particles and associated magnetic fields. Despite decades of intensive research, we still do not fully understand certain basic questions such as how the solar wind is heated and accelerated. The solar wind fills the heliosphere and interacts with the magnetospheres of the Earth and the other planets, as well as with the outer boundaries of the heliosphere and the local interstellar medium, creating a variety of phenomena that still need explanation.

This book is one of five volumes in the International Association of Geomagnetism and Aeronomy (IAGA) Special Book Series. It is dedicated to the science covered by IAGA Division IV, *Solar Wind and Interplanetary Field*, and features topics from the interior of the Sun to the outermost regions of the heliosphere.

The concept of the book series was developed simultaneously with the preparation of the 11th Scientific Assembly of IAGA that took place on August 23–30, 2009 in Sopron, Hungary. This provided an added benefit, namely, the opportunity to seek leading experts that could both give review presentations in Sopron and also provide articles for the book. We also selected a few summary articles written by the conveners of Division IV sessions that describe the most relevant results presented in Sopron in context with recent advancements in the field.

All the chapters of this book were invited, and represent a selection of overview papers. The book contains some long chapters which summarize the current knowledge of the field. These major review papers are complemented with invited reviews of special topics presented in Sopron. A few short chapters of special interest were also included in this volume.

Our aim was to provide a comprehensive view of the state of knowledge of the physics of the phenomena of the Sun, the solar wind, and the heliosphere. This book focuses on the new insights that have been obtained through space- and ground-based observations as well as modeling of the relationship between the Sun and its environment. The new insights cover the detailed structure of the Sun, solar corona, magnetic field interactions, coronal heating, flares, accelerated particles, coronal mass ejections, shocks, and several other topics.

This volume is a representation of how our advances in observations of space plasmas, space weather, and the Sun-Earth connection have been key in improving the modeling of the Sun, the solar wind, the heliosphere, and our understanding of its various physical processes. The current and future solar missions featured in the book will continue to provide new discoveries in all key plasma regions of the heliosphere and challenge our existing ideas to help us develop a more realistic picture of the relevant physics.

We would like to express our thanks to the authors who have prepared the chapters for this book, and also to all scientists involved in IAGA Division IV. Without their drive for solar and heliophysics research this book would not have been possible.

Finally, the editors would like to thank the referees of the chapters included in the book for their dedication to the details and their gracious acceptance of short deadlines: Markus J. Aschwanden, Gemma D. R. Attril, Eugene H. Avrett, Adriaan A. Van Ballegooijen, Ofer Cohen, Anthea J. Coster, Steven R. Cranmer, Horst Fichtner, Katya Georgieva, Stephen W. Kahler, Justin C. Kasper, Kelly E. Korreck, Alexander Lazarian, Bo Li, Nicholas A. Murphy, Divya Oberoi, John C. Raymond, Manfred Schuessler, Leonard Strachan, Manuel Vázquez, and Gary P. Zank.

Cambridge, Massachusetts
La Laguna, Tenerife
November 2010

Mari Paz Miralles
Jorge Sánchez Almeida

Contents

Contributors

Martin Asplund Max-Planck-Institut für Astrophysik, Garching, Germany, asplund@mpagarching.mpg.de

Tom Bogdan NOAA Space Weather Prediction Center, Boulder, CO, 80305 USA, tom.bogdan@noaa.gov

Iver H. Cairns School of Physics, University of Sydney, Sydney, NSW 2006, Australia, i.cairns@physics.usyd.edu.au

Sébastien Couvidat W.W. Hansen Experimental Physics Laboratory, Stanford University, Stanford, CA 94305, USA, couvidat@stanford.edu

Joseph M. Davila Goddard Space Flight Center, Greenbelt, MD 20771, USA, Joseph.M.Davila@nasa.gov

Steven Dewitte Institut Royal de Météorologie de Belgique, B-1180 Brussels, Belgique, Steven.Dewitte@meteo.be

Ruth Esser Institute of Science and Technology, University of Tromsø, NO–9037 Tromsø, Norway, Ruth.Esser@uit.no

Horst Fichtner Institut für Theoretische Physik IV, Ruhr-Universität Bochum, 44780 Bochum, Germany, hf@tp4.rub.de

Antoinette B. Galvin University of New Hampshire, Durham, NH 03824, USA, toni.galvin@unh.edu

Nat Gopalswamy NASA Goddard Space Flight Center, Greenbelt, MD 20771, USA, nat.gopalswamy@nasa.gov

Nicolas Grevesse Centre Spatial de Liège and Institut d'Astrophysique et de Géophysique, Université de Liège, Liège, Belgium, nicolas.grevesse@ulg.ac.be

Mike Hapgood Rutherford Appleton Laboratory, Didcot, UK, hapgood@stst.ac.uk

Andrew L. Haynes School of Mathematics & Statistics, University of St Andrews, St Andrews, Fife, KY16 9SS, Scotland, UK, andrew@mcs.st-and.ac.uk

Shinsuke Imada Institute of Space and Astronautical Science, Japan Aerospace Exploration Agency, 3-1-1 Yoshinodai, Sagamihara-shi, Kanagawa 229-8510, Japan, imada.shinsuke@jaxa.jp

Hiroaki Isobe Kyoto University, Kyoto, Japan, isobe@kwasan.kyoto-u.ac.jp

H. Gordon James Communications Research Centre Canada, Ottawa, Ontario, Canada K2H 8S2, gordon.james@crc.ca

Paul J. Kellogg University of Minnesota, Minneapolis, MN, USA, pauljkellogg@gmail.com

Karel Kudela IEP SAS, Watsonova 47, 040 01 Kosice, Slovakia, kkudela@upjs.sk

Leonid L. Lazutin Skobeltsyn Institute of Nuclear Physics, Moscow University, Moscow, Russia, lazutin@dec1.npi.msu.ru

Bo Li University of Sydney, Sydney, NSW 2006, Australia, boli@physics.usyd.edu.au

Øystein Lie-Svendsen Norwegian Defence Research Establishment, NO–2027 Kjeller, Norway, Oystein.Lie-Svendsen@ffi.no

Jon A. Linker Predictive Science, Inc., San Diego, CA 92121-2910, USA, linkerj@predsci.com

Roberto Lionello Predictive Science, Inc., San Diego, CA 92121-2910, USA, lionel@predsci.com

W. William Liu Canadian Space Agency, 6767 route de l'Aéroport, Saint-Hubert, Québec, Canada, william.liu@space.gc.ca

Quanming M. Lu CAS Key Laboratory of Basic Plasma Physics, School of Earth and Space Sciences, University of Science and Technology of China, Hefei, Anhui, 230026, China, qmlu@ustc.edu.cn

Rhona C. Maclean School of Mathematics & Statistics, University of St Andrews, St Andrews, Fife, KY16 9SS, Scotland, UK, rhonam@mcs.st-and.ac.uk

Dalmiro J. F. Maia Porto University, Porto, Portugal, dmaia@fc.up.pt

Milan Maksimovic CNRS and LESIA, Observatoire de Paris, 92195 Meudon, France, milan.maksimovic@obspm.fr

David M. Malaspina University of Colorado, Boulder, CO, USA, David.Malaspina@colorado.edu

Richard G. Marsden ESA-ESTEC, 2200AG Noordwijk, The Netherlands, Richard.Marsden@esa.int

Mauro Messerotti INAF-Astronomical Observatory of Trieste, Loc. Basovizza n. 302, 34012 Trieste, Italy; Department of Physics, University of Trieste, Via A. Valerio n. 2, 34127 Trieste, Italy; INFN-Trieste Division, Via A. Valerio n. 2, 34127 Trieste, Italy, messerotti@oats.inaf.it

Zoran Mikić Predictive Science, Inc., San Diego, CA 92121-2910, USA, mikicz@predsci.com

Mari Paz Miralles Harvard-Smithsonian Center for Astrophysics, 60 Garden Street, Cambridge, MA 02138, USA, mmiralles@cfa.harvard.edu

Daniel Müller ESA-GSFC, Greenbelt, MD, USA, dmueller@esa.nascom.nasa.gov

Clare E. Parnell School of Mathematics and Statistics, University of St Andrews, St Andrews, Fife, KY16 9SS, Scotland, UK, clare@mcs.st-and.ac.uk

Spiros Patsourakos Department of Physics, Section of Geophysics, University of Ioannina, GR 451 10 Ioannina, Greece, spatsour@cc.uoi.gr

Giannina Poletto INAF - Arcetri Astrophysical Observatory, Largo Fermi, 5, 50125 Firenze, Italy, poletto@arcetri.astro.it

Matthias Rempel HAO, National Center for Atmospheric Research, Boulder, CO, USA, rempel@hao.ucar.edu

John D. Richardson Kavli Center for Astrophysics and Space Science, Massachusetts Institute of Technology, Cambridge, MA, USA, jdr@space.mit.edu

Pete Riley Predictive Science, Inc., San Diego, CA 92121-2910, USA, pete@predsci.com

A. Jacques Sauval Observatoire Royal de Belgique, Brussels, Belgium, jacques.sauval@oma.be

Jorge Sánchez Almeida Instituto de Astrofísica de Canarias, E38205 La Laguna, Tenerife, Spain, jos@iac.es

Werner Schmutz Physikalisch-Meteorologisches Observatorium Davos, World Radiation Center, CH-7260 Davos Dorf, Switzerland, werner.schmutz@pmodwrc.ch

Pat Scott Department of Physics and Oskar Klein Center for Cosmoparticle Physics, University of Stockholm, Stockholm, Sweden, pat@fysik.su.se

Dastgeer Shaikh Center for Space Plasma and Aeronomic Research and Physics Department, University of Alabama, Huntsville, AL 35899, USA, dastgeer.shaikh@uah.edu

Toshifumi Shimizu Institute of Space and Astronautical Science, Japan Aerospace Exploration Agency, 3-1-1 Yoshinodai, Sagamihara-shi, Kanagawa 229-8510, Japan, shimizu.toshifumi@isas.jaxa.jp

Hendrik C. Spruit Max Planck Institute for Astrophysics, 85741 Garching, Germany, henk@mpa-garching.mpg.de

Toshio Terasawa Institute for Cosmic Ray Research, University of Tokyo, 5-1-5 Kashiwa-no-Ha, Kashiwa City Chiba, 277-8582, Japan, terasawa@icrr.u-tokyo.ac.jp

Barbara J. Thompson Goddard Space Flight Center, Greenbelt, MD 20771, USA, Barbara.J.Thompson@nasa.gov

Gérard Thuillier LATMOS-CNRS, Bp3, 91371 Verrières-le Buisson, France, gerard.thuillier@latmos.ipsl.fr

Viacheslav S. Titov Predictive Science, Inc., San Diego, CA 92121-2910, USA, titovv@predsci.com

Eino Valtonen Department of Physics and Astronomy, Space Research Laboratory, University of Turku, Turku, NSW FI-20014, Finland, eino.valtonen@utu.fi

N. Vilmer LESIA-Paris Observatory, 92195 Meudon Cedex, France, nicole.vilmer@obspm.fr

Igor S. Veselovsky Kobeltsyn Institute of Nuclear Physics, Moscow State University, Moscow, 119992, Russia, veselov@dec1.sinp.msu.ru

Andrew W. Yau Department of Physics and Astronomy, University of Calgary, Calgary, AB, Canada T2N1N4, yau@phys.ucalgary.ca

Gary P. Zank Center for Space Plasma and Aeronomic Research and Physics Department, University of Alabama, Huntsville, AL 35899, USA, garyp.zank@gmail.com

Part I
Introduction

Chapter 1

The Sun, the Solar Wind, and the Heliosphere

Mari Paz Miralles and Jorge Sánchez Almeida

Abstract We describe the aims and contents of the book entitled "The Sun, the Solar Wind, and the Heliosphere". This is a volume in the IAGA Special Book Series dedicated to the science covered by IAGA Division IV, *Solar Wind and Interplanetary Field*. The book features review articles on topics from the interior of the Sun to the outermost regions of the heliosphere. In addition, we highlight some of the results presented during the Division IV symposia at the 11th Scientific Assembly of IAGA in Sopron, Hungary, which was planned simultaneously with this book.

1.1 Purpose and Contents of the Book

As editors, our commitment with Springer was setting up a volume representing the state of the art of the science covered by the International Association of Geomagnetism and Aeronomy (IAGA) Division IV, *Solar Wind and Interplanetary Field*. This volume surpasses our original expectations. It contains a collection of contributions by top experts addressing and reviewing a variety of topics included under the umbrella of the division. It covers subjects that extend from the interior of the Sun to the heliopause, and from the study of physical processes in the Sun and the solar wind plasma to space weather forecasts. Obviously, not all topics are treated with the same depth, and the reader will probably notice, for example, that chapters on solar surface magnetism and neutral-plasma interactions for planets are missing. However, the list is fairly complete. We asked the authors to spell out arguments in depth, imposing only limited constraints on the length of the manuscripts. As a reward, many contributions are comprehensive and detailed and, therefore, expected to be references in the field for years to come.

The volume has been divided in 7 parts, including this introduction, loosely following the organization of the Division IV scientific sessions during the 11th Scientific Assembly of IAGA held in Sopron, Hungary. Part II on the *Solar Interior* describes our current views of the interior of the Sun through numerical modeling (Rempel) and observations (Couvidat). The physical processes that maintain the large scale flows in the Sun are critical for understanding the magnetic cycle that, in the end, produces and modulates the heliosphere. One of the contributions is specifically devoted to critically revising some of the established views of the solar magnetic cycle (Spruit). The revised solar metallicity stands out as one of the *few* problems of the standard solar model. The contribution of Grevesse et al. explain the bases for the revision, and the problems it poses.

Part III on the *Solar Atmosphere* collects works dealing with the part of the heliosphere closest to the solar surface, where coronal mass ejections (CMEs) are triggered and where the solar wind originates. Imada et al. summarize the physical conditions for magnetic reconnection to occur, and they discuss Hinode observations of magnetic reconnection in the chromosphere. CMEs are at the basis of geomagnetic storms, therefore, understanding their onset and early evolution is crucial for space weather forecasting. The topic is reviewed by Patsourakos in the context of the new *STEREO* mission. The origin of the solar wind, from historic developments to the current

M.P. Miralles (✉)
Harvard-Smithsonian Center for Astrophysics,
60 Garden Street, Cambridge, MA 02138, USA
e-mail: mmiralles@cfa.harvard.edu

M.P. Miralles, J. Sánchez Almeida (eds.), *The Sun, the Solar Wind, and the Heliosphere*, IAGA Special Sopron Book Series 4, DOI 10.1007/978-90-481-9787-3_1, © Springer Science+Business Media B.V. 2011

understanding, is treated in a balanced contribution by Esser and Lie-Svendsen. Developments toward a predictive three-dimensional magnetohydrodynamic model of the solar corona and the solar wind using the observed photospheric magnetic fields as boundary conditions are presented by Lionello et al.

Part IV on the *Heliosphere* treats the realm of the outermost regions of the solar system. Galvin summarizes the solar wind observations taken with *STEREO* at 1 AU from 2007 to 2009. This review provides a new perspective on the origins and propagation of in situ signatures of the solar-wind stream interfaces and the interplanetary coronal mass ejections (ICMEs) during the recent solar cycle minimum. Terasawa reviews the nonthermal particle acceleration of collisionless shocks in the heliosphere such as those formed ahead of CMEs and the Earth's bow shock, around corotating interaction regions, and at the solar wind termination shock. New knowledge about the heliospheric outer boundaries, regions between the solar wind and the local interstellar cloud, is presented by Richardson. The data from the journey of *Voyagers* 1 and 2 across the solar wind termination foreshock, the termination shock, and the heliosheath is reviewed.

Part V on *Heliophysical Processes* updates our understanding on the basic physical processes that play key roles in shaping the heliosphere. Recent developments in the understanding of the magnetic topology involved in 3D reconnection are reviewed by Parnell et al. Current sheets and their observational aspects are analyzed in depth in one of the contributions (Poletto). Another article treats solar energetic particles (SEPs), which may be produced by flares or accelerated up to relativistic energies by shock waves driven by CMEs (Valtonen). The heliosphere and planetary magnetospheres provide a test bed to explore many fundamental questions in plasma physics, in particular the role played by the coupling of different spatial and temporal scales in determining the structure and dynamics of space plasmas. Shaikh et al. discuss several examples of this interplay between scales in a variety of space plasma environments. The state of the magnetosphere and the information that can be derived from the study of cosmic rays is discussed by Kudela and Lazutin. Messerotti presents various aspects of the heliospheric radio emission processes employing an ontological approach.

Part VI on *Radio Emissions* from the Sun and the interplanetary medium are used to better understand the plasma conditions of energy release sites and particle acceleration regions in the solar corona, the solar wind, and the interplanetary medium. Type II and III solar radio bursts, and emissions from the Earth's foreshock are reviewed by Li et al. Kellogg and Malaspina report on recent progress in understanding the processes of linear and nonlinear conversion of Langmuir waves to electromagnetic radiation. A review of radio images of solar flares, CMEs, and electron beams in the corona and in the interplanetary medium is presented by Vilmer. Finally, the overview by Cairns is a comprehensive review of the field of solar system shocks that summarizes in detail the observation and theory of type II coronal and interplanetary bursts and the 2–3 kHz emissions from the outer heliosphere.

ILWS (International Living With a Star) and CAWSES (Climate and Weather of the Space-Earth System) are two major international initiatives that coordinate observations and modeling to improve our understanding of the space environment and its impacts on life and society. Part VII on *Coordinated Science in the Sun-Earth System* contains a description of activities carried out within the two initiatives (Fichtner and Liu). Various space missions being developed in and around the initiative are described in detail: the *Solar Orbiter* mission (Marsden and Müller), *e-POP* mission (Yau and James), the *Picard* investigation (Thuillier et al.). In addition, the new international Space Weather Initiative (ISWI) program is designed to promote the observation, understanding, and prediction of near-Earth space weather phenomena. Davila et al. summarize this program and the collaboration with the other two initiatives named above.

1.2 Solar Wind and Interplanetary Field Topics Addressed in Sopron

The 11th Scientific Assembly of IAGA was held in Sopron, Hungary, on 23–30 August 2009, and was simultaneously planned with the IAGA series of books that includes the present volume. Although this book is by no means the proceedings of the assembly corresponding to our sessions, we have decided to briefly attest what happened there and highlight some key scientific results presented at the meeting. Our IAGA Division IV had six splinter meetings and a Reporter reviews session. Leading experts giving invited reviews were asked to contribute to this volume, therefore, their works are not generally mentioned here, and we refer to their contributions for details. In addition, several of the conveners of the symposia have prepared their own summaries and/or introductions, and we directly refer to then to report on the science of their corresponding sessions.

The different symposia and their contents were as follows.

Symposium IV.01, New Solar and Interplanetary Results. This symposium covered new results using data from space and ground-based observatories, as well as theory and modeling of different aspects of the Sun and the heliosphere.

One of the topics discussed at the Sopron meeting was the impact of the differences and similarities between the minimum of solar activity that persisted from 2007 to 2009 and the prior one (1996–1997). In particular, the role of the solar interior in determining solar cycle characteristics was highlighted. The search for internal signatures of the sunspot cycle is of prime importance in the quest to understand the origin of solar magnetism. Gizon (2009) reviewed current advances toward retrieving a detailed 3D view of the solar interior through local helioseismology. Mursula (2009) commented on the changes of several solar and heliospheric parameters (cosmic rays, solar irradiance, heliospheric and solar polar magnetic fields) during the extended ending of solar cycle 23 and the ongoing solar minimum. The presenter suggested that solar cycle 23 may not be so exceptional after all, when looking at long-term solar activity records. Yi and Oh (2009) discussed the implications of the weaker solar magnetic fields and the decrease of the solar dipole tilt angles at the ending of solar cycle 23 and the delayed onset of solar cycle 24, and its implications for global climate. Finally, Muraközy and Ludmány (2009) presented several features of the solar north-south hemispheric asymmetries using sunspot data from solar cycles 12 to 23. These authors concluded that the alternating hemispheric precedence could be considered as a long-term torsional oscillation of the solar cycle. Despite significant advances, our understanding of the solar magnetism is still incomplete. However, the abundance of new magnetic structures makes them potentially important to understand the global magnetic properties of the Sun. Sánchez Almeida et al. showed the presence of many small vortex flows throughout the photosphere, which may stir and heat the corona (Bonet et al. 2008).

In the solar atmosphere, explosive phenomena such as solar flares and CMEs producing coronal disturbances and shock waves were discussed. Vršnak reported on observational signatures of large-scale coronal MHD shocks and the relationship between shocks, CMEs, and flares (e.g., Vršnak and Cliver 2008). In addition, Cliver (2009) argued for Moreton waves driving CMEs looking at an event on 6 December 2006, using He I λ 10830, Hα, and *TRACE* imaging data. Nitta et al. (2009) employed new *STEREO* observations to revisit the CME-flare relationship during solar cycle 23. Their goal was to understand the effect of the flare process on the initiation and subsequent dynamics of CMEs. Santos and Büchner (2009) presented a 3D numerical simulation of a flaring active region associated with a CME. These authors studied how the magnetic energy is stored and released to produce these powerful events in the solar atmosphere. Asikainen et al. (2009) analyzed hemispheric and longitudinal asymmetries in CME occurrence using *SOHO*/LASCO data in solar cycle 23 in the context of known asymmetries in solar activity.

Erupting solar prominences or filaments have been associated with CMEs. These filaments consist of cool and dense plasmas near equilibrium in the corona. Foullon et al. (2009) discussed the use of ultra-long-period oscillations in EUV filaments to forecast prominence eruptions.

Coronal plasma temperatures and abundances were also discussed. Shibasaki (2009) proposed a mechanism of temperature dependent plasma flows to explain steady plasma upflows observed in soft X-rays in open magnetic field regions. Ofman (2009) presented a model of coronal quiescent streamers to explain why their structures differ significantly depending on whether they are observed in heavy ions or hydrogen lines.

Recent *Hinode* and *STEREO* observations have reported a variety of phenomena in the solar atmosphere over a range of coronal plasma temperatures. While it is widely believed that jets are produced by magnetic reconnection, the mechanisms responsible for jet initiation and propagation are not still completely understood. Miralles (2009) showed the first discovered ultraviolet counterparts in the extended corona of the hot X-ray jets resolved by *Hinode*/XRT. This provides evidence that the jet material travels throughout the corona and reaches *SOHO*/UVCS heights, at least 1.7 $R_\odot$, in the acceleration region of the fast solar wind.

During a CME, the eruption of a magnetic structure is associated with the formation of a large-scale current sheet (e.g., Vršnak et al. 2009). Ko et al. (2009) presented a first attempt to model the UV line emission in a post-CME current sheet, and discussed its observational consequences.

Interplanetary CMEs (ICMEs) are the most important cause of magnetic storms at the Earth's magnetosphere. Kilpua presented new results of ICMEs from the *STEREO* observatory, i.e., observing the ICMEs from two different perspectives. When an ICME exhibits a large magnetic field, rotation in the magnetic field direction, and a low proton plasma β, it is classified as a magnetic cloud (MC) and is interpreted as the signature of a magnetic flux rope (Kilpua et al. 2009). Lynnyk and Vandas (2009) modeled a full set of magnetic clouds using force-free cylindrical flux ropes, and concluded that the expansion model provides better fits. Dal Lago et al. (2009) used the *STEREO*/SECCHI and *SOHO*/LASCO instruments to study the evolution of a CME from 1.4 $R_\odot$ to beyond 32 $R_\odot$, and estimated its travel time to Venus.

The solar wind is an important source of geomagnetic disturbances, especially in the declining phase of the solar cycle as seen in solar cycle 23. Coronal holes are sources of fast solar wind, predominantly associated with solar polar regions. But coronal holes also appear at low-latitudes forming high-speed streams (see, e.g., Miralles et al. 2001, 2004, 2006) that can interact with slow streams and create disturbances nearth Earth. Luo et al. (2009) presented a new index to forecast the solar wind speed from the brightness of 284 Å EUV emission images from *SOHO*/EIT.

During the last few decades, in situ observations have provided information about the particles and wave processes in the solar wind (e.g. Marsch et al. 1982a, b; Goldstein et al. 1995; Marsch 2006). As the solar wind expands from the Sun toward the orbit of the Earth, electrons and protons cool less rapidly than adiabatically expected and supplementary heating mechanisms are needed to account from these departures. Kiran and Shah (2009) presented a new calculation of the power dissipated through Alfvén waves in the solar wind, which is in better agreement with observations than previous estimates, especially for the slow solar wind streams (Kiran et al. 2006).

Depressions in the interplanetary magnetic field (IMF) were presented by Erdös and Tátrallyay (2009). These authors compared magnetic field measurements from the four *CLUSTER* spacecraft separated between 500 km and 10,000 km, to conclude that magnetic holes are likely to be frozen in the solar wind and may be formed in sheet-like structures. Kuznetsova (2009) analyzed data from the interplanetary magnetic field, the solar wind speed, and sunspot number to show that the Sun and the Moon may both influence the temporal changes of the solar wind parameters.

The Earth's bow shock is the region where the solar wind is decelerated from supersonic to subsonic speeds before being deflected around the Earth. The length scales of various processes occurring at the bow shock have been studied by the *CLUSTER* mission. Kis et al. (2009) analyzed the properties of the energetic ions in front of the Earth's bow shock using *CLUSTER*/CIS particle data for different solar wind conditions, whereas Kovács et al. (2009) monitored the plasma turbulence upstream and downstream of the Earth's bow shock using *CLUSTER* magnetic data. These authors examined the intermittency level of the plasma fluctuations and the varying solar wind parameters.

McComas et al. (2009) provided an overview of the Interstellar Boundary Explorer (*IBEX*) mission and some of the early results from its first all-sky maps. *IBEX* is meant to study the interaction between the heliosphere and the interstellar medium around us by imaging energetic neutral atoms emanating from this region. In particular, these authors found a bright ribbon of emission that was not predicted by any prior models or theories. This feature appears to be ordered by the interaction of the heliosphere with the local interstellar magnetic field.

Symposium IV.02, From Micro- to Macro-scales in the Heliosphere and Magnetospheres. Coupling complexity, class of problem or system that consist of significantly different scales, regions, or particle populations, is well represented by physical processes occurring throughout the heliosphere and magnetosphere. The contribution by Shaikh, Veselovsky, Lu, and Zank provided an introduction to the topic, which includes highlights of some of the contributions made during meeting. We refer to their work for details.

Symposium IV.03, Universal Heliophysical Processes. The International Heliophysical Year program has been promoting investigations on the universality of physical processes such as shocks, particle acceleration, dynamos, magnetic reconnection, magnetic flux ropes, plasma-neutral matter interactions, turbulence, and several others. The paper by Gopalswamy provides a balanced and accurate account of what happened during this symposium. We refer the interested reader to his contribution that follows this introduction.

Symposium IV.04, Advances in Coordinated Sun-Earth System Science Through CAWSES and ILWS. The Climate and Weather of the Sun-Earth System and the International Living with a Star programs were created in order to conduct interdisciplinary research in a plasma system as large and complex as the heliosphere. We refer to the contribution by Fichtner and Liu in this volume for details. Rather than a traditional summary, it contains a discussion on the need for interdisciplinary research and coordinated international cooperation to approach the Sun-Earth system science.

Symposium IV.05, Neutral-Plasma Interactions for Planets, Moons, Asteroids, and Comets. This symposium focused on recent progress in the space plasma physics of solar system bodies without measurable dynamo magnetic fields. The interactions of Mars, Venus, and Titan with the solar wind were featured. Results from *Mars Express*, *Venus Express*, and from *Cassini* were given.

Symposium IV.06, Solar and Interplanetary Radio Emissions. This symposium discussed recent developments concerning non-thermal radio emissions from the Sun and the heliosphere, with links to solar energetic particles events and/or CMEs. As it happened with session IV.04, the contribution by Li et al. can be regarded as a balanced introduction to solar and interplanetary radio emissions, therefore, we have preferred to place it as the first article in the corresponding section of this volume.

Symposium IV.07, Reporter Reviews. This symposium contained reviews of recent advances, both theoretical and observational, on the Sun, solar wind, and heliosphere. Given by active researchers, these reviews covered a comprehensive range of topics in a manner that is accessible to researchers from other IAGA divisions, while offering synthesis and context to Division IV scientists. All these reviews were by invitation only. Because the organization of the meeting in Sopron and the planning of this book were simultaneous, the reporters in this session were chosen with the idea of their contributions to this volume. Grevesse, Spruit, Esser, and Richardson were the invited reporters in this session. All delivered extensive overview papers to this volume, therefore, we refer their contributions for details.

References

Asikainen T, Mursula K, Robbrecht E, Berghmans D (2009) 11th scientific assembly, "Our planet moving in space", Sopron, Hungary, 23–30 Aug 2009, p 35, abstract 401-MON-01430-1088

Bonet JA, Márquez I, Sánchez Almeida J, Cabello I, Domingo V (2008) Astrophys J Lett 687:L131–L134

Cliver E (2009) 11th scientific assembly, "Our planet moving in space", Sopron, Hungary, 23–30 Aug 2009, p 33, abstract 401-MON-01145-0547

Dal Lago A, Inhester B, Antunes Vieira LE (2009) 11th scientific assembly, "Our planet moving in space", Sopron, Hungary, 23–30 Aug 2009, p 50, abstract 401-TUE-01130-1203

Erdös G, Tátrallyay M (2009) 11th scientific assembly, "Our planet moving in space", Sopron, Hungary, 23–30 Aug 2009, p 53, abstract 401-TUE-01430-1238

Foullon C, Verwichte E, Nakariakov VM (2009) 11th scientific assembly, "Our planet moving in space", Sopron, Hungary, 23–30 Aug 2009, p 36, abstract 401-MON-01445-1210

Gizon L (2009) 11th scientific assembly, "Our planet moving in space", Sopron, Hungary, 23–30 Aug, 2009, p 29, abstract 401-MON-00830-1224

Goldstein ML, Roberts DA, Matthaeus WH (1995) Ann Rev Astron Astrophys 33:283–326

Kilpua EKJ, Liewer PC, Farrugia C, Luhmann JG, Möstl C, Li Y, Liu Y, Lynch BJ, Russell CT, Vourlidas A, Acuna MH, Galvin AB, Larson D, Sauvaud JA (2009) Solar Phys 254:325–344

Kiran Z, Shah HA, Qureshi MNS, Murtaza G (2006) Solar Phys 236:167–183

Kiran Z, Shah HA, 11th scientific assembly, "Our planet moving in space", Sopron, Hungary, 23–30 Aug 2009, p 53, abstract 401-TUE-01415-0035

Kis A, Scholer M, Klecker B, Lucek EA, Réme H (2009) 11th scientific assembly, "Our planet moving in space", Sopron, Hungary, 23–30 Aug 2009, p 56, abstract 401-TUE-01645-0326

Ko Y-K, Raymond JC, Vršnak B (2009) 11th scientific assembly, "Our planet moving in space", Sopron, Hungary, 23–30 Aug 2009, p 47, abstract 401-TUE-00945-0555

Kovács P, Heilig B, Csontos A (2009) 11th scientific assembly, "Our planet moving in space", Sopron, Hungary, 23–30 Aug 2009, p 56, abstract 401-TUE-01630-0338

Kuznetsova T (2009) 11th scientific assembly, "Our planet moving in space", Sopron, Hungary, 23–30 Aug, 2009, p 56, abstract 401-TUE-01615-0234

Luo B, Zhong Q, Liu S, Gong J (2009) 11th scientific assembly, "Our planet moving in space", Sopron, Hungary, 23–30 Aug 2009, p 53, abstract 401-TUE-01400-0663

Lynnyk A, Vandas M (2009) 11th scientific assembly, "Our planet moving in space", Sopron, Hungary, 23–30 Aug 2009, p 50, abstract 401-TUE-01115-1058

Marsch E, Muhlhauser KH, Schwenn R, Rosenbauer H, Neubauer FM (1982a) J Geophys Res 87:35

Marsch E, Muhlhauser KH, Schwenn R, Rosenbauer H, Phillip W, Neubauer FM (1982b) J Geophys Res 87:52

Marsch E (2006) Living Rev Solar Phys 3:1

McComas DJ, Allegrini F, Bochsler P, Bzowski M, Christian ER, Crew GB, DeMajistre R, Fahr H, Fichtner H, Frisch PC et al (2009) Science 326(5955):959

Miralles MP (2009) 11th scientific assembly, "Our planet moving in space", Sopron, Hungary, 23–30 Aug 2009, p 46, abstract 401-MON-00900-1307

Miralles MP, Cranmer SR, Kohl JL (2006)In: Lacoste H, Ouwehand L (eds) SOHO-17: 10 years of SOHO and beyond. ESA, Noordwijk, The Netherlands, ESA SP-617, 15.1

Miralles MP, Cranmer SR, Panasyuk AV, Romoli M, Kohl JL (2001) Astrophys J 549:L257

Miralles MP, Cranmer SR, Kohl JL (2004) Adv Space Res 33:696

Muraközy J, Ludmány A (2009) 11th scientific assembly, "Our planet moving in space", Sopron, Hungary, 23–30 Aug 2009, p 29, abstract 401-MON-00955-1252

Mursula K (2009) 11th scientific assembly, "Our planet moving in space", Sopron, Hungary, 23–30 Aug 2009, p 29, abstract 401-MON-00915-1094

Nitta NV, Aschwanden MJ, Lemen JR, Wuelser J-P, Zarro DM (2009) 11th scientific assembly, "Our planet moving in space", Sopron, Hungary, 23–30 Aug, 2009, p 35, abstract 401-MON-01400-1010

Ofman L (2009) 11th scientific assembly, "Our planet moving in space", Sopron, Hungary, 23–30 Aug 2009, p 38, abstract 401-MON-01630-1053

Santos JC, Büchner J (2009) 11th scientific assembly, "Our planet moving in space", Sopron, Hungary, 23–30 Aug 2009, p 35, abstract 401-MON-01415-0315

Shibasaki K (2009) 11th scientific assembly, "Our planet moving in space", Sopron, Hungary, 23–30 Aug 2009, p 38, abstract 401-MON-01600-0018

Vršnak B, Cliver EW (2008) Solar Phys 253:215–235

Vršnak B, Poletto G, Vujić E, Vourlidas A, Ko Y-K, Raymond JC, Ciaravella A, Žic T, Webb DF, Bemporad A, Landini F, Schettino G, Jacobs C, Suess ST (2009) A&A 499:905–916

Yi Y, Oh SY (2009) 11th scientific assembly, "Our planet moving in space", Sopron, Hungary, 23–30 Aug, 2009, p 29, abstract 401-MON-00930-0502

Chapter 2

Universal Heliophysical Processes

Nat Gopalswamy

Abstract The physical processes in the heliospace are a direct consequence of the Sun's mass and electromagnetic emissions. There has been enormous progress in studying these processes since the dawn of the space age half a century ago. The heliospace serves as a great laboratory to study numerous physical processes, using the vast array of ground and space-based measurements of various physical quantities. The observational capabilities collectively form the Great Observatory to make scientific investigations not envisioned by individual instrument teams. The International Heliophysical Year (IHY) program has been promoting scientific investigations on the universality of physical processes such as shocks, particle acceleration, dynamo, magnetic reconnection, magnetic flux ropes, plasma-neutral matter interactions, turbulence, and several other topics. This chapter highlights scientific deliberations on these and related topics that took place during the IAGA session on "Universal Heliophysical Processes" in Sopron, Hungary. The session featured several invited and contributed papers that focused on observations, theory and modeling of the universal heliophysical processes.

2.1 Introduction

The Sun strongly influences a region of space around it to a distance of about 100 AU by virtue of its mass and electromagnetic emissions. In-situ and remote-sensing observations combined with theory and modeling efforts have helped accumulate a wealth of knowledge on the heliospace. The magnetized plasma, energetic particles, and the electromagnetic radiation from the Sun interact with the neutral and charged fluid envelopes that surround the planets and their moons. Neutral material from the interstellar medium and charged particles of galactic origin (cosmic rays) also enter the heliospace resulting in additional physical processes. The interaction between the solar wind and the interstellar neutral matter results in pickup ions. The magnetic enhancements in the solar wind caused by solar eruptions deflect the galactic cosmic rays, reducing their severity in impacting Earth. The International Geophysical Year (IGY) during 1957–1958 provided a major impetus to the development of space science. During 2007–2009, the International Heliophysical Year (IHY 2007) was conducted to take stock of the enormous achievements that the world scientific community has accumulated since 1957 (see Chapter 28 by Davila et al. this volume, for an overview of the IHY program and its continuation as the International Space Weather Initiative (ISWI)). The IAGA session on "Universal Heliophysical Processes" in Sopron, Hungary provided a forum to discuss these achievements. Invited and contributed papers addressed issues related to the variability of the Sun from its interior to its atmosphere and in the extended heliospace.

N. Gopalswamy (✉)
NASA Goddard Space Flight Center, Greenbelt, MD 20771, USA
e-mail: nat.gopalswamy@nasa.gov

M.P. Miralles, J. Sánchez Almeida (eds.), *The Sun, the Solar Wind, and the Heliosphere*, IAGA Special Sopron Book Series 4, DOI 10.1007/978-90-481-9787-3_2,

2.2 Solar Interior, Dynamo, and the Solar Cycle

The mass and electromagnetic emissions from the Sun are ultimately related to the generation of energy in the solar core and its transport by radiative and convective processes in the solar interior. The energy generation in the solar core is primarily due to thermonuclear reactions that convert hydrogen to helium. Such a conversion is consistent with the stellar structure inferred from helioseismology, a field that developed rapidly with its origin in the discovery of the 5-min solar oscillations (Leighton and Brophy 1961). The interior structure and dynamics of the Sun have been established to a very high degree of accuracy by studying the acoustic waves trapped inside the Sun. The solution to the solar neutrino problem (Ahmad et al. 2001) also confirmed the standard solar model established by helioseismology many years earlier. Couvidat (Chapter 4, this volume) summarized the current developments in the global and local helioseismology branches. The local helioseismology is concerned with sunspot regions, where one observes intense magnetic fields produced at the base of the convection zone and transported to the surface. The sunspot regions are the source of the most violent eruptions whose consequences can be felt throughout the heliosphere.

2.2.1 *Solar Activity Cycles*

The generation and maintenance of solar magnetism by the differential rotation of the Sun and the turbulent convection in its outer layers is central to most of the energetic phenomena that take place in the solar atmosphere. The dynamo mechanism involves the conversion of the poloidal field into toroidal field and vice versa resulting in the 11-year sunspot cycle and the 22-year magnetic cycle (see e.g., Charbonneau 2005, for a review). The sunspot regions are the source of coronal mass ejections (CMEs) and flares that lead to hazardous space weather. The largest number of sunspots appears on the Sun during the solar maximum, when the likelihood of energetic eruptions is very high. Prediction of the sunspot number is therefore of great practical importance and provides a metric of our understanding of solar magnetism. Petrovay (2009) reviewed empirical and theoretical regularities and patterns underlying the sunspot cycle and discussed recent model-based prediction attempts for solar cycle 24. Petrovay also discussed the possible mechanisms that explain the amplitude of the solar activity modulated over a longer time-scale (~90 yr) resulting in epochs of suppressed solar activity (grand minima). There is widespread interest in grand minima after Eddy (1976) discovered the Maunder minimum and its implications for Earth's climate.

There have been attempts, both based on empirical and dynamo models, to predict solar cycles. The sunspot number predictions for cycle 23 occupy a huge range above and below the observations for cycle 23 (see e.g., Brajša et al. 2009). Using a combined method, Brajša et al. (2009) predicted the strength and the epoch of cycle 24: the sunspot number will be in the low 80s, occurring around the year 2012. Verbanac et al. (2009) presented the details of this prediction method: (1) the calculation of the asymmetry of the duration of the ascending and descending solar cycle parts, (2) the correlation of the relative sunspot numbers in and around solar activity minima and the following activity maxima and (3) the method of the autoregressive moving average model (ARMA) applied to the relative sunspot number data measured up to now. They utilized various data sets that comprise of yearly, corrected yearly, monthly and smoothed monthly relative sunspot number values.

2.2.2 *Solar Dynamo and Grand Minima*

Recognizing solar activity cycles from various proxy data has greatly enhanced our understanding of solar variability. Demetrescu et al. (2009) used annual means of measured and reconstructed solar, heliospheric, and magnetospheric parameters to infer solar activity signatures at the Hale magnetic (MC) and Gleissberg cycle (GC) timescales. They reconstructed available open solar flux, modulation strength, cosmic ray flux, and total solar irradiance data back to 1700; solar wind parameters (speed and density) and the magnitude of the heliospheric magnetic field at 1 AU were reconstructed back to 1870; time series of geomagnetic activity indices (aa, IDV, IHV), going back to 1870, were also considered. Simple filtering procedures (successive 11-, 22-, and 88-year running averages and differences between them) and scaling by the standard deviation from the average value for the common interval covered by the data show that the long-discussed variation in the twentieth century (a pronounced increase since ~1900, followed by a depression in the 1960s and a new, slower, increase) seen in the 11-year averages of the analyzed parameters, is a result of the superposition in data of solar activity signatures MC and GC timescales (see Fig. 2.1). This leads to the conclusion that the MC and GC signals are quite similar in all studied parameters, pointing to a common pacing source, the solar dynamo.

2.3 Solar Eruptions and Their Interplanetary Consequences

It is possible to recognize distinct activity phenomena associated with the toroidal and poloidal fields of the Sun. Flares, CMEs, and irradiance variations are connected to the toroidal fields, while the high-speed solar wind from solar poles is linked to the poloidal field. Both sets of phenomena have important consequences on Earth. Thus the variations in the action of the solar dynamo can result in varying influences on the geoeffectiveness and terrestrial climate (Georgieva 2009). Flares, high-speed solar wind, and CMEs produce space weather effects, while irradiance variation produces climate effects. CMEs impact on Earth's magnetosphere and result in intense geomagnetic storms, while high-speed solar wind can produce moderate and weak storms. Flares produce short-term ionospheric disturbances that can affect radio communication and navigations and accelerate solar energetic particles (SEPs) of the impulsive type. CMEs on the other hand drive fast mode MHD shocks that accelerate gradual SEPs, which are potentially hazardous to space-based technological systems and humans in space.

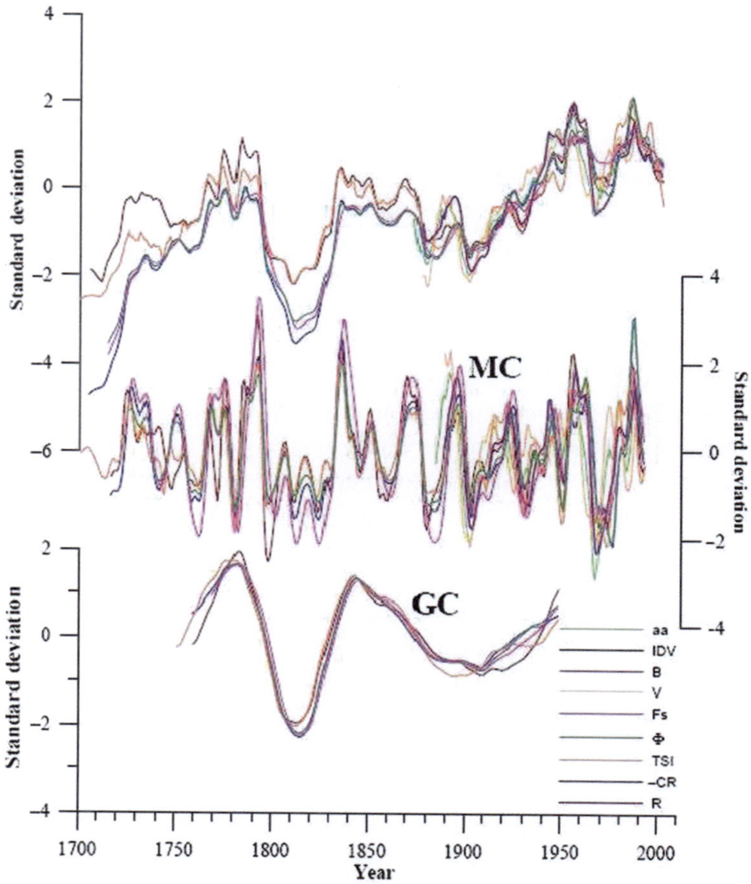

Fig. 2.1 11-year averages (*top*), the Hale magnetic cycle (MC) signal (*middle*) and the Gleissberg (GC) signal (*bottom*) in various data: aa - geomagnetic activity index, IDV - geomagnetic interdiurnal variability index, B - magnitude of the heliospheric magnetic field; V - solar wind speed; Fs - open solar flux; *Φ* - heliospheric modulation strength; TSI - total solar irradiance; CR - cosmic ray flux (reconstructed count rate of the standard neutron monitor); R - sunspot number

CMEs originate from closed magnetic regions that possess free magnetic energy. Active regions produce the most energetic CMEs because of the large free energy that can be stored in them. One of the important indicators of free energy is the active region helicity. The helicity generation is thought to be closely connected with the toroidal and poloidal fields (Longcope et al. 1998; Choudhuri et al. 2004). Based on the estimates of magnetic helicity in interplanetary flux ropes, Sung et al. (2009) showed that the CME kinetic energy and the flux rope helicity are closely related. Thus, the active region helicity and the flux rope structure play a significant role in understanding the generation and interplanetary consequences of CMEs.

There is plenty of observational evidence for flux rope structure in CMEs and their interplanetary counterparts (ICMEs). The flux rope structure was inferred from the solar wind magnetic field data (see e.g., Burlaga et al. 1981). The coronagraphic observations from the Solar and Heliospheric Observatory (SOHO) reveal flux rope structure (see e.g., Chen et al. 1997). Linton (2009) compared the CME flux rope structure with the plasmoids observed in the Earth's magnetotail, though the two structures are of vastly different spatial scale. With this comparison, Linton concluded that flux rope formation is a universal space physics phenomenon and that the physical mechanisms responsible for flux rope formation occur over a wide range of plasma conditions wherever current sheets exist. The current sheets in the solar atmosphere are of particular interest since they are supposed to indicate the production of CME flux ropes. Poletto (Chapter 15, this volume) reviewed the observational evidence of current sheets throughout the solar atmosphere.

One of the important consequences of a flux rope structure is that we can predict which portion of the flux rope is likely to interact with Earth's magnetosphere. For example, if the flux rope axis is in the ecliptic plane, then the leading field is pointed to the north or south, while the trailing field has the opposite sense. South-pointing field is necessary for reconnection with the magnetospheric field, which points to the north. Another advantage is the possibility that the flux rope axis is parallel to the neutral line in the solar active region (or the axis of the filament overlying the neutral line). Observations indicate that there are significant deviations between the flux rope axis and the active region neutral line, suggesting that the field orientations rotate by up to 160° with respect to the magnetic orientation of the CME source region on the Sun. The cause of this rotation it is not well understood. Török et al. (2009) reported on a systematic study of CME rotation mechanisms, based on numerical MHD simulations of a flux rope CME model. They focused on two mechanisms: (i) the conversion of flux-rope twist into writhe and (ii) the interaction of the flux-rope current with the component of the ambient coronal magnetic field along the flux rope. By varying the initial flux rope twist and the shear angle of the ambient magnetic field, they found a continuous range of flux rope rotations between ~20 and ~140° in the simulations. Strong rotations of more than ~100° require both high flux rope twist (i.e., the occurrence of the helical kink instability) and the presence of a significant shear of the ambient field with respect to the initial flux rope orientation.

Occasionally, linking the active region helicity to that of the interplanetary flux rope originating from the active region may not be straightforward. Chandra et al. (2010) discussed a CME from NOAA AR 10501 that was associated with a magnetic cloud (MC) of positive magnetic helicity on 20 November, 2003. However, the active region had a negative helicity, opposite to that of the MC. They reconciled the disparity using the observation that a smaller region within the active region had emerging flux with a positive helicity, which might explain the positive helicity in the MC.

2.4 CME-Driven Shocks and Related Phenomena

The heliospace plasma is an ideal laboratory for collisionless shock physics, because it supports a variety of shocks, ranging from mini-bow shocks ahead of the lunar magnetic anomaly to large scale shocks in front of CMEs (Terasawa Chapter 12, this volume). The dissipation processes at collisionless shocks naturally result in the production of nonthermal particles. CME-driven shocks are of particular interest because they accelerate electrons and ions throughout the heliospace. Shock waves near the Sun produce type II radio bursts and release SEPs (see Chapter 16 by Valtonen this volume). Type II bursts represent one of the several different radio emission processes that occur in the heliospace (Messerotti Chapter 19, this volume). When shocks impact the magnetosphere, they produce the storm sudden commencements, which are signatures of magnetospheric compression.

2.4.1 EUV Wave Transients

The CME-driven shocks also manifest as EUV wave transients, which are thought to be bow waves enveloping CMEs. Veronig (2009) reviewed the current status of large-scale EUV waves (also known as "EIT waves" and "coronal Moreton waves") in the solar corona that are launched in association with solar flares and CMEs. In particular she discussed the physical nature of the wave: blast wave, driven-wave, or non-waves. Based on the wave kinematics and dynamics derived from high-cadence observations by the EUVI instruments onboard the twin STEREO spacecraft, she concluded that the coronal wave reveals deceleration, indicative of a freely propagating MHD wave after an initial driving by the expanding flanks of the associated CME. When EIT waves are associated with type II radio bursts, one infers that the EIT wave is in fact a fast mode shock.

2.4.2 Radio-Loud and Radio-Quiet Shocks

The fact that shocks are important source of energetic particles in the heliosphere can be inferred in a number of ways. Type II radio bursts are the earliest signature of particle acceleration by CME-driven shocks. Type II bursts begin when the CMEs are typically at a heliocentric distance of 1.5 Rs (Gopalswamy et al. 2009b) indicating that electrons accelerated to an energy of up to ~10 keV escape from the CME-driven shock front. Type II bursts also serve as an indicator of particle acceleration by shocks throughout the inner heliosphere. SEPs are released typically at larger heliocentric distances (~5–7 Rs), where a combination of circumstances favor SEP acceleration and release (CMEs reaching peak speeds and Alfven speed starts declining). This spatial domain overlaps with the fields of view of white-light coronagraphs, so shock signatures can be discerned in the coronagraphic images as the diffuse feature ahead of the bright structures. Gopalswamy et al. (2009a) described the diffuse feature as the compressed sheath ahead of the flux rope in the 2005 January 15 CME (see Fig. 2.2). The extent of the diffuse feature differed substantially when shocks with (radio-loud) and without type II bursts (radio-quiet) were considered. They investigated the widths of the flux rope and the surrounding disturbance for 13 radio-quiet and 47 radio-loud interplanetary shocks associated with limb CMEs. The sky-plane widths of the surrounding disturbance were much larger for the case of radio-loud shocks (284° compared to 197° for the radio-quiet shocks).

2.4.3 Extended Shocks and SEPs

The large extent of the shock surrounding energetic CMEs has important consequences in the heliosphere. Malandraki et al. (2009) reported energetic particle

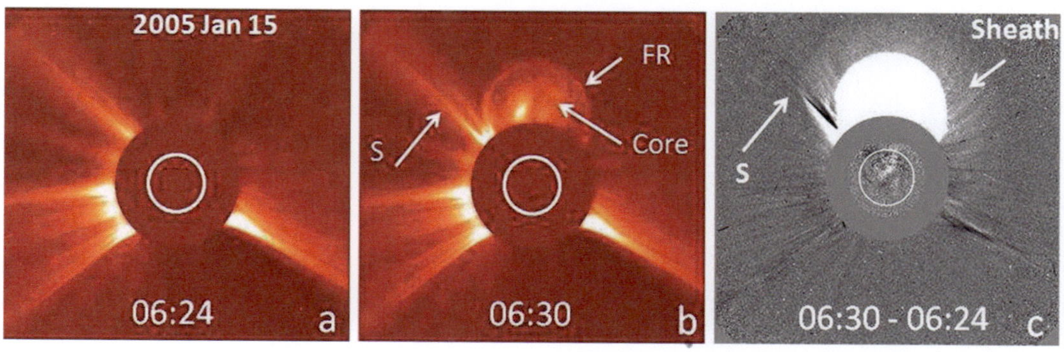

Fig. 2.2 A flux rope CME (2005 January 15) observed by SOHO/LASCO with the surrounding shock disturbance. (*left*) Pre-CME corona at 6:24 UT, (*middle*) the CME with flux rope (FR) and prominence core and a kink (S) in the streamer outside the flux rope, and (*right*) the difference between the images at 06:30 and 06:24 showing the diffuse material (sheath) surrounding the flux rope (from Gopalswamy, 2010)

observations by Ulysses during the period of isolated but intense solar activity from active region 10,930 in December 2006. The heliocentric distance of Ulysses was 2.8 AU, located at 72° southern heliographic latitude immersed in the high-speed solar wind from the southern polar coronal hole. The particle event was associated with an X9.0 flare from S07E79 on December 5, being the largest among a set of 4 X-class events. Three forward shocks from these eruptions were observed at Ulysses, even though Ulysses was located at W120. The shocks were not followed by discernible ejecta suggesting that the shocks were much more extended than the ejecta. The onset and decay profiles of the event for the 8–19 MeV proton channel were relatively smooth. The event had a "clean" onset, occurring in a period nearly devoid of solar wind structures and with relatively low pre-event intensities (see Fig. 2.3). The pitch-angle distributions of energetic particles were near-isotropic. The particle flow directions were along the field and there was no evidence for any net flow across the field lines, so there was little cross-field diffusion close to the observer. The magnetic field in the fast solar wind is much more turbulent, so particle propagation should be more difficult leading to significant scattering. The rise-time of the SEP event at STEREO-B was found to be faster than at Ulysses (although this corresponds to an eastern poorly connected event as viewed from the Earth), suggesting a more diffusive transport to high latitudes than to the near-Earth STEREO-B. The relatively simple structure of the heliosphere in December 2006 when Ulysses was in the southern polar wind seem to exclude the possibility that magnetic field lines originating at low-latitudes reached Ulysses. The energetic particles observed as large SEP events over the south pole of the Sun were released when the propagating coronal waves reached high latitude magnetic field lines connected to Ulysses. Based on the observations available, however, cross-field diffusion closer to the Sun cannot be definitely excluded.

2.4.4 Geospace Consequences of Solar Eruptions

CMEs have important consequences in geospace by direct impact (geomagnetic storms) and via SEPs, which can be trapped in Earth's radiation belt over long periods of time. SEPs can also penetrate Earth's atmosphere at high latitudes leading to ozone depletion. The properties of magnetosphere can also be modified dur-

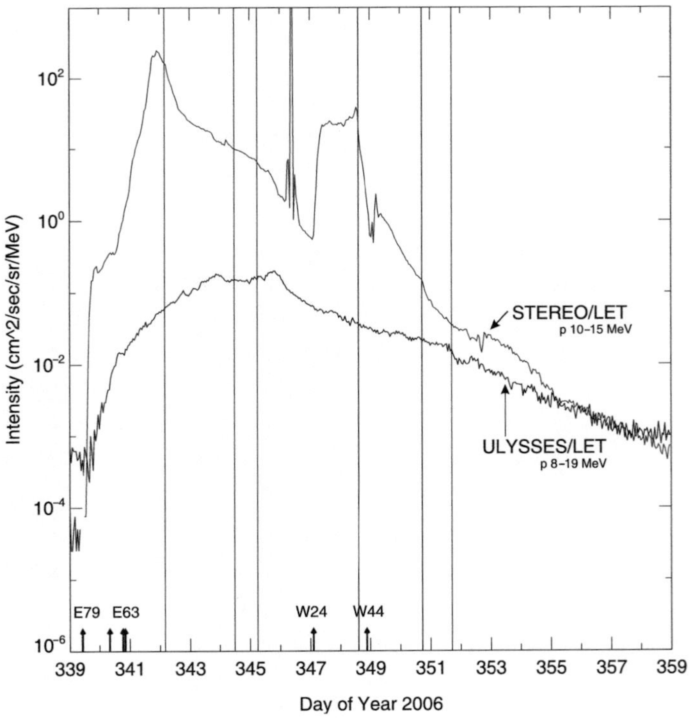

Fig. 2.3 Hourly averages of proton intensities at similar energy intervals measured in and out of the ecliptic plane by the COSPIN/LET and STEREO/LET experiments onboard Ulysses (*lower trace*) and STEREO-B (*upper trace*) during the December 2006 period. courtesy: O.E. Malandraki (IAA/NOA)

ing geomagnetic storms. Kudela and Lazutin (Chapter 18, this volume) reviewed studies on the cosmic ray anisotropy observed at neutron monitors and muon telescopes just before the onset of some geomagnetic storms. They also discussed changes of geomagnetic cut-off, structure of the cosmic ray transmissivity function and the asymptotic directions for various geomagnetic field models during strong geomagnetic storms. They reported preliminary results of a study on the solar proton capture into radiation belts using coronas data and explored the use of different geomagnetic field models that can fit the observed trapped particle profiles in different local time sectors. Furthermore, they pointed out that measurements of energetic neutral emissions (gammas and neutrons) near the Earth or on the ground can serve as indicators of acceleration processes on solar surface.

The first interplanetary structure that impacts Earth's magnetosphere is the shock ahead of the ICME, which is recognized in the ground-based magnetometer data as a sudden increase in the horizontal component of Earth's magnetic field known as the geomagnetic storm sudden commencement (SC). The preliminary impulse (PI) of SCs appears as a negative impulse (PRI) at the afternoon high latitude and dayside geomagnetic equator, and a positive impulse (PPI) at morning high latitude and nightside geomagnetic equator. The temporal and spatial variations of the PI are explained by means of a magnetosphere-ionosphere current system, composed of the ionospheric Hall and Pedersen currents at high latitudes and the Pedersen current amplified by the Cowling effect at the equator. The ionospheric currents are driven by the dusk-to-dawn electric field impressed from the magnetosphere. The electric field then propagates to low latitude, driving westward/eastward currents in the day/nightside equatorial ionosphere. Although the observations indicate instantaneous transmission of the electric field to the equator, a complete confirmation required high time resolution and better data coverage. Kikuchi et al. (2010) analyzed PI events using 1-s sampled magnetometer data recorded at high to equatorial latitudes on both day–and nightside. They found the PI to start simultaneously with the temporal resolution of 1 s at all latitudes and local times. The D-component deflections at mid latitudes were consistent with the Pedersen currents connecting the field-aligned currents with the equatorial PI currents. The instantaneous development of the PI currents is consistent with their scenario based on the TM0 mode wave in the Earth-ionosphere waveguide.

Balan et al. (2009) studied the changes in the Earth's ionosphere due to the passage of the CMEs of 07–11 November 2004. The observed changes include (1) the direct response of the high latitude ionosphere, (2) the development of a rare super double geomagnetic storm, (3) the development of strong daytime eastward prompt penetration electric field (PPEF) events in the longitudes that were in daytime sectors during the main phases of both super storms, (4) the strong F3 layer with large density depletions around the equator in the longitudes of the PPEF events, and (5) the large positive/negative ionospheric storms at low to mid latitudes. Using the physics based model SUPIM, they evaluated the relative importance of diffusion, daytime eastward PPEF and neutral wind on the equatorial plasma fountain and positive ionospheric storms. The plasma fountain was found to rapidly develop into a super fountain and the equatorial ionization anomaly (EIA) crests shifted to higher than normal latitudes during the PPEF event both in the presence and absence of neutral winds. However, the super fountain became stronger with less poleward turning of the plasma flux vectors and the EIA crests became stronger than normal in the presence of an equatorward wind. The equatorward wind reduced (or stopped) the downward velocity component due to diffusion and raised the ionosphere to high altitudes of reduced chemical loss. These mechanical effects of the equatorward wind accumulated the plasma brought by the super fountain, and hence strengthened the EIA crests and produced positive ionospheric storms; the wind need not be a storm-time wind, although stronger wind could lead to stronger ionospheric storms.

2.5 Solar Wind Processes

Solar wind represents another important mass emission process from the Sun, which is rather steady in contrast to that during CMEs. The large number of observations and theoretical studies of the solar wind since its original prediction by Parker (1958) have established the basic picture that the wind is driven by heating of the solar corona. Lie-Svendsen (2009) emphasized the need to treat the corona and wind as one tightly coupled system. He also noted that the solar mass loss is controlled by the amount of coronal heating and by the energy flow between the chromosphere and corona. He also highlighted the following key results from SOHO/UVCS observations: (i) protons and heavy ions are much hotter than electrons, (ii) the wind is rapidly accelerated within a few solar radii, and (iii) the ion temperature perpendicular to the magnetic field is much higher than the parallel temperature. From these observations he concluded that the solar wind is driven mainly by proton heating close to the Sun, perhaps through ion cyclotron waves.

2.5.1 Evolution of Solar Wind Properties

The twin spacecraft of the STEREO mission have helped understand the temporal evolution of the solar wind plasma properties using the electron and ion measurements. Opitz et al. (2009) compared the proton bulk velocity measurements made by the PLASTIC instruments on STEREO A and B after adjusting for the theoretical time lag. They obtained the correlation coefficients for the proton bulk velocity as a function of the time difference between two ejections of the plasma parcel from the solar source as 0.95 and 0.85 for time lags of 0.5 and 2 days, respectively (Fig. 2.4a). They performed a similar correlation analysis of the electron core densities measured by the two STEREO/IMPACT instruments (Opitz et al. 2010b). The correlation coefficients were 0.80 and 0.65 for time lags of 0.5 and 2 days, respectively (Fig. 2.4b). The correlation was lower for the electron density than for the proton bulk velocity, though for both parameters the solar wind can be considered as persistent over more than 2 days. Opitz et al. (2010a) extrapolated the solar wind bulk velocity measurements for different in-ecliptic heliospheric positions, which were validated using on-site measurements at Venus (VEX), Earth (SOHO) and Mars (MEX). They also showed that the solar wind is accelerated or decelerated at stream interfaces during its radial propagation.

2.5.2 Solar Wind Turbulence

The solar wind also offers a unique laboratory for studying turbulent plasma processes. Of particular interest is the turbulence at MHD scales because it strongly affects several aspects of the solar wind: generation and heating of the wind, particle acceleration, and cosmic-ray propagation. Alfvenic fluctuations in the solar wind are also linked to geomagnetic activity. Bavassano (2009) reviewed the solar wind observations at different heliocentric distances and latitudes that have helped understand the mechanisms driving the solar wind turbulence. Gary (2009) presented their idea on the role of whistler fluctuations in the dissipation regime of solar wind turbulence. They proposed the whistler fluctuations as an alternative hypothesis to describe short-wavelength turbulence in the solar wind, because the kinetic Alfven waves may not contribute to a power law spectrum. Particle-in-cell simulation showed that the whistler cascade yields steep power-law power spectra consistent with observations. Vörös and Leubner (2009) reported on the occurrence of magnetic turbulence in the solar wind on the basis

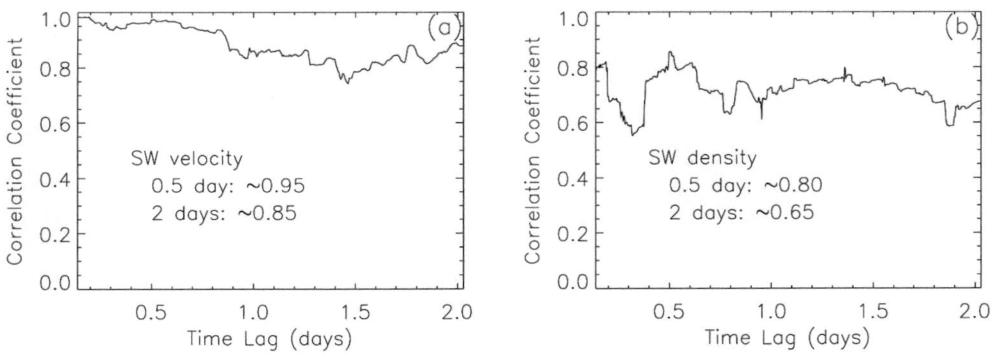

Fig. 2.4 Temporal evolution of the solar wind proton bulk velocity (**a**) and electron core density (**b**) obtained by correlating the STEREO A and B measurements

of a wavelet approach. They concluded that the solar wind turbulence is strongly localized and associated with large-scale structures, in particular shocks. The quiet inter-shock periods in between the localized turbulent fluctuations were found to be unrelated to turbulence or turbulent intermittency. They also pointed out that the misinterpretations caused by the stationarity assumption in the recent literature are significant and therefore suggested a revision of the basic idea of stationarity in solar wind turbulence.

2.5.3 Reconnection Exhaust in the Solar Wind

The reconnection events, known to be central to many energetic phenomena in the heliospace, have recently been found to occur in the solar wind. Lavraud et al. (2009) reported on the analysis of one such event in the heliospheric current sheet (HCS) crossed by WIND, ACE and STEREO (see Fig. 2.5). Although only WIND and ACE provided good ion flow data in accord with a reconnection exhaust, the magnetic field bifurcation typical of such exhausts was clearly observed at all spacecraft. Unambiguous strahl mixing within the exhaust was consistent with the sunward flow deflection observed at WIND and ACE and thus with the formation of closed magnetic field lines within the exhaust with both ends attached to the Sun. The strong downward flow deflection was consistent with the exhaust and X-line orientations obtained from minimum variance analysis (MVA) at each spacecraft so that the X-line was almost along the GSE Z-axis and duskward of all the spacecraft. The observation of strahl mixing in extended and intermittent layers outside the exhaust by STEREO A and B was con-

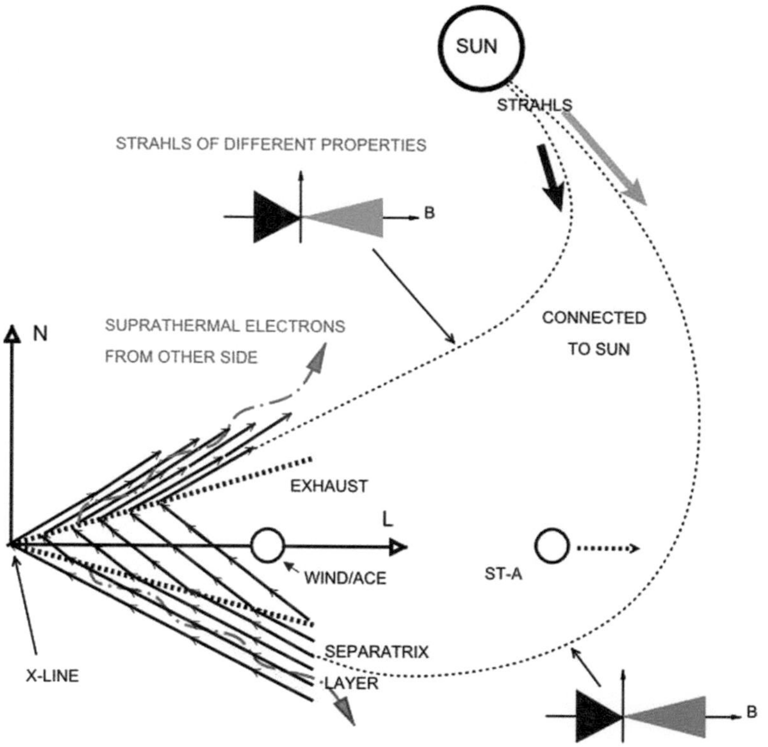

Fig. 2.5 (*Lower left part*) Idealized projection of asymmetric reconnection in the solar wind (in the L-N plane formed by the directions of minimum (N) and maximum (L) magnetic field variance). The field line kinks at each exhaust boundary accelerate the plasmas they intercept. The exhaust boundaries are locally open, rotational-like discontinuities. Suprathermal electrons can flow from one side to the other (*curvy dash-dotted lines*), so as to form electron separatrix layers. (*Right part of figure*) Spacecraft such as Wind/ACE and STEREO-A, located on the sunward side of the reconnection site, are on newly closed field lines and thus observe the mixing of suprathermal electrons from the two sides of the reconnecting current sheet. The strahls from the two sides may be of different strengths (*dark and light grey arrows at the Sun*). From Lavraud et al. (2009)

sistent with the formation of electron separatrix layers surrounding the exhaust. This event also provided further evidence that balanced parallel and anti-parallel suprathermal electron fluxes are not a necessary condition for identification of closed field lines in the solar wind. In the present case the origin of the imbalance simply was the mixing of strahls of substantially different strengths from a different solar source each side of the HCS. The inferred exhaust orientations and distances of each spacecraft relative to the X-line show that the exhaust was likely non-planar, following the Parker spiral orientation. Finally, the separatrix layers and exhausts properties at each spacecraft suggest that the magnetic reconnection X-line location and/or reconnection rate were variable in both space and time at such large scales.

2.5.4 Influence of the Solar Wind on Energetic Particles

The solar wind plays also plays an important role in establishing the global characteristics of MeV particles in the inner heliosphere. Two aspects of this relation were reported by Kecskeméty et al. (2009): the decay rate after solar energetic particle events and the energy spectrum during quiet solar activity. As convection and adiabatic deceleration play a dominant role in particle propagation at these energies, the characteristic time constant of exponential-form decays should be inversely proportional to solar wind speed resulting in an exponential time (t) profile $J(t) \propto E^{-\gamma} exp[-4V(1+\gamma)t/3r]$ with V denoting solar wind speed, γ the slope of the energy (E) spectrum, and r the heliocentric distance. The characteristic decay time was found to be valid in more than 50% of events in SOHO and Ulysses simultaneous time profiles; the V and r dependence was also confirmed. Based on IMP-8, SOHO, and Ulysses data, the quiet-time MeV proton fluxes were found to correlate with solar activity, and in particular, with the solar wind speed. The comparison of particle fluxes and solar wind speed during the solar activity minima of 1976–1977 and 1986–1987 indicated that whereas fluxes of both quiet time low energy protons and GCR were lower in 1986–1987, the corresponding solar wind speeds were higher. The two recent minima, 1996–1997, and 2006–2007 were also compared using SOHO data. The closest correlation was found between 4.5–20 MeV proton fluxes and solar wind pressure (see Fig. 2.6, including quiet periods from the last two minima), the lowest proton fluxes appeared at lowest plasma pressures.

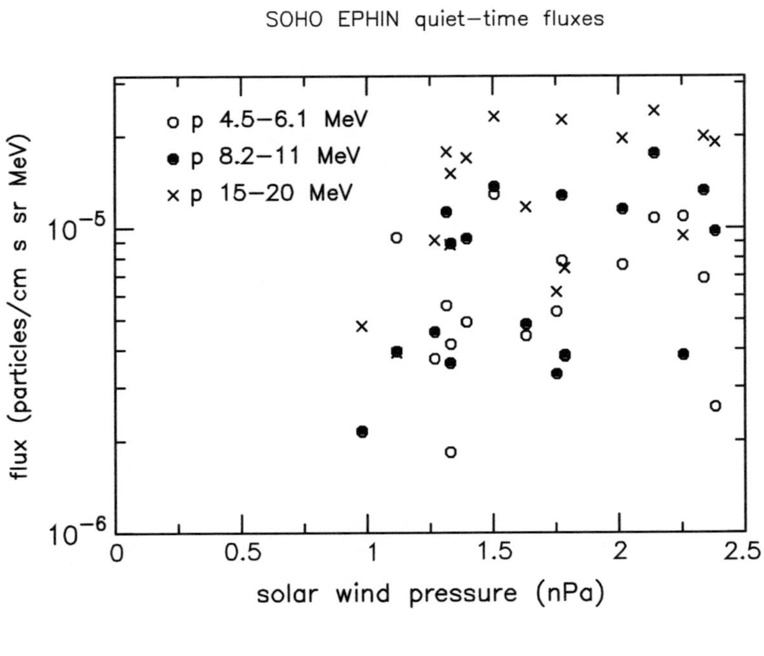

Fig. 2.6 Scatter plot between solar wind pressure and quiet-time particle flux for three different energy ranges

2.5.5 Solar Wind in the Outer Heliosphere

The character of the solar wind undergoes a significant transformation in the outer heliosphere owing to the greater influence exerted by the interstellar medium. Using observations from multiple spacecraft distributed throughout the heliosphere and a multi-fluid MHD model, Wang and Richardson (2009) traced the propagation of interplanetary CMEs (ICMEs) and the shocks driven by them from 1 AU to the location of Voyager 2. They showed that the ICME shock relationship is not simple as one observes at 1 AU: successive merging and interaction of relatively small interplanetary shocks could form a well-developed strong forward shock beyond 30 AU. They showed that the strong shock (speed jump ~100 km/s) of 2001 October 16 was due to the merging of a series of shocks observed at Earth in April 2001. They also examined the characteristics of the termination shock in detail, making use of multiple shock crossing of Voyager 2 in August 2007. For two crossings events, the flow was found to be still supersonic with respect to the thermal ions downstream of the termination shock, probably due to the fact that most of the solar wind energy is transferred to pickup ions. They concluded that the solar wind in the outer heliosphere is fundamentally different from that in the inner heliosphere, since the influence of the local interstellar source becomes significant.

2.5.6 Solar Wind Structure and Cosmic-Ray Modulation

One of the implications of the solar wind and the magnetic field it carries is its impact on the galactic cosmic rays that enter the heliosphere. Ferreira (2009) described the current status of numerical model computations that compare well with spacecraft observations at various energies. The basis of these calculations is the transport equation, which is solved using realistic transport parameters, model of the heliosphere (including the asymmetry caused by the relative motion between the local interstellar medium and the Sun), and the heliospheric magnetic field. Time-dependent modulation was also discussed to show that drift effects together with propagating diffusion barriers are responsible for modulation over a solar cycle.

2.6 Concluding Remarks

The summary provided in this chapter gives only a birds eye view of a limited number of heliophysical processes. However, the topics mentioned are of current interest both from physics and practical points of view. The origin of solar wind and CMEs continue to occupy the forefront of solar-terrestrial research because they have been identified as the primary sources of space weather. The past decade has seen an enormous growth in research focusing on space weather not only because of its scientific research, but also because of the increased dependency of the humans on space based technology, which is vulnerable to space weather. With the recent launch of the Solar Dynamics Observatory (SDO), there will be more focus on the source of mass emission using the unprecedented SDO imagery coupled with what is already available from ground and spacebased assets. While the past decades have seen man made instruments making in situ measurements all the way to the edge of the heliospace, the current decade will see in situ measurements made from close to the Sun where the mass emission begins using the Solar Probe Plus mission.

Acknowledgements The author thanks Drs. K. Kecskeméty, B. Vršnak and B. J. Thompson for their help in convening the IAGA session. He also thanks Dr. P. Mäkelä for proof reading the manuscript.

References

Ahmad QR, Allen RC, Andersen TC et al (2001) Phys Rev Lett 87:071301

Bavassano B (2009) 1st Scientific Assembly, "Our Planet Moving in Space", Sopron, Hungary, 23–30 Aug, 2009, p. 116, abstract 403-SAT-00830-0189

Balan N, Shiokawa K, Otsuka Y, Kikuchi T (2009) 11th IAGA Scientific Assembly, "Our Planet Moving in Space", Sopron, Hungary, 23–30 Aug, 2009, p. 119, abstract 403-SAT-01100-0065

Brajša R, Wöhl H, Hanslmeier A et al (2009) Astron Astrophys 496:855

Burlaga L, Sittler E, Mariani F, Schwenn R (1981) J Geophys Res 86:6673

Chandra R, Pariat E, Schmieder B, Mandrini CH, Uddin W (2010) Sol Phys 261:127

Charbonneau P (2005) Living Rev Solar Phys 2:2

Chen J, Howard RA, Brueckner GE et al (1997) Astrophys J Lett 490:L191

Choudhuri AR, Chatterjee P, Nandy D (2004) Astrophys J Lett 615:L57
Demetrescu C, Dobrica V, Maris G (2009) IAGA abstract #403-FRI-O1045-0292
Eddy JA (1976) Science 192:1189
Ferreira S (2009) 11th IAGA Scientific Assembly, "Our Planet Moving in Space", Sopron, Hungary, 23–30 Aug, 2009, p. 124, abstract 403-SAT-01630-0842
Gary SP (2009) IAGA abstract #403-SAT-O0915-1144
Georgieva K (2009) IAGA abstract #403-FRI-O1030-0324
Gopalswamy N, Mäkelä P, Xie H, Yashiro S (2009a) IAGA abstract #403-FRI-O1415-1318
Gopalswamy N, Thompson WT, Davila JM et al (2009b) Sol Phys 259:227
Gopalswamy N (2010) in Heliophysical Processes, ed. N. Gopalswamy, S.S. Hasan, A. Ambastha, Springer, p. 53
Kecskeméty K, Daibog E, Zeldovich M, Logachev Y (2009) IAGA abstract #403-SAT-O1530-0530
Kikuchi T, Shinbori A, Tsuji Y et al 2010, abstract #403-SAT-01130-0601
Lavraud B, Gosling JT, Rouillard AP et al (2009) Sol Phys 256:379
Leighton MM, Brophy JA (1961) J Geol 69:1
Lie-Svendsen O (2009) 1st Scientific Assembly, "Our Planet Moving in Space", Sopron, Hungary, 23–30 Aug, 2009, p. 107, abstract 403-FRI-01530-1024
Linton M (2009) 11th Scientific Assembly, "Our Planet Moving in Space", Sopron, Hungary, 23–30 August, 2009, p. 102, abstract 403-FRI-001115-0470
Longcope DW, Fisher GH, Pevtsov AA (1998) Astrophys J 507:417
Malandraki OE, Marsden RG, Lario D et al (2009), IAGA abstract #403-SATO1445-0335
Opitz A, Fedorov A, Wurz P et al (2010a) Sol Phys, 264:337
Opitz A, Karrer R, Wurz P et al (2009) Sol Phys 256:365
Opitz A, Sauvaud J-A, Fedorov A et al (2010b) Sol Phys, DOI: 10.1007/s11207-010-9613-5
Parker, E. N. 1958, Astrophys. J., 128, 664
Petrovay K (2009) 11th IAGA Scientific Assembly, "Our Planet Moving in Space", Sopron, Hungary, 23–30 Aug, 2009, p. 99, abstract 403-FRI-00930-1046
Sung S, Marubashi K, Cho K et al (2009) Astrophys J 699:298
Török T, Kliem B, Berger MA, Démoulin P, van Driel-Gesztelyi L (2009) IAGA abstract #403-FRI-O1145-0274
Verbanac G, Brajsa R, Cliver E et al (2009) IAGA abstract #403-FRI-O1100-1025
Veronig AM (2009) 11th IAGA Scientific Assembly, "Our Planet Moving in Space", Sopron, Hungary, 23–30 Aug, 2009, p. 105, abstract 403-FRI-01345-0277
Vörös Z, Leubner M (2009) IAGA abstract #403-SAT-O0900-0793
Wang C, Richardson D (2009) 11th IAGA Scientific Assembly, "Our Planet Moving in Space", Sopron, Hungary, 23–30 Aug, 2009, p. 124, abstract 403-SATO1600-0153

Part II
The Solar Interior

Chapter 3

Solar Convection Zone Dynamics

Matthias Rempel

Abstract A comprehensive understanding of the solar magnetic cycle requires detailed modeling of the solar interior including the maintenance and variation of large scale flows (differential rotation and meridional flow), the solar dynamo and the flux emergence process connecting the magnetic field in the solar convection zone with magnetic field in the photosphere and above. Due to the vast range of time and length scales encountered, a single model of the entire convection zone is still out of reach. However, a variety of aspects can be modeled through a combined approach of 3D MHD models and simplified descriptions. We will briefly review our current theoretical understanding of these processes based on numerical models of the solar interior.

3.1 Introduction

The solar convection zone comprises the outer most 30% of the solar radius and contains about 2% of the total solar mass. Due to a density variation of more than 6 orders of magnitude a variety of different physical regimes are encountered. While fluid motions are highly subsonic (Ma $\approx 10^{-4}$) and strongly influenced by rotation near the base of the convection zone, they turn supersonic in the photosphere and the influence of rotation diminishes. The pressure scale height varies between about 50 Mm at the base of the convection one and about 100 km in the photosphere of the sun. As a consequence a comprehensive model of the entire convection zone is currently out of reach and different aspects have to be modeled independently. The deep convection zone up to about 10–20 Mm beneath the photosphere can be modeled most efficiently using the anelastic approach which is filtering out sound waves, but is fully considering the compressibility in the stratification (Glatzmaier 1984). The upper most parts of the convection zone require fully compressible MHD (see e.g. Nordlund et al. 2009, for a recent review). While most anelastic models of the solar interior are global models with computational domains covering an entire shell between two radii (or at least a shell segment), MHD models of the outer parts of the convection zone typically focus on details in rectangular computational domains.

Apart from 3D MHD models adapted to the different physical regimes a variety aspects have been modeled based on simplified models, such as the mean field approach. Here the focus is on the large scales, while the effects of unresolved turbulence is parametrized. Non-linear terms in the momentum, energy and induction equations lead to non-vanishing second order correlation terms of small scale quantities that act as drivers for large scale flows or as turbulent induction effects for the large scale magnetic field. The decomposition into large and small scale properties and the arising correlation terms driving large scale flows are the strength and the weakness of this approach at the same time. On one hand the computational expense is decreased by orders of magnitude allowing for simulations covering long time scales as well as exploring wide parameter ranges, on the other hand the results are heavily dependent on parametrization of the second order correlation terms. For a comprehensive

M. Rempel (✉)
HAO, National Center for Atmospheric Research, Boulder, CO, USA
e-mail: rempel@hao.ucar.edu

M.P. Miralles, J. Sánchez Almeida (eds.), *The Sun, the Solar Wind, and the Heliosphere*, IAGA Special Sopron Book Series 4, DOI 10.1007/978-90-481-9787-3_3, © Springer Science+Business Media B.V. 2011

description of mean field theory we refer to Rüdiger and Hollerbach (2004).

3.2 Differential Rotation and Meridional Flow

Differential rotation is the consequence of angular momentum transport in the solar convection zone. Starting with a decomposition of the turbulent velocity field into fluctuating and (axisymmetric) mean flows $v = \langle v \rangle + v'$ leads to the following terms in the angular momentum flux (neglecting magnetic terms for simplicity):

$$\langle F_i \rangle = r \sin\theta \rho \, (\underbrace{\langle v_i{}' v_\phi{}' \rangle}_{\text{Reynolds stress}} + \underbrace{\langle v_i \rangle \Omega \, r \sin\theta}_{\text{Meridional flow}}) \, . \tag{3.1}$$

Here the Reynolds stress describes the transport due to correlations of fluctuating turbulent velocity components, while the Meridional flow describes the transport due to large scale coherent mean flows in the $r - \theta$ plane. Angular momentum transport through Reynolds stresses requires the presence of rotation and anisotropy and expressions for these transport terms have been derived within the mean field approach by Durney and Spruit (1979); Hathaway (1984) and more recently by Kitchatinov and Rüdiger (1993) using a quasi-linear approach (see also Kitchatinov and Rüdiger (2005) for an improved representation). In 3D simulations the influence of rotation on convection leads to a preferential north-south alignment of convection cells (Gilman 1979; Miesch et al. 2000; Brun and Toomre 2002; Miesch et al. 2008). The consequence is a dominance of east-west motions over north-south motions. By means of the Coriolis force eastward (faster rotating) flows are deflected equatorward, while westward (slower rotating) flows are deflected poleward, leading on average to an equatorward transport of angular momentum.

The profile of differential rotation cannot be determined on the basis of angular momentum transport processes alone. As stationary state requires beside vanishing divergence of the total angular momentum flux also a force balance in the meridional plane between Coriolis, centrifugal, buoyancy and pressure forces. The latter is most conveniently expressed by (follows from ϕ-component of vorticity equation):

$$r \sin\theta \frac{\partial \Omega^2}{\partial z} = \frac{g}{c_p \, r} \frac{\partial s}{\partial \theta} \tag{3.2}$$

Helioseismic observations by Thompson et al. (1996) show clearly a differential rotation profile with contours of constant Ω inclined by about 25° with respect to the rotation axis (deviation from Taylor-Proudman state). It turns out that avoiding the Taylor-Proudman state is a key problem for a theoretical understanding of solar differential rotation. While early models attempted to achieve this by assuming large viscosities ("Taylor-number puzzle" after Brandenburg et al. (1990)), Kitchatinov and Rüdiger (1995) showed that an alternative solution of this problem can be given if the anisotropic convective energy transport is considered, leading to a pole-equator temperature difference of about 10 K. Anisotropic convective energy transport is automatically considered in global 3D simulations, but in many cases it turns out to be insufficient for obtaining solar-like differential rotation.

Recently Rempel (2005) showed that coupling between the tachocline and convection zone can also provide the latitudinal entropy variation needed to explain the observed profile of solar differential rotation. A typical solution from that model is shown in Fig. 3.1, displaying differential rotation (a), corresponding entropy perturbation (b) and the stream function of the meridional flow (c). Panel (d) shows for comparison the profile of differential rotation obtained if the effects of the entropy perturbation displayed in (b) are neglected. Inclusion of this effect through the bottom boundary condition in 3D models allows also for more solar-like differential rotation in 3D convection models (Miesch et al. 2006).

While there is a general agreement that thermal effects are essential for solar-like differential rotation, it is still unclear whether the required latitudinal entropy variation is a consequence of anisotropic convective energy transport, imposed by the tachocline, or a combination of both.

How does the meridional flow come into play here? A stationary solution requires that the divergence of the angular momentum flux Eq. (3.1) vanishes. While in very special situations the Reynolds stress can be divergence free on its own, in general a contribution from the meridional flow is required to close the system. It

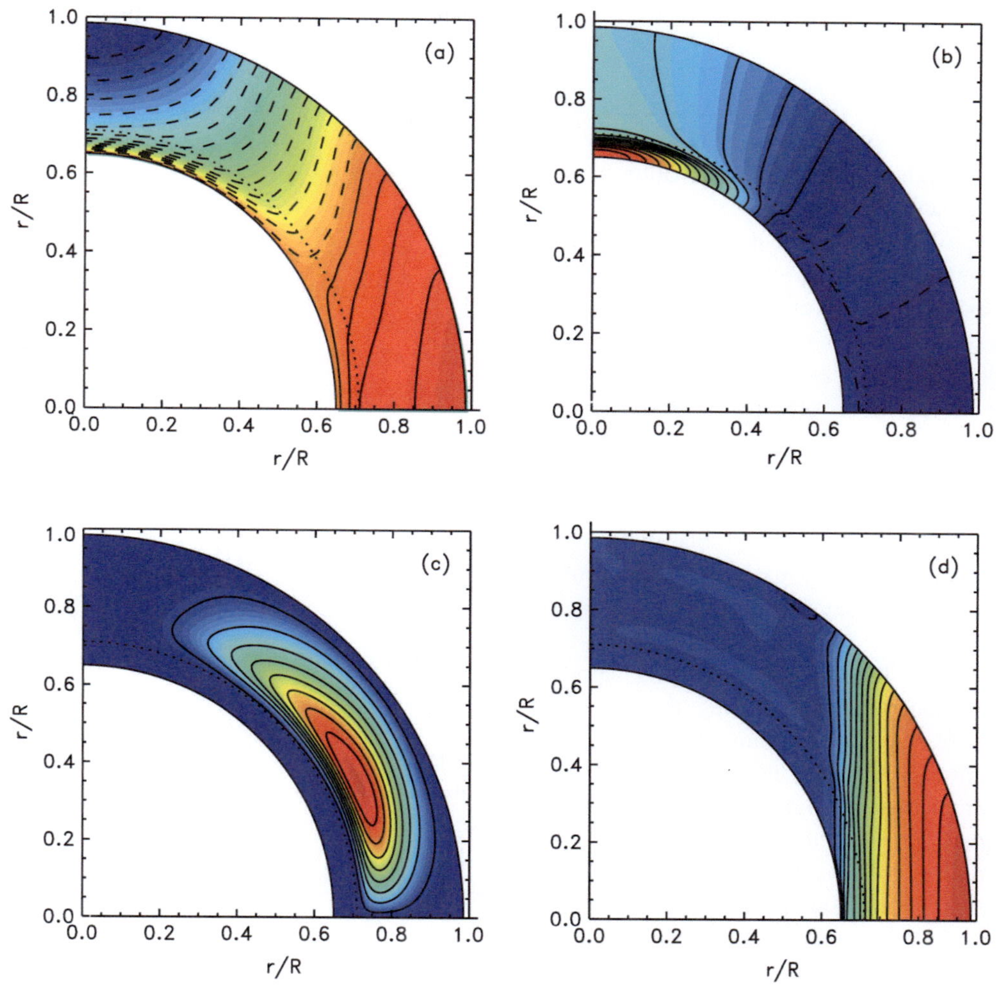

Fig. 3.1 Contour plots of differential rotation (**a**), entropy perturbation (**b**) and stream function of meridional flow (**c**) using the mean field model of Rempel (2005). Panel (**d**) shows the differential rotation profile obtained using the same parametrization of the Reynolds stress but neglecting the effects of baroclinicity

turns out that primarily the component of the Reynolds stress that transports angular momentum parallel to the axis of rotation influences most strongly the direction of the meridional flow. If the transport of angular momentum is inward directed, the resulting meridional flow is poleward at the surface and equatorward near the bottom of the convection zone. While this is found in most mean field models such as Kitchatinov and Rüdiger (1995), 3D simulations present a more complicated situation. Early models with lower resolution (Brun and Toomre 2002) typically show multicellular flow pattern, while a recent high resolution run (Miesch et al. 2008) shows a single flow cell (poleward at top, equatorward at bottom of CZ) in the bulk of the convection zone. To which degree these results are converged with respect to numerical resolutions remains to be seen in the future.

Differential rotation shows also cyclic variations known as torsional oscillations, which point toward a close relation to the solar magnetic cycle. We refer here to Howe (2009) and Brun and Rempel (2009) for reviews of observations as well as theoretical models for the time varying component of solar differential rotation.

3.3 Solar Dynamo

Similar to models of differential rotation and meridional flow we discussed in Section 3.2, also the solar dynamo is modeled through a combination of mean field models and 3D simulations. Currently mean field models of the solar dynamo are the only models that provide dynamo solutions that are compatible with basic cycle features and can be evolved over time scales much longer than a cycle. However, as already stated above, these models are heavily dependent on parametrization of turbulent induction processes and cannot provide an explanation from first principles. On the other hand, 3D MHD simulations describe currently only aspects of the dynamo process, a comprehensive model of a solar dynamo with features compatible with the basic dynamo constraints derived from the solar butterfly diagram is still an open challenge.

Regardless of the adopted modeling approach the primary uncertainties regarding the underlying dynamo process are similar. Many of these uncertainties result directly from our limited ability to model processes from first principles and the rather sparse observational constraints on the solar interior. The best known ingredient is differential rotation (Ω-effect) due to observational constraints from helioseismology on the mean profile and variation of differential rotation (Howe 2009). But even the exact knowledge of the differential rotation profile is not sufficient to determine whether radial velocity gradients at the base of the convection zone (tachocline) or latitudinal gradients in the bulk of the convection zone play the major role in the generation of toroidal magnetic fields, since this would require knowledge of the detailed distribution of poloidal field in the convection zone. Even less known are the processes related to the regeneration of poloidal field (α-effect). A third unknown are the transport processes of magnetic flux in the convection zone. Since in general the locations where the α-effect and Ω-effect operate do not coincide, transport of magnetic flux inbetween these regions is crucial for a coherent operation of the large dynamo in the convection zone.

It is beyond the scope of this chapter to review all the possible dynamo scenarios which have been considered and we refer to Charbonneau (2005) for further reading. In the following three subsections we will briefly discuss some of the key uncertainties and open questions.

3.3.1 Role of Tachocline

Soon after helioseismology revealed the detailed structure of differential rotation in the solar interior (Thompson et al. 1996) it was suggested that the base of the convection zone with its strong radial shear layer (tachocline) is a likely location for the solar dynamo (production of strong toroidal field by shear). In addition, the stable stratification found in the solar overshoot region at the base of the convection zone allows for storage of magnetic field over time scales comparable to the solar cycle. Both aspects are crucial since simulations of rising flux in the convection zone as well as studies of magnetic stability in the solar overshoot regions (see Section 3.4 for further detail) point toward a rather strong toroidal magnetic field of 10^5 G at the base of the convection zone. More recently the role of the tachocline for the organization and amplification of large scale toroidal field has been also seen in global 3D MHD simulations of the solar dynamo (Browning et al. 2006). However, Brown et al. (2009) presented simulations of solar like stars at faster rotation rates, which point toward the possibility that substantial magnetic field can be produced and maintained within the convection zone in near equatorial regions. It is currently not clear to which degree this result can be also relevant for the solar rotation rate.

While most models of flux emergence point toward a field strength of 10^5 G at the base of the convection zone, it is far from trivial to amplify field to this strength solely through differential rotation. Dynamo models that include non-linear feedbacks consistently (Rempel 2006) lead to an upper limit more around 10^4 G, similar values are also found in most 3D simulations such as (Browning et al. 2006). Whether this discrepancy can be bridged through an alternative field amplification mechanism (e.g. harvesting potential energy of the stratification as proposed by Rempel and Schüssler (2001)) or the possibility that also initially weaker magnetic field from the bulk of the convection zone can lead to active region formation is currently an open question.

3.3.2 Regeneration Process of Poloidal Field

The details of the processes rebuilding the poloidal field from toroidal field are still very uncertain. In the meanfield language these processes are formally described as α-effect and in the context of solar dynamo models the following 3 classes of α-effects are typically considered: (1) Helical turbulence, (2) MHD shear flow instabilities in the tachocline, (3) Rising flux tubes in the convection zone (Babcock-Leighton). While all these processes are likely to contribute, their amplitude and spacial distribution is not known well enough to clearly quantify their individual role.

Furthermore recent research also points toward highly non-linear and also time dependent α-effects resulting from additional constraints due to conservation of magnetic helicity (Brandenburg and Sandin 2004). Indirect constraints on the operation of the α-effect might be gained from helicity fluxes observable in the photosphere and above.

The only α-effect contribution that is directly constrainable through observations is the Babcock-Leighton α-effect (Babcock 1961; Leighton 1969), which has been used in most of the recent flux-transport dynamo models (Dikpati and Charbonneau 1999; Dikpati et al. 2004; Rempel 2006). The Babcock-Leighton α-effect is based on the flux emergence process leading to the formation of active regions, the key ingredient is the systematic tilt resulting from the action of the Coriolis force twisting the rising flux tube. While it is possible to construct dynamo models entirely based on the Babcock-Leighton α-effect, these models lead in general to rather strong polar fields at the surface in contradiction with observations, unless a strong magnetic diffusivity gradient and additional contributions from α-effects at the base of the convection zone are considered (Dikpati et al. 2004).

3.3.3 Transport of Magnetic Flux in Convection Zone

Traditionally most models considered only turbulent transport in the convection zone, which can be decomposed (in the meanfield language) into diffusive transport (turbulent diffusion) but also advection like transport in form of turbulent pumping. The latter has been also studied extensively through 3D MHD simulations (Tobias et al. 1998, 2001). If magnetic field becomes sufficiently strong magnetic buoyancy drives additional transport in terms of rising flux bundles that can lead to the formation of active regions on the visible surface (see Section 3.4 for more detail). Additional to these processes magnetic flux can be transported by the large scale meridional flow in the convection zone. Dynamo models based primarily on the latter are called flux transport dynamos and were first introduced by Choudhuri et al. (1995) and Dikpati and Charbonneau (1999) and have been developed further by several groups since then.

The attraction of flux transport dynamos comes primarily from the fact that a meridional flow which is poleward at the top and equatorward at the bottom of the convection zone gives a very robust explanation for the equatorward propagation of the activity in the course of the solar cycle. In addition the poleward flow in the near surface levels in combination with the systematic tilt angle of sunspot groups leads automatically to the correct phase relation between toroidal and poloidal field. However, as pointed out by Schüssler (2005), the phase relation is primarily a consequence of the tilt angle of active regions and in that sense only a weak constraint on dynamo processes in the solar interior. For meridional flow speeds consistent with surface observations and an extrapolation based on mass conservation for the deeper layers, these models also yield dynamo periods in agreement with the solar cycle.

Overall the flux transport picture is currently one of the most successful scenarios for the large scale solar dynamo, but (as many other models) it is based on two strong assumptions which cannot be proven from first principles: (1) The meridional flow is dominated by one flow cell with poleward flow close to the surface layers and equatorward flow at the base of the convection zone. (2) Turbulent transport processes are sufficiently weak to allow advection effects to dominate. While meanfield models of differential rotation and meridional flow typically lead to the required flow patterns (see e.g. Küker and Rüdiger 2005), the situation is more complicated in 3D simulations. Most of the earlier models at moderate resolution lead to multi-cellular flows (Brun and Toomre 2002), in contrast a more recent model at higher resolution shows a flow pattern dominated by a single cell in the convection (Miesch et al. 2008). Overall the situation cannot be considered as converged yet. The amplitude of turbulent transport estimated from simple mixing length arguments is typically 1–2 orders of magnitude larger than the values required for the flux transport picture. Since turbulent transport is in general more complicated than a simple diffusive transport this aspect needs to be studied in more detail through 3D simulations taking into account the presence of large scale flows and the full non-linearity of the problem.

3.4 Flux Emergence Process

It is generally accepted that sunspots form from magnetic field rising from the base of the convection zone to the surface (see reviews by Moreno-Insertis 1997; Fisher et al. 2000; Fan 2004, and further references therein), Solar dynamo models as presented in Section 3.3 focus on the large scale evolution of magnetic field and cannot address detailed processes such as the flux emergence process leading to the formation of sunspots on the visible surface of the sun. The latter is primarily a consequence of limited numerical resolution. Nevertheless, studying flux emergence is integral to our understanding of solar magnetism, since it allows us to connect the solar dynamo to observational constraints on the magnetic field structure in the solar photosphere. To date the flux emergence process has been studied decoupled from dynamo models using a variety of different approaches.

3.4.1 *Flux Emergence in Lower Convection Zone*

Early work was based on the thin flux tube approximation (Choudhuri and Gilman 1987; Fan et al. 1993, 1994; Moreno-Insertis et al. 1994; Schüssler et al. 1994; Caligari et al. 1995). These studies concluded that the overall properties of active regions, such as the low latitude of emergence, latitudinal trend in tilt angles as well as asymmetries between leading and following spots can be understood on the basis of rising flux tubes, provided the initial field strength at the base of the convection zone is around 100 kG. This conclusion was also consistent with stability considerations of flux in a subadiabatic overshoot region (Ferriz-Mas and Schüssler 1993, 1995).

Based on two-dimensional MHD simulations it was early realized by Schüssler (1979) that untwisted magnetic flux tubes cannot rise coherently and fragment. It was shown later by Moreno-Insertis and Emonet (1996); Emonet and Moreno-Insertis (1998) that this fragmentation can be alleviated provided that flux tubes have enough initial twist.

More recently also 3D MHD simulations of rising flux tubes based on the anelastic approximation have become possible (Fan 2008) and give support for results from earlier simulations based on the thin flux tube approximation. It was however found by Fan (2008) that there is a very delicate balance between the amount of twist required for a coherent rise and the amount of twist allowed to be in agreement with observations of sunspot tilt angles (twist with the observed sign produces a tilt opposite to the effect of Coriolis forces on rising tubes).

The simulations presented above consider the flux emergence process decoupled from convection. First attempts to address flux emergence in global simulations of the convection zone were made recently by Jouve and Brun (2007, 2009). Understanding the interaction of emerging flux with the ambient convective motions in the convective envelope is a crucial step toward more realism; however, currently the focus on the global scale limits the resolution required to resolve this interaction in detail. Substantial progress will likely happen in the next decade with increase in computing power.

3.4.2 *Flux Emergence in Upper Convection Zone*

Another challenge is understanding the last stages of the flux emergence process in the near surface layers (last 10–20 Mm). All of the models presented above exclude the upper most 10–20 Mm since the adopted approximations (thin flux tube, anelastic) loose their validity and also the steep decrease of pressure scale height and increase in convective velocities would lead to very stringent resolution and time step constraints.

The upper most layers of the convection zone require fully compressible MHD and also the inclusion of radiative processes is necessary if a detailed comparison with the available high resolution observations is desired (Cheung et al. 2007, 2008, 2010). While the primary modeling focus in the deep convection zone lies on large scale properties of active regions, near surface simulations focus on the detailed interaction of emerging flux with convection on the scale of granulation and below. One of the major open questions concerns the re-amplification process of magnetic flux into coherent sunspots from flux that has risen through a convection zone with a density stratification of six orders of magnitude.

Recently MHD simulations with radiative transfer also provided a breakthrough in our understanding of sunspot fine structure in the photosphere such as umbral dots, penumbral filaments, light bridges and the Evershed flow in terms of a common magneto-convection process modulated by inclination angle and field strength (Schüssler and Vögler 2006; Heinemann et al. 2007; Rempel et al. 2009a, b).

3.4.3 Open Questions, Connection to Dynamo Models

While we have seen strong progress in modeling the flux emergence process over the past decades, we do not have at this point a fully consistent model. The latter is a consequence of the fact that different aspects are modeled independently due to computational constraints. As a consequence there are some "missing links" between different models, which have to be addressed in the future through a more coherent coupling of models. Here we mention just a few of the open questions: (1) As pointed out before most models of emerging magnetic flux require an initial field strength of about 10^5 G at the base of the convection zone to be consistent with observational constraints. On the other hand the majority of dynamo models falls short of such values, more typical are 10^4 G. (2) Due to the strong density stratification in the convection zone even magnetic field with initially 10^5 is weakened to sub kG field strength in the upper most layers of the convection during the emergence process. It is currently an open question if such weak field can get re-amplified to sunspot field strength. Near surface simulations start very often from 10 kG field about 5 Mm beneath the photosphere (Cheung et al. 2008) to overcome the influence of convective motions. (3) Rising Ω-shaped flux tubes in the deep convection zone form typically as low wave number instability ($m = 1$ and $m = 2$ modes are preferred). In the near surface layers such low wave numbers should lead to much faster diverging motions in bipolar sunspot groups as observed (due to magnetic tension) if sunspots stay connected to their deep roots. A possible dynamical "disconnection" mechanism has been proposed by Schüssler and Rempel (2005), but it is also unclear if sunspots are sufficiently stable if they are rather shallow.

3.5 Summary

We presented here a brief summary of the state of the art of modeling of dynamical processes in the solar convection zone with focus on differential rotation/meridional flow, the large scale solar dynamo and the flux emergence process. We see currently in this field a dramatic change from more simplified models toward large 3D MHD simulations, primarily driven by the strong increase in computing resources. At the same time the field suffers from very limited observational constraints on processes in the solar interior. Progress in the future will heavily rely on improving and exploiting helioseismic constraints and also on coupling models to allow for a check of consistency. In the near terms the latter is likely to be most successful for models of the flux emergence process.

Acknowledgements The National Center for Atmospheric Research is sponsored by the National Science Foundation. This research was partially funded through NASA award number NNH09AK14I.

References

Babcock HW (1961) ApJ 133:572

Brandenburg A, Sandin C (2004) A&A 427:13

Brandenburg A, Tuominen I, Moss D, Ruediger G (1990) Sol Phys 128:243

Brown BP, Browning MK, Miesch MS, Brun AS, Toomre J (2009) ArXiv e-prints

Browning MK, Miesch MS, Brun AS, Toomre J (2006) ApJ 648:L157
Brun AS, Rempel M (2009) Space Sci Rev 144:151
Brun AS, Toomre J (2002) ApJ 570:865
Caligari P, Moreno-Insertis F, Schüssler M (1995) ApJ 441:886
Charbonneau P (2005) Living Rev Solar Phys 2:2
Cheung MCM, Schüssler M, Moreno-Insertis F (2007) A&A 461:1163
Cheung MCM, Schüssler M, Tarbell TD, Title AM (2008) ApJ 687:1373
Cheung MCM, Rempel M, Title AM, Schüssler M (2010) ApJ 720:233–244, arXiv:1006.4117
Choudhuri AR, Gilman PA (1987) ApJ 316:788
Choudhuri AR, Schüussler M, Dikpati M (1995) A&A 303:L29
Dikpati M, Charbonneau P (1999) ApJ 518:508
Dikpati M, de Toma G, Gilman PA, Arge CN, White OR (2004) ApJ 601:1136
Durney BR, Spruit HC (1979) ApJ 234:1067
Emonet T, Moreno-Insertis F (1998) ApJ 492:804
Fan Y (2004) Living Rev Solar Phys 1:1
Fan Y (2008) ApJ 676:680
Fan Y, Fisher GH, Deluca EE (1993) ApJ 405:390
Fan Y, Fisher GH, McClymont AN (1994) ApJ 436:907
Ferriz-Mas A, Schüssler M (1993) Geophys Astrophys Fluid Dyn 72:209
Ferriz-Mas A, Schüssler M (1995) Geophys Astrophys Fluid Dyn 81:233
Fisher GH, Fan Y, Longcope DW, Linton MG, Pevtsov AA (2000) Sol Phys 192:119
Gilman PA (1979) ApJ 231:284
Glatzmaier GA (1984) J Comput Phys 55:461
Hathaway DH (1984) ApJ 276:316
Heinemann T, Nordlund Å, Scharmer GB, Spruit HC (2007) ApJ 669:1390
Howe R (2009) Living Rev Solar Phys 6:1
Jouve L, Brun AS (2007) Astron Nachr 328:1104
Jouve L, Brun AS (2009) ApJ 701:1300
Kitchatinov LL, Rüdiger G (1993) A&A 276:96
Kitchatinov LL, Rüdiger G (1995) A&A 299:446
Kitchatinov LL, Rüdiger G (2005) Astron Nachr 326:379
Küker M, Rüdiger G (2005) Astron Nachr 326:265
Leighton RB (1969) ApJ 156:1
Miesch MS, Brun AS, De Rosa ML, Toomre J (2008) ApJ 673:557
Miesch MS, Brun AS, Toomre J (2006) ApJ 641:618
Miesch MS, Elliott JR, Toomre J, Clune TL, Glatzmaier GA, Gilman PA (2000) ApJ 532:593
Moreno-Insertis F (1997), Mem Soc Astron Italiana 68:429
Moreno-Insertis F, Caligari P, Schüssler M (1994) Sol Phys 153:449
Moreno-Insertis F, Emonet T (1996) ApJ 472:L53+
Nordlund Å, Stein RF, Asplund M (2009) Living Rev Solar Phys 6:2
Rempel M (2005) ApJ 622:1320
Rempel M (2006) ApJ 647:662
Rempel M, Schüssler M (2001) ApJ 552:L171
Rempel M, Schüssler M, Cameron RH, Knölker M (2009a) Science 325:171
Rempel M, Schüssler M, Knölker M (2009b) ApJ 691:640
Rüdiger G, Hollerbach R (2004) The magnetic universe: geophysical and astrophysical dynamo theory (ISBN 3-527-40409-0). Wiley-VCH, Weinheim
Schüssler M (1979) A&A 71:79
Schüssler M (2005) A&A 439:749
Schüssler M, Caligari P, Ferriz-Mas A, Moreno-Insertis F (1994) A&A 281:L69
Schüssler M, Rempel M (2005) A&A 441:337
Schüssler M, Vögler A (2006) ApJ 641:L73
Thompson MJ, Toomre J, Anderson ER, Antia HM, Berthomieu G, Burtonclay D, Chitre SM, Christensen-Dalsgaard J, Corbard T, De Rosa M, et al (1996) Science 272:1300
Tobias SM, Brummell NH, Clune TL, Toomre J (1998) ApJ 502:L177
Tobias SM, Brummell NH, Clune TL, Toomre J (2001) ApJ 549:1183

Chapter 4

Solar Oscillations: Current Trends

Sébastien Couvidat

Abstract Oscillations observed at the surface of the Sun through Doppler shifts of spectral lines or through irradiance measurements are used by solar physicists to investigate subsurface physical processes. This is the purpose of helioseismology. Seminal results were obtained since the inception of this research field in 1962, regarding both the global structure and dynamics of the Sun: e.g., detection of the presence of the tachocline—a region of intense radial shear in the rotation velocity, at the base of the convection zone—, measurement of the depth of this convection zone, measurement of the flow velocity of meridional circulation, of the torsional oscillations, determination of the abundance of photospheric helium, of the solar neutrino fluxes, and of the solar rotation profile... Moreover, we now access the local properties—structure and flows—of magnetized regions (sunspots) and supergranules, using oscillations that are not global. In this chapter we briefly summarize some of these results and emphasize the current trends in both global and local helioseismology.

4.1 Helioseismology: A Brief Overview

The first definite observations of solar oscillations in 1962 (Leighton et al. 1962) were understood as the solar surface response to standing waves trapped in the solar interior at the beginning of the 1970s (Ulrich 1970; Leibacher and Stein 1971). With the identification of ridges in the $k_h - \omega$ diagram (Deubner 1975), where k_h is the horizontal wavenumber and ω is the angular frequency, came the confirmation that solar oscillations exhibit a modal structure. They are therefore conveniently described in term of spherical harmonics: the spherical component of the eigenfunction of such an oscillation mode is described by the degree ℓ (the number of nodal lines on the sphere) and the azimuthal order m of a spherical harmonics. The radial component of the eigenfunction is described by the (radial) order n (roughly speaking the number of nodes in the radial direction).

Once the interest of these standing waves as probes to the solar interior was established, the field of global helioseismology was born and first inversions of the radial solar rotation (Duvall et al. 1984) and the radial sound-speed profile (Christensen-Dalsgaard et al. 1985) were performed. Indeed, both the structure and dynamics of the Sun can be studied with helioseismology. Since the beginning of the 1990s, the emphasis is put on local-helioseismology techniques, namely time-distance helioseismology (Duvall et al. 1993), ring-diagram analysis (Hill 1988), and acoustic holography (Lindsey and Braun 1990). These techniques allow a study of local structures, like sunspots or supergranules, and also, e.g., of possible asymmetries between the northern and southern solar hemispheres. In this chapter we remind the reader of a few results of global and local helioseismology, and we mention some of the current trends of these fields. The reader is invited to consult the recent review (Chaplin and Basu 2008) for further details regarding the prospects of global helioseismology.

S. Couvidat (✉)
W.W. Hansen Experimental Physics Laboratory, Stanford University, Stanford, CA 94305, USA
e-mail: couvidat@stanford.edu

M.P. Miralles, J. Sánchez Almeida (eds.), *The Sun, the Solar Wind, and the Heliosphere*, IAGA Special Sopron Book Series 4, DOI 10.1007/978-90-481-9787-3_4,

4.2 A Few Historical Results

After the confirmation of the modal structure of global solar oscillations, physicists envisaged their potential for measuring physical properties of the interior of the Sun (e.g. Christensen-Dalsgaard and Gough 1976; Severny et al. 1979).

Indeed, helioseismic data were soon utilized to improve the 1D solar models: helioseismology gives us access to, for instance, the sound-speed, density, and first adiabatic exponent (Γ_1) radial profiles. These profiles demonstrated, among others, the positive impact that inclusion of the microscopic diffusion of elements has on solar models (e.g. the review Christensen-Dalsgaard 2002). Moreover, the helioseismic (or observed) sound-speed and density profiles allowed the production of seismic solar models which confirmed the deficit of solar neutrinos observed on Earth (e.g. Shibahashi and Takata 1996; Turck-Chièze et al. 2001; Couvidat et al. 2003a) and thus helped solve the solar neutrino puzzle (described in Bahcall and Ulrich 1988). Figure 4.1 shows relative-difference profiles for the sound speed and density and for various solar models of 2001, including a standard model with microscopic diffusion and a seismic model. The good agreement between the seismic sound-speed profile and the profiles predicted by the solar models at that time is striking, and is a direct consequence of the improvement in the modeling of the Sun that resulted from helioseismic constraints.

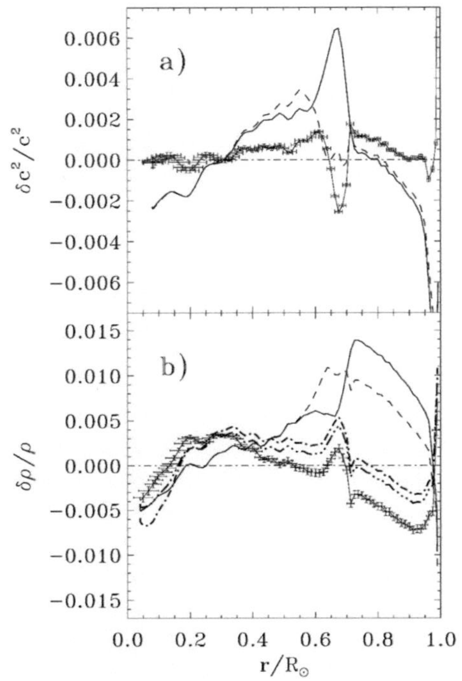

Fig. 4.1 Figure from (Turck-Chièze et al. 2001). Relative differences between (**a**) the square of the sound speed and (**b**) the density, deduced from helioseismic data, and those of different solar models. These solar models are a reference standard model (*solid line*), the model Btz of (Brun et al. 1999) (*dashed line*) which includes mixing in the tachocline, and a seismic model (*solid line with error bars*). On panel (**b**) two other models were added for which some nuclear reaction rates have been modified compared to their values in the standard model

The seismic sound-speed profile was also used to determine the location of the base of the convection zone, at about 0.7 $R_\odot$ (where $R_\odot = 6.9599 \times 10^5$ km is the solar radius), thanks to a change in the slope of this profile produced by the transition from a predominantly radiative to a predominantly convective mode of energy transportation (e.g. Gough 1986; Christensen-Dalsgaard et al. 1991; Basu and Antia 1997).

Similarly, the abundance of helium at the solar surface, which cannot be measured spectroscopically due to the noble-gas properties of this element, was obtained from the Γ_1 profile: Γ_1 drops sharply near the solar surface in the zones of ionization of helium, and the amplitude of this drop depends on the abundance of this element (e.g. Gough 1984; Christensen-Dalsgaard and Pérez Hernández 1991; Dziembowski et al. 1991; Vorontsov et al. 1991).

Finally, the inversion of the rotation profile as a function of depth and co-latitude (e.g. Brown et al. 1989; Schou et al. 2002) showed the existence of a region of intense radial shear in the rotation velocity at the base of the convection zone: the tachocline (whose thickness was measured by, e.g., Corbard et al. 1998, at about 0.05 $R_\odot$). This tachocline is now widely thought to be the seat of the solar dynamo that produces an intense toroidal magnetic field from a poloidal one and is responsible for the solar activity cycle.

4.3 Current Research Trends in Global Helioseismology

4.3.1 *Search for Gravity Modes*

Unlike the acoustic waves (p modes), which are excited in, and mainly sample, the convection zone, the gravity waves (g modes) are mostly sensitive to physical

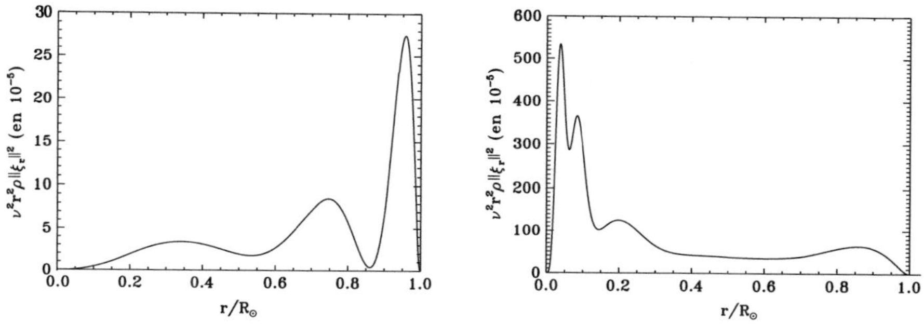

Fig. 4.2 Radial kinetic energy density as a function of the fractional solar radius for a "typical" acoustic mode (p mode) of degree $\ell = 2$ and radial order $n = 2$ (*left panel*), and a gravity mode (g mode) of degree $\ell = 2$ and radial order $n = -2$ (*right panel*)

phenomena inside the solar radiative zone—especially the solar core—and are evanescent in the convection zone. Figure 4.2 shows how the radial kinetic energy density of a typical p mode peaks near the solar surface, while for a typical g mode it peaks in the solar core. Therefore, g modes are mainly sensitive to, and can inform us about, physical processes inside this nuclear core. Their detection is an arduous problem, considering the current estimate of the upper limit of their amplitude at the solar surface: about 10 mm/s (Appourchaux et al. 2000). The search for these elusive g modes has been going on for decades (e.g. Delache and Scherrer 1983). Looking at the power spectral density of Doppler-velocity time series from the GOLF (Global Oscillation at Low Frequencies) instrument on board the SOHO spacecraft, Turck-Chièze et al. (2004) found some statistically significant peaks, potentially g modes at the level of a few mm/s. Recently, García et al. (2007) used a different approach: they focused on the asymptotic properties of low-degree high-order g modes to successfully detect their presence, with the caveat that this technique does not give access to individual modes but only to the general behaviour of a specific group of modes. These results of the solar physics group at Saclay, France, favor a solar core rotating faster than the rest of the radiative zone.

4.3.2 Rotation of the Solar Core

With only p modes detected with certainty so far, it is difficult to extract the rotation rate in the solar core. Indeed, inversions of the internal solar rotation using low-degree p modes reach only about 0.25 $R_\odot$ (e.g. Eff-Darwich et al. 2002; Couvidat et al. 2003b; Thompson et al. 2003; García et al. 2004), and give somewhat different results at that depth: usually either showing a core rotating slightly more slowly than the rest of the radiative interior (which rotates like a solid body), or at about the same rate. Inversions showing a faster rotating core are rare. Detecting more low-degree low-order p modes and with a better accuracy in their frequency determination would help, but the breakthrough is expected to come from the unambiguous detection of—even a few—g modes, as is clearly shown by (Mathur et al. 2007).

4.3.3 Problem of Solar Abundances

The abundances of chemical elements at the solar surface have recently been revised (e.g. Asplund et al. 2005) using 3D atmosphere models, a NLTE treatment whenever possible, and better defined spectral lines (thus getting rid of some blends): in particular the abundances of C, N, O, and Ne were reduced by at least 35% compared to previous estimates (Grevesse and Noels 1993). This led to a reduction in the photospheric Z/X ratio (where Z is the mass fraction of metals, and X is the mass fraction of hydrogen) from 0.0245 (from Grevesse and Noels 1993) to 0.0165. These new abundances create a discrepancy with helioseismic results: the solar seismic sound-speed profile becomes significantly different from the profile predicted by the solar models based on these new abundances, and the base of the convection zone in these models is displaced. To reconcile models and seismic observations, attempts have been made at increasing the radiative opacities (e.g. Basu and Antia 2004) by as much as 20%, increasing the abundance of metals in the radiative interior through enhanced microscopic diffusion (e.g. Guzik et al. 2005), or other more exotic solutions (such as mixing by gravity waves). None of these attempts has been fully satisfying so far.

4.3.4 *Variations with the Solar Activity Cycle*

It is well established that p-mode frequencies increase with the solar activity (e.g. Libbrecht and Woodard 1990; Elsworth et al. 1990; Rabello-Soares et al. 2008). Higher frequencies increase more than lower ones, suggesting that the perturbation responsible for these changes is located mainly near the solar surface. However, the physical mechanism producing this variation is still unclear.

Deeper in the Sun, Howe et al. (2000) found a periodic variation in the solar rotation rate at the base of the convection zone. The period is about 1.3 years near the equator, and 1 year at higher latitudes. The amplitude of the variation, 12 nHz at most out of a rotation rate of 450 nHz, is small.

Near the tachocline, Christensen-Dalsgaard et al. (2004) searched for evidence of (variable) prograde jets, as predicted by certain dynamo models to stabilize the strong toroidal field against a poleward slip.

Another example of variation with the solar cycle are the torsional oscillations: latitudinal bands of faster and slower than average flows, migrating with the solar cycle. First described by Howard and LaBonte (1980), they were studied with helioseismology by Ulrich (2001). Today, inversions of helioseismic data allow for an inference of the depth dependence of these torsional oscillations (e.g. Zhao and Kosovichev 2004; Howe et al. 2005).

Meridional circulation is also an instance of large scale flow probably linked to the solar dynamo and being investigated with helioseismology (e.g. Giles et al. 1997).

Everything connected to the solar dynamo and the solar activity cycle is fair game for helioseismologists.

4.3.5 *Variations of the Solar Radius*

Even though a part of the variations with the solar cycle, we treat the possible change in the solar radius in this separate section. This variation of radius with solar activity has already been investigated using various methods, like Danjon astrolabes, which produced conflicting results. Helioseismology can help: f-mode (surface gravity waves) frequencies are sensitive to the radius $R_\odot$ (e.g. Schou et al. 1997; Antia 1998; Dziembowski and Goode 2005). Based on an integral relation (described in Dziembowski and Goode 2004) between f-mode frequencies and fractional radius, i.e. the radial distance to solar center divided by $R_\odot$, (Lefebvre and Kosovichev 2005) found that near the solar surface (from 0.99 $R_\odot$ to 1 $R_\odot$) the change of fractional radius is in anti-phase with the solar activity, while deeper (from 0.97 $R_\odot$ to 0.99 $R_\odot$) it is in phase. The meaning of this result is not yet understood.

4.4 Current Research Trends in Local Helioseismology

Here we only present a few results of local helioseismology, wich uses local acoustic or surface gravity waves. These waves are quickly damped during their journey inside the Sun and do not have a global reach: they do not interfere constructively in the "horizontal" direction, and therefore do not qualify as global modes.

4.4.1 *Structure and Dynamics of Sunspots*

The first inversion of the sound-speed perturbations below an active region was presented in Kosovichev et al. (2000). Since then, numerous articles have dealt with this topic (e.g. Jensen et al. 2001; Couvidat et al. 2004). Figure 4.3 shows an example of an inversion result for the sound speed below a sunspot. This plot exhibits a two-region structure: a decrease in the sound speed compared to the quiet Sun, immediately below the solar surface, followed by an increase in sound speed in deeper layers. Similarly, Zhao et al. (2001) applied the time-distance formalism to derive the flow velocities underneath a sunspot.

These groundbreaking results are controversial, due to the numerous assumptions involved in the sound-speed and flow-velocity inversion process, and that might not be valid in presence of the strong magnetic field of sunspots (e.g. Braun and Birch 2006; Couvidat and Rajaguru 2007). To emphasize but one example, it is known that acoustic waves crossing a sunspot are partly converted into slow and fast MHD waves

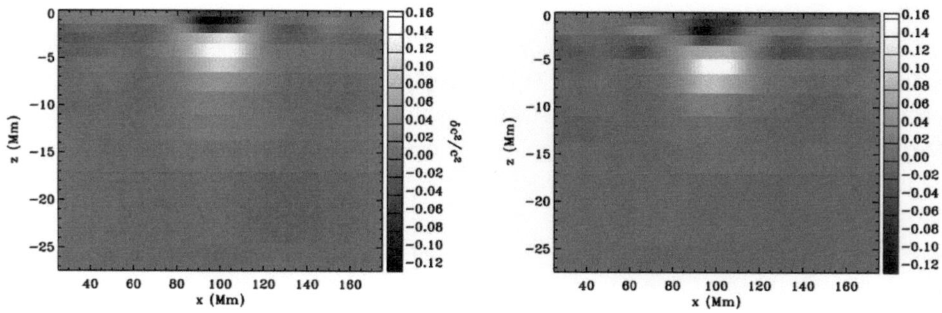

Fig. 4.3 Figure from (Couvidat et al. 2006). Vertical cut in the inversion results for the squared sound-speed perturbations—relative to the quiet Sun—and for the sunspot NOAA 8243 of June 1998, observed by MDI. *Left*: Inversion using Born-approximation sensitivity kernels. *Right*: Inversion using ray-path approximation sensitivity kernels

(e.g. Crouch and Cally 2005) which is not taken into account in the inversion of travel times of wavepackets in local helioseismology.

4.4.2 Remote Sensing of the Far-Side of the Sun

Acoustic holography was used to detect the presence of sunspots on the far-side of the Sun (Lindsey and Braun 1990). This is a first step in the long-term space-weather goal of forecasting solar flares and other disturbances that affect our technological societies more and more. Similarly, Zhao (2007) developed the same far-side imaging capabilities using time-distance helioseismology. The basic concept behind far-side imaging is that active regions produce a negative travel-time perturbation of several seconds—relative to the quiet Sun—for the specific wavepackets considered, which travel from the far side to the near side (e.g. González Hernández et al. 2007). We can detect this perturbation long before the active regions become visible.

4.4.3 Supergranulation as a Travelling Wave

Using f modes to measure the horizontal divergence of the flows in supergranules, (Gizon et al. 2003) found that supergranulation undergoes oscillations and supports waves with periods of 6–9 days. These waves are predominantly prograde, which explains the apparent super-rotation of supergranules. The authors suggest that this interesting result could be the beginning of the use of supergranular waves to probe the upper convection zone.

4.4.4 Improvement of Travel-Time Measurement Methods, of Sensitivity-Kernel Calculations, and of Inversion Methods

A lot of work aims at improving local-helioseismology techniques. New travel-time definitions for wavepackets are created (Gizon and Birch 2002, 2004), while the sensitivity of these wavepackets to different kinds of perturbations (sensitivity expressed as 3D functions called sensitivity kernels) is being investigated in a manner less approximate than the ray-path—geometrical optics—method (Birch et al. 2001, 2004). The inversion techniques are being scrutinized and improved (e.g. Couvidat et al. 2005). There is also an attempt at understanding the validity of Doppler-velocity measurements (the primary data of helioseismology) in active regions (e.g. Rajaguru et al. 2007). Finally an emphasis is put on using MHD simulations of the solar convection zone to test local-helioseismology techniques and assumptions (e.g. Hanasoge et al. 2007, 2008; Parchevsky et al. 2008).

4.5 Some Future Instruments

Several instruments with helioseismic capabilities have recently been launched, or will be launched in the upcoming years.

The Helioseismic and Magnetic Imager (HMI) instrument was launched on board the Solar Dynamics Observatory (SDO) satellite in February 2010. This NASA mission succeeds to the SOHO spacecraft. HMI is an improved version of MDI (Michelson Doppler Imager): with two 4,096 × 4,096 CCDs, HMI measures both the Doppler velocity at the solar surface and the vector magnetic field. The spatial resolution is 0.5″ in full-disk mode, and the Doppler velocity data are obtained every 45 s.

Another successful instrument on board SOHO, GOLF, will hopefully be succeeded by GOLF-NG (e.g. Turck-Chièze et al. 2009). GOLF-NG will measure the sodium spectral line D1 at 8 points in the blue wing, and 8 points in the red wing, using the same resonant scattering cell technique as GOLF. A prototype of GOLF-NG has already been assembled and is currently working in the Canary Islands.

PICARD (e.g. Thuillier 2005), a CNES project, is planned for launch in 2010. Three instruments on board PICARD are of interest to helioseismologists: they will measure the solar spectral irradiance at different wavelengths, and the solar radius.

Finally, Solar Orbiter is an ESA project whose goal is the observation of the high latitudes of the Sun, and whose launch should occur sometime after 2015.

4.6 Conclusion

During the past four decades helioseismology provided astrophysicists with a flurry of spectacular results, and there is no sign of abatement. More sophisticated helioseismology instruments have recently been launched, or are soon to be launched, and will provide us with higher quality data, needed to further our understanding of the structure and dynamics of the Sun.

Acknowledgements This work was supported by the NASA Grant NAS5-02139 (HMI).

References

Antia HM (1998) Estimate of solar radius from f-mode frequencies. Astron Astrophys 330:336–340

Appourchaux T, Frohlich C, Andersen B, Berthomieu G, Chaplin JW, Elsworth Y et al (2000) Observational upper limits to low-degree solar g-modes. Astrophys J 538:401–414

Asplund M, Grevesse N, Sauval AJ (2005) The solar chemical composition. In: Proceedings of a symposium held 17–19 June, 2004 in Austin, Texas. ASP Conference Series 336, Astronomical Society of the Pacific, San Francisco, pp 25–38

Bahcall JN, Ulrich RK (1988) Solar models, neutrino experiments, and helioseismology. Rev Mod Phys 60:297–372

Basu S, Antia HM (1997) Seismic measurement of the depth of the solar convection zone. MNRAS 287:189–198

Basu S, Antia HM (2004) Constraining solar abundances using helioseismology. Astrophys J 606:L85–L88

Birch AC, Kosovichev AG, Price GH, Schlottmann RB (2001) The accuracy of the born and ray approximations in time-distance helioseismology. Astrophys J 561:L229–L232

Birch AC, Kosovichev AG, Duvall TL Jr (2004) Sensitivity of acousticwave travel times to sound-speed perturbations in the solar interior. Astrophys J 608:580–600

Braun DC, Birch AC (2006) Observed frequency variations of solar p-mode travel times as evidence for surface effects in sunspot seismology. Astrophys J 647:L187–L190

Brown TM, Christensen-Dalsgaard J, Dziembowksi WA, Goode P, Gough DO, Morrow CA (1989) Inferring the sun's internal angular velocity from observed p-mode frequency splittings. Astrophys J 343:526–546

Brun AS, Turck-Chièze S, Zahn JP (1999) Standard solarmodels in the light of new helioseismic constraints. II. Mixing below the convective zone. Astrophys J 525:1032–1041

Chaplin WJ, Basu S (2008) Perspectives in global helioseismology and the road ahead. Solar Phys 251:53–75

Christensen-Dalsgaard J (2002) Helioseismology. Rev Mod Phys 74:1073–1129

Christensen-Dalsgaard J, Gough DO (1976) Towards a heliological inverse problem. Nature 259:89–92

Christensen-Dalsgaard J, Pérez Hernández F (1991) Influence of the upper layers of the sun on the p-mode frequencies. Lect Notes Phys 388, 43–50

Christensen-Dalsgaard J, Duvall TL Jr, Gough DO, Harvey JW, Rhodes EJ Jr (1985) Speed of sound in the solar interior. Nature 315:378–382

Christensen-Dalsgaard J, Gough DO, Thompson MJ (1991) The depth of the solar convection zone. Astrophys J 378:413–437

Christensen-Dalsgaard J, Corbard T, Dikpati M, Gilman PA, Thompson MJ (2004) Detection of jets and associated toroidal fields in the solar tachocline. In: Proceedings of the SOHO 14/GONG 2004 workshop, 12–16 July 2004, pp 376–380. ESA SP-559, New Haven

Corbard T, Berthomieu G, Provost J, Morel P (1998) Inferring the equatorial solar tachocline from frequency splittings. Astron Astrophys 330:1149–1159

Crouch AD, Cally PS (2005) Mode conversion of solar p-modes in non-vertical magnetic fields. Solar Phys 227:1–26

Couvidat S, Turck-Chièze S, Kosovichev AG (2003a) Solar seismic models and the neutrino predictions. Astrophys J 599:1434–1448

Couvidat S, Garcia RA, Turck-Chièze S, Corbard T, Henney CJ, Jiménez-Reyes S (2003b) The rotation of the deep solar layers. Astrophys J 597:L77–L79

Couvidat S, Birch AC, Kosovichev AG, Zhao J (2004) Three-dimensional inversion of time-distance helioseismology data: ray-path and fresnel-zone approximations. Astrophys J 607:554–563

Couvidat S, Gizon L, Birch AC, Larsen RM, Kosovichev AG (2005) Time-distance helioseismology: inversion of noisy correlated data. Astrophys J Suppl Ser 158: 217–229

Couvidat S, Birch AC, Kosovichev AG (2006) Three-dimensional inversion of sound speed below a sunspot in the born approximation. Astrophys J 640:516–524

Couvidat S, Rajaguru SP (2007) Contamination by surface effects of time-distance helioseismic inversions for sound speed beneath sunspots. Astrophys J 661:558–567

Delache P, Scherrer PH (1983) Detection of solar gravity mode oscillations. Nature 306:651–653

Deubner F-L (1975) Observations of low wavenumber nonradial eigenmodes of the sun. Astron Astrophys 44:371–375

Duvall TL Jr, Dziembowski WA, Goode PR, Gough DO, Harvey JW, Leibacher JW (1984) Internal rotation of the sun. Nature 310:22–25

Duvall TL Jr, Jefferies SM, Harvey JW, Pomerantz MA (1993) Time-distance helioseismology. Nature 362:430–432

Dziembowski WA, Goode PR (2004) Helioseismic probing of solar variability: the formalism and simple assessments. Astrophys J 600:464–479

Dziembowski WA, Goode PR (2005) Sources of oscillation frequency increase with rising solar activity. Astrophys J 625:548–555

Dziembowski WA, Pamiatnykh AA, Sienkiewicz R (1991) Helium content in the solar convective envelope from helioseismology. MNRAS 249:602–605

Eff-Darwich A, Korzennik SG, Jiménez-Reyes SJ (2002) Inversion of the internal solar rotation rate. Astrophys J 573:857–863

Elsworth Y, Howe R, Isaak GR, McLeod CP, New R (1990) Variation of low-order acoustic solar oscillations over the solar cycle. Nature 345:322–324

García RA, Corbard T, Chaplin WJ, Couvidat S, Eff-Darwich A, Jiménez-Reyes SJ et al (2004) About the rotation of the solar radiative interior. Solar Phys 220:269–285

García RA, Turck-Chièze S, Jiménez-reyes SJ, Ballot J, Pallé PL, Eff-Darwich A et al (2007) Tracking solar gravity modes. Science 316:1591–1593

Giles PM, Duvall TL, Scherrer PH, Bogart RS (1997) A subsurface flow of material from the Sun's equator to its poles. Nature 390:52–54

Gizon L, Birch AC (2002) Time-distance helioseismology: the forward problem for random distributed sources. Astrophys J 571:966–986

Gizon L, Birch AC (2004) Time-distance helioseismology: noise estimation. Astrophys J 614:472–489

Gizon L, Duvall TL, Schou J (2003) Wave-like properties of solar supergranulation. Nature 421:43–44

González Hernández I, Hill F, Lindsey C (2007) Calibration of seismic signatures of active regions on the far side of the sun. Astrophys J 669:1382–1389

Gough DO (1984) Towards a solar model. Mem Soc Astron Italiana 55:15–35

Gough DO (1986) Asymptotic sound-speed inversions. In: Proceedings of the NATO advanced research workshop, Cambridge, England, 17–21 June 1985, pp 125–140. D. Reidel Publishing Co., Dordrecht

Grevesse N, Noels A (1993) Cosmic abundances of the elements. In: Origin and evolution of the elements: proceedings of a symposium in honour of H. Reeves, held in Paris, 22–25 June 1992, pp 15–25. Cambridge University Press, Cambridge

Guzik JA, Watson LS, Cox AN (2005) Can enhanced diffusion improve helioseismic agreement for solar models with revised abundances? Astrophys J 627:1049–1056

Hanasoge SM, Duvall TL Jr, Couvidat S (2007) Validation of Helioseismology through Forward Modeling. Astrophys J 664:1234–1243

Hanasoge SM, Couvidat S, Rajaguru SP, Birch AC (2008) Impact of locally suppressed wave sources on helioseismic traveltimes. MNRAS 391:1931–1939

Hill F (1988) Rings and trumpets – Three-dimensional power spectra of solar oscillations. Astrophys J 333:996–1013

Howard R, LaBonte BJ (1980) The sun is observed to be a torsional oscillator with a period of 11 years. Astrophys J 239:L33–L36

Howe R, Christensen-Dalsgaard J, Hill F, Komm RW, Larsen RM, Schou J, Thompson MJ, Toomre J (2000) Dynamic variations at the base of the solar convection zone. Science 287:2456–2460

Howe R, Christensen-Dalsgaard J, Hill F, Komm RW, Schou J, Thompson MJ (2005) Solar convection-zone dynamics, 1995–2004. Astrophys J 634:1405–1415

Jensen JM, Duvall TL Jr, Jacobsen BH, Christensen-Dalsgaard J (2001) Imaging an emerging active region with helioseismic tomography. Astrophys J 553:L193–L196

Kosovichev AG, Duvall TL Jr, Scherrer PH (2000) Time-distance inversion methods and results. Solar Phys 192:159–176

Lefebvre S, Kosovichev AG (2005) Changes in the subsurface stratification of the sun with the 11-year activity cycle. Astrophys J 633:L149–L152

Leibacher JW, Stein RF (1971) A new description of the solar five-minute oscillation. Astrophys Lett 7:191–192

Leighton RB, Noyes RW, Simon GW (1962) Velocity fields in the solar atmosphere. I. Preliminary report. Astrophys J 135:474–499

Libbrecht KG, Woodard MF (1990) Solar-cycle effects on solar oscillation frequencies. Nature 345:779–782

Lindsey C, Braun DC (1990) Helioseismic imaging of sunspots at their antipodes. Solar Phys 126:101–115

Mathur S, Turck-Chièze S, Couvidat S, Garcia RA (2007) On the characteristics of the solar gravity mode frequencies. Astrophys J 668:594–602

Parchevsky KV, Zhao J, Kosovichev AG (2008) Influence of nonuniform distribution of acoustic wavefield strength on time-distance helioseismology measurements. Astrophys J 678:1498–1504

Rabello-Soares MC, Korzennik SG, Schou J (2008) Variations of the solar acoustic high-degree mode frequencies over solar cycle 23. Adv Space Res 41:861–867

Rajaguru SP, Sankarasubramanian K, Wachter R, Scherrer PH (2007) Radiative transfer effects on Doppler measurements as sources of surface effects in sunspot seismology. Astrophys J 654:L175–L178

Schou J, Kosovichev AG, Goode PR, Dziembowski WA (1997) Determination of the sun's seismic radius from the SOHO Michelson Doppler Imager. Astrophys J 489: L197–L200

Schou J, Howe R, Basu S, Christensen-Dalsgaard J, Corbard T, Hill F et al (2002) A comparison of solar p-mode parameters

from the Michelson Doppler Imager and the Global Oscillation Network Group. Astrophys J 567:1234–1249
Severny AB, Kotov VA, Tsap TT (1979) Solar oscillations and the problem of the internal structure of the sun. Astron Zh 56:1137–1148
Shibahashi H, Takata M (1996) A seismic solarmodel deduced from the sound-speed distribution and an estimate of the neutrino fluxes. PASJ 48:377–387
Thompson MJ, Christensen-Dalsgaard J, Miesch MS, Toomre J (2003) The internal rotation of the sun. Annu Rev Astron Astrophys 41:599–643
Thuillier G (2005) Status of solar global properties measurements. PICARD mission. Mem Soc Astron Italiana 76:901–906
Turck-Chièze S, Couvidat S, Kosovichev AG, Gabriel AH, Berthomieu G, Brun AS et al (2001) Solar neutrino emission deduced from a seismic model. Astrophys J 555:L69–L73
Turck-Chièze S, Garcia RA, Couvidat S, Ulrich RK, Bertello L, Varadi F et al (2004) Looking for gravity-mode multiplets with the GOLF experiment aboard SOHO. Astrophys J 604:455–468
Turck-Chièze S, Lamy P, Carr C, Carton PH, Chevalier A, Dandouras I et al (2009) The DynaMICCS perspective. Exp Astron 23:1017–1055
Ulrich RK (1970) The five-minute oscillations on the solar surface. Astrophys J 162:993–1002
Ulrich RK (2001) Very long lived wave patterns detected in the solar surface velocity signal. Astrophys J 560:466–475
Vorontsov SV, Baturin VA, Pamyatnykh AA (1991) Seismological measurement of solar helium abundance. Nature 349:49–51
Zhao J (2007) Time-distance imaging of solar far-side active regions. Astrophys J 664:L139–L142
Zhao J, Kosovichev AG (2004) Torsional oscillation, meridional flows, and vorticity inferred in the upper convection zone of the sun by time-distance helioseismology. Astrophys J 603:776–784
Zhao J, Kosovichev AG, Duvall TL Jr (2001) Investigation of mass flows beneath a sunspot by time-distance helioseismology. Astrophys J 557:384–388

Chapter 5

Theories of the Solar Cycle : A Critical View

Hendrik C. Spruit

Abstract Some established views of the solar magnetic cycle are discussed critically, with focus on two aspects at the core of most models: the role of convective turbulence, and the role of the "tachocline" at the base of the convection zone. The standard view which treats the solar cycle as a manifestation of the interaction between convection and magnetic fields is shown to be misplaced. The main ingredient of the solar cycle, apart from differential rotation, is instead buoyant instability of the magnetic field itself. This view of the physics of the solar cycle was already established in the 1950s, but has been eclipsed mathematically by mean field turbulence formalisms which make poor contact with observations and have serious theoretical problems. The history of this development in the literature is discussed critically. The source of the magnetic field of the solar cycle is currently assumed to be located in the "tachocline": the shear zone at the base of the convection zone. While the azimuthal field of the cycle is indeed most likely located at the base of the convection zone, it cannot be powered by the radial shear of the tachocline as assumed in these models, since the radiative interior does not support significant shear stresses. Instead, it must be the powered by the latitudinal gradient in rotation rate in the convection zone, as in early models of the solar cycle. Possible future directions for research are briefly discussed.

H.C. Spruit (✉)
Max Planck Institute for Astrophysics, 85741 Garching, Germany
e-mail: henk@mpa-garching.mpg.de

5.1 The Role of Convective Turbulence

For a star to generate a self-sustained magnetic field, it is sufficient that it rotate differentially. This differs from the traditional view of dynamos in stars, which holds that in addition to the shear flow due to differential rotation, a small scale velocity field has to be imposed in order to "close the dynamo cycle", thus enabling a selfsustained field independent of initial conditions. Convection can provide such a velocity field, and in fact convection has become such an integral part of thinking about dynamos in stars that the subject of "stellar magnetic fields" has been almost synonymous with "convective dynamos" for decades (for reviews see e.g. Weiss 1981–1997; Rüdiger and Hollerbach 2004; Tobias 2005, for recent texts Brandenburg 2009; Jones et al. 2009; Charbonneau 2005). Whether or not such a dynamo process can take place in principle is a separate matter. From the observations it is evident, however, that it is not the way the solar cycle works. Instead, as I will argue below, the cycle operates on dynamic instability of the magnetic field itself. Convection plays only an indirect role, namely by maintaining the differential rotation of the envelope from which the cycle derives its energy.

5.1.1 Mechanism of the Solar Cycle as Inferred from Observations

The common ingredient in all dynamo models such as those for the Earth's magnetic field or the solar cycle is the generation of a toroidal (azimuthally directed) field by stretching ("winding-up") of the lines of a poloidal

M.P. Miralles, J. Sánchez Almeida (eds.), *The Sun, the Solar Wind, and the Heliosphere*, IAGA Special Sopron Book Series 4, DOI 10.1007/978-90-481-9787-3_5, © Springer Science+Business Media B.V. 2011

field (e.g. Elsasser 1956). This is "the easy part". It produces a field that increases in strength linearly with time and is proportional to the imposed initial field. To produce a cyclic, self-sustained field as observed there must be a second step that turns some of the toroidal field into a new poloidal component, which is again wound up, completing a field-amplification cycle that becomes independent of initial conditions. The particular process by which the new poloidal field is generated distinguishes the models from each other. In early models of the solar cycle that were popular in their time (Babcock 1961, 1963; Leighton 1969) observations of the emergence of active regions were used to infer the nature of the process responsible for this key step in the dynamo cycle.

These models proposed that the increasing toroidal field eventually becomes unstable, erupting to the surface to form the observed active regions (Cowling 1953; Elsasser 1956; Babcock 1961, see sketches in Figs. 2, 5). The equatorward drift of the main zone of activity reflects the latitude dependence of the time taken for the toroidal field to reach the point of buoyant instability (Babcock 1961, 1963). This is illustrated with a simple model in Fig. 5.1. In this sketch, a uniform poloidal field is assumed to be stretched passively by the latitudinal differential rotation as observed on the surface of the Sun. Helioseismic observations (see review by Howe, 2009) show that this pattern of rotation also holds to a fair approximation inside the convection zone.

The azimuthal field becomes unstable to buoyant rise when a critical strength of $\sim 10^5$ G is reached

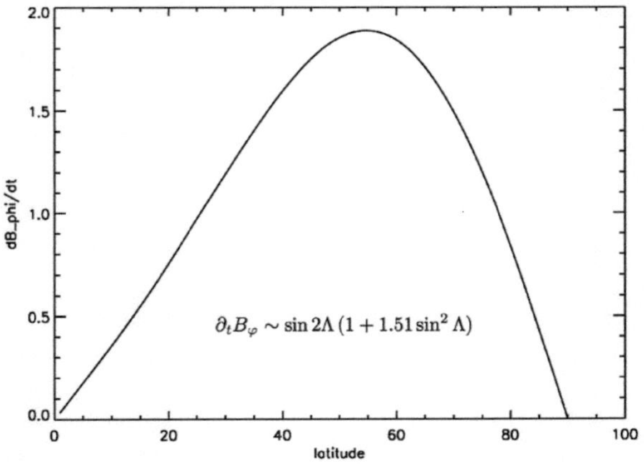

Fig. 5.1 Rate of increase of the azimuthal field strength as a function of heliographic latitude, due to the observed differential rotation acting on an assumed uniform poloidal field

(Schüssler et al. 1994). This happens first at the latitude where the rate of increase of the field is largest, around a latitude of 60° in the simple model of Fig. 5.1. This agrees with observations (Altrock 2010), though initially only small-scale magnetic activity without sunspots is produced. As time progresses, the field also becomes unstable at lower latitudes, producing an equatorward drift of the zone of activity. For reasons unknown, sunspots form only below a latitude of around 40°. As Fig. 5.1 implies, Babcock's model also predicts a poleward propagating branch. Such a branch (but without sunspots) is actually present on the Sun (the "poleward rush", Leroy and Trellis, 1974; Altrock 2010). Its observational status and interpretation are not entirely clear, however.

The process of emergence of an active region has been studied in great detail for more than a century. A small patch of fragmented magnetic fields with mixed polarities appears and expands as more flux emerges (Fig. 5.3). The surroundings of this patch remain unaffected by this process. The mix of polarities then separates into two clumps, the polarities traveling in opposite directions to their destination, ignoring the convective flows in the region.

This striking behavior is the opposite of diffusion. To force it into a diffusion picture, one would have to reverse the arrow of time. Instead of opposite polarities decaying by diffusing into each other, they segregate out from a mix. The MHD equations are completely symmetric with respect to the sign of the magnetic field, however. There are no flows (no matter how complex) that can separate fields of different signs out of a mixture. This rules out a priori all models attempting to explain the formation of sunspots and active regions by turbulent diffusion. For recent such attempts, which actually ignore the observations they are trying to explain, see Kitiashvili et al. (2010), Brandenburg et al. (2010). The observations, instead, demonstrate that the orientation and location of the polarities seen in an active region must already be have been present in the initial conditions: in the layers below the surface from which the magnetic field traveled to the surface.

The fragmented state near the surface in the early stages of the eruption process is only temporary. The intuitive "rising tree" picture (Zwaan, 1978) illustrates this (Fig. 5.4). The observed fragmentation and subsequent formation of spots from a horizontal strand of magnetic field below the surface has recently been

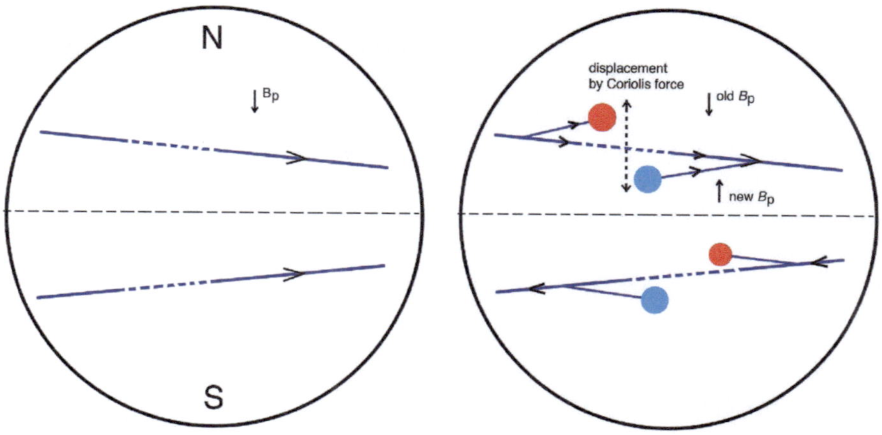

Fig. 5.2 Closing of the dynamo cycle by active region emergence. *Left*: sub-surface field produced by stretching of a poloidal field $\mathbf{B}_p$ by differential rotation (equator rotates faster). Coriolis forces during emergence of a stretch of the field (broken) to the surface causes displacements of the footpoints, observed at the surface as "tilt" of the active regions (*circles*). At depth, this produces a new poloidal field component of opposite sign

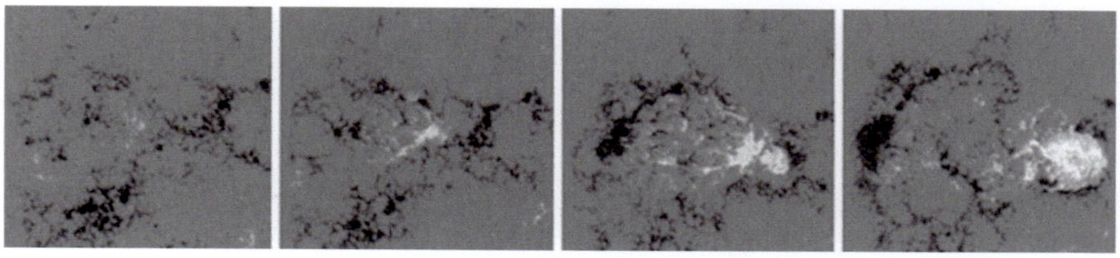

Fig. 5.3 Sequence (time from left to right) showing the emergence of an active region at the solar surface observed with the Hinode satellite. The opposite magnetic polarities (vertical component of the field) are shown in *black* and *white*. For a movie of this sequence see http://science.nasa.gov/headlines/y2007/images/trilobite/Hinode_lower.mov

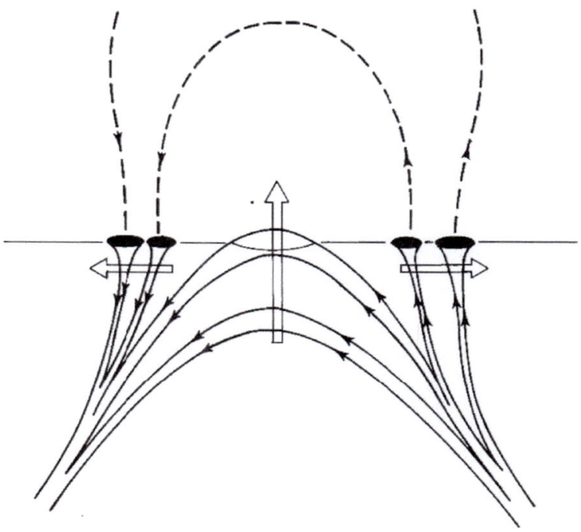

Fig. 5.4 "Rising tree" sketch to explain the phenomenology of an active region emergence. (From Zwaan, 1978)

reproduced in striking realism in full 3-D radiative MHD simulations (Cheung et al. 2008; Rempel, this volume).

The axes of active regions are observed to be tilted with respect to the east-west direction. This was attributed to the action of Coriolis forces during the emergnece process, and identified with the generation of the new poloidal field component that closes the dynamo cycle by Leighton (1969, see sketch in Fig. 5.2).

5.1.2 Later Developments

Models like Leighton's thus made a direct connection between observations of the active regions that make up the solar cycle and the functioning of the cycle as a whole. One might have expected that this natural state of affairs would have led to a further development of the theoretical ideas in continued contact with the observations. But this has not been the case.

Instead, the development of these ideas has been eclipsed for several decades by the parallel development of turbulent mean field formalisms for the solar cycle. These ideas postulated mathematically tractable equations which were claimed to represent the physics of the interaction between magnetic fields and convection in some statistical sense. They relied on theoretical assumptions like cascades in wavenumber space, correlation functions to represent the interaction between magnetic fields and flows, and an assumed separation of length scales between mean fields and fluctuations. Just looking at the data as described above, it is difficult see how a separation would be accomplished. What is more, the data themselves already contain more detailed and more critical information on the functioning of the cycle than is present in mean field models. The dominance of these formalisms in the astrophysical literature (thousands of papers) has led to a particularly sterile theoretical view of the solar cycle, supported neither by a sound theoretical foundation of the equations used nor making much contact with the observations.

In addition, it has had the effect of obscuring an important fact, namely that no turbulence needs to be imposed at all for dynamo action to take place. A system that is completely laminar in the absence of magnetic fields can produce dynamo action from shear and magnetic instability alone (cf. Spruit 2002). A well studied and very successful example of such a dynamo process is the MRI turbulence observed in numerical simulations of accretion disks (e.g. Hawley et al. 1996). The models by Babcock and Leighton are just another example where magnetic instability is the key element in closing the field amplification cycle. These kinds of magnetic cycle are intrinsically nonlinear (i.e. not "kinematic" in dynamo parlance): their functioning depends on the finite amplitude of the field generated. This is because the time scale of the magnetic instabilities that close the dynamo cycle depends on field strength.

The conditions for self-sustained field generation to occur by differential rotation and instability alone, the properties of the magnetic field produced in this way, and its observable consequences all reflect the nature of the magnetic instability involved. In the case of the solar cycle: the properties of magnetic buoyancy. It is sometimes argued that such a process brings about an "alpha effect", so that one just has to use a set of equations that incorporate such an effect. Neither the fact that a poloidal field component can appear by a process changing the direction of an initially toroidal field, however, nor the fact that turbulent mean field equations contain a term describing such an effect, are justifications for using these equations. An understanding of the solar cycle, or any other dynamo process, requires physics to be found out first, rather than assumed in some parametrized form. The idea that insight about the solar cycle can be obtained from the solutions of such models has been an impediment to real progress, however tempting the equations may have looked.

Justification for this critical view is found in the history of ideas about the solar cycle; this is done in the following section. I briefly discuss there how mean field thinking has led to a systematic disconnect between theory and observations. In all likelihood this would not have been necessary if the observations and their interpretation in models such as Leighton's (1969), had been taken more serious.

5.2 Failure of Convective Dynamos Models of the Solar Cycle

The turbulent view of magnetic field generation in convective stellar envelopes holds that the generation of a new poloidal field from the toroidal field produced by differential rotation should be seen as due to the effect of convection acting on a magnetic field in a rotating fluid (Parker 1955, 1979; Steenbeck et al. 1966; Weiss 1981). A consequence of this model is therefore that dynamo action takes place throughout the convection zone. In this model the equatorward drift of the main belt of activity during the cycle reflected the (at the time unknown) *radial* gradient in the rotation rate, not the observed latitudinal differential rotation that were key to the Babcock and Leighton models. This is because mean field dynamo equations naturally lead to dynamo waves traveling in a direction perpendicular to surfaces of constant rotation rate. The observed equatorward drift of activity during the cycle therefore required the rotation rate to have a predominantly radial gradient, with rotation increasing inward. Finally, it was noted that the Lorentz force limits the action of convection on magnetic fields when the field strength reaches equipartition with the kinetic energy of the convective "eddies" (e.g. Proctor and Weiss 1982). This predicts that the field in the solar envelope should be an intermittent turbulent field, with intrinsic field strengths of a few thousand Gauss.

5.2.1 Predictions

Mean field dynamo models thus made three testable predictions. (They are found in many of the texts of the 1970s and 1980s, where they appear mostly as accepted consensus rather than as testable predictions):

- The dynamo action takes place through interaction with turbulent convection,
- The rotation rate in the convection zone depends mainly on radius, it increases with depth.
- The field strength does not exceed equipartition with convective energy densities (few thousand Gauss).

These predictions have never agreed with the phenomenology of the solar cycle very well. One of the important observations is Hale's polarity law: the fact that magnetic fields are not present on the surface in a random "turbulent" form, but appear in a strikingly systematic way, as bipolar active regions oriented east-west, with one of the polarities systematically leading (in the direction of rotation). On top of the east-west orientation, the leading polarity is systematically shifted towards the equator compared with the following polarity (Joy's law, see sketch in Fig. 2). To circumvent these observations, dynamo theories ignore heliographic longitude (of sunspot locations, for example); parameters of the model are then adjusted to fit the remaining data (the "butterfly diagram"). Given the reduced nature of these data and the degrees of freedom of the models, this process is usually successful. The price paid is that most observations of active region phenomenology have to be declared irrelevant, when in fact they provide the most telling evidence about the operation of the cycle. This attitude has remained an integral part of mainstream dynamo thinking.

Attempts have been made to reconcile the magnetic eruption view of active regions with the role of convective turbulence in the mean field dynamo view. Weiss (1964) proposed that magnetic fields rise from the interior, but that turbulence takes over in bringing about observations like the formation of sunspots. In this view, sunspots would form by random walk of magnetic fields in a turbulent flow. This proposal thus kept Cowling's view of active regions as emerging from below, but effectively discarded the observational evidence that led to this idea in the first place, namely the formation of sunspots. Meyer et al. (1974) explicitly repeated the view that sunspots form by random walk of magnetic field lines in convective turbulence.

This has been challenged by observers, who noted that sunspots do not form randomly but in a strikingly deterministic way, as described above (e.g. Zwaan 1978). The motion of active region magnetic fields independent of and opposing surface flows is documented by virtually all observations of active region formation (e.g. Tarbell et al. 1990; Strous et al. 1996). The consequence, namely that the magnetic field itself, rather than convective turbulence, forms active regions has always contradicted the role of convection assumed in mean field models.

Since then, helioseismic measurements of the internal rotation have shown the second prediction to be wrong as well: the differential rotation is in fact mainly in latitude. The radial gradient is weak, and where it is present it is mostly of the wrong sign.

The third prediction can be tested somewhat more indirectly by making use of the many clues given by observations of active regions. A major step forward in the interpretation of this phenomenology are the simulations of flux bundles rising from the base of the convection by Moreno-Insertis (1986), D'Silva and Choudhuri (1993), Fan et al. (1994), Schüssler et al. (1994). In these simulations, a horizontal (azimuthal, zonal) bundle of magnetic field lines at the base of the convection zone is allowed to become unstable and rise to the surface. The degrees of freedom in these calculations are the magnetic flux of the bundle (set by the value observed for a typical active region), and its initial field strength. The results show that several key characteristics of active regions can be reproduced simultaneously by such magnetic flux loops emerging from the interior of the convection zone: the time scale for emergence of an active region, the heliographic latitude range of emergence, and the degree of tilt of active region axes. For all these phenomena, agreement between simulations and observation points to the same value of the field strength: about 10^5 G. This is also the field strength at which instability is predicted to set in (Schüssler et al. 1994). It can thus be identified with Babcock's (then still unquantified) critical field strength.

Within this picture, a further piece of evidence that would otherwise be a disconnected observation finds its natural place. After formation of a spot, its position drifts a bit in latitude and longitude ("proper motion"), with a random component superimposed on a systematic drift. The random component varies quasiperiodically on a time scale of a few days, decaying with time (e.g. Herdiwijaya et al. 1997). For a field strength of 10^5 G at the base, the Alfvén travel time along the flux strand from the base to the surface is around 3 days. The random proper motions of spots are thus neatly interpreted as reflecting the "settling" of a sunspot to its equilibrium position after the eruption process is completed. [In addition to this random component there is a systematic drift in longitude, corresponding to the increasing separation between the two polarities. This was explained as due to the tension in the sub-surface magnetic field by van Ballegooijen (1982), and reproduced in simulations of rising flux tubes (Caligari et al. 1995)].

5.2.2 Assessment of the Turbulent Convective Dynamo View

The success of the rising flux tube simulations was not immediately seen as a threat to mean field models. In line with the status of active region phenomenology in mean field models, the eruption process would simply be of marginal significance: it would just be some secondary manifestation of the mean field dynamo operating in deeper layers. The success of the simulations, if it is not accidental, has ominous consequences, however, since the agreement with each of the observations only holds if the field strength in the deep interior of the convection zone is around 10^5 G. The energy density in such a field is at least two orders of magnitude larger than the kinetic energy of the convective turbulence invoked in mean field models. If taken serious as a diagnostic, the observed mode of emergence of active regions thus implies that the third prediction also fails.

The situation is more serious, however, since at this field strength the rising flux tubes are so strong that convective turbulence can have little effect on them. The positions of active regions on the surface must consequently correspond reasonably with those of their anchors at the base, also explaining the regularity with which active regions follow Hale's and Joy's laws, and the proper motions of sunspots mentioned above. Active regions as seen at the surface are therefore not a manifestation of a convective mean field dynamo, even if it were to exist somewhere in the convection zone. This is a major setback for this theory, since observations of active regions are the dominant source of information we have about the solar cycle, and virtually the only source ever used for mean field parameter fitting.

Considering these spectacular failures of the turbulent mean field dynamo paradigm for the solar cycle it is useful to reflect for a moment how it could have survived for several decades, and is still going strong (cf. Rüdiger and Hollerbach 2004; Brandenburg 2005, 2009; Tobias 2005; Jones et al. 2009; Charbonneau 2005). Major problems like the lack of a stable theoretical foundation for the equations used, and the lack of connection with most of the relevant observations should have been reasons to pause and reflect on the basis of the enterprise. Instead such problems, when faced at all, were usually countered with these arguments:

(i) in a complicated problem one has to start somewhere,
(ii) the key to a dynamo cycle is the recreation of a poloidal component, and this is included in the mean field dynamo equation.

The first argument was perfectly reasonable at the time of the formulation of mean field electrodynamics, (cf. Parker 1955; Steenbeck et al. 1966), but half a century hence it is beginning to wear a bit thin.

The second argument is essentially a semantic kludge. At a sufficiently abstract level, the nature of the solar cycle as a combination of winding up and an alpha-effect need not be contested. What should have been considered more critically, however, is the question how much more the use of mean field dynamo formalisms provides, above just the alpha-effect that was already put in as an assumption from the beginning. The applied mathematical attractions of solving this equation in countless variations of geometry and parameters appears to have led to a misplaced sense of reality. The failure of the theory to show demonstrable progress by providing increasing contact with observations should have been a warning here. The mean field dynamo model for the solar cycle is best regarded as a mirage. Remarkably, it still keeps a substantial community busy searching for an oasis.

The theoretical basis of mean field electrodynamics has always been problematic. The series expansions used to derive mean field equations, for example are known to diverge unconditionally at large Reynolds numbers. In the mean time, it is now also being called into question by the results from high resolution numerical simulations of magnetic fields driven by imposed small scale forces. From these simulations it is becoming clear that large scale fields do not appear from small scales as expected (e.g. Cattaneo and Hughes 2009), at least not under the generic conditions where the mean field dynamo equation was applied. The occurrence of *small scale* dynamos (i.e. the exponential growth of magnetic energy on small scales) in the interaction of a magnetic field with turbulence is now also somewhat in doubt. Standard wisdom (the "proper" view, Brandenburg 2010), that an appropriately complex velocity field is sufficient to produce a selfsustained small-scale magnetic field turns out to be incorrect. Selfsustained fields have been found in such flows when the viscosity is larger than the magnetic diffusivity (magnetic Prandtl number $P_m > 1$, e.g. Schekochihin et al. 2004). For the case $P_m < 1$ however, conflicting results are reported on the presence or absence of selfsustaining fields, depending also on numerical method. In the case $P_m > 1$, the actual operation mechanism of the field amplification does not agree with conventional ideas based on cascades in wavenumber space (Schekochihin et al. 2004). Kinematic models of small-scale dynamo action, i.e. models using an imposed velocity field, which have guided much of previous thinking, do not provide guidance in this context since they are equivalent to assuming an infinite magnetic Prandtl number.

5.3 Tachocline Dynamos

A somewhat older idea to reconcile mean field models with flux emergence observations is that the dynamo works as a turbulent mean field dynamo near the base of the convection zone (Galloway and Weiss 1981; Parker 1993). After its discovery, the narrow shear zone below the convection zone, the tachocline, was quickly identified as a region of choice to operate a mean field dynamo. It contained the strong radial gradient needed in mean field models to produce the drift of active latitudes during the cycle (e.g. Dikpati 2006 and references therein), and required no conceptual adjustments to the models developed before.

This idea, however does not make physical sense. It assumes that the shear zone can be exploited in the same way as the shear between two moving plates in the laboratory. Turbulence generated by the shear exerts stress on both plates, the energy put into the system by the work done against this boundary stress can be tapped to maintain turbulence and a magnetic field. In the Sun, stresses can be maintained in the convection zone by the rapid momentum exchange due to convective flows. On the other side, however, in the stable stratification on the interior side of the tachocline, the stress that can be supported by fluid motions is many orders of magnitude weaker, since the stratification is very stable on this side (in terms of the buoyancy N^2, the interior is 10^6 times more stable than the convection zone is unstable). This means that the analogy with shear maintained between two moving plates is incorrect (resembling the Zen exercise to clap with one hand).

If the analogy is incorrect, what is then the cause of the tachocline? The tachocline is more appropriately treated as a shear flow driven only by the latitudinal dependence of the rotation rate on the convective side, with a free-slip surface on the interior side. The velocities in the tachocline can just be an "imprint" into the interior of the differential rotation with latitude in the convection zone. This has been proposed early on after the discovery of the tachocline by Spiegel and Zahn (1992), who studied the long-term evolution of the rotation in the interior under the effect of a (weak) viscous stress. A more complete analysis of this problem, which stresses the importance of baroclinicity and thermal diffusion, is the "gyrotropic pumping" model of McIntyre (2007). In a model by Forgács-Dajka and Petrovay (2001) the tachocline is also seen as an imprint of the convection zone on the interior, but this model involves turbulence of unspecified origin inside the tachocline.

The radial gradient in the tachocline is thus useless for driving a dynamo, since it does not support any significant stress with which a field could be amplified. The latitudinal gradient in the tachocline can of course be tapped, but this is no different from what the bulk of the convection zone can do. Its shallow imprint into the interior does not add much, and defeats the original idea of using the (radial) tachocline shear to drive a dynamo.

5.4 New Directions

5.4.1 Compromises

The mean field paradigm holds that, after ignoring or averaging out most of the surface observations, the bit that is left is still a useful validation of the theory. This point of view is still popular in the literature on the solar cycle. Another view appears to follow more of an "adiabatic adjustment" approach: attempts are made to incorporate elements like the disregarded observations mentioned in Section 5.1.1, and physics like buoyant instability of the magnetic field into mean field turbulence, as gradual adjustments of the formalism. This view thus attempts to accommodate intrinsically incompatible elements into a mean field approach without questioning its status as the underlying fundamental theory.

The kind of compromises this leads to looks ugly. In one such attempt to conciliate mean field theory with observations, the emergence of strong magnetic flux tubes is in fact acknowledged to account for phenomena observed at the surface. However, the surface fields are then seen as a separate phenomenon, not representative of the real solar cycle. The real cycle takes place, unseen, somewhere in the convection zone in the manner demanded by the mean field dynamo equations (e.g. Tobias 2009). Another proposal (Brandenburg 2005) postulates the existence of a shallow surface layer (a few Mm depth). This layer contains the puzzling observations, again in some form of mean field dynamo, while at the same time shielding the turbulent mean field dynamo happening below it from our view. "Turbulent pumping" is advanced as achieving this. Observations of active region emergence (Fig. 5.3) that refute such ideas (however vague) are ignored.

5.4.2 Weak Fields

A more progressive view within the convective turbulence category is the idea (Durney et al. 1993; Cattaneo, 1999) that turbulent interaction between field and convection is observed at the surface of the Sun in the form of the so-called weak or inner-network fields. These appear as fields of mixed polarity, short life times and low intrinsic strength (Martin 1988). These properties are more in line with a priori expectations about turbulent fields.

This proposal bypasses the question what causes the strong fields observed as spots and active regions: it only looks at the weak field component and assumes it has a different origin. It is not clear, however, whether the weak fields are really an independent phenomenon: they might just be a part of the solar cycle as seen in active regions. The weak fields might represent either a small scale tail in the distribution of emerging flux units, or some kind of "debris" from the fragmentation of larger units during the decay phase of active regions. Finally, they might be related to the "annealing step" by which old flux disappears again from the convection zone (Section 5.4.4). In the latter case, they would be part of the *decay* of magnetic fields produced in the cycle, rather than an amplification process.

5.4.3 Numerical Simulations

Convincing 3-D numerical simulations of a solar cycle "from scratch" are likely to remain out reach for the indefinite future. This is because of the well-known problems of dynamical range in length and time scale intrinsic to a convective stellar envelope. A brute-force simulation of the entire convection zone would have to resolve time scales as short as a few seconds in the photospheric layers and above, as well as years to cover the duration of a cycle. Corresponding length scales range from a kilometer at the surface to a solar radius. Extrapolating Moore's law with a constant doubling time of 1.5 years, the computing resources needed for a simulation at this resolution would become available 100 years from now (Schüssler 2008). Existing "global" simulations of the convection zone or its magnetic cycle are possible only by leaving out key parts of the physics. Usually, the top layers where most of the dynamic range in length- and time scales is located, are left out. Conclusions drawn from such simulations are unlikely to be very meaningful, since the easier case of a convective envelope without magnetic fields is already known to produce results that bear no relation to observations when these surface layers are left out from the simulation. Individual aspects of the cycle still provide interesting unsolved conceptual problems, however, that may be addressed in isolation before realistic numerical simulations are attempted on more global scales. Two such problems are discussed in the next subsections.

5.4.4 The Annealing Step, "Turbulent Diffusion"

The most challenging problem may well be finding a satisfactory description for the process by which the mass of buoyant vertical flux tubes resulting from a cycle's worth of eruptions gets "annealed" back into a simpler configuration. As the eruption of active regions from the toroidal field proceeds during the cycle, an ever increasing number of magnetic strands develops connecting the surface with the base of the convection zone (cf. Fig. 5.5). The sections of field remaining at the base are sufficient to provide the toroidal field of the next cycle, but this picture does not explain how the clutter of strands connecting to the surface gets simplified from one cycle to the next. In dynamo parlance, this is the "turbulent diffusion" step. The difficulty here is that appeal to traditional convective "turbulent diffusion" will not work (even if the concept itself is accepted), since the fields are now much stronger than equipartition with convection (at least near the base of the convection zone where this annealing has to take place).

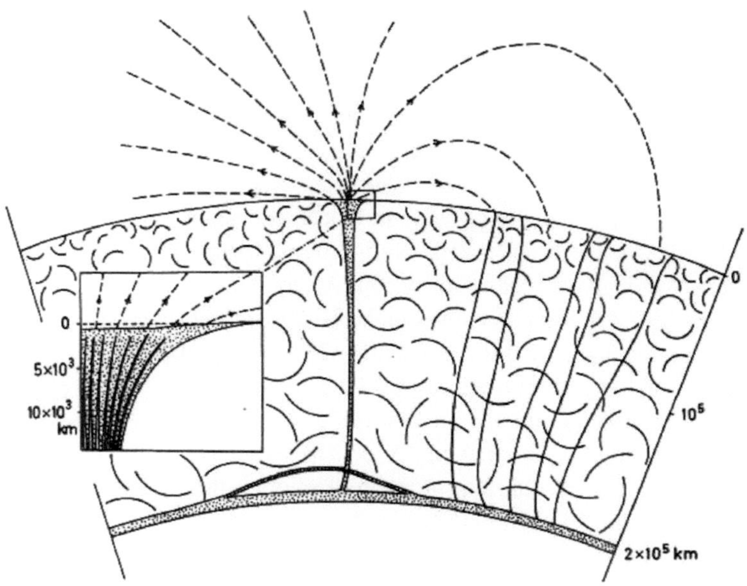

Fig. 5.5 Vertical cut through an active region illustrating the connection between a sunspot at the surface and its origins in the toroidal field layer at the base of the convection zone. (From Spruit and Roberts 1983)

At the moment, it is not clear how long this annealing process takes, or even by which mechanism. For a discussion relating to this problem, see Parker (2009). Perhaps the "residence time" of dispersed active regions is substantially longer than the cycle length? In that case a large amount of small scale mixed-polarity magnetic flux should be present distributed over the solar surface. Such a component could remain undetected except at very high spatial resolution. In fact, recent observations of mixed-polarity fields in quiet regions of the solar surface with Hinode and the Swedish 1-m solar telescope appear to show evidence in this direction (cf. Pietarila Graham et al. 2009).

Extrapolating this thought further, it might even be that such a component is relevant for total solar irradiance (TSI, the solar energy flux received at earth), since small scale magnetic fields are known to produce a net brightening of the solar surface (Spruit 1977). If this is the case, the small but systematic decrease of TSI during the current extended minimum, below the previous shorter minima, might be indicative of the decrease of surface magnetic flux in the course of the annealing process.

5.4.5 Thermodynamics

A central question concerns the thermodynamics of fields of $\sim 10^5$ G at the base of the convection zone. For the field to be wound up quietly over several years before becoming unstable, it has to reside in a stable buoyant equilibrium near the base of the convection zone. In the absence of such a stable equilibrium, the field would rise to the surface on a time scale of weeks (as it actually does in the emergence of a new active region). Magnetic pressure produces buoyancy, and this has to be compensated for equilibrium to hold. Neutral buoyancy through density equilibrium requires significantly lower temperatures in the field than in its environment. For a field strength of 10^5 G, the required amount of temperature reduction is some 100 times larger than canonical temperature fluctuations in a mixing length model of convection.

To solve this equilibrium problem, it is often postulated that flux tubes erupt from stably stratified layers below the base of the convection zone. Though this recognizes the buoyancy problem, it does not actually help much since it begs the question how the magnetic field got to this location in the first place (in particular: on a time scale less than the solar cycle).

5.5 Conclusions

Observations of active region phenomenology, most of them already old and well-established, show that the solar cycle operates on buoyant instability of the magnetic field itself rather than the conventional view based on interaction with convection. This puts us back to ideas developed half a century ago. Significant steps forward, however are the direct 3-D, radiative numerical MHD simulations which are now beginning to make contact with some of the classical observations. Though these simulations cannot deal with the cycle as a whole, their success in reproducing limited aspects such as the emergence of magnetic flux discussed above, or the observed structure of sunspots (Heinemann et al. 2007; Scharmer et al. 2008; Rempel et al. 2009) give confidence for the future. At the same time they clean the table by eliminating a number of dead-end views on the solar cycle, some of which considered well-established thus far.

At the same time, a number of unsolved questions appear that are specific for the picture of a magnetic cycle operating on buoyant instability. Some of these questions are unlikely to be answered from first principles or numerical simulations. Clues taken from observations may well play an important role in making progress in figuring out the physics relevant for these questions. As the history of the subject shows, however, taking observational clues serious will require one to jettison the turbulent mean field baggage that has impeded the development of a sensible theory of the solar cycle for so long. This process would be assisted by healthy skepticism on the part of the observational community. In fact, it is rather surprising how easily observers have acquiesced in the past to the treatment of their data by mean field theories ("sorry but your observations are just turbulence, they have to be averaged out").

References

Altrock RC (2010) arXiv:1002.2401
Babcock HW (1961) ApJ 133:1049

Babcock HW (1963) Ann Rev Astron Astrophys 1:41
Brandenburg A (2005) ApJ 625:539
Brandenburg A (2009) Plasma Phys Control Fusion 51:124043
Brandenburg A (2010) MNRAS 401:347
Brandenburg A, Kleeorin N, Rogachevskii I (2010) Astron Nachr 331:5
Caligari P, Moreno-Insertis F, Schussler M (1995) ApJ 441:886
Cattaneo F (1999) ApJ 515:L39
Cattaneo F, Hughes DW (2009) MNRAS 395:L48
Charbonneau P (2005) Living Rev Solar Phys, http://solarphysics.livingreviews.org/Articles/lrsp-2005-2/
Cheung MCM, Schüssler M, Tarbell TD, Title AM (2008) ApJ 687:1373
Cowling TG (1953) In: Kuiper, G (ed) The sun. University of Chicago Press, Chicago, IL, Chapter 8
Deforest CE, Lamb DA, Berger T, Hagenaar H, Parnell C, Welsch B (2008) AGU Spring Meeting Abstracts, 1
Dikpati M (2006), Adv Space Res 38:839
D'Silva S, Choudhuri AR (1993) A&A 272:621
Durney BR, De Young DS, Roxburgh IW (1993) Sol Phys 145:207
Elsasser WM (1956) Rev Mod Phys 28:135
Fan Y, Fisher GH, McClymont AN (1994) ApJ 436:907
Forgács-Dajka E, Petrovay K (2001) Sol Phys 203:195
Galloway DJ, Weiss NO (1981) ApJ 243:945
Gilman PA (2005) Astron Nachr 326:208
Hawley JF, Gammie CF, Balbus SA (1996) ApJ 464:690
Heinemann T, Nordlund Å, Scharmer GB, Spruit HC (2007) ApJ 669:1390
Herdiwijaya D, Makita M, Anwar B (1997) Pub Astron Soc Jpn 49:235
Howe R (2009) Living Rev Solar Phys 6:1, arXiv:0902.2406
Jones CA, Thompson MJ, Tobias SM (2009) Space Sci Rev 152, 591. doi:10.1007/s11214-009-9579-5
Kitiashvili IN, Kosovichev AG, Wray AA, Mansour NN (2010) ApJ 719:307, arXiv:1004.2288v1
Leighton RB (1969) ApJ 156:1
Leroy JL, Trellis M (1974) A&A 35:283
Martin SF (1988) Sol Phys 117:243
McIntyre ME (2007) In: Hughes DW et al (eds) The solar tachocline.CUP, Cambridge, p 183
Meyer F, Schmidt HU, Wilson PR, Weiss NO (1974) MNRAS 169:35
Moreno-Insertis F (1986) A&A 166:291
Parker EN (1955) ApJ 122:193
Parker EN (1979) Cosmical magnetic fields. Clarendon Press, Oxford
Parker EN (1993) ApJ 408:707
Parker EN (2009) Space Sci Rev 144:15
Pietarila Graham J, Danilovic S, Schüssler M (2009) ApJ 693:1728
Proctor MRE, Weiss NO (1982) Rep Prog Phys 45:1317
Rempel M, Schüssler M, Cameron RH, Knölker M (2009) Science 325:171
Rüdiger G, Hollerbach R (2004) The magnetic universe : geophysical and astrophysical dynamo theory. Wiley-VCH, Weinheim
Scharmer GB, Nordlund Å, Heinemann T (2008) ApJ 677:L149
Schekochihin AA, Cowley SC, Taylor SF, Maron JL, McWilliams JC (2004) ApJ 612:276
Schüssler M (2008) 12th European solar physics meeting, Freiburg, Germany, held 8–12 September 2008. Online at http://espm.kis.uni-freiburg.de/, p 1.1, 12, 1
Schüssler M, Caligari P, Ferriz-Mas A, Moreno-Insertis F (1994) A&A 281:L69
Spiegel EA, Zahn J-P (1992) A&A 265:106
Spruit HC (1977) Sol Phys 55:3
Spruit HC (2002) A&A 381:923
Spruit HC, Roberts B (1983) Nature 304:401
Steenbeck M, Krause F, Rädler K-H (1966) Z Naturforsch A 21:369
Strous LH, Scharmer G, Tarbell TD, Title AM, Zwaan C (1996) A&A 306:947
Tarbell T., Ferguson, S., Frank, Z., Shine, R., Title, A., Topka, K., & Scharmer, G. (1990). IAU Symp 138:147
Tobias S (2005) Advances in astronomy. In: Thompson JMT (ed). Royal Society series on advances in science, vol 1. Imperial College Press, London, p 355
Tobias SM (2009) Space Sc Rev 144:77
van Ballegooijen AA (1982) A&A 113:99
Vrabec D (1974) In: Athay RG (ed) Chromospheric fine structure, IAU Symp. Reidel, Dordrecht 56, p 221
Weiss NO (1964) MNRAS 128:225
Weiss NO (1981) JGR 86:11689
Weiss NO (1989) ASSL 156: Accretion disks and magnetic fields in astrophysics, p 11
Weiss NO (1993) ASSL 183: Physics of Solar and Stellar Coronae, p 541
Weiss NO (1994) In: Proctor MRE, Gilbert AD (eds) Lectures on solar and planetary dynamos, ISBN 0 521 46142 1 and ISBN 0 521 46704 7. Cambridge University Press, Cambridge, p 59
Weiss NO (1997) In: Castagnoli G, Provenzale A (eds) Past and present variability of the solar-terrestrial system, "Enrico Fermi": course CXXXIII.IOS, Oxford, p 325
Zwaan C (1978) Sol Phys 60:213

Chapter 6

The New Solar Composition and the Solar Metallicity

Nicolas Grevesse, Martin Asplund, A. Jacques Sauval, and Pat Scott

Abstract We review the current status of our knowledge of the chemical composition of the sun and present a redetermination of the solar abundances of all available elements. These new results have very recently been published by Asplund et al. (2009). The basic ingredients of this work, the main results and their implications are discussed.

6.1 Historical Introduction

The first quantitative analysis of the composition of the solar atmosphere was done by H. N. Russell in 1929 (Russell 1929). The name of Russell is associated with a large number of pioneer researches in astrophysics during the first half of last century, for example, the Hertzsprung-Russell diagram, as well as with a series of basic works in atomic spectroscopy, for example the LS or Russell-Saunders coupling. Using eye estimates of the solar line intensities and a very simple one temperature model of the solar photosphere, Russell succeeded to derive the abundances of 56 elements. Russell's mixture was used for quite a long time by all astronomers; it had a very great impact on the whole astronomical community showing the overwhelming abundance of hydrogen in the Universe. Russell's abundance distribution already showed the main remarkable features related to nuclear properties that have been used by the theories of nucleosynthesis to explain the origin of the chemical elements.

After this pioneer work, new more sophysticated solar system abundance analyzes, often called cosmic abundances, not only used the sun, but also the earth crust and, especially, meteorites. Actually, other members of the solar system and, particularly, the meteorites (Sections 6.4 and 6.7.3), could be as good indicators as the sun itself of the solar composition. These new works resulted from progress in the quality of the solar spectra, in the modeling of the photosphere, in the atomic data needed and in the meteoritic composition analyzes. Probably the most important analyzes that had important impacts on the whole astronomical society are the pioneer works of Goldschmidt (1938), essentially based on the earth crust and meteorites, of Unsöld (1948), with a new analysis of the solar photospheric abundances and of Suess and Urey (1956), who made a new refined analysis of the composition of meteorites combined with solar and stellar data for the elements absent in meteorites. This last work was really the basis of nucleosynthesis, the theory of the formation of the chemical elements, developed by Burbidge et al. (1957) and Cameron (1957). Very important basic solar abundance researches have also been done by Goldberg et al. (1960), who made a new extensive analysis of the solar abundances of many elements, by Cameron (1973) with a new critical analysis of all the elements combining solar and meteoritic data and by Ross and Aller (1976), who applied the technique of spectrum synthesis to determine the abundances of the elements in the solar photosphere for a few elements and made a critical evaluation of the results available in the literature. This last review also summarizes the main past analyzes and introduces to

N. Grevesse (✉)
Centre Spatial de Liège and Institut d'Astrophysique et de Géophysique, Université de Liège, Liège, Belgium
e-mail: nicolas.grevesse@ulg.ac.be

M.P. Miralles, J. Sánchez Almeida (eds.), *The Sun, the Solar Wind, and the Heliosphere*, IAGA Special Sopron Book Series 4, DOI 10.1007/978-90-481-9787-3_6, © Springer Science+Business Media B.V. 2011

the various techniques used to derive solar abundances (Section 6.3).

More detailed reviews of the evolution of the solar system chemical composition can be found in Suess and Urey (1956) for the period covering the first half of the twentieth century and in Ross and Aller (1976), Grevesse and Sauval (1998) and Lodders et al. (2009) for more recent works.

In the present chapter we shall mainly concentrate on the evolution during the last two decades that finally lead to the present solar composition (Asplund et al. 2009).

6.2 Interest of Solar Abundances

We give here after some of the reasons why the accurate knowledge of the solar chemical composition is a key data not only for solar and solar system analyzes but for the whole astrophysical and geophysical community.

- It is obvious that if we want to model the sun we need to know its precise chemical composition. The role of elements like Mg, Si and Fe as electron donors in the photosphere is well known, as well as the crucial contributions to the opacity of Fe in the central solar layers and of O and Ne just below the convective zone.
- Studying solar abundances allows one to analyze the structure as well as the physical processes in the various solar layers. We discover, for example, that there are unexpected chemical composition variations in the layers below the convective zone as well as in the outermost solar layers (Sections 6.3 and 6.7.4).
- The sun, being the best known star, has always been considered as the typical star, the STANDARD star, to which other stars are compared.
- From comparisons between stellar and solar compositions, we learn a lot concerning the structure and evolution of the stars as well as of the galaxies and the universe.
- The standard solar composition is also the basic data to be reproduced by nucleosynthesis theories.
- The sun is the ideal laboratory where new methods, for example 3D models, are first tested, before to be applied to other stars.
- The sun is also unique because chemical composition data can be derived from various types of matter in different objects of the solar system like the earth, moon, planets, comets, meteorites. Comparing these data gives interesting informations on the formation and evolution of these objects as well as of the solar system as a whole.
- If the meteorites are also a very precise source of solar system abundances (Sections 6.4 and 6.7.3), they have lost a great fraction of the most volatile but also the most abundant elements like H, He, C, N, O and Ne. The abundances of these very important elements can therefore only be derived from solar analyzes.

6.3 Sources of Solar Chemical Composition Data

Because of its proximity, the sun is unique. Solar abundances can be obtained by very different techniques and for very different types of solar matter, from the interior to the outermost coronal layers.

Calibration of solar interior models and helioseismology, the analysis of the solar oscillations, allow one to derive the solar abundance of He and to estimate the metallicity in the inner solar layers. Using spectroscopy in a large wavelength range, from X-rays to infrared, gives us informations on the chemical composition of the outer solar layers, from the photosphere to the corona, including sunspots for a few elements only. Particle collection techniques from space allow one to measure the abundances of the elements in the solar wind (SW) and solar energetic particles (SEP). We could also cite gamma ray spectroscopy of solar flares as well as lunar soils that record the past chemical history of the sun.

These various measurements, concerning very different solar layers with very different physical conditions, have shown unexpected variations of the chemical composition, from one type of solar matter to another one as well as with time. Below the convection zone, we observe element migration, often called diffusion. During the solar lifetime, the convection zone reservoir has lost about 12% of its helium and 10% of all the heavier elements. In the outer layers, we notice a very variable composition. Elements with first ionization potentials (FIP) lower than about 10 eV appear to be enhanced relative to their photospheric values whereas elements with higher FIP show about their photospheric values, with the exception of He which is under-abundant in the outer solar layers.

The solar photosphere is without any doubt the layer for which we have the largest number of data. A large number of photospheric spectra of very high quality from the visible to the infrared are available. Such spectra are used since quite a long time to derive solar abundances. They show a very large number of spectral lines of a large number of elements from Li to Th, whereas the other sources of solar abundances only concern a limited number of elements. Just above the convection zone, the photosphere is well mixed and does not show, on the contrary to other layers, any chemical composition variation with space and/or time. Furthermore, the heterogeneous structure of this layer can now be modeled with great realism (Section 6.5). For these many reasons, the abundances of the elements derived from the photospheric spectrum will form the basis of the solar chemical composition.

A few elements however do not show any spectral line in the photospheric spectrum for purely spectroscopic reasons and not because of their abundances. Helium has already been mentioned here above (see also Section 6.6.1). Fluorine and chlorine are estimated from their molecular lines in sunspots spectra. The other noble gazes, Ne (Section 6.6.3), Ar, Kr and Xe are estimated from their presence in the solar wind and also from interpolations between neighboring elements.

6.4 Recent Analyzes

Global analyzes of the solar chemical composition during the last 20 years (Anders and Grevesse 1989; Palme and Beer 1993; Grevesse and Noels 2003; Grevesse and Sauval 1998; Lodders 2003; Lodders et al. 2009) are essentially based on critical reviews of the literature covering direct solar analyzes as well as meteoritic results.

It became clear with the very largely cited work by Anders and Grevesse (1989) that the photospheric and meteoritic abundances agree. However, not all meteorites are concerned by this agreement: only the very rare CI carbonaceous chondrites obey to this rule. Only five such meteorites are known with a total mass of about 30 kg. They are the least fractionated among the various types of meteorites, having preserved the bulk composition of their parent planitesimals and containing the largest amount of volatile elements. However, the most abundant and most volatile elements like H, He, C, N, O, Ne, Ar, Kr, Xe,... are either absent or largely depleted in these CI meteorites. Therefore the solar abundances of these very important elements, because of their large abundances, can only be derived from the sun itself. It is clear that the two methods, solar photosphere and meteorites, complement each other very well.

The main characteristics among these tables of solar abundances is the low evolution of the accuracy of the solar photospheric results essentially due to the lack of more accurate atomic and molecular data, particularly, the transition probabilities.

The situation has however changed recently with the rather severe downward revision of the abundances of a few elements, including the very important O, C and N (Asplund et al. 2005). There are three main reasons for these changes: the use of a 3D hydrodynamical solar model atmosphere, the relaxation of the assumption of local thermodynamic equilibrium (LTE) and the improvements in the atomic and molecular data.

6.5 The New Analysis

In our new analysis (Asplund et al. 2009), we have redetermined the abundances of all the elements present in the photospheric spectrum. We also discuss the abundances of elements like He, Ne, Ar, Kr, Xe, ..., which are not present in the photospheric spectrum.

The basic ingredients of this new analysis are discussed here after.

- We used a new 3D hydrodynamical solar model atmosphere instead of the classical 1D models of the photosphere used since many decades.
- We have made a very careful and very demanding selection of the spectral lines.

- We have replaced the often used LTE hypothesis by non-LTE analyzes, when possible.
- In the cases of C, N and O, we have used all the indicators of the abundances, atoms as well as molecules.
- And we have made a careful choice of the atomic and molecular data among the available data.

The realism of the new 3D models seems well established. Details concerning the construction as well as the various tests of these models can be found in the papers cited by Asplund et al. (2005, 2009), Nordlund et al. (2009) and Trampedach, Asplund, Hayek and Collet (private communication). The new 3D model we use reproduces very well the topology of the heterogeneous solar upper layers, the observed shapes (widths, shifts and asymmetries) of the spectral lines, the observed spectral distribution of the flux and of the intensity and the center-to-limb variation of the intensity versus wavelength, as well as the observed wings of the hydrogen lines. 1D models generally fail to reproduce these observations. With the 3D models, spectral line shapes are very well reproduced without the help of fake parameters like micro- and macroturbulence that have to be used together with 1D models. Furthermore, with the 3D model, molecules and atoms, as well as lines of different intensities or excitation energies, lead to results in pretty good agreement; this is not the case with 1D models.

The new very demanding selection of spectral lines we have made is not trivial. Actually we wanted to avoid, as much as possible, the use of spectral lines that are blended. Including such lines leads, without any doubt, to an increase in the abundance scatter and skews the results to higher abundances. So we examined very carefully each line we have retained, looking carefully for hidden blends which show up in the widths and/or shapes of the lines.

Relaxing the LTE hypothesis is also a great step forward because many spectral lines of important elements, for example the important O I lines, can only be successfully interpreted in non-LTE. NLTE analyzes are however very demanding in the number and quality of atomic data like transition probabilities, cross-sections for collisions with electrons and with neutral hydrogen atoms. Such data are unfortunately only available for a limited number of atoms and ions.

In deriving the abundances of C, N and O, we have used all the indicators, atoms as well as molecules. We have shown that molecules are as good indicators of the abundances as atoms because they are not more sensitive to temperature than the very high excitation permitted lines of C I, N I and O I, traditionally used in abundance analyzes.

Finally the careful choice of atomic and molecular data is obvious but not trivial. Of course one has to use the most precise of these data. These atomic and molecular data are not only the obvious and well known transition probabilities but also, for example, the cross sections for collisions with the neutral hydrogen atoms and, in some cases, the partition functions themselves, that have to be carefully selected.

6.6 Results and Discussion

Our new solar abundances are given in Table 6.1, in the usual astronomical scale relative to hydrogen where $\log N_H = 12.00$. They are compared with the very recent meteoritic values of Lodders et al. (2009).

In order to estimate the uncertainties of the photospheric results, we have added, to the statistical errors derived from the scatter of the results and traditionally used to represent the uncertainty of the solar results, possible systematic errors introduced by the model (mean atmospheric stratification and homogeneities) and by the physical processes (NLTE vs. LTE).

One of the main characteristics of our 3D results is that all indicators of the abundances lead to the same results and no dependence is observed with the strength or excitation energy of the lines. This is not the case with 1D models which lead both to large variations of the abundance results among the indicators as well as to dependencies with the above parameters. For C and O, for example, we do not have any more differences between the abundance results obtained from the low excitation forbidden lines formed in LTE, the very high excitation permitted lines formed in NLTE and the various molecular transitions from the visible to the infrared as discussed in Section 6.6.2.

The other main and very important characteristic of our new photospheric results is that they

Table 6.1 Element abundances, from Asplund et al. (2009), in the present-day solar photosphere relative to hydrogen, where $\log N_H = 12.00$. We also give meteoritic data from Lodders et al. (2009). Indirect photospheric estimates are marked with [...]

	Elements	Photosphere	Meteorites		Elements	Photosphere	Meteorites
1	H	12.00	8.22 ± 0.04	44	Ru	1.75 ± 0.08	1.76 ± 0.03
2	He	$[10.93 \pm 0.01]$	1.29	45	Rh	0.91 ± 0.10	1.06 ± 0.04
3	Li	1.05 ± 0.10	3.26 ± 0.05	46	Pd	1.57 ± 0.10	1.65 ± 0.02
4	Be	1.38 ± 0.09	1.30 ± 0.03	47	Ag	0.94 ± 0.10	1.20 ± 0.02
5	B	2.70 ± 0.20	2.79 ± 0.04	48	Cd		1.71 ± 0.03
6	C	8.43 ± 0.05	7.39 ± 0.04	49	In	0.80 ± 0.20	0.76 ± 0.03
7	N	7.83 ± 0.05	6.26 ± 0.06	50	Sn	2.04 ± 0.10	2.07 ± 0.06
8	O	8.69 ± 0.05	8.40 ± 0.04	51	Sb		1.01 ± 0.06
9	F	4.56 ± 0.30	4.42 ± 0.06	52	Te		2.18 ± 0.03
10	Ne	$[7.93 \pm 0.10]$	-1.08	53	I		1.55 ± 0.08
11	Na	6.24 ± 0.04	6.27 ± 0.02	54	Xe	$[2.24 \pm 0.06]$	-1.93
12	Mg	7.60 ± 0.04	7.53 ± 0.01	55	Cs		1.08 ± 0.02
13	Al	6.45 ± 0.03	6.43 ± 0.01	56	Ba	2.18 ± 0.09	2.18 ± 0.03
14	Si	7.51 ± 0.03	7.51 ± 0.01	57	La	1.10 ± 0.04	1.17 ± 0.02
15	P	5.41 ± 0.03	5.43 ± 0.04	58	Ce	1.58 ± 0.04	1.58 ± 0.02
16	S	7.12 ± 0.03	7.15 ± 0.02	59	Pr	0.72 ± 0.04	0.76 ± 0.03
17	Cl	5.50 ± 0.30	5.23 ± 0.06	60	Nd	1.42 ± 0.04	1.45 ± 0.02
18	Ar	$[6.40 \pm 0.13]$	-0.46	62	Sm	0.96 ± 0.04	0.94 ± 0.02
19	K	5.03 ± 0.09	5.08 ± 0.02	63	Eu	0.52 ± 0.04	0.51 ± 0.02
20	Ca	6.34 ± 0.04	6.29 ± 0.02	64	Gd	1.07 ± 0.04	1.05 ± 0.02
21	Sc	3.15 ± 0.04	3.05 ± 0.02	65	Tb	0.30 ± 0.10	0.32 ± 0.03
22	Ti	4.95 ± 0.05	4.91 ± 0.03	66	Dy	1.10 ± 0.04	1.13 ± 0.02
23	V	3.93 ± 0.08	3.96 ± 0.02	67	Ho	0.48 ± 0.11	0.47 ± 0.03
24	Cr	5.64 ± 0.04	5.64 ± 0.01	68	Er	0.92 ± 0.05	0.92 ± 0.02
25	Mn	5.43 ± 0.04	5.48 ± 0.01	69	Tm	0.10 ± 0.04	0.12 ± 0.03
26	Fe	7.50 ± 0.04	7.45 ± 0.01	70	Yb	0.84 ± 0.11	0.92 ± 0.02
27	Co	4.99 ± 0.07	4.87 ± 0.01	71	Lu	0.10 ± 0.09	0.09 ± 0.02
28	Ni	6.22 ± 0.04	6.20 ± 0.01	72	Hf	0.85 ± 0.04	0.71 ± 0.02
29	Cu	4.19 ± 0.04	4.25 ± 0.04	73	Ta		-0.12 ± 0.04
30	Zn	4.56 ± 0.05	4.63 ± 0.04	74	W	0.85 ± 0.12	0.65 ± 0.04
31	Ga	3.04 ± 0.09	3.08 ± 0.02	75	Re		0.26 ± 0.04
32	Ge	3.65 ± 0.10	3.58 ± 0.04	76	Os	1.40 ± 0.08	1.35 ± 0.03
33	As		2.30 ± 0.04	77	Ir	1.38 ± 0.07	1.32 ± 0.02
34	Se		3.34 ± 0.03	78	Pt		1.62 ± 0.03
35	Br		2.54 ± 0.06	79	Au	0.92 ± 0.10	0.80 ± 0.04
36	Kr	$[3.25 \pm 0.06]$	-2.23	80	Hg		1.17 ± 0.08
37	Rb	2.52 ± 0.10	2.36 ± 0.03	81	Tl	0.90 ± 0.20	0.77 ± 0.03
38	Sr	2.87 ± 0.07	2.88 ± 0.03	82	Pb	1.75 ± 0.10	2.04 ± 0.03
39	Y	2.21 ± 0.05	2.17 ± 0.04	83	Bi		0.65 ± 0.04
40	Zr	2.58 ± 0.04	2.53 ± 0.04	90	Th	0.02 ± 0.10	0.06 ± 0.03
41	Nb	1.46 ± 0.04	1.41 ± 0.04	92	U		-0.50 ± 0.03
42	Mo	1.88 ± 0.08	1.94 ± 0.04				

are significantly smaller, for the most abundant elements like C, N, O, Ne and Fe, than those recommended in the widely used compilations of Anders and Grevesse (1989) and Grevesse and Sauval (1998) (Fig. 6.1). They are generally only somewhat smaller for the other elements as seen from Fig. 6.1. If we now compare with the recent compilation of Asplund et al. (2005), already based on the same rules as the present one, but for a few elements only, the new results, for C, N and O, are about 10% larger. This is essentially due to the present use of a new, still more realistic version, of the 3D model.

Details concerning these new analyzes are given in Asplund et al. (2009). Still more detailed accounts will

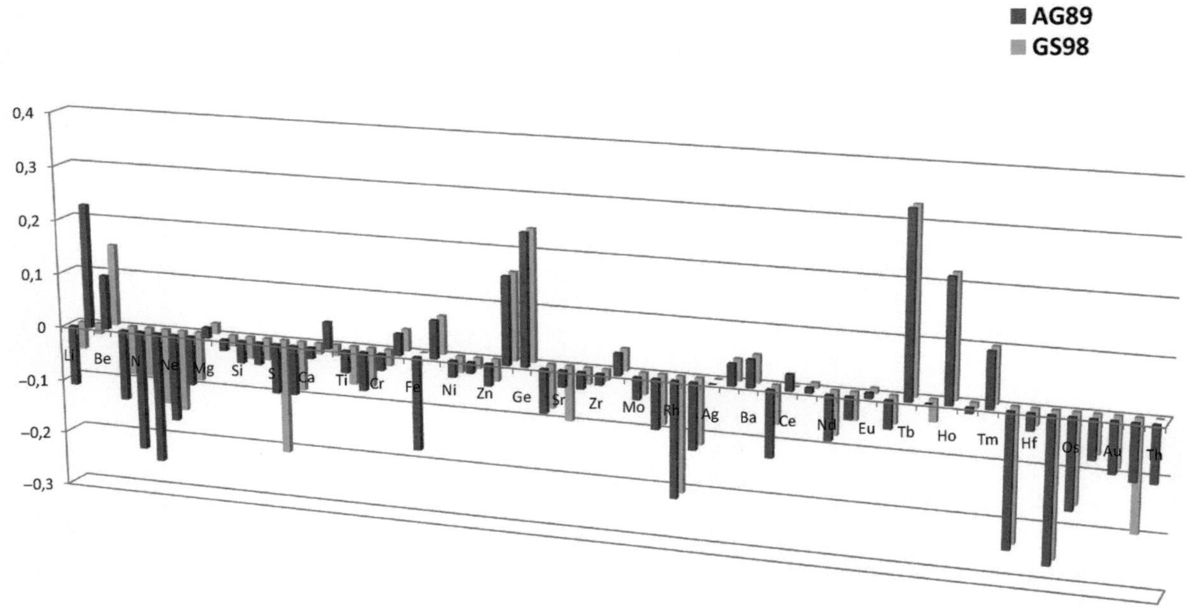

Fig. 6.1 Comparison of the present photospheric results (Asplund et al. 2009) with those of Anders and Grevesse (1989, AG89) and Grevesse and Sauval (1998, GS98). Plotted are the differences (This work − AG89) and (This work − GS98)

be given in a series of forthcoming papers on the solar abundances of C, N, O, the light elements (Na to Ca), the iron group elements and the neutron-capture elements (Cu to Th).

6.6.1 The Solar Abundance of Helium

Helium has been discovered in a solar spectrum obtained during an eclipse in 1868, about 30 years before to be identified on earth in 1895. These helium lines are however very high excitation lines that are produced in active centers like prominences. The He abundance derived from these analyzes as well as from the far UV coronal lines are very uncertain and range from $N_{He}/N_H = 0.065$–0.085 with uncertainties of order 15–25%.

Solar wind and solar energetic particles show a very variable and rather low value of N_{He}/N_H when compared with values observed spectroscopically in hot stars and in the interstellar medium in the solar neighborhood.

The present solar He abundance in the outer convective zone is actually derived from the analysis of the solar oscillations. Helioseismology allows one to derive a very accurate value of Y, the He content by mass, in the convective zone, Y = 0.2485 ± 0.034 (Basu and Antia 2004) (see also Asplund et al. 2009, for a short discussion). This value corresponds to $N_{He}/N_H = 0.085$ or $\varepsilon_{He} = 10.93$ in the astronomical logarithmic scale.

6.6.2 The Abundances of Oxygen and Nitrogen

Caffau et al. (2008, 2009) found abundances of O and N larger than our values, also using their own 3D model, but applied to the forbidden and permitted atomic lines only. A full discussion of the reasons why the results of Caffau et al. (2008) are larger than ours, would be out of the scope of the present review. We suggest three main reasons for explaining these differences. We estimated the contributions of the atomic and molecular lines that blend the forbidden O I lines by deriving them in a purely empirical way, independent of the photospheric model and of the abundances. We better derived the important NLTE effects on the high excitation O I lines by estimating very precisely the cross sections of the collisions with the hydrogen atoms from center-to-limb observations of a sample of these lines. And finally the equivalent widths of Caffau et al. (2008) for the main permitted O I lines are systematically larger than our values and also larger than all previously published values. These various reasons explain why Caffau et al. (2008) find a solar abundance of O, only based on the O I lines, $\varepsilon_O = 8.76 \pm 0.07$, 0.07 dex i.e. 17% larger than our result, $\varepsilon_O = 8.69 \pm 0.05$.

The larger N abundance found by Caffau et al. (2009) is certainly due to their selection of N I lines, some of the near infrared lines they retained being blended. This explains why, because of these blends, their result is somewhat larger than ours ($\varepsilon_N = 7.86$ vs. 7.83) and, especially, why the dispersion of their results is much larger than ours (0.12 dex vs. 0.05 dex).

The differences due to the use of different 3D models by the two groups are however rather small, of order 5–10% at most.

6.6.3 The Abundance of Neon

As for He, and for the same purely spectroscopic reasons, no Ne line (this is true as well for all the noble gazes) is present in the photospheric spectrum, although Ne is the fifth most abundant element in the sun. The solar Ne abundance has to be inferred from UV and X-ray spectroscopy of various types of coronal matter including solar flares as well as from direct measurements in the solar wind and solar energetic particles.

The interpretation of these data is however complicated by the so-called first ionization potential (FIP) effect mentioned in Section 6.3. In the upper solar atmosphere, elements with FIP smaller than about 10 eV are systematically enhanced compared to their photospheric abundances whereas the elements with larger FIP are in principle not enhanced. Furthermore, this enhancement is very variable and even for high FIP elements like O(13.6eV) and Ne(21.6eV), the ratio Ne/O is seen to vary and the variations are to some extent related to the "activity" of the matter that is observed.

We therefore choose to use the ratio Ne/O observed for quiet sun matter, $N_{Ne}/N_O = 0.175 \pm 0.031$, as the value to derive the photospheric Ne abundance from our photospheric O abundance. We have however to point out that the uncertainty of the so obtained solar Ne abundance is rather large, 0.10 dex. A full discussion of the solar Ne problems as well as of the other noble gases, Ar, Kr and Xe, is given in Asplund et al. (2009).

6.7 Implications of the New Solar Chemical Composition

6.7.1 The New Solar Metallicity

With the solar chemical composition we recommend in Table 6.1, the mass fractions of hydrogen, X, helium, Y, and metals, Z, the metallicity, become X = 0.7380, Y = 0.2485 and Z = 0.0134 with Z/X = 0.0181. The solar metallicity is not any more the canonical 2% value recommended by Anders and Grevesse (1989), and largely used by astronomers, but a much smaller value of 1.34%. The reason for this decrease is easily understood if one recalls that the main contributors to Z are, by order of decreasing contribution, O, 42.9%, C, 17.7%, Fe, 9.7%, Ne, 9.4%, Mg, 5.3%, N, 5.2%, Si, 5.0%, ... If we look at Fig. 6.1, we see also from Section 6.6 that the new abundances of the main contributors to Z have decreased by quite a large amount when compared with older largely used values. See also Section 6.7.4.

6.7.2 The New Solar Bulk Composition

In order to derive the bulk composition of the sun, 4.56 Gyrs ago, we have to take into account the effects of diffusion at the bottom of the convective zone (Section 6.3) that very slowly impoverish the convective zone and the photosphere. The values for the bulk composition are therefore somewhat larger than the present day values given in Table 6.1. The increase to be applied to the photospheric values of Table 6.1 is 0.04 dex for the heavier elements and 0.05 dex for He; the X_0, Y_0 and Z_0 become therefore 0.7154, 0.2703 and 0.0142 respectively, with $Z_0/X_0 = 0.0198$. See also Section 6.7.4.

6.7.3 Comparison with Meteorites

A comparison between our photospheric results and the recent meteoritic values of Lodders et al. (2009) is shown in Fig. 6.2. For the elements which show an uncertainty of the photospheric abundance less than 25%, the agreement is perfect, the mean difference photosphere-meteorites being 0.00 ± 0.05 dex. A few elements however show rather large disagreements. We do believe that the photospheric results for these few exceptions (Co, Rb, Rh, Ag, Hf, W and Pb) are to be blamed for unidentified blends or NLTE effects that are impossible to estimate because of the lack of the required atomic data.

A remark has to be made concerning the accuracy of the meteoritic results. In Table 6.1, the uncertainties of these data are the values found from meteoritic analyzes that are made in a scale relative to Silicon. In order to place the meteoritic abundances, measured relative to Si, on the astronomical scale relative to H, the meteoritic data are re-normalized such that the meteoritic and photospheric abundances of Si agree. Therefore strictly speaking an additional uncertainty of 0.03 dex, the uncertainty of the photospheric abundance of Si, should be incorporated in the uncertainties of the meteoritic abundances of Lodders et al. (2009) given in Table 6.1. The uncertainties of order 0.01–0.04 dex, reported for most of the meteoritic values, are more likely of order 0.03–0.05 dex, still lower however than most of the photospheric values.

6.7.4 The Sun, "Back to Normal"?

In the past, the sun appeared to be metal-rich compared with observations of the solar neighborhood. With our new solar abundances, the sun is "back to normal". There is now a good agreement between the sun and the results of the most recent analyzes of solar-type stars, OB stars, H II regions and the interstellar medium in the solar neighborhood. This result of the comparison sun-solar neighborhood might perhaps eventually suggest that the chemical evolution of our galaxy was not very efficient during the last 4.5 Gyrs. More details are given in Asplund et al. (2009).

Very recently however Meléndez et al. (2009) convincingly showed that our Sun has a somewhat different chemical composition than solar twins that are, in all other aspects, very similar to the Sun. Actually the Sun shows a depletion of about 20% of the refractory elements relative to the volatile elements in comparison with the solar twins. This peculiarity is also found in solar analogs (G0–G5) known to have close-in giant planets while the solar analogs without such planets show the solar abundance pattern. This has

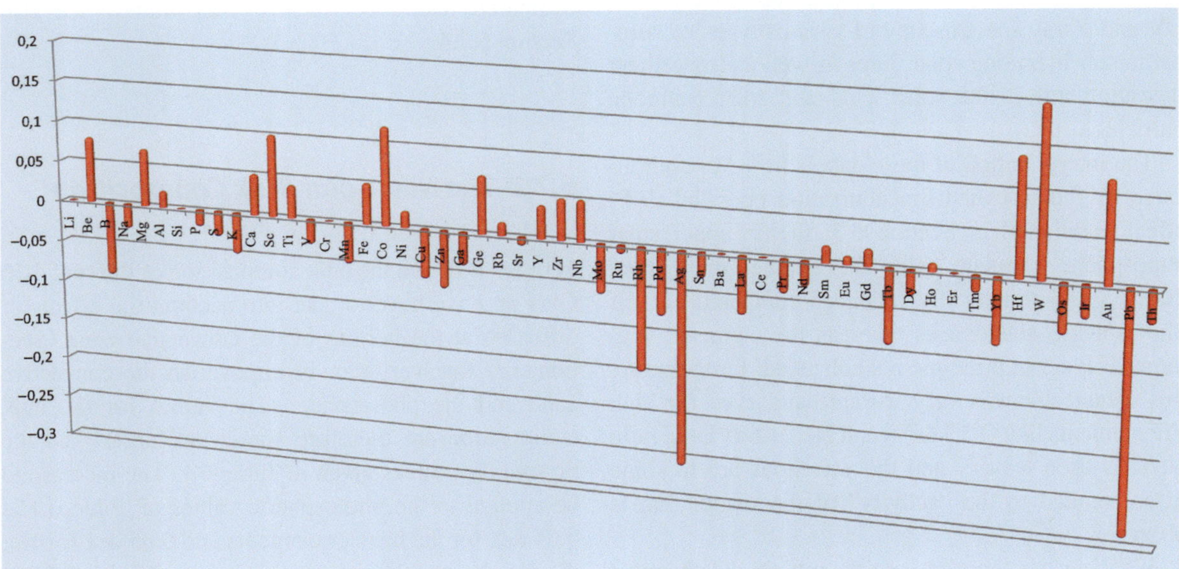

Fig. 6.2 Differences between our photospheric results (Asplund et al. 2009) and the meteoritic abundances of Lodders et al. (2009)

been confirmed by Ramírez et al. (2009) and further discussed by Meléndez et al. (2010): these authors suggest that these peculiarities in the solar chemical composition can be explained as the signatures of the formation of terrestrial planets like our own Earth.

It might also well be that the convective zone (CZ) and the photospheric metal content is naturally lower than the value in the radiative zone below the CZ, as suggested very recently by Nordlund (private communication). From comparisons of the amount of metals locked in planets and available in the CZ reservoir, Nordlund concludes that the metallicity in the CZ, as mentioned in Section 6.7.1, could be lower than in the deeper radiative layers. This had already been speculated by Haxton and Serenelli (2008). See also Section 6.7.5.

6.7.5 The Solar Standard Model Problem

One problem remains unsolved: the disagreement between the predictions by solar standard model (SSM) and the results of the observations using helioseismology, even if the situation is somewhat alleviated owing to our new abundances of O, C and Ne, somewhat larger than the ones of Asplund et al. (2005). It is well known that since the heavy element content of the sun has been revised downward by Asplund et al. (2005) (Sections 6.4 and 6.6), the excellent agreement between SSM predictions and determinations of the solar inner layer structure by helioseismology obtained with older but larger solar heavy element abundances, has been very seriously compromised. This problem essentially comes from the reduction of the opacity caused by the decreased abundances of elements that contribute much to the opacity in the inner solar layers, like C, O and Ne in the layers just below the CZ and Fe and Mg in the central layers. Actually, careful analysis of the very broad range of solar oscillations allows one to determine with high accuracy the sound speed profile in the radiative layers below the CZ, the depth of this CZ as well as its helium content (Section 6.6.1). With the new solar abundances, the SSM cannot any more predict values in agreement with those deduced from the observations. In order to reconcile SSM predictions and helioseismic observations the opacity should be enhanced by about 5% in the central solar layers and by about 15% at the base of the convective zone as suggested in the very recent detailed discussion of these problems by Serenelli et al. (2009). We also note that solar models calculated with a metallicity, below the CZ, larger than the present photospheric metallicity (Sections 6.7.1 and 6.7.4) do not succeed to solve the here above mentioned problem (Castro et al. 2007).

6.7.6 Miscellaneous

We have however to be very cautious when using our new solar abundances for other stars. Of course the new solar abundances alter the cosmic yardstick i.e. the comparison solar to stellar abundances where the sun is often taken as the standard. But one has to be very careful when comparing our 3D solar results with corresponding results for late-type stars derived with classical 1D model atmospheres. Ideally, differential analyzes of the stars and the sun have to be performed and our 3D solar results should then only be used to place the results of such differential analyzes on an absolute scale.

Without any doubt, our new solar abundances of Table 6.1 will have important impacts on stellar structure and evolution modeling.

6.8 Conclusions

We have presented and discussed the results of our new analysis of the solar chemical composition (Asplund et al. 2009). The abundances of all the elements present in the photospheric spectrum have been redetermined using the basic ingredients given in Section 6.5. In this work, we also discuss the abundances of elements like He, Ne, Ar, Kr, Xe, ..., which are not present in the photospheric spectrum.

The end product is the first comprehensive and homogeneous solar abundance analysis since many decades.

The use of new 3D models of the photospheric layers, instead of the classical 1D models used for about 50 years, of course plays a role in the new photospheric results but the other basic rules (Section 6.5) we apply, in addition to the 3D models, also play crucial roles.

The new photospheric abundances are generally lower to much lower, for some elements like C, N, O and Ne, than previously adopted values as discussed in Sections 6.5 and 6.6, and as shown in Fig. 6.1.

The impacts of these new results are summarized in Section 6.7. They lead to a lower solar metallicity, $Z = 0.0134$, much lower than the canonical $Z = 0.02$ generally used everywhere. But the disagreement between the Solar Standard Model predictions and the data derived from helioseismology is still present although somewhat alleviated.

For many elements, unfortunately, the uncertainties in the photospheric values are still very large essentially because of the uncertainties in the atomic and molecular data and of the lack of data for many elements making adequate NLTE analyzes impossible. Therefore, progress in atomic and molecular spectroscopy are, without any doubt, the keys for allowing solar and stellar spectroscopists to make more accurate determinations of chemical compositions. For a large number of elements, we need more accurate transition probabilities, more data for the cross-sections for collisions with electrons and with atomic hydrogen atoms, more accurate molecular data, ...

Acknowledgements N.G. thanks the editors for their invitation and IAGA for support.

References

Anders E, Grevesse N (1989) Geochim. Cosmochim. Acta 53:197

Asplund M, Grevesse N, Sauval AJ (2005) In: Barnes TG III, Bash FN (eds) Cosmic abundances as records of stellar evolution and nucleosynthesis, ASP Conference Series 336. ASP, San Francisco, p 25

Asplund M, Grevesse N, Sauval AJ, Scott P (2009) Annu Rev Astron Astrophys 47:481

Basu S, Antia HM (2004) Astrophys J Lett 606:L85

Burbidge EM, Burbidge GR, Fowler WA, Hoyle F (1957) Rev Mod Phys 29:547

Caffau E, Ludwig HG, Steffen M, Ayres TR, Bonifacio P, Cayrel R, Freytag B, Plez B (2008) Astron Astrophys 488:1031

Caffau E, Maiorca E, Bonifacio P, Faraggiana R, Steffen M, Ludwig HG, Kamp J, Busso M (2009) Astron Astrophys 498:877

Cameron AGW (1957) Pub Astron Soc Pac 69:201

Cameron AGW (1973) Space Sci Rev 62:9

Castro M, Vauclair S, Richard O (2007) Astron Astrophys 463:755

Goldberg L, Müller EA, Aller LH (1960) Astrophys J Supp Ser 45:1

Goldschmidt VM (1938) Skrifter Norske Videnskaps-Akad., OlsoI Math-Naturv. Klasse, No.4

Grevesse N, Noels A (2003) In: Prantzos N, Vangioni-Flam E, Cassé M (eds) Origin and evolution of the elements. Cambridge University Press, Cambridge, p 561

Grevesse N, Sauval AJ (1998) Space Sci Rev 85:161

Haxton WC, Serenelli AM (2008) Astrophys J 687:678

Lodders K (2003) Astrophys J 591:1220

Lodders K, Palme H, Gail HP (2009) In: Trumper JE (ed) Landolt-Börnstein, New series, Astronomy and astrophysics, vol VI/4B, Chap. 4.4, Springer, Berlin, Heidelberg, New York, p 560

Meléndez J, Asplund M, Gustafsson B, Young D (2009) Astrophys J 704:L66

Meléndez J, Asplund M, Gustafsson B, Young D, Ramírez I (2010) In: Cunha K, Spite M, Barbuy B (eds) Chemical abundances in the Universe: connecting first stars to planets, IAU Symposium No. 265, Cambridge University Press, Cambridge, p 412

Nordlund A, Stein R, Asplund A (2009) Living Rev Solar Phys 6:1

Palme H, Beer H (1993) In: Voigt HH (ed) Landolt-Börnstein, Group IV, Astronomy and astrophysics, vol 3. Springer, Berlin, p 196

Ramírez I, Meléndez J, Asplund M (2009) Astron Astrophys 508:L17

Ross JE, Aller LH (1976) Science 191:1223

Russell HN (1929) Astrophys J 70:11

Serenelli AM, Basu S, Ferguson JW, Asplund A (2009) Astrophys J 705:L123

Suess HE, Urey HU (1956) Rev Mod Phys 28:53

Unsöld A (1948) Z Astrophys 24:306

Part III
The Solar Atmosphere

Chapter 7

Magnetic Reconnection in the Solar Atmosphere Observed by Hinode

Shinsuke Imada, Hiroaki Isobe, and Toshifumi Shimizu

Abstract One of the most famous rapid energy conversion mechanisms in space is a magnetic reconnection. The general concept of a magnetic reconnection is that the rapid energy conversion from magnetic field energy to thermal energy, kinetic energy or non-thermal particle energy. The understanding of rapid energy conversion rates from magnetic field energy to other energy is the fundamental and essential problem in the space physics. One of the important goals for studying magnetic reconnection is to answer what plasma condition/parameter controls the energy conversion rates. Recently, solar atmosphere has been focused as a space laboratory for magnetic reconnection because of its variety in plasma condition. So far considerable effort has been devoted toward understanding the energy conversion rates of magnetic reconnection, and various typical features associated with magnetic reconnection have been observed in the solar atmosphere with the modern spacecraft/ground base telescopes. In this chapter, we first introduce the variety of plasma condition/parameter in solar atmosphere. Later, we discuss what we can learn about a magnetic reconnection from the *Hinode* observation.

7.1 Introduction

Magnetic reconnection has been discussed as one of the important mechanisms for the plasma heating and particle acceleration in astrophysical plasma, because the magnetic field energy can be rapidly released to the plasma during reconnection. One of major aspects of magnetic reconnection is the rapid energy conversion of stored free magnetic energy to kinetic energy, thermal energy, non-thermal particle energy and wave/turbulence energy. These energy conversions are fundamental and essential to understand the dynamical behavior of plasma not only in the solar atmosphere (e.g., Tsuneta et al. 1992; Masuda et al. 1994; Ohyama and Shibata 1998; Yokoyama et al. 2001) but also in the Earth's magnetosphere (e.g., Hones 1979; Nagai et al. 1998, 2001; Baumjohann et al. 1999; Øieroset et al. 2002; Imada et al. 2005, 2007a, 2008b), laboratory (e.g., Baum and Bratenahl 1974; Ono et al. 1988; Ji et al. 1998; Yamada et al. 1997), or other astronomical objects. One of the important goals for studying magnetic reconnection is what plasma condition/parameter, such as reconnection rate or plasma-β, controls the rates of the energy conversion (Fig. 7.1). Thus we need to observe the energy conversion rates of magnetic reconnection in various plasma conditions. Recently, solar atmosphere has been focused as a space laboratory for magnetic reconnection because of its variety in plasma condition. Actually, with the solar atmosphere, we can cover from high plasma β (> 1) to low plasma β (< 1), weakly ionized to fully ionized, and collisional to collisionless plasma. Observing magnetic reconnection in various plasma conditions is physically important in the category of not only solar physics but also other plasma physics. Thus observing entire solar atmosphere, from photosphere to corona, is very important to understand the characteristics of magnetic reconnection in various plasma conditions.

S. Imada (✉)
Institute of Space and Astronautical Science, Japan Aerospace Exploration Agency, 3-1-1 Yoshinodai, Sagamihara-shi, Kanagawa 229-8510, Japan
e-mail: imada.shinsuke@jaxa.jp

M.P. Miralles, J. Sánchez Almeida (eds.), *The Sun, the Solar Wind, and the Heliosphere*, IAGA Special Sopron Book Series 4, DOI 10.1007/978-90-481-9787-3_7,

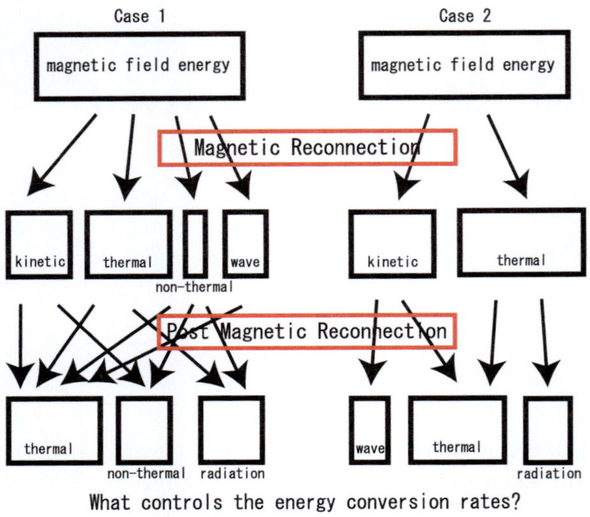

Fig. 7.1 Concept of magnetic reconnection study

7.2 Typical Plasma Parameters in the Solar Atmosphere

In this section, we introduce the variety of typical plasma parameters in the solar atmosphere. Figure 7.2 shows the temperature and electron density as a function of height from photosphere. We can clearly see the variety of temperature/electron density from photosphere to corona. Table 7.1 shows the typical plasma parameters as a function of location in the solar atmosphere, where h: height from photosphere, N_e: electron density, N_H: neutral Hydrogen density, T_e: electron temperature, B: magnetic field, V_S: sound velocity, V_A: Alfven velocity, V_{the}: electron thermal velocity, ω_{ce}: electron cyclotron frequency, ω_{ci}: proton cyclotron frequency, ν_e: electron collision frequency, ν_{e0}: electron neutral collision frequency, c/ω_{pe}: electron inertia length, c/ω_{pi}: ion inertia length, λ_p: pressure scale height, and λ_{mfp}: electron mean free path, Plasma-β: gas pressure/magnetic pressure. It is often useful to have a simpler formula, which already incorporates the numerical values of any constants include in the formula as well as possible unit conversion factors. Thus we write down these formulas below (Equations 7.1–7.8). We can calculate the numerical values of all plasma parameters in Table 7.1 with these simple formulas. We can clearly see the high variation of typical plasma parameter in the solar atmosphere in Table 7.1.

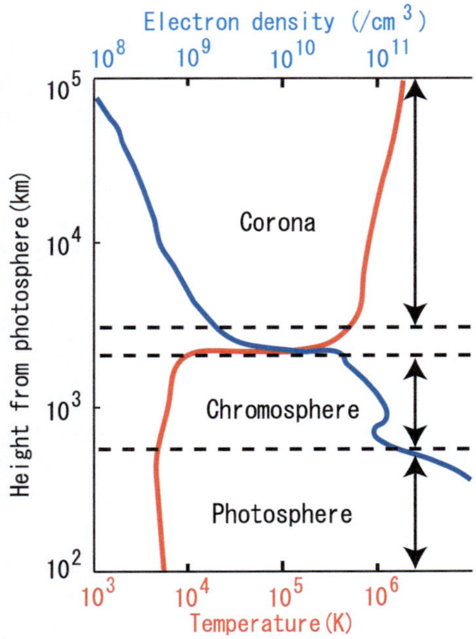

Fig. 7.2 Temperature and density as a function of height from photosphere

Let us discuss five of non-dimensional parameters which critically decide plasma behavior. Plasma-β represents the ratio between gas pressure and magnetic pressure. Generally plasma behaves as charged-gas in the low plasma-β condition. On the other hand, plasma behaves as neutral-gas in the high plasma-β condition. Most parts of the chromosphere/corona have a low plasma-β of $\beta < 1$, but are sandwiched between the higher values $\beta > 1$ in the photosphere and outer corona. Thus the magnetic field is dynamically important in the chromosphere/corona but not in the photosphere/outer corona. Actually, plasma-β is related to the ratio between sound velocity and Alfven velocity. The phase velocity of Alfven wave is faster than that of sound wave in most part of chromosphere/corona, though the phase velocity of sound wave is faster than that of Alfven wave in photosphere or outer corona. Therefore, the most information is carried by Alfven wave in chromosphere or corona, though the most information is carried by sound wave in photosphere or outer corona. The ratio between N_e and $N_H + N_e$ decides whether neutral or charged particles is main component. In the photosphere and chromosphere, N_H is sufficiently larger than N_e. On the other hand, neutral particles are negligible in the corona and outer corona. Actually the plasma becomes fully ionized at the sharp transition from chromosphere to coronal temperature. Thus the main component

Table 7.1 The plasma parameter in the solar atmosphere

	Photosphere	Chromosphere	Corona	Outer corona
h (km)	0	$\sim 2 \times 10^3$	$\sim 10^4$	$\sim 10^6$
N_e (/cc)	$\sim 10^{14}$	$\sim 10^{10}$	$\sim 10^9$	$\sim 10^7$
N_H (/cc)	$\sim 10^{17}$	$\sim 10^{12}$	~ 0	~ 0
T_e (K)	6×10^3	$\sim 7 \times 10^3$	$\sim 10^6$	$\sim 10^6$
B (G)	$\sim 10^3$	$\sim 10^2$	$\sim 10^1$	$\sim 10^{-1}$
V_S (km/s)	$\sim 9 \times 10^0$	$\sim 1 \times 10^1$	$\sim 1 \times 10^2$	$\sim 1 \times 10^2$
V_A (km/s)	$\sim 7 \times 10^0$	$\sim 2 \times 10^2$	$\sim 7 \times 10^2$	$\sim 7 \times 10^1$
V_{the} (km/s)	$\sim 3 \times 10^2$	$\sim 3 \times 10^2$	$\sim 4 \times 10^3$	$\sim 4 \times 10^3$
ω_{ce} (Hz)	$\sim 3 \times 10^9$	$\sim 3 \times 10^8$	$\sim 3 \times 10^7$	$\sim 3 \times 10^5$
ω_{ci} (Hz)	$\sim 2 \times 10^6$	$\sim 2 \times 10^5$	$\sim 2 \times 10^4$	$\sim 2 \times 10^2$
ν_e (Hz)	$\sim 5 \times 10^9$	$\sim 6 \times 10^5$	$\sim 7 \times 10^1$	$\sim 7 \times 10^{-1}$
ν_{e0} (Hz)	$\sim 1 \times 10^8$	$\sim 2 \times 10^3$	~ 0	~ 0
c/ω_{pe} (km)	$\sim 5 \times 10^{-7}$	$\sim 5 \times 10^{-5}$	$\sim 2 \times 10^{-4}$	$\sim 2 \times 10^{-3}$
c/ω_{pi} (km)	$\sim 2 \times 10^{-5}$	$\sim 2 \times 10^{-3}$	$\sim 7 \times 10^{-3}$	$\sim 7 \times 10^{-2}$
λ_p (km)	$\sim 3 \times 10^2$	$\sim 3 \times 10^2$	$\sim 5 \times 10^4$	$\sim 5 \times 10^4$
λ_{mfp} (km)	$\sim 6 \times 10^{-8}$	$\sim 6 \times 10^{-4}$	$\sim 6 \times 10^1$	$\sim 5 \times 10^3$
Plasma-β	$\sim 2 \times 10^0$	$\sim 2 \times 10^{-3}$	$\sim 7 \times 10^{-2}$	$\sim 7 \times 10^0$
$N_e/(N_H + N_e)$	$\sim 1 \times 10^{-3}$	$\sim 1 \times 10^{-2}$	$\sim 1 \times 10^0$	$\sim 1 \times 10^0$
λ_p/λ_{mfp}	$\sim 4 \times 10^9$	$\sim 6 \times 10^5$	$\sim 8 \times 10^2$	$\sim 9 \times 10^0$
$\omega_{ce}/(\nu_e + \nu_{e0})$	$\sim 6 \times 10^{-1}$	$\sim 5 \times 10^2$	$\sim 4 \times 10^5$	$\sim 4 \times 10^5$
$\omega_{ci}/(\nu_e + \nu_{e0})$	$\sim 3 \times 10^{-4}$	$\sim 3 \times 10^{-1}$	$\sim 2 \times 10^2$	$\sim 2 \times 10^2$

in the photosphere/chromosphere is neutral-particle, although charged-particle is main in corona and outer corona region. Note that the ratio between N_e and $N_H + N_e$ do not decide whether plasma behaves as charged-particle or neutral gas, because the ionized component can drag the neutral component in the case that ion-neutral collisions are enough frequent. The ratio between λ_p and λ_{mfp} represents the ratio between the collision-scale and typical macro-fluid-scale. In the photosphere/chromosphere/corona, the electron mean free path is significantly small compared with the pressure scale height. Thus plasma behaves collisional plasma in the macro-scale. Actually, all neutral and charged particles are fully interacted by collision in the macro-scale. On the other hand, in the outer corona the electron mean free path is comparable to the pressure scale height. Thus the interaction by collision is weak in the outer corona. The ratio between gyro and collision frequency represents the collisionality in the micro-scale. In the case that the gyro frequency is enough larger than collision frequency, microscopic effects might be taken place. Therefore it seems that anomalous resistivity can work in the corona or outer corona. On the other hand, it may not work in the photosphere/chromosphere. It is generally believed that the fast Petschek-type reconnection will take place with localized anomalous resistivity. On the other hand, the slow Sweet-Parker type reconnection will take place in the case of no localized anomalous resistivity. Therefore, in classically slow magnetic reconnection takes place in the photosphere/chromosphere, and fast magnetic reconnection take place in the corona/outer corona. Recently, fast MHD reconnection in the limit of uniform low resistivity is intensively discussed (e.g. Lazarian and Vishniac 1999; Loureiro et al. 2007; Kowal et al. 2009; Zweibel and Yamada 2009; Samtaney et al. 2009; Bhattacharjee et al. 2009; Cassak et al. 2009) and needed more discussion. It is important to reveal fast MHD reconnection in the limit of uniform low resistivity in both of observational and theoretical viewpoint.

We can summarize this section as follows (see Table 7.2): photospheric plasma behaves as collisional (high collisionality) neutral gas (high plasma-β and low ionization), chromospheric plasma behaves as collisional (high collisionality) semi-charged gas (low plasma-β but low ionization), coronal plasma behaves as semi-collisionless (collisional in macro-scale but collisionless in micro-scale) charged gas (low plasma-β and high ionization), outer coronal plasma behaves as collisionless (low collisionality) semi-charged gas (high ionization but high plasma-β).

Table 7.2 Typical characteristics of plasma in solar atmosphere

	Photosphere	Chromosphere	Corona	Outer corona
Dominant pressure	Gas	Magnetic field	Magnetic field	Gas
Dominant component	Neutral	Neutral	Charged	Charged
Collisionality	Collisional	Collisional	semi-Collisionless	Collisionless

$$\omega_{ce} = \frac{eB}{2\pi m_e c} = 2.80 \times 10^7 \left(\frac{B}{10\ \mathrm{G}}\right) \mathrm{Hz} \tag{7.1}$$

$$\omega_{ci} = \frac{eB}{2\pi m_i c} = 1.52 \times 10^4 \left(\frac{B}{10\ \mathrm{G}}\right) \mathrm{Hz} \tag{7.2}$$

$$V_S = (\gamma kT/m_i)^{\frac{1}{2}} = 1.17 \times 10^2 \left(\frac{T_e}{1\ \mathrm{MK}}\right)^{\frac{1}{2}} \mathrm{km/s} \tag{7.3}$$

$$V_A = \frac{B}{\sqrt{4\pi n m_i}} = 6.89 \times 10^2 \left(\frac{n}{10^9/\mathrm{cc}}\right)^{-\frac{1}{2}} \left(\frac{B}{10\ \mathrm{G}}\right) \mathrm{km/s} \tag{7.4}$$

$$V_{the} = (kT_e/m_e)^{\frac{1}{2}} = 3.89 \times 10^3 \left(\frac{T_e}{1\ \mathrm{MK}}\right)^{\frac{1}{2}} \mathrm{km/s} \tag{7.5}$$

$$\nu_e = \frac{4\sqrt{2\pi} n \ln \Lambda e^4}{3\sqrt{m_e}(kT_e)^{3/2}} = 6.58 \times 10^1 \left(\frac{n}{10^9/\mathrm{cc}}\right) \left(\frac{T}{1\ \mathrm{MK}}\right)^{-3/2} \left(\frac{\ln \Lambda}{18}\right) \mathrm{Hz} \tag{7.6}$$

$$\nu_{e0} = n_0 \sigma_s (kT_e/m_e)^{1/2} = 1.51 \times 10^3 \left(\frac{n_0}{10^{12}/\mathrm{cc}}\right) \left(\frac{T_e}{7 \times 10^3\ \mathrm{K}}\right)^{\frac{1}{2}} \mathrm{Hz} \tag{7.7}$$

$$\frac{c}{\omega_{pe}} = c\left(\frac{4\pi n_e e^2}{m_e}\right)^{-\frac{1}{2}} = 1.68 \times 10^{-4} \left(\frac{n_e}{10^9/\mathrm{cc}}\right)^{-\frac{1}{2}} \mathrm{km} \tag{7.8}$$

$$\frac{c}{\omega_{pi}} = c\left(\frac{4\pi n_i e^2}{m_i}\right)^{-\frac{1}{2}} = 7.21 \times 10^{-3} \left(\frac{n_i}{10^9/\mathrm{cc}}\right)^{-\frac{1}{2}} \mathrm{km} \tag{7.9}$$

$$\beta = \frac{8\pi n k T}{B^2} = 6.95 \times 10^{-2} \left(\frac{n}{10^9/\mathrm{cc}}\right) \left(\frac{T}{1\ \mathrm{MK}}\right) \left(\frac{B}{10\ \mathrm{G}}\right)^2 \tag{7.10}$$

7.3 Magnetic Reconnection in the Solar Corona

Magnetic reconnection in the solar corona has been discussed as one of the popular mechanisms of a solar flare in the category of solar physics. Various features expected from the magnetic reconnection model have been confirmed by modern observations. These include a cusp-like structure in X-ray images (e.g., Tsuneta et al. 1992), non-thermal electron acceleration (e.g., Masuda et al. 1994), chromospheric evaporation (e.g., Teriaca et al. 2003), reconnection inflow (e.g., Yokoyama et al. 2001), and plasmoid ejection (e.g., Ohyama and Shibata 1998). Recently the *Hinode* spacecraft was launched (Kosugi et al. 2008). It is a Japanese mission collaborating with US and UK with three instruments on board; the Solar Optical Telescope (SOT), the X-Ray Telescope (XRT) and the EUV Imaging Spectrometer (EIS). SOT measures the three components of the magnetic field of the photosphere and the dynamics of the lower atmosphere. XRT observes the dynamics of the corona by imaging, and EIS measures the flows and dynamics from the transition region to the corona. After their first light, Hinode has been revealing many new solar flare aspects. There are already more than ten papers about one of the most popular flare event which has been occurred on 2006 December 13th (see Table 7.3). In this section

Table 7.3 Published papers of 2006 December 13 solar flare

Energy category	Stage	Content	Telescope	Authors
Magnetic field	Whole	B configuration	SOT	Guo et al. (2008)
	Main	Reconnection E	SOT	Jing et al. (2008a)
	Whole	B configuration	SOT	Jing et al. (2008b)
	Whole	B photosphere	SOT	Kubo et al. (2007)
	Whole	Helicity injection	SOT	Magara and Tsuneta (2008)
	Whole	B configuration	XRT	Su et al. (2007)
	Whole	Sunspot decay	SOT	Tan et al. (2009)
	Whole	B photosphere	SOT	Wang et al. (2008)
	Whole	Helicity injection	SOT	Zhang et al. (2008)
	Whole	B photosphere	MDI	Zhang et al. (2007)
Kinetic	Main	Ejection and shock	EIS	Asai et al. (2008)
	Main	Microwave burst	Ground	Chen (2008)
Thermal	Main	Density and B	Ground	Yan et al. (2007)
Non-thermal	–	Energy spectrum	Ground	Abbasi et al. (2008)
	Main	White light flare	SOT	Isobe et al. (2007)
	–	SEP	In-situ	Li et al. (2009)
	Main	Electron acceleration	RHESSI	Minoshima et al. (2009)
	Main	Spectral index	Ground	Ning (2008)
	–	Energy spectrum	Ground	Vashenyuk et al. (2008)
Wave/turbulence	Whole	Intermittency	SOT	Abramenko et al. (2008)
	Whole	Line broadening	EIS	Harra et al. (2009)
	Main	Line broadening	EIS	Imada et al. (2008a)
	Post	Radio oscillation	Ground	Tan et al. (2007)
Others	Post	Dimming	EIS	Imada et al. (2007b)
	Post	CME	EIS	Jin et al. (2009)
	Main	Sunspot oscillation	SOT	Kosovichev and Sekii (2007)
	Post	CME	In-situ	Liu et al. (2008)
	Pre	Filament rotation	EIS	Williams et al. (2009)

we briefly review what we understand about magnetic reconnection in solar corona from the Hinode observation of 2006 December 13th flare event, and what we still do not understand. We also discuss further possibility of the flare observation in the future.

One of the important parameter for the magnetic reconnection in solar corona is how much energy is stored in the coronal magnetic fields. The stored energy is generally estimated from the continuous observation of magnetic filed in the photosphere. Kubo et al. (2007) qualitatively discussed by using continuous photospheric magnetic fields observations in the flare productive active region NOAA 10930 with the SOT aboard the Hinode spacecraft during 2006 December 6–19. They found that the sunspot rotated and created enormous shears before the flare. The free energy was seemed to be stored in coronal magnetic field by the sunspot rotating motions. Furthermore, Magara and Tsuneta (2008) calculated the magnetic helicity in a flare-productive active region and found that the helicity increases very rapidly before the flare. One of the most interesting their findings is that the magnetic helicity became saturated in the late pre-flare phase. This quantitative analysis is very important for the estimation of stored energy before magnetic reconnection (top in Fig. 7.1). Jing et al. (2008a) discussed the local magnetic reconnection rate from the flare ribbons separation speed and local magnetic fields on this event using the high-resolution SOT data. They also discussed the correlation between the ribbon separation speed and magnetic field strength around the maximum phase of the flare. They concluded that the ribbon separation is weakly and negatively correlated with magnetic field strength. An empirical relation of $V_r \propto B_n^{-0.15}$ is obtained at the flare peak time. These studies are important for understanding the energy release rate in each time during the magnetic reconnection. Therefore, we can observationally evaluate the stored magnetic field energy and its release rate which is the top part (pre reconnection stage) of Fig. 7.1.

To understand the energy conversion rate of magnetic reconnection we have to evaluate the thermal, kinetic, non-thermal, and wave/turbulence energy converted from the stored magnetic energy in the corona. Unfortunately the estimation of thermal energy in this flare is very difficult, because the main part of flare loops was saturated with excess photons from the bright post flare loops. The flare have occurred in the initial operation phase of *Hinode*. Thus the exposure time is too long for the flare observation. Recently, Imada et al. (2009) proposed the new method to estimate ion temperature without ionization equilibrium assumption in solar corona by using a pair of emission lines of different atomic species. This method is useful to estimate the thermal energy in magnetic reconnection region. Generally, kinetic energy can be evaluated from doppler velocity (line-of-sight velocity) obtained by spectroscopic observation in corona. Asai et al. (2008) discussed the two types of strongly blueshifted emission lines of this flare observed with EIS. They concluded that one was related to a plasmoid ejection seen in soft X-rays, and the other was associated with the faint arc-shaped ejection seen in soft X-rays. The velocity of plasmoid is faster than the sound speed (a few hundred km s^{-1}), but slower than the Alfven speed ($\sim$1,000 km s^{-1}). This speed is consistent with the prediction of MHD simulation. The faint soft X-ray ejection is thought to be a magnetohydrodynamic (MHD) fast-mode shock wave. Although we have much information about fast flows associated with solar flare, we cannot detect the reconnection outflow itself yet. Non-thermal energy in solar flare can be estimated from hard X-ray (HXR) observation. Minoshima et al. (2009) discussed the electron acceleration site and mechanism by using the magnetic field observation at photosphere (SOT) and HXR in corona (*RHESSI*). They found that the HXR sources are located at the region where horizontal magnetic fields change direction. The region is interpreted as the footpoint of magnetic separatrix. They concluded that these observations as evidence of electron acceleration near the magnetic separatrix and injection parallel to the field line. Actually, we can estimate the non-thermal energy with these HXR observations. On the other hand, these HXR emissions come from not reconnection region but footpoints. Generally, further acceleration or de-acceleration can be taken place in the course of propagating from reconnection region to footpoints. For the wave/turbulence energy can be estimated from the excess of line width obtained from spectroscopic observation in corona. Imada et al. (2008a) studied the characteristics of the non-Gaussian line profile in and around a flare arcade. They successfully distinguished the symmetric broad line profile from the distorted broad line profile by *M*—the additional component contribution—defined in their paper. From their study, the distorted/symmetric broad line profiles were preferentially observed in new/old flare loops, respectively. Their results may indicate that the old flare loops store the turbulent plasma. Note that their discussion is related to not reconnection region but post flare loops.

The *Hinode* observation of magnetic reconnection in solar corona can be summarized as follows. Both of the stored magnetic field energy in corona before magnetic reconnection (top part of Fig. 7.1) and the most part of energy post reconnection stage (bottom part of Fig. 7.1) can be estimated in detail. On the other hand, there is not enough observational knowledge of the physical parameters in the reconnection region (middle part of Fig. 7.1) for these events. The inflow into the reconnection region, the temperature of the plasma in the reconnection region, and the temperatures and densities of the plasma jets predicted by reconnection, have not been quantitatively measured in sufficient. EIS on Hinode may provide some answers if solar cycle 24 ever produces a solar maximum. But it is important to answer why the most observation cannot detect the predicted flow or temperature in the reconnection region.

One of the reasons why we cannot observe inside the magnetic reconnection region is due to its darkness. Generally we can see the bright cusp-like structure during solar flare, although the reconnection region is faint/blind. Thus it is plausible that the throughput of current telescope is not sufficient to observe inside the reconnection region. Further, the stray/scattered lights from the flare loops are so strong. Therefore, it is possible that the stray/scattered lights from the flare loops hide the reconnection region. One may think that the temperature in the reconnection region is enough higher than that of cusp-like flare loops. Thus the wavelength of emission from reconnection region is different from flare loops. However, this is not entirely true. Magnetic reconnection causes rapid heating. Thus ionization cannot reach to the equilibrium stage. Figure 7.3 shows the non-ionization equilibrium calculation at the down stream of Petschek

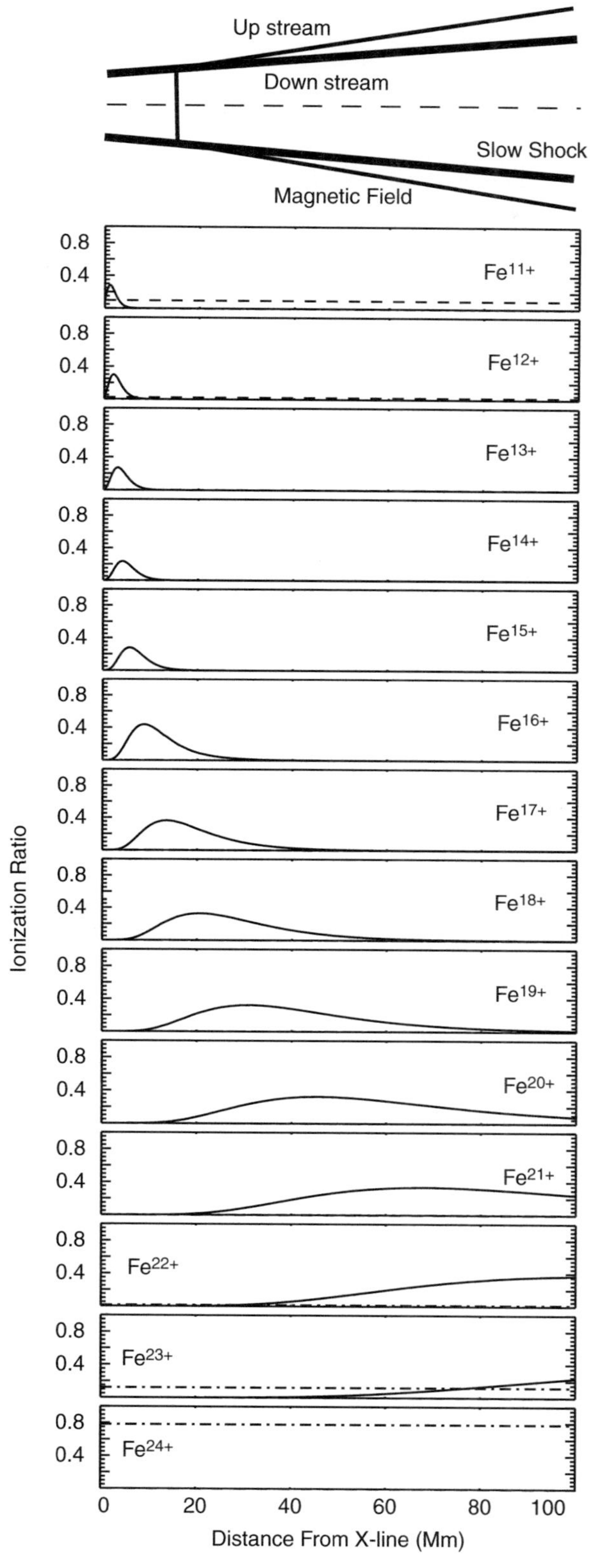

Fig. 7.3 Non-ionization equilibrium calculation at the down stream of magnetic reconnection region

type magnetic reconnection. At the top panel show the schematic picture of reconnection region. The ionization rate in the center of reconnection region as a function of distance from X-type neutral line (dashed line in the top) are plotted from Fe^{+11} to Fe^{+24}. Table 7.4 shows the parameter of slow-mode shock in this calculation, where B_0, N_0, $T_{0,1}$, β_0, $\theta_{0,1}$ show the magnetic field, density, temperature, plasma-β, shock angle (subscripts 0 and 1 represent upstream and downstream parameters of the slow-mode shock, respectively). We stop our calculation $\sim$100 Mm, because the distance from X-type neutral line to flare loops generally seems to be less than 100 Mm. Note that in this calculation thermal conduction are not taken into account. The dashed/dot-dashed lines show the ionization rate of ionization equilibrium condition for 1 MK/40 MK plasma, respectively. From this result we can clearly see that plasma cannot reach the ionization equilibrium in the downstream of slow-mode shock. The typical temperature and density of flare loops are 10 MK and 10^{11}/cc, and the dominant ionization is from Fe^{+18} to Fe^{+21}. Thus the wavelength of emission from reconnection region is not so much different from flare loops. The density is higher for two orders, and the emissions from flare loops are higher for four orders than that of reconnection region. Therefore high-throughput and reducing the stray/scattered light is quite important for observing magnetic reconnection region. One of the solutions is that the new telescope do not use without pre-filter. Without pre-filter we can achieve not only high-throughput but also reducing the stray/scattered light from flare loops. Using occulter to reduce the stray/scattered light from bright flare loops is also one of the possibilities.

7.4 Magnetic Reconnection in the Solar Chromosphere

One of the most important *Hinode* findings is that magnetic reconnection may frequently occur in the solar chromosphere. Shibata et al. (2007) found the ubiquitous presence of chromospheric anemone jets outside sunspots in active regions. Their typical velocity is 10–20 km/s. These small jets have an inverted Y-shape, similar to the shape of X-ray anemone jets in the corona. They concluded that these features imply the magnetic reconnection in the chromosphere.

Table 7.4 Slow-mode shock condition in the calculation

B_0	N_0	T_0	β_0	θ_0	θ_1	outflow	T1
12 G	4.2×10^8/cc	1 MK	0.01	85°	2.3°	1150 km/s	40 MK

Actually, *Hinode* have found many evidences of magnetic reconnection in the chromosphere (e.g., Katsukawa et al. 2007; Nishizuka et al. 2008; Shimizu et al. 2009). One of the important question about magnetic reconnection in the solar chromosphere is "Can fast magnetic reconnection (independent on resistivity) take place?". Current our understanding which is based on the *Hinode* observation is "Yes". Unfortunately, *Hinode* observes the solar chromosphere by not spectro/polarimetric observation but Ca II H imaging. The time cadence of the observation is fast, but we do not have velocity, density, or magnetic field information from *Hinode* observation. We need more information in the chromosphere to conclude the question. Although we do not have sufficient information about magnetic reconnection in chromosphere, it is worthy to discuss the possibility of fast magnetic reconnection in chromosphere. We already mentioned in Section 7.2 that anomalous resistivity may not work in the chromosphere. Therefore, it seems that fast magnetic reconnection cannot take place in the chromosphere. One of the important differences of plasma condition between solar corona and chromosphere is ion-neutral collision. The collision between ions and neutrals cause ambipolar diffusion (e.g., Zweibel 1988, 1989; Brandenburg and Zweibel 1994; Chiueh 1998). Recently, Isobe et al. (in preparation) discussed the effect of ambipolar diffusivity in the current sheet of chromosphere. The effect of ambipolar diffusivity is large in the solar chromosphere because of their mild ion-neutral collision. According to their results, the bursty magnetic reconnection can take place in the chromosphere. The magnetic reconnection in the chromosphere is still not clear in both of observational and theoretical viewpoint.

7.5 Summary

We briefly introduced a variety of typical plasma parameter in the solar atmosphere, and mentioned the importance of comparison study of magnetic reconnection in different plasma conditions. Especially we claimed that it is important to discuss magnetic reconnection not only in the solar corona but also in the solar chromosphere. We also briefly reviewed a series of *Hinode* studies about 2006 December 13 X- Class flare. *Hinode* observation is revealing new aspects of solar flare. One of the most important advantages in Hinode observation is quantitative study. Many physical models or assumptions can be discussed by not only morphological study but also quantitative study. On the other hand, we still do not have some important observation such as direct measurement of reconnection outflow. We have to consider why we still do not have such important observations. Next solar maximum is coming soon.

Acknowledgements Hinode is a Japanese mission developed and launched by ISAS/JAXA, collaborating with NAOJ as a domestic partner, NASA and STFC (UK) as international partners. Scientific operation of the Hinode mission is conducted by the Hinode science team organized at ISAS/JAXA. This team mainly consists of scientists from institutes in the partner countries. Support for the post-launch operation is provided by JAXA and NAOJ (Japan), STFC (U.K.), NASA (U.S.A.), ESA, and NSC (Norway).

This work was supported by the Grant-in-Aid for Young Scientists Start-up (21840062) and by the Grant-in-Aid for Creative Scientific Research "The Basic Study of Space Weather Prediction" (17GS0208, Head Investigator: K. Shibata) from the Ministry of Education, Science, Sports, Technology, and Culture (MEXT) of Japan.

References

Abbasi R et al (2008) Intermittency in the photosphere and corona above an active region. Astrophys J 689(1):L65–L68

Abramenko V et al (2008) Solar energetic particle spectrum on 2006 December 13 determined by IceTop. Astrophys J 681(2):1669–1676

Asai A et al (2008) Strongly blueshifted phenomena observed with Hinode EIS in the 2006 December 13 solar flare. Astrophys J 685(1):622–628

Baum PJ, Bratenahl A (1974) Spectrum of turbulence at a magnetic neutral point. Phys Fluids 17:1232–1235

Baumjohann W et al (1999) Substormdipolarization and recovery. J Geophys Res 104: 24995–25000

Bhattacharjee A et al (2009) Fast reconnection in high-Lundquist-number plasmas due to the plasmoid Instability. Phys Plasmas 16(11):112102–112102-5

Brandenburg A, Zweibel EG (1994) The formation of sharp structures by ambipolar diffusion. Astrophys J 427(2): L91–L94

Cassak PA et al (2009) Scaling of Sweet-Parker reconnection with secondary islands. Phys Plasmas 16(12): 120702

Chen B, Yan Y (2008) Short-lived absorptive Type III-like microwave bursts as a signature of fragmented electron injections. Astrophys J 689(2):1412–1420

Chiueh T (1998) Ambipolar diffusion–driven tearing instability in a steepening background magnetic field. Astrophys J 494:90

Guo Y et al (2008) 3D magnetic field configuration of the 2006 December 13 flare extrapolated with the optimization method. Astrophys J 679(2):1629–1635

Harra LK et al (2009) Coronal nonthermal velocity following helicity injection before an X-class flare. Astrophys J 691(2):L99–L102

Hones EW (1979) Transient phenomena in themagnetotail and their relation to substorm. Space Sci Rev 23:393–410

Imada S et al (2005) Average profiles of energetic and thermal electrons in the magnetotail reconnection regions. Geophys Res Lett 32(9):L09101

Imada S et al (2007a) Energetic electron acceleration in the downstreamreconnection outflow region. J Geophys Res 112(A3):A03202

Imada S et al (2007b) Discovery of a temperature-dependent upflow in the plage region during a gradual phase of the X-class flare. Pub Astron Soc Jpn 59(SP3):S793–S799

Imada S et al (2008a) Non-Gaussian line profiles in a large solar flare observed on 2006 December 13. Astrophys J 679(2):L155–L159

Imada S et al (2008b) The dawn-dusk asymmetry of energetic electron in the Earth's magnetotail: observation and transportmodels. J Geophys Res 113(A11):A11201

Imada S et al (2009) Ion temperature and non-thermal velocity in a solar active region: using emissionlines of different atomic species. Astrophys J Lett 705(2):L208–L212

Isobe H et al (2007) Flare ribbons observed with G-band and FeI 6302 filters of the solar optical telescope on board Hinode. Pub Astron Soc Jpn 59(SP3):S807–S813

Ji H et al (1998) Experimental test of the Sweet-Parker model of magnetic reconnection. Phys Rev Lett 80(15):3256–3259

Jin M et al (2009) Coronal mass ejection induced outflows observed with Hinode/EIS. Astrophys J 702(1): 27–38

Jing J et al (2008a) Spatial distribution of magnetic reconnection in the 2006 December 13 solar flare as observed by Hinode. Astrophys J 672(1):L73–L76

Jing J et al (2008b) Changes of magnetic structure in three dimensions associated with the X3.4 flare of 2006 December 13. Astrophys J 676(1):L81–L84

Katsukawa Y et al (2007) Small-scale jetlike features in penumbral chromospheres. Science 318(5856):1594

Kosovichev AG, Sekii T (2007) Initial observations of sunspot oscillations excited by solar flare. Astrophys J 670(2): L147–L149

Kosugi T et al (2007) The Hinode (Solar-B) mission: an overview. Solar Phys 243(1):3–17

Kowal G et al (2009) Numerical tests of fast reconnection in weakly stochastic magnetic fields. Astrophys J 700(1): 63–85

Kubo M et al (2007) Hinode observations of a vectormagnetic field change associated with a flare on 2006 December 13. Pub Astron Soc Jpn 59(SP3):S779–S784

Lazarian A, Vishniac ET (1999) Reconnection in a weakly stochastic field. Astrophys J 517(2):700–718

Li C et al (2009) Solar source of energetic particles in interplanetary space during the 2006 December 13 event. Astron Astrophys 503(3):1013–1021

Liu Y et al (2008) A comprehensive view of the 2006 December 13 CME: from the Sun to interplanetary space. Astrophys J 689(1): 563–571

Loureiro et al (2007) Instability of current sheets and formation of plasmoid chains. Phys Plasmas 14(10):100703

Magara T, Tsuneta S (2008) Hinode's observational result on the saturation of magnetic helicity injected into the solar atmosphere and its relation to the occurrence of a solar flare. Pub Astron Soc Jpn 60(5):1181–1189

Masuda S et al (1994) A loop-top hard X-ray source in a compact solar flare as evidence for magnetic reconnection. Nature 371(6497):495–497

Minoshima T et al (2009) Multiwavelength observation of electron acceleration in the 2006 December 13 flare. Astrophys J 697(1):843–849

Nagai T et al (1998) Structure and dynamics of magnetic reconnection for substorm onsets with Geotail observations. J Geophys Res 103:4419–4440

Nagai T et al (2001) Geotail observations of the Hall current system: evidence of magnetic reconnection in the magnetotail. J Geophys Res 106:25929–25949

Ning Z (2008) Microwave and hard X-ray spectral evolution for the 13 December 2006 solar flare. Solar Phys 247(1):53–62

Nishizuka N et al (2008) Giant chromospheric anemone jet observed with Hinode and comparison with magnetohydrodynamic simulations: evidence of propagating Alfven waves and magnetic reconnection. Astrophys J 683(1):L83–L86

Ohyama M, Shibata K (1998) X-ray plasma ejection associated with an impulsive flare on 1992 October 5: physical conditions of X-ray plasma ejection. Astrophys J 499:934

Øieroset M et al (2002) Evidence for electron acceleration up to 300 keV in the magnetic reconnection diffusion region of Earth's magnetotail. Phys Rev Lett 89(19):195001

Ono Y et al (1988) Relaxation phenomena in the high-temperature S-1 spheromak. Phys Rev Lett 61(25):2847–2850

Samtaney R et al (2009) Formation of plasmoid chains in magnetic reconnection. Phys Rev Lett 103(10):105004

Shibata K et al (2007) Chromospheric anemone jets as evidence of ubiquitous reconnection. Science 318(5856) 1591

Shimizu T et al (2009) Hinode observation of the magnetic fields in a sunspot light bridge accompanied by long-lasting chromospheric plasma ejections. Astrophys J Lett 696(1):L66–L69

Su Y et al (2007) Evolution of the sheared magnetic fields of two X-class flares observed by Hinode/XRT. Pub Astron Soc Jpn 59(SP3):S785–S791

Tan B et al (2007) The microwave pulsations and the tearing modes in the current-carrying flare loops. Astrophys J 671(1):964–972

Tan C et al (2009) Evolution of optical penumbral and shear flows associated with the X3.4 flare of 2006 December 13. Astrophys J 690(2):1820–1828

Teriaca L et al (2003) Solar and heliospheric observatory/coronal diagnostic spectrograph and ground-based observations of a two-ribbon flare: spatially resolved signatures of chromospheric evaporation. Astrophys J 588(1):596–605

Tsuneta S et al (1992) Observation of a solar flare at the limb with the YOHKOH Soft X-ray Telescope. Pub Astron Soc Jpn 44(5):L63–L69

Wang H et al (2008) Study of magnetic channel structure in active region 10930. Astrophys J 687(1):658–667

Vashenyuk EV et al (2008) Characteristics of relativistic solar cosmic rays during the event of December 13, 2006. Geomagnet Aeron 48(2):149–153

Williams DR et al (2009) Evidence from the extreme-ultraviolet imaging spectrometer for axial filament rotation before a large flare. Pub Astron Soc Jpn 61(3):493–497

Yamada M et al (1997) Study of driven magnetic reconnection in a laboratory plasma. Phys Plasmas 4(5): 1936–1944

Yan Y et al (2007) Diagnostics of radio fine structures around 3 GHz with Hinode data in the impulsive phase of an X3.4/4B flare event on 2006 December 13. Pub Astron Soc Jpn 59(SP3):S815–S821

Yokoyama T et al (2001) Clear evidence of reconnection inflow of a solar flare. Astrophys J 546(1):L69–L72

Zhang J et al (2007) Interaction between a fast rotating sunspot and ephemeral regions as the origin of the major solar event on 2006 December 13. Astrophys J 662(1):L35–L38

Zhang Y et al (2008) Correlation between the sharp variation of the transport rate of magnetic helicity and solar eruptive events. Astrophys J 682(2):L133–L136

Zweibel EG (1988) Ambipolar diffusion drifts and dynamos in turbulent gases. Astrophys J 329(1):384–391

Zweibel EG (1989) Magnetic reconnection in partially ionized gases. Astrophys J 340(1):550–557

Zweibel EG, Yamada M (2009) Magnetic reconnection in astrophysical and laboratory plasmas. Annu Rev Astron Astrophys 47(1):291–332

Chapter 8

Constraining the Initiation and Early Evolution of CMEs with SECCHI on STEREO

Spiros Patsourakos

Abstract One of the major objectives of the STEREO mission is to understand how and where Coronal Mass Ejections (CMEs) are accelerated. We review here the status of our knowledge from SECCHI CME observations in the period 2007–2009 about various topics concerned with the early stages of CMEs. This includes: impulsively accelerated CMEs, the flare-CME relationship, EUV dimmings, EUV waves and eruptive prominences. We will show how the unique characteristics of this mission and most importantly the distinct viewpoints and the high image cadence allowed for significant progress in the above mentioned areas. We also discuss potential future uses of SECCHI data.

8.1 Introduction

One of the major problems in solar physics today is the understanding of Coronal Mass Ejections (CMEs). Significant progress into our understanding of CME onsets and of their impact into the inner corona has been made by the Solar and Heliospheric Observatory SOHO; see for instance the reviews Pick et al. (2006), Schwenn et al. (2006). The availability of data from the Solar Terrestrial Relations Observatory (STEREO) offers new powerful ways of looking at and studying CME onsets and their early evolution which continuously progresses our understanding. The STEREO mission was launched in late 2006 (Kaiser et al. 2008), and consists of two almost identical spacecraft traveling ahead (STA) and behind (STB) Earth's orbit. The separation, SC_{sep}, between STA and STB increases at a rate of $\approx 45^{\circ}$ of year.

For CME onset studies extreme ultra-violet (EUV) and total brightness white-light coronagraph (WLC) images are used from the Extreme Ultraviolet Imaging Telescope (EUVI; Wuelser et al. 2004) and the COR1 coronagraph (Thompson et al. 2003) respectively, of the Sun-Earth Connection Coronal and Heliospheric Investigation (SECCHI; Howard et al. 2008) instrument suite onboard STEREO. EUVI observes the entire solar disk and the corona up to 1.4 $R_{\odot}$ in four EUV channels (171, 195, 284, 304 Åwith nominal cadence of 2.5, 10, 20 and 10 min respectively) while COR1 has a field of view from 1.5–4 $R_{\odot}$ and a nominal cadence of 10 min.

STEREO observations excel over SOHO observations in three important domains: (1) higher cadence, (2) multi-temperature coverage and (3) multi-viewpoint coverage. High cadence is required in order to temporally resolve very rapidly varying phenomena (e.g., impulsively accelerated CMEs): this increases the number of useable measurements and allows to separate various facets of the same event. SOHO Extreme ultra-violet Imaging telescope (EIT;) observations have a rather low cadence of $\approx$ 12 min. Multi-temperature coverage is needed in order to differentiate between density and temperature effects in various observed transients (e.g., are observed intensity changes in a EUV wave a temperature or a density effect?) and to pinpoint their thermal structure (isothermal or multi-thermal?). EIT takes observations at a regular cadence of about 12 min at only one temperature (*approx* 1.6 MK, the 195 Å channel),

S. Patsourakos (✉)
Department of Physics, Section of Geophysics, University of Ioannina, GR 451 10 Ioannina, Greece
e-mail: spatsour@cc.uoi.gr

M.P. Miralles, J. Sánchez Almeida (eds.), *The Sun, the Solar Wind, and the Heliosphere*, IAGA Special Sopron Book Series 4, DOI 10.1007/978-90-481-9787-3_8,

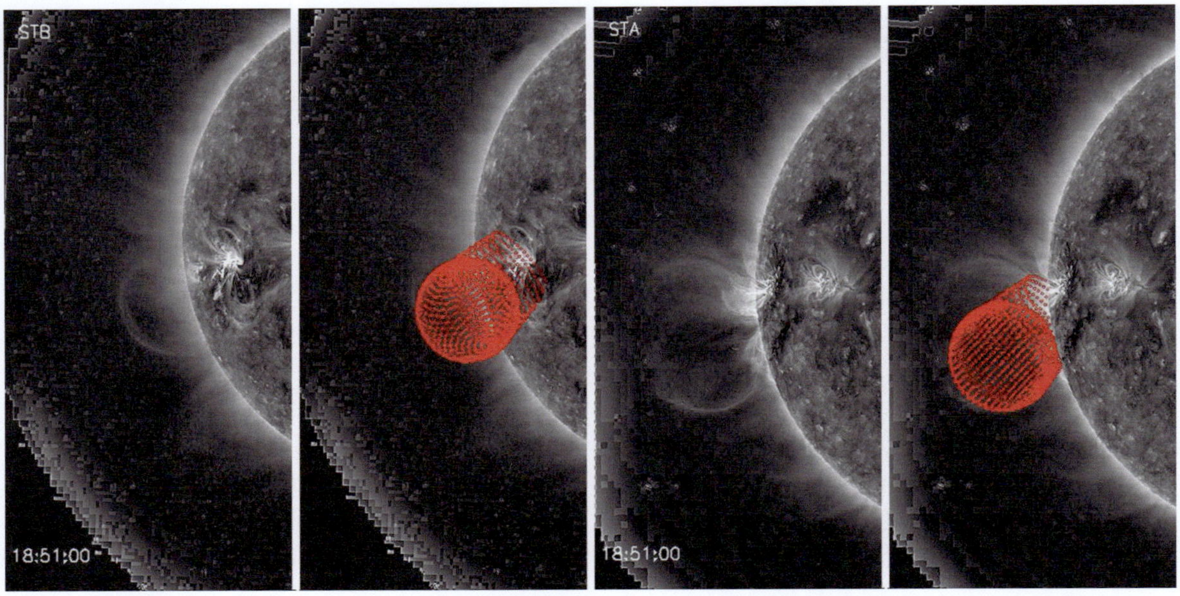

Fig. 8.1 Simultaneous STA and STB fitting of the CME bubble of the 25 March 2008 CME with a 3D model for 18:51:00 UT during that day. STB view and modeling (wireframe) first two panels; STA view and modeling (wireframe) last two panels. From Patsourakos et al. (2010)

while it takes observations at all channels only every four hours. Finally, before STEREO all observations employed a single viewpoint. They could thus have suffered from projection effects in the optically-thin coronal medium, which could be the source of significant ambiguity for the interpretation of observations. Moreover, having only one viewpoint gives access to projected instead of true quantities (e.g., heights, shapes). The situation is clearly improved with the two distinct STEREO viewpoints.

With this chapter we will discuss and put into perspective several new SECCHI results regarding CME initiation and early evolution.

8.2 Bubble Morphology of Impulsively Accelerated CMEs

SECCHI observations started to provide significant clues on the genesis of impulsively accelerated CMEs (e.g., Aschwanden et al. 2009a; Patsourakos et al. 2010). One of the few examples so far was the flare-CME event which took place during 25 March 2008, when SC_{sep} was $\approx 47°$. The CME started when a set of low-lying loops erupted and formed a bubble (e.g. Fig. 8.1). The bubble size increased dramatically in a matter of 15 min at a speed of almost 1,000 kms^{-1}. Both the high cadence of the EUVI measurements and their multi-viewpoint aspect allow the temporal resolution of the evolution of the bubble and determine its 3D shape. An example of the employed procedure is given in Fig. 8.1. The 3D geometrical model of Thernisien et al. (2009) was used to simultaneously fit the observed bubble on both STA and STB. We found that the eruption induced deflections of ambient structures in the corona (e.g., the curved structures north of the fitted bubble in the STA view of Fig. 8.1). These structures were rooted at the AR periphery, and not at its core, as revealed from the STB view.

Preliminary results Patsourakos et al. (2010) showed that the bubble is a true 3D object (a sphere) and not a 2D one (i.e., an arcade of loops). This means that it could represent an early instance of a magnetic flux rope, which is thought to be the basic structural element of most CMEs when they reach the coronagraph field of view, as demonstrated by recent STEREO observations and 3D geometrical modeling of several CMEs (Thernisien et al. 2009). We finally found that the EUV cavity initially expanded in a non self-similar manner since it moved faster in the lateral than in the radial direction.

8.3 CME-Flare Relationship

The high-cadence and multi-viewpoint SECCHI observations solidify the tight association between flares and CMEs. Previous SOHO studies using data from the C1 large angle spectroscopic coronagraph (LASCO) have shown that the impulsive phase of CMEs is synchronized with the impulsive phase of the associated flare as observed in soft X-rays by GOES (Zhang et al. 2001). This underlines the existence of a common physical mechanism between the flare and the CME. A common problem of the above studies was their relatively low cadence which did not allow us to temporally resolve the early acceleration profiles of very impulsively accelerated CMEs close to the solar surface. EUVI with its higher cadence represents a significant improvement in this area.

Temmer et al. (2010) compared the acceleration profiles of several impulsively accelerated CMEs observed in high-cadence EUVI movies (e.g. the 31 December 2007 CME when SC_{sep} was $\approx 44°$) with the Hard X-rays (HXRs) of the associated flares observed with RHESSI. A close temporal correlation between the acceleration profile of the early CME and HXRs was found. Similar results were obtained by Cheng et al. (2010) who studied the build-up, initiation, and acceleration of the 26 April 2006 CME, observed when SC_{sep} was $\approx 49°$). Besides a close temporal relationship between the acceleration profile of the CME and the associated HXRs Cheng et al. (2010) found a close relationship with the reconnecting flux (estimated from flare ribbon motions and MDI magnetograms) as well.

The above results were interpreted as indicative of a close feedback between the acceleration profile and HXR energy release (i.e. reconnection). The upward CME acceleration brings new flux to reconnect into the current sheet underneath the erupting flux rope but at the same time reconnection in this current sheet feeds new poloidal flux into the erupting flux which maintains the self Lorentz force, thereby sustaining the CME acceleration.

The later result was recently further substantiated by Krucker et al. (2010) who studied the partially occulted solar flare which took place on 31 December 2007. The Krucker et al. (2010) study is complementary to the study of Temmer et al. (2010) of the same event in that they compare RHESSI and EUVI imaging. The flare footpoints were behind the east limb as seen from the Earth which allowed to observe faint HXR coronal sources with RHESSI, whereas EUVI observed the early development of the associated CME. Krucker et al. (2010) observed that a faint HXR sources at the bottom of the large CME bubble observed by EUVI. This provides a "visual" indication of a reconnection site in the wake of the eruption which could play a significant role into the closely correlated flare-CME dynamics.

One the first and most extensively studied STEREO CME events is that of the 19 May 2007 with a SC_{sep} of $\approx 9°$. Magnetic field extrapolations have showed that the event had a multi-polar configuration (Li et al. 2008). Using anaglyph EUVI movies these authors found that about 1 h before the event brightenings occurred along the separator that split different flux systems. Slowly rising transient brightenings at the upper edge of the associated filament were also observed. The pre-event transient activity along the separator gives support to processes (such as magnetic breakout) which require multi-polar magnetic configurations for initiating the CME.

Finally Aschwanden et al. (2009a) performed a large statistical study of more than 180 flare events observed by EUVI. They found that a large fraction of the observed flares 79% show impulsive EUV emission (occurring simultaneously with the HXRs) and 73% show delayed EUV emission.

The former originated from heated, highly sheared filaments whereas the latter were from post-flare loop systems and arcades. The impulsive EUV emission originating from the normally low-lying filaments is a clue that the associated CMEs are triggered deep in the atmosphere.

8.4 EUV Dimmings

EUV dimmings are intensity depletions associated with CME onsets, and are believed to be an evacuation of plasma. Aschwanden et al. (2009b) made the first comprehensive measurements using EUVI data of the mass contained in EUV dimmings and compared these estimates to CME mass (of the same dimming-CME events) determined from WLC data.

The EUV dimmings in the 3 EUVI coronal channels (171, 195, 284 Å) were approximated with a geometrical model describing the erupting volume, together with temperature-density models for the background-foreground corona and for the CME as observed in the EUV. Fitting of the observed intensity depletions for each dimming event supplied an estimate of the associated mass $m_{dimming}$ which was supposedly removed by the eruption. For the same events the true (projection-free) total CME mass m_{CME} from WL observations of the mature CMEs when they have reached the outer corona observed by the COR2 coronagraphs on both STA and STB was determined (Colaninno and Vourlidas 2009).

It was found that $m_{dimming}/m_{CME} = 1.1 \pm 0.2$. Therefore most of CME mass should originate in the low active region corona which is removed when a CME happens there. This result could have significant implications for the understanding of EUV waves (see Section 8.5). It basically shows that is not neccessary that large areas in the low corona to open during the eruption. Otherwise, WLC CMEs would have possibly contained much larger masses than observed.

However, it turns out that *not all* CMEs correspond to a dimming. This was already noticed from SOHO observations (e.g., Bewsher et al. 2008). With the exception of a few studies (e.g., Reinard and Biesecker 2009) the common suggestion was that a lack of a dimming signature meant the CME was backsided. In such cases the source region of those CMEs and therefore the dimmings were not accessible from the single viewpoint SOHO observations.

This hypothesis was recently tested with STEREO observations (Robbrecht et al. 2009). During the period 31-May–2-June 2008 a very slow CME was observed to ascend from the east limb of STA. The kinematics and mass of the CME were typical of slow CMEs (speeds not exceeding ≈ 600 kms^{-1} and masses of several times 10^{15} gr). SC_{sep} was $\approx 50°$ for this event which means that the source region of the CME should have been seen face-on from STB. A candidate source region area in STB was determined by using the direction and the width of the CME; the source region covered quiet Sun and there were no active regions on the solar disk. Close inspection of EUVI movies in all channels from STB showed no evidence of large-scale dimmings around the postulated source region; nothing but general quiet Sun small-scale activity which can be observed anywhere on the Sun. The lack of any large scale dimming was interpreted as indication of a large size (and initial height) of the erupting structure; given the very rapid drop-off of density and therefore EUV intensity with height, CMEs with large initiation heights would not lead to an appreciable intensity depletion. The inferred large initiation height of this CME was corroborated by the large size of the erupting cavity which gave rise to the CME as seen from STB. Tomographic reconstructions of the Differential Emission Measure distributions for such cavities showed they contain less dense but somehow hotter plasmas than the surrounding corona (Vásquez et al. 2009). It remains to be seen if there are more such CMEs with no EUV disk signature: preliminary inspection of EUVI movies of other slow events show they are not uncommon.

8.5 EUV Waves

One of the most interesting and unexpected discoveries of EIT on SOHO was the detection of large-scale EUV disturbances traveling over significant fractions of the solar disk (e.g., Thompson et al. 1998, 1999).

Before the launch of STEREO their origin was strongly debated. They were first interpreted as fast-mode coronal Moreton waves triggered by the eruption (e.g., Thompson et al. 1999). This interpretation was then strongly contested by several authors (e.g., Delannée 2000; Delannée et al. 2008; Attrill et al. 2007a). These authors suggested that EUV waves are not really waves but rather the footprints or the low coronal lateral extension of the associated CMEs, i.e. therefore pseudo-waves. Let us note here that it exists a rather extensive list of EUV waves theories (see for example the review of Wills-Davey and Attrill 2009). However, they can be broadly categorized into wave and pseudo-wave theories. A significant part of the difficulties for deciding the nature of EUV waves originated from some limitations of the EIT observations used in their study: low cadence and single-viewpoint coverage did not allow full resolution of their early evolution and it was more difficult to unambiguously determine the structures associated with the wave and with the CME. These are limiting factors for deciding whether EUV waves are true waves or pseudo-waves.

A major criticism for the fast-mode wave interpretation was the relatively low speeds in several of the observed waves (e.g. Wills-Davey et al. 2007). However this could be at least partially due to be related to the low cadence of the EIT observations. High cadence

(2.5 min) EUVI observations of the EUV wave of 19 May 2007 observed at a small SC_{sep} of $\approx 9°$ showed that its initial speed could have been underestimated by a factor of almost 2 by using a reduced EIT-like cadence of 10 min (Long et al. 2008; Veronig et al. 2008, see also Ma et al. 2009 for a similar result for the 7 December 2007 wave). These high cadence observations showed that the wave experienced a deccelaration from ≈ 400 kms^{-1} to ≈ 200 kms^{-1}. This is consistent with a freely-propagating wave distributed over a volume which increases with time. Moreover, Gopalswamy et al. (2009) found for the same event that the wave was reflected off an adjacent equatorial coronal hole. Reflection of fast-mode waves is expected when they reach regions of steep gradients of high magneto-acoustic speed like coronal holes.

The first STEREO observations of EUV waves at large separation were reported in Patsourakos et al. (2009a) for an EUV wave which took place in 7 December 2007, when SC_{sep} was $\approx 44°$. Multi-temperature almost simultaneous observations of the wave in all EUVI channels, showed that the wave was most prominent in the 195 channel and was less prominent in the 171 and 284 channels. The wave was associated with a rather moderate intensity increase of up to 30% with respect to pre-event values. A very scant signature was noticed in the mainly chromospheric/transition region 304 channel.[1] Triangulations of the wave front location allowed to determine a height of about 90 Mm to be determined. Similar wave heights (80–100 Mm) were found by Kienreich et al. (2009) from the analysis of the EUV wave of 13 February 2009. A different method, comparing the wave front location from two distrinct viewpoints was used. This is comparable to the hydrostatic scale-height at coronal temperatures (75 Mm for a temperature of 1.5 MK). These results indicate that the wave was a small (linear) perturbation of the ambient quiet Sun over which it propagated. High-cadence movies for this event showed that a set of slowly rising loops from the heart of the erupting active region underwent a sudden jump and in a matter of 2.5 min the first wave front in the AR periphery was seen. This suggests that the impulsive loop expansion associated with CME onset produced the observed wave. Forward 3D geometrical modeling of the corresponding CME as seen by the COR1 coronagraphs showed that the projection of the best-fit model on the solar surface was not consistent with the location and size of the co-temporal EUV wave.

However we had to wait until early 2009 to be in a position to perform the ultimate test for the nature of EUV waves (Patsourakos 2009b; Kienreich et al. 2009; Cohen et al. 2009). During 13 February 2009 SC_{sep} was $\approx 90°$ and a EUV Wave – CME event took place. Its source region was at disk center as seen from STB; therefore it was observed at the east limb of STA. This quadrature configuration is the ideal one for studying EUV waves. This is since the STB could follow the wave almost over the entire solar disk, whereas the STA view allowed to monitor the low coronal lateral and radial expansion of the associated early CME together with its impact on the surrounding corona. Single viewpoint observations of EUV waves from SOHO tend to be optimized for either the wave (disk observations) or the CME (limb observations), with the exception of events that occurred at the limb and expanded onto the disk. The contribution of STEREO allows for comparison of fully face-on disk and CME-limb observations

The early CME in the low corona observed by EUVI on STA had the form of an erupting bubble. The analysis of Patsourakos (2009b) and Kienreich et al. (2009) found that after about 20 min into the eruption, the wave (as seen in STB) and the CME bubble (as seen in STA) started to *decouple*. The wave attained a significantly larger size than the bubble (e.g., Fig. 8.2) and their sizes evolved at different rates. Indeed Kienreich et al. (2009) measured the lateral expansion speeds of the wave and of the early CME and found that the CME was expanding slower than the wave. Moreover, Patsourakos (2009b) fitted a 3D geometrical model to composite EUVI and COR1 images of the CME cavity and found it smaller than a similar 3D model applied to the EUV wave. This decoupling between the wave and the CME bubble clearly demonstrate that they correspond to *different* physical entities.

Moreover, high cadence STA 171 movies, showed that the erupting bubble induced deflections of ambient coronal structures (Patsourakos 2009b). The outer extensions of these deflections propagated at

[1] The 304 and 284 waves both looked very diffuse. This means that possibly the dominant contribution to the 304 channel signal was not from the cool He II line but from the coronal Si XI line.

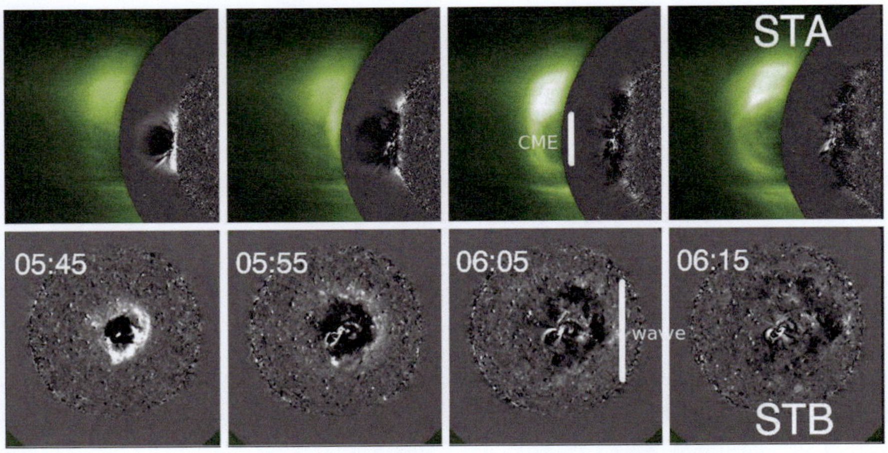

Fig. 8.2 Simultaneous views of the EUV Wave - CME event of 13 February 2009 as seen from above (*lower panel*) and from the side (*upper panel*). Inner field of view images are running difference to emphasize the wave. Note the decoupling, i.e. different spatial scales of the EUV bubble (STA view) and of the EUV wave (STB and STA views) after ≈ 05:55. From Patsourakos (2009b)

increasing distances at both sides of the erupting bubble; their locations approximatively matched the latitudinal extension of the wave at the same time. Patsourakos (2009b) interpretted these observations to mean a wave was hitting these structures and setting them into oscillations. At the same time the erupting CME bubble was significantly smaller and confined in the heart of the associated active region where deep dimmings were formed. Similar phenomena, at even larger volumes were observed by COR1 in the form of deflections of coronal streamers bounding the eruption site. These deflections are a "smoking-gun" of a wave acting over much larger volumes and areas in the corona than the CME itself occupies.

If one considers the deflected EUV and WLC structures as an integral part of the erupting flux itself then the lateral extend of the CME could roughly match the longitudinal extend of the wave (see for example Cohen et al. 2009 for this event and Attrill et al. 2007b, 2009; Dai et al. 2010, for other STEREO events).

Both the decoupling between wave and CME bubble and the observed deflections clearly indicate that EUV waves are true wave phenomena (Patsourakos 2009b; Kienreich et al. 2009). This behavior was observed in several high cadence EUVI movies of impulsively accelerated CMES with an associated EUV wave (e.g., 31 December 2007, 2 January 2008, 25 March 2008, 18 February 2009). The 3D analysis of deflections for several other events (e.g., Aschwanden et al. 2009a; Verwichte et al. 2009) allowed for the first time to determine their polarization, something not possible before with single-viewpoint observations. It was found that these oscillations represent horizontally polarized fast magnetoacoustic kink mode waves (Verwichte et al. 2009).

We wish to stress here that the results above strictly apply to EUV waves occurring during solar minimum: very simple coronal landscape dominated by quiet Sun with one or few active regions present on the disk and waves propagating over a significant fraction of the solar disk (i.e. global waves). When any of the above conditions is not met, i.e. multiple active regions exist (i.e. solar maximum conditions), or existed in a previous rotation, or disturbances not covering global scales pseudo-wave theories could be well applicable (i.e. reconnection and/or plasma compression at the sides of the CME). Moreover we stress here that pseudo-waves could possibly explain the observations of the expanding cavities (giving rising to core dimmings) within the host active regions (e.g. the intensity front seen in the core of the active region of Fig. 8.2 before the wave and the CME decouple). Therefore it could be well that both wave and non-wave components could co-exist (e.g. Zhukov and Auchère 2004; Cohen et al. 2009). What really matters then, is the proper application of each component.

A possible example of a pseudo-wave is reported by Zhukov et al. (2009). They observed a faint wave-like distrurbance associated with a slow prominence eruption which took place during 8 December 2007.

The faint wave was confined relatively close to the host active region, and did not reach global scales as in other events (e.g., the 7 December 2007 global wave described in Patsourakos et al. (2009a) and Ma et al. (2009) which was indeed launched from the same active region as the Zhukov et al. (2009) event). Therefore it is well possible that the EUV wave seen for this event was nothing else that the footprint of erupting prominence.

8.6 Eruptive Prominences

STEREO observations at relatively small separation angles allowed the 3D reconstructions of erupting prominences.

Gissot et al. (2008) used a cross-correlation method on STA-STB image pairs which supplied the true height of the erupting prominence of the 19 May 2007 event. They found that the west part of the prominence was ascending faster than the east, therefore an *assymetric* eruption took place. Heights in the range 20–80 Mm were found. This adds an extra degree of complexity into the modeling of such events. Liewer et al. (2009) used a tie-pointing method on STA-STB image pairs to reconstruct the same erupting prominence. They found that the erupting prominence was located above the flare ribbons, therefore confirming the standard CME-flare model, and the causal relationship and mapping between the eruption (prominence) and its footpoints (ribbons).

Finally Thompson et al. (2009) reported on prominence rotation during the 9 April 2008 event (SC_{sep} of $\approx 48°$). Matching several threads of the erupting prominence in the 304 channel from both STA and STB data Thompson et al. (2009) showed that the prominence rotated roughly by 120° in the EUVI field of view with respect to some estimates of its original orientation. MHD simulations have shown that various physical mechanisms could account for the observed rotation. These include moderately twisted (kink-unstable) but weakly sheared configurations as well as moderately sheared but weakly twisted (kink-stable) configurations. Also breakout magnetic topologies (multipolar) can give rise to rotation in the inner corona (Lynch et al. 2009). The observed rotation is significant ($>100°$). This may hold significant implications for Space Weather since the original orientation of the filament does not neccessarily tells us much about the its intermediate (low coronal) orientation. The latter could be related to the angle at which an Earth-directed CME will hit the magnetosphere. This study could be extended to more events to see if such big rotations are common.

Certainly not all prominences exhibit such big rotations, for instance the erupting prominence observed during 9 May 2009 with a SC_{sep} of $\approx 7°$ did not exhibit significant rotation (Bemporad 2009).

8.7 Conclusions and Outlook

The main new SECCHI findings on CME onsets and early evolution discussed in the previous sections are :

- Impulsively accelerated CMEs can start from very deep in the corona and take the form of bubble;
- Observations of the early stages of CME development in the EUV show that the EUV is a true 3D feature;
- The flare-CME synchronization holds even for very impulsively accelerated events;
- EUV dimmings (when existing) contain significant part of the CME mass;
- Not all CMEs have EUV disk signatures (e.g., dimmings) and can go unnoticed;
- Solar minimum EUV waves that reach global scales on the Sun are real waves and not the CME footprints;
- The early stages of CME expansion disturb much larger volumes in the corona that what it occupies: it induces deflections of off-limb coronal structures and launches global waves on the disk;
- 3D reconstructions unambiguously demonstrated that erupting prominences can exhibit substantial rotation in the low corona (before entering coronagraph field of view).

All of the above represent significant developments providing new insight into CME onsets. However there still exists much to be done with existing and future STEREO data. As seen in the previous sections observations at small or moderate separation angles can be used to determine the 3D coordinates of observed features like coronal loops, prominences, CMEs. Having the 3D coordinates we can then determine several geometrical properties of the observed structures like their height, width, shape and their temporal evolution. This will allow us to check the validity of various instability criteria for eruptions (e.g., critical height, twist, shear, footpoint seperation).

On the other hand, observations at larger separations cannot supply 3D reconstructions like above. They however allow us to view the same phenomenon from two dramatically different viewpoints like the observations of Section 8.5, where the CME was simultaneously viewed edge-on and face-on. The EUV wave was only one application of such a data set. One might also employ such datasets to understand the 3D geometry and evolution of coronal sigmoids (e.g., are they a single flux tube or two *J*'s), the relationship between the evolution of EUV dimmings (from the disk view) and of the CME ascent (from the limb view of the same area with the other spacecraft) etc.

SECCHI promises to deepen our understanding of magnetic flux ropes which are widely accepted to be the building block of most CMEs once they fully develop in the coronagraph field of view. However, we dont know if such structures pre-exist before or form during an eruption. The answer to this question has profound implications for determining which is the initiation mechanism (if a single one exists) of CMEs. Determining how the 3D geometry of pre-CME and CME structures (i.e., CME bubbles) evolves with time during eruptions could hold important clues.

Acknowledgements I'm grateful to my colleagues Angelos Vourlidas, Eva Robbrecht, Bernhard Kliem, Yi-Ming Wang, Arnaud Thernisien and Guillermo Stenborg for many discussions on various topics discussed here. I wish to thank the organizers of the IAGA 2009 meeting and Kuen Ko in particular for giving me the opportunity to give this chapter. Special thanks go the referee for helpful comments/suggestions. The SECCHI data used here were produced by an international consortium of the Naval Research Laboratory (USA), Lockheed Martin Solar and Astrophysics Lab (USA), NASA Goddard Space Flight Center (USA), Rutherford Appleton Laboratory (UK), University of Birmingham (UK), Max−Planck−Institute for Solar System Research (Germany), Centre Spatiale de Liège (Belgium), Institut d Optique Théorique et Appliqueé (France), and Institut dAstrophysique Spatiale (France).

References

Aschwanden MJ, Wuelser JP, Nitta NV, Lemen JR (2009a) Solar flare and CME observations with STEREO/EUVI. Solar Phys 256:3–40

Aschwanden MJ, Nitta NV, Wuelser J, Lemen JR, Sandman A, Vourlidas A, Colaninno RC (2009b) First measurements of the mass of coronal mass ejections from the EUV dimming observed with stereo EUVI A and B spacecraft. Astrophys J 706:376–392

Attrill GDR, Harra LK, van Driel-Gesztelyi L, Démoulin P. (2007a) Coronal "wave": magnetic footprint of a coronal mass ejection? Astrophys J 656:L101–L104.

Attrill GDR, Harra LK, van Driel-Gesztelyi L, Démoulin P, Wüslser J-P (2007b) Coronal "wave": a signature of the mechanism making CMEs large-scale in the low corona? Astron Nachr 328:760–763

Attrill GDR, Engell AJ, Wills-Davey MJ, Grigis P, Testa P (2009) Hinode/XRT and STEREO observations of a diffuse coronal "wave"-coronal mass ejection-dimming event. Astrophys J 704:1296–1308.

Bemporad A (2009) Stereoscopic reconstruction from STEREO/EUV imagers data of the three-dimensional shape and expansion of an erupting prominence. Astrophys J 701:298–305.

Bewsher D, Harrison RA, Brown DS (2008) The relationship between EUV dimming and coronal mass ejections. I. Statistical study and probability model. Astron Astrophys 478:897–906

Cheng X, Ding MD, Zhang J (2010) A study of the build-up, initiation, and acceleration of 2008 April 26 coronal mass ejection observed by STEREO. Astrophys J 712:1302–1310

Cohen O, Attrill GDR, Manchester WB, Wills-Davey MJ (2009) Numerical simulation of an EUV coronal wave based on the 2009 February 13 CME event observed by STEREO. Astrophys J 705:587–602

Colaninno RC, Vourlidas A (2009) First determination of the true mass of coronal mass ejections: a novel approach to using the two STEREO viewpoints. Astrophys J 698:852–858

Dai Y, Auchère F, Vial J-C, Tang YH, Zong WG (2010) Large-scale extreme-ultraviolet disturbances associated with a limb coronal mass ejection. Astrophys J 708:913–919

Delannée C (2000) Another view of the EIT wave phenomenon.Astrophys J 545:512–523

Delannée C, Török T, Aulanier G, Hochedez J-F (2008) A new model for propagating parts of EIT waves: a current shell in a CME. Solar Phys 247:123–150

Gissot SF, Hochedez J-F, Chainais P, Antoine J-P (2008) 3D reconstruction from SECCHI-EUVI images using an optical-flow algorithm: method description and observation of an erupting filament. Solar Phys 252:397–408

Gopalswamy N, Yashiro S, Temmer M, Davila J, Thompson WT, Jones S, McAteer RTJ, Wuelser J-P, Freeland S, Howard RA (2009) EUV wave reflection from a coronal hole. Astrophys J 691:L123–L127

Howard RA et al (2008) Sun Earth connection coronal and heliospheric investigation (SECCHI). Space Sci Rev 136:67–115

Kaiser ML, Kucera TA, Davila JM, St Cyr, OC, Guhathakurta M, Christian E (2008) The STEREO mission: an introduction. Space Sci Rev 136:5–16

Kienreich IW, Temmer M, Veronig AM (2009) STEREO quadrature observations of the three-dimensional structure and driver of a global coronal wave. Astrophys J 703:118–122

Krucker S, Hudson HS, Glesener L, White SM, Masuda S, Wuelser J-P, Lin RP (2010) Measurements of the Coronal acceleration region of a solar flare. Astrophys J 714:1108

Li Y, Lynch BJ, Stenborg G, Luhmann JG, Huttunen KEJ,Welsch BT, Liewer PC, Vourlidas A (2008). The solar-magnetic field and coronal dynamics of the eruption on 2007 May 19. Astrophys J 681:L37–L40

Liewer PC, de Jong EM, Hall JR, Howard RA, Thompson WT, Culhane JL, Bone L, van Driel-Gesztelyi L (2009) Stereoscopic analysis of the 19 May 2007 erupting filament. Solar Phys 256:57–72

Long DM, Gallagher PT, McAteer RTJ, Bloomfield DS (2008) The kinematics of a globally propagating disturbance in the solar corona. Astrophys J 680:L81–L84

Lynch BJ, Antiochos SK, Li Y, Luhmann JG, DeVore CR (2009) Rotation of coronal mass ejections during eruption. Astrophys J 697:1918–1927

Ma S, Wills-Davey MJ, Lin J, Chen PF, Attrill GDR, Chen H, Zhao S, Li Q, Golub L (2009). A new view of coronal waves from STEREO. Astrophys J 707:503–509

Patsourakos S, Vourlidas A, Wang YM, Stenborg G, Thernisien A (2009a) What is the nature of EUV waves? First STEREO 3D observations and comparison with theoreticalmodels. Solar Phys 259:49–71

Patsourakos S, Vourlidas A (2009b). "Extreme ultraviolet waves" arewaves: first quadrature observations of an extreme ultraviolet wave from STEREO. Astrophys J 700:182–186

Patsourakos S, Vourlidas A, Kliem B (2010) Toward understanding the early stages of an impulsively accelerated coronal mass ejection. Astron and Astrophys, in press, available online in http://arxiv.org/abs/1008.1171

Pick M, et al (2006) Multi-wavelength observations of CMEs and associated phenomena. report of working group F. Space Scie Rev 123:341–382

Reinard AA, Biesecker DA (2009) The relationship between coronal dimming and coronal mass ejection properties. Astrophys J 705:914–919

Robbrecht E, Patsourakos S, Vourlidas A (2009) No trace left behind: STEREO observation of a coronalmass ejectionwithout low coronal signatures. Astrophys J 701:283–291

Schwenn R et al (2006) Coronal observations of CMEs. Report of working group A. Space Sci Rev 123:127–176

Temmer M, Veronig AM, Kontar EP, Krucker S, Vrsnak B (2010) Combined STEREO/RHESSI study of CME acceleration and particle acceleration in solar flares. Astrophys J 712:1410–1420

Thernisien A, Vourlidas A, Howard RA (2009) Forward modeling of coronal mass ejections using STEREO/SECCHI data. Solar Phys 256:111–130

Thompson BJ, Plunkett SP, Gurman JB, Newmark JS, St Cyr OC, Michels DJ (1998) SOHO/EIT observations of an Earth-directed coronal mass ejection on May 12, 1997. Geophys Res Lett 25:2465–246

Thompson BJ, Gurman JB, Neupert WM, Newmark JS, Delaboudinière J-P, St Cyr OC, Stezelberger S, Dere KP, Howard RA, Michels DJ (1999) SOHO/EIT observations of the 1997 April 7 coronal transient: possible evidence of coronal moretonwaves. Astrophys J 517:L151–L154

Thompson WT, et al (2003) COR1 inner coronagraph for STEREOSECCHI. society of Photo-Optical Instrumentation Engineers (SPIE) Conference series 4853, 1–11

Thompson WT, Kliem B, Toeroek T (2009) 3D reconstruction of an erupting prominence. AAS/Solar Physics Division Meeting 40, #21.11

Vásquez AM, Frazin RA, Kamalabadi F (2009) 3D temperatures and densities of the solar corona via multi-spacecraft EUV tomography: analysis of prominence cavities. Solar Phys 256:73–85

Veronig AM, Temmer M, Vrşnak B (2008) High-cadence observations of a global coronal wave by STEREO EUVI. Astrophys J 681:L113–L116

Verwichte E, Aschwanden MJ, Van Doorsselaere T, Foullon C, Nakariakov VM (2009) Seismology of a large solar coronal loop from EUVI/STEREO observations of its transverse oscillation. Astrophys J 698:397–404

Wuelser J-P et al (2004) EUVI: the STEREO-SECCHI extreme ultraviolet imager. Society of Photo-Optical Instrumentation Engineers (SPIE) Conference series 5171, 111–122

Wills-Davey MJ, Attrill GDR (2009) EIT waves: a changingunderstanding over a solar cycle. Space Sci Rev 149: 325–353

Wills-Davey MJ, DeForest CE, Stenflo JO (2007) Are "EIT waves" fast-mode MHD waves? Astrophys J 664: 556–562

Zhang J, Dere KP, Howard RA, Kundu MR, White SM (2001) On the temporal relationship between coronal mass ejections and flares. Astrophys J 559:452–462

Zhukov AN, Auchère F (2004) On the nature of EIT waves, EUV dimmings and their link to CMEs. Astron Astrophys 427:705–716

Zhukov AN, Rodriguez L, de Patoul J (2009) STEREO/SECCHI observations on 8 December 2007: evidence against the wave hypothesis of the EIT wave origin. Solar Phys 259: 73–85

Chapter 9

Funnels and the Origin of the Solar Wind

Ruth Esser and Øystein Lie-Svendsen

Abstract A short overview of the historic development of the idea of a funnel shaped geometry in the lower solar atmosphere (chromosphere, transition region and lower corona) is presented as an introduction. It is followed by a short presentation of the early prediction of solar wind expansion by Parker. A large number of observations and theoretical studies have since helped to refine this basic picture. We know the wind is driven by heating of the corona. Modelling has shown that in a magnetically "open" corona most of the energy deposited is lost in the solar wind. Hence the wind is not merely an "evaporating tail" of a hydrostatic plasma, rather the corona and wind should be regarded as one tightly coupled system. An overview of observations that are crucial for our present understanding of the wind is given. It is followed by a discussion of the energy flow between corona and chromosphere which determines the solar wind mass flux. Once the basic physical understanding of the solar wind is established we will present some recent studies which introduce funnel type geometries and investigate their effect on the properties of the solar wind.

9.1 Introduction: History of Funnel Geometry

In earlier solar wind models it was assumed that all plasma heating occurred below the base of the corona, and they ignored the energy transported by thermal conduction from the corona down to the transition region. The modelling of the solar wind and the modelling of the solar atmosphere (chromosphere, transition region, and corona) were carried out completely independently from each other. Withbroe (1988) presented the first solar wind model, termed "radiative energy balance" model, and discussed in detail the importance for the solar wind of the energy balance between the chromosphere and corona (see also Hammer 1982a, b).

The idea of a funnel type geometry in the source region of the solar wind was developed in models of the lower solar atmosphere and corona which did not include the solar wind. In the standard model describing the magnetic structure of the chromosphere-transition region-corona system of the quiet Sun, it was assumed that all magnetic field lines were constricted to the boundaries of supergranules, the network lanes, by supergranular flows (Gabriel 1976). These field lines, emerging from the photosphere, were predicted to expand rapidly with height and to fill the whole corona uniformly. This geometry is shown in Fig. 9.1.

All energy necessary to maintain a transition region was thought to be provided by conductive flow from the corona (back heating). A sketch of this scenario is shown in Fig. 9.2. Also presented in this figure are the height above the photosphere and typical densities and temperatures. These numbers are from Vernazza et al. (1981), Model F.

The "standard" funnel model (Gabriel 1976) is in agreement with observations showing that ultraviolet emission from the transition region is concentrated in network lanes, indicating that only the transition region in funnels is subject to back heating from the corona.

R. Esser (✉)
Institute of Science and Technology, University of Tromsø, NO–9037 Tromsø, Norway
e-mail: Ruth.Esser@uit.no

M.P. Miralles, J. Sánchez Almeida (eds.), *The Sun, the Solar Wind, and the Heliosphere*, IAGA Special Sopron Book Series 4, DOI 10.1007/978-90-481-9787-3_9, © Springer Science+Business Media B.V. 2011

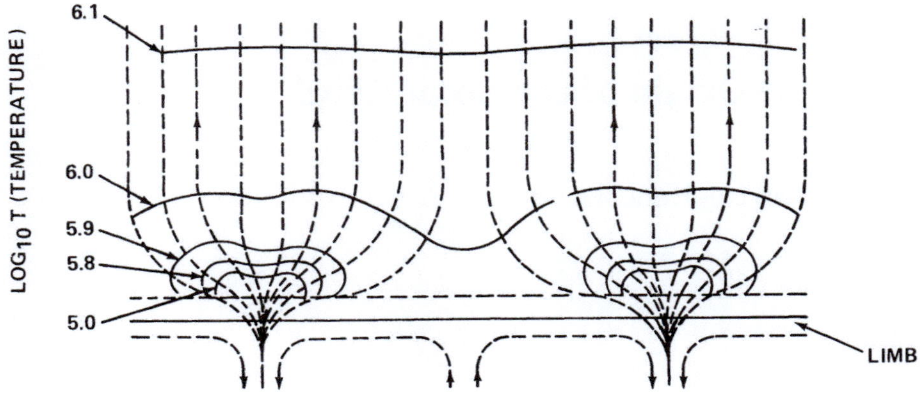

Fig. 9.1 Funnel geometry according to (Gabriel 1976)

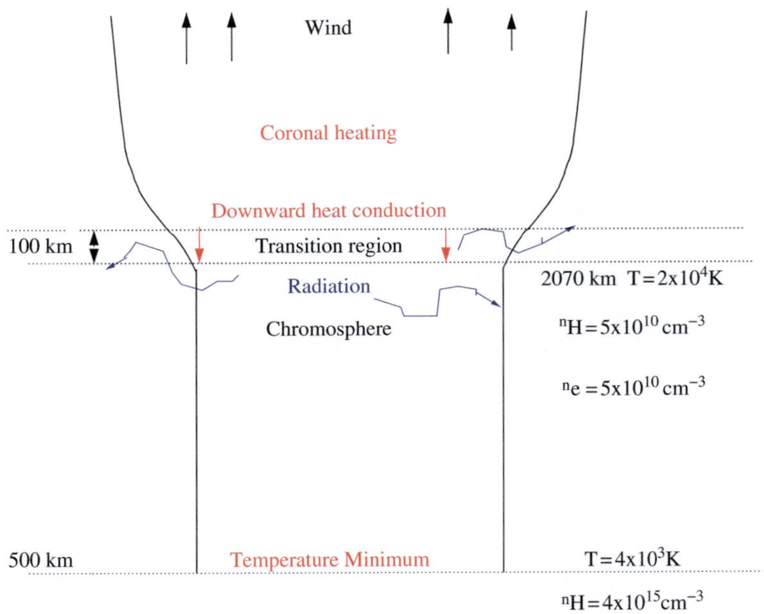

Fig. 9.2 Sketch of a back heating model of the chromosphere, transition region, and corona. Numbers are from Vernazza et al. (1981), Model F

It also predicts the right emission measure at temperatures $T \approx 10^6$ K. At lower temperatures the discrepancy between predicted and observed emission measure is, however, significant, as can be seen in Fig. 9.3. This discrepancy led several authors to the conclusion that the structures that contain the cooler transition region must be distinct from the ones containing the hotter transition region (Athay 1982).

These studies and the close inspection of photospheric magnetograms showing that even in so-called unipolar quiet regions at least 10% of the flux has the polarity opposite to the dominant one, led Dowdy and coworkers (1986) to conclude that network lanes contain both open and closed magnetic structures, as shown in Fig. 9.4. In this scenario the hotter coronal plasma is contained in the open magnetic field regions which are similar to the coronal funnels in the standard model. The cooler material is assumed to be mostly contained in low-lying loops. Since the loops are disconnected from the corona, they must be heated internally, not by back heating. Due to the mixture of open and closed field regions, the funnels in this geometry

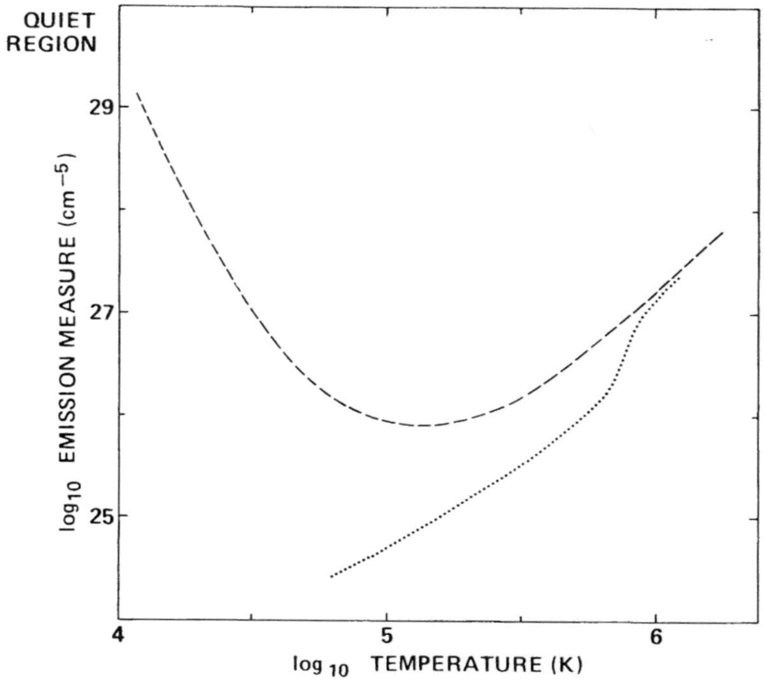

Fig. 9.3 Discrepancy between observed (*dashed curve*) and calculated emission measure (*dotted curve*) (from Gabriel 1976)

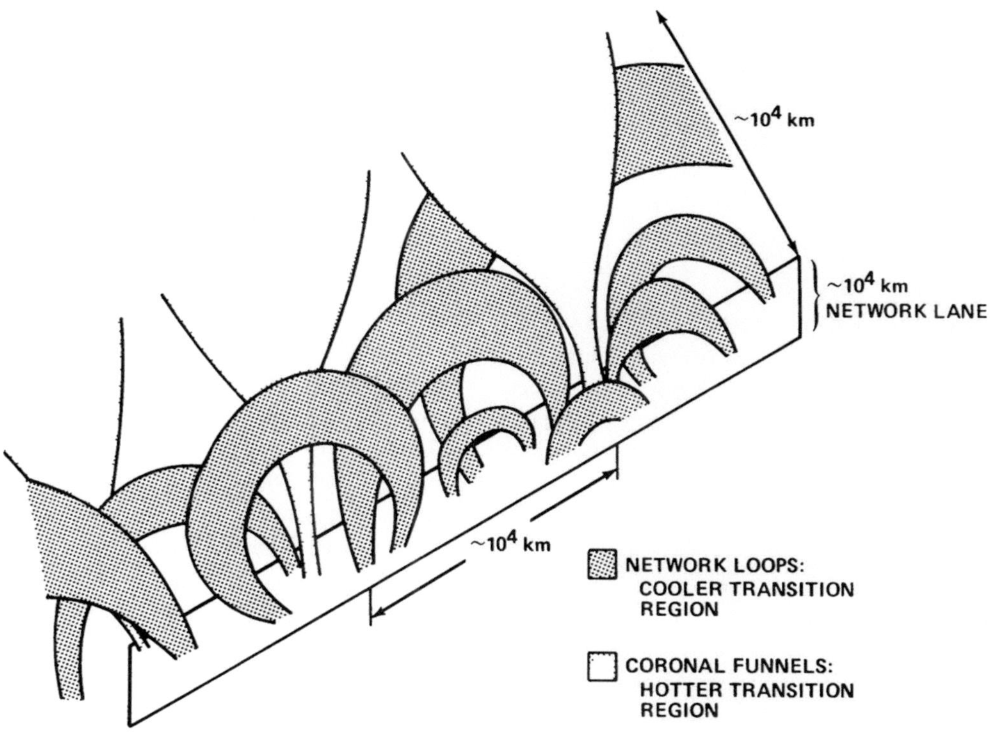

Fig. 9.4 Funnel geometry suggested by Dowdy et al. (1986)

are more constricted than in the standard model. The constriction factor was estimated to be of order 100 (that is, the funnels cover roughly 1% of the surface area) (Dowdy et al. 1986).

Since the solar wind was considered independently from the lower solar atmosphere before Hammer and Withbroe's work (Hammer 1982a, b; Withbroe 1988), the funnel geometry by Dowdy et al. (1986) was not applied to the solar wind expansion until much later. The wind was assumed to start somewhere in the corona and expand radially, or by a factor of 5–7 faster than radially. This factor was implied by observations by Munro and Jackson (1977).

In the following we shall present the basic ideas that let to the earliest prediction of the existence of the solar wind by Parker (1958) (Section 9.2). We summarize coronal and in situ solar wind observations and the constraints they put on solar wind expansion (Section 9.3). In Section 9.4 we discuss some basic physical principles that apply to the chromosphere, transition region, corona, and solar wind system. We shall then discuss some basic results when a funnel-type geometry is assumed for the solar wind. We shall also discuss a scenario where funnels and loops interact through reconnection as in the ideas originally presented by Fisk (2003).

9.2 Early History of the Solar Wind Theory

Parker (1958) predicted the existence of the supersonic solar wind before it was discovered experimentally. His arguments for the inevitability of the solar wind was based on rather simple physical principles. Assuming that the coronal plasma was in hydrostatic equilibrium (hence no supersonic wind), the force balance is everywhere that gravity is balanced by the pressure gradient force of the electron-proton plasma:

$$G_N \frac{M_\odot m_p}{r^2} n = -\frac{d(2nkT)}{dr}. \qquad (9.1)$$

Here G_N is the gravitational constant, $M_\odot$ the mass of the Sun, m_p the proton mass, r the radial distance from the Sun, n the electron (= proton) density, k Boltzmann's constant, and T the temperature. Assuming furthermore that the heat flux is constant beyond a distance r_0 in a spherically symmetric solar atmosphere,

$$r^2 q(r) = -r^2 \kappa_0 T^{5/2} \frac{dT}{dr} = r_0^2 q(r_0), \qquad (9.2)$$

where κ_0 is the electron heat conduction coefficient, and that $T(r \to \infty) \ll T_0 \equiv T(r_0)$, leads to a finite pressure at infinity,

$$P(r = \infty) = P_0 \exp\left(-\frac{7}{5}\frac{T_{\rm esc}}{T_0}\right), \qquad (9.3)$$

where

$$T_{\rm esc}(r_0) \equiv G_N \frac{M_\odot m_p}{2kr_0}. \qquad (9.4)$$

Inserting typical coronal parameters this leads to a pressure ratio at infinity between the solar wind and interstellar pressure of the order

$$\frac{P(r = \infty)}{P(\text{interstellar gas})} \sim 10^8. \qquad (9.5)$$

There is hence an imbalance between the pressures of the solar atmosphere and interstellar medium of many orders of magnitude. It can be shown that this is still true even for a plasma with a very small, subsonic, outflow speed, a solar breeze. Hence the gas cannot be in hydrostatic equilibrium nor can it be a breeze with subsonic flow, it must be flowing with a supersonic speed.

Assuming an isothermal plasma, Parker (1958) predicted the wind speed as a function of coronal temperature, shown in Fig. 9.5. To achieve a high speed wind, the coronal temperature should be of order 2.5×10^6 K.

Although this first, very simple, model obtained the essential behaviour of the solar wind, it had of course several shortcomings, the most notable being the assumption of an isothermal plasma in the solar wind acceleration region. (Maintaining a constant temperature with distance requires continuous heating of the plasma.)

For the temperatures needed to achieve a high speed wind ($\sim 2.5 \times 10^6$ K) an isothermal model predicts a mass flux that is much larger than what is observed. This can easily be shown for a spherically symmetric wind with constant temperature since the mass flux may then be derived analytically,

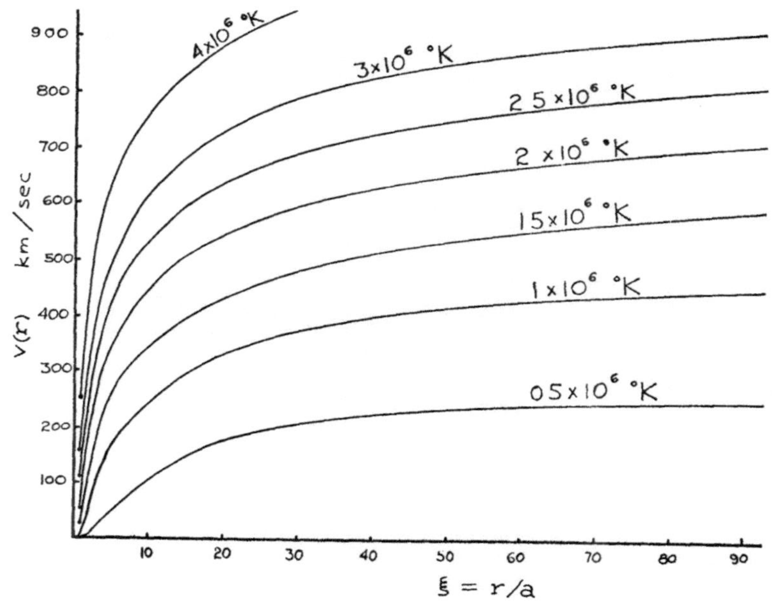

Fig. 9.5 Solar wind speed as a function of distance and for different assumed temperatures in Parker's original isothermal model with a being the radius of the corona taken to be 10^{11} cm $\approx 1.44\,R_\odot$ (Parker 1958)

$$(nu)_E = \frac{1}{4} e^{3/2} \left(\frac{R_\odot}{1\ \mathrm{AU}} \right)^2 n_0 \sqrt{\frac{2kT}{m_p}} \left(\frac{T_{\mathrm{esc}}}{T} \right)^2 e^{-T_{\mathrm{esc}}/T} \tag{9.6}$$

where $(nu)_E$ is the proton flux density at Earth, n_0 is the coronal density, and T_{esc} is defined in (9.4). In the Parker model, and indeed in most solar wind models to date, the coronal density and temperature are fixed input parameters to the model. Choosing a coronal density $n_0 = 10^{14}$ m^{-3}, close to the observed density, and $r_0 = R_\odot$, leading to $T_{\mathrm{esc}}(R_\odot) \approx 1.2 \times 10^7$ K, and $T = 2.5 \times 10^6$ K, we obtain a mass flux of order 10^{14} m^{-2} s^{-1}, almost 2 orders of magnitude higher than what is observed.

Furthermore, according to (9.6) the solar wind mass flux should be extremely sensitive to coronal temperature when $T \ll T_{\mathrm{esc}}$ due to the $\exp(-T_{\mathrm{esc}}/T)$ factor. This is illustrated in Fig. 9.6, which also shows that even with an isothermal wind a temperature of order 10^6 K is needed in order to reproduce the observed solar wind flux.

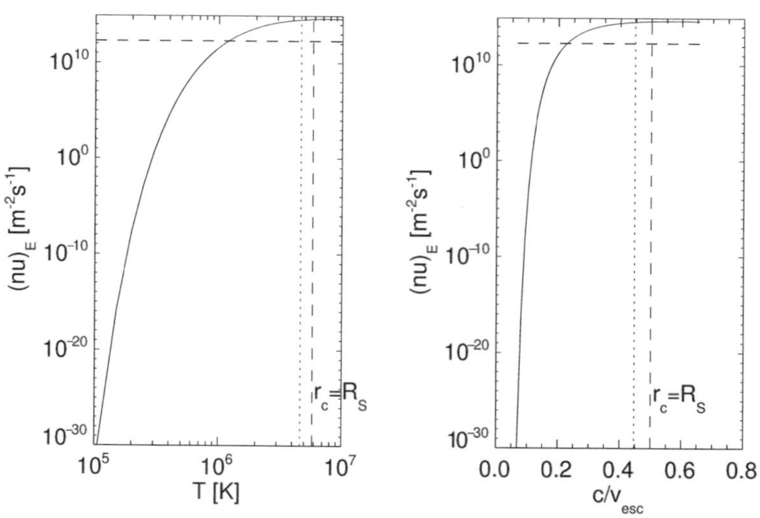

Fig. 9.6 Proton flux density at Earth in an isothermal wind as a function of temperature, from (9.6), using $n_0 = 1 \times 10^{14}$ m^{-3}. The *horizontal dashed line* indicates the observed solar wind flux

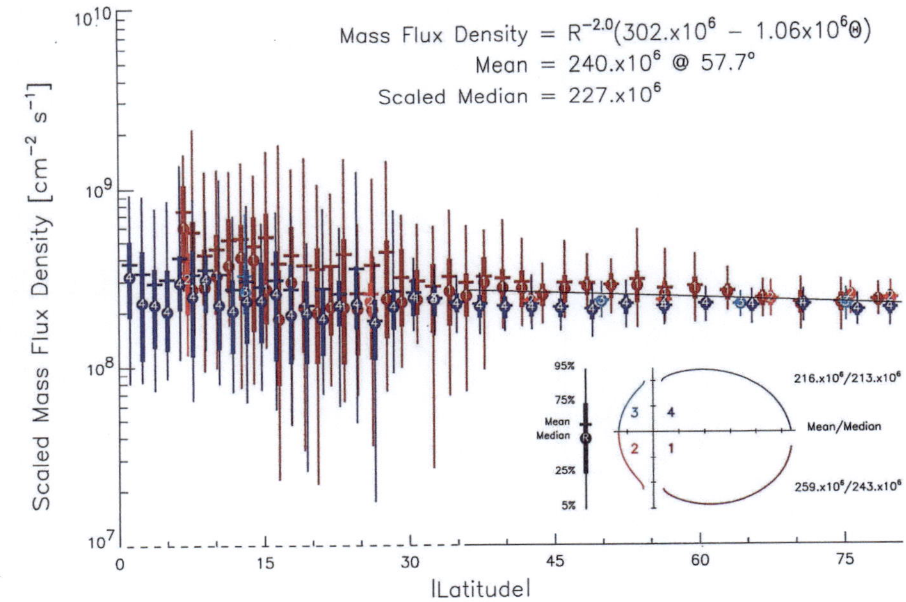

Fig. 9.7 Ulysses observations of the solar wind particle flux density scaled to 1 AU (McComas et al. 2000)

Even if a more realistic fall off of the coronal temperature is taken into account, the predicted mass flux varies by orders of magnitude (e.g. Esser et al. 1986). The observed mass flux, on the other hand, varies far less than predicted by these models (see, e.g., Fig. 9.7), which means either that the temperature in the inner corona is extremely constant or there is some mechanism that regulates the mass flux. We return to this "mass flux problem" in Section 9.4.3.

9.3 Coronal and Interplanetary Solar Wind Observations

The mass flux observed by Ulysses is shown in Fig. 9.7 (from McComas et al. 2000). The mass flux in the low-latitude slow wind varies by roughly a factor 10, whereas in the high-latitude high-speed wind it is almost constant, varying by only a few percent.

Temperatures observed in the corona in a polar coronal hole region are shown in Fig. 9.8. Electron temperatures are difficult to observe and there are relatively few examples; the observations shown in Fig. 9.8 are from the SUMER instrument on SOHO (Wilhelm et al. 1998). They were derived from intensity measurements of the Mg IX line pair at 706 and 750 Å at distances between 1.03 and 1.3 $R_{\odot}$. The derived coronal hole electron temperature is less than 10^6 K. Electron temperatures at greater heights in the corona have so far not been observed directly. Charge states observed in situ in the solar wind carry information on coronal electron temperatures. However, the charge states depend on a number of additional coronal plasma parameters, such as density and flow speed, and the modeled electron temperaures are very sensitive to the assumptions made in the calculations (e.g. Ko et al. 1997). They can therefore not be considered true measurements of coronal electron temperatures. Also shown in Fig. 9.8 are proton temperatures and the temperatures of magnesium and oxygen ions, all derived from spectral line widths measurements. Most of these observations are from UVCS, see the figure for references. The proton temperatures are about 3–4 times higher than electron temperatures. The oxygen temperatures are significantly higher than the proton temperaure with a maximum of order 2×10^8 K. Observations of heavy ions other than O have so far not been carried out except close to the Sun (e.g., Tu et al. 1998).

Observations together with model studies show strong evidence for temperature anisotropies of both the protons and O ions. The temperatures shown in the figure are the perpendicular ion temperatures, the par-

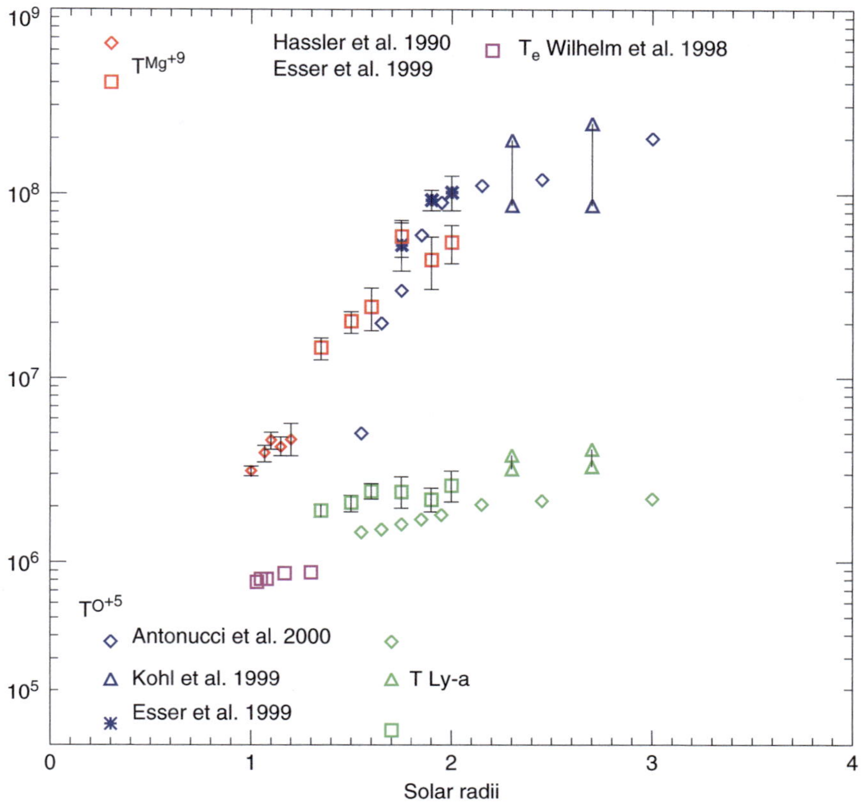

Fig. 9.8 Observed coronal hole temperatures, where *green symbols* denote hydrogen (Lyα), *blue symbols* O^{+5}, *red symbols* Mg^{+9} ions, and *purple squares* the electron temperature

allel components are close to the electron temperature (Cranmer et al. 1999, 2008).

A summary of observed coronal electron densities is shown in Fig. 9.9. These observations are derived from white light intensity measurements and in a few cases from spectral line ratios (see references in the figure for details). Electron densities can be used to place some rough limits on the solar wind outflow speed. An example is shown in Fig. 9.10, together with a summary of other observations that place limits on the flow speed, mostly derived from UVCS and one example of Interplanetary Scintillation Observations is also shown (see references in the figure for details). These observations indicate a rather rapid acceleration close to the sun. This rapid acceleration indicates the presence of an efficient heating mechanism in the near sun region.

In the original calculations by Parker (see Fig. 9.5) this rapid acceleration is also present and is a result of the assumed constant temperature in this model.

Note, however, that the low observed electron temperatures in Fig. 9.8 in Parker's isothermal wind model do *not* lead to rapid acceleration and can only produce

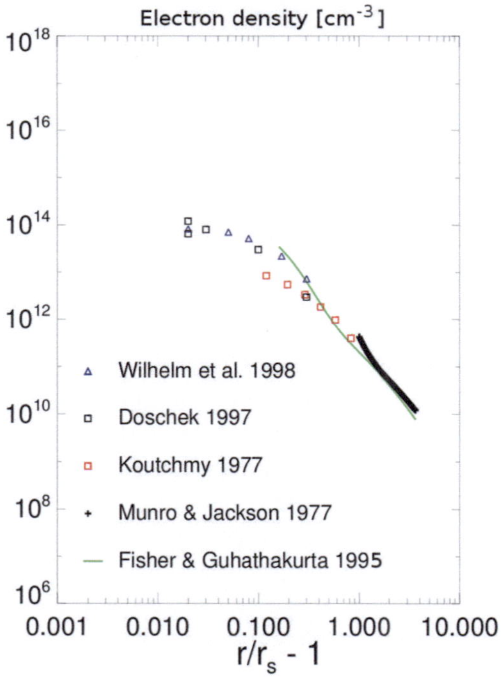

Fig. 9.9 Electron densities derived from coronal white light and line intensity ratios (see references for details)

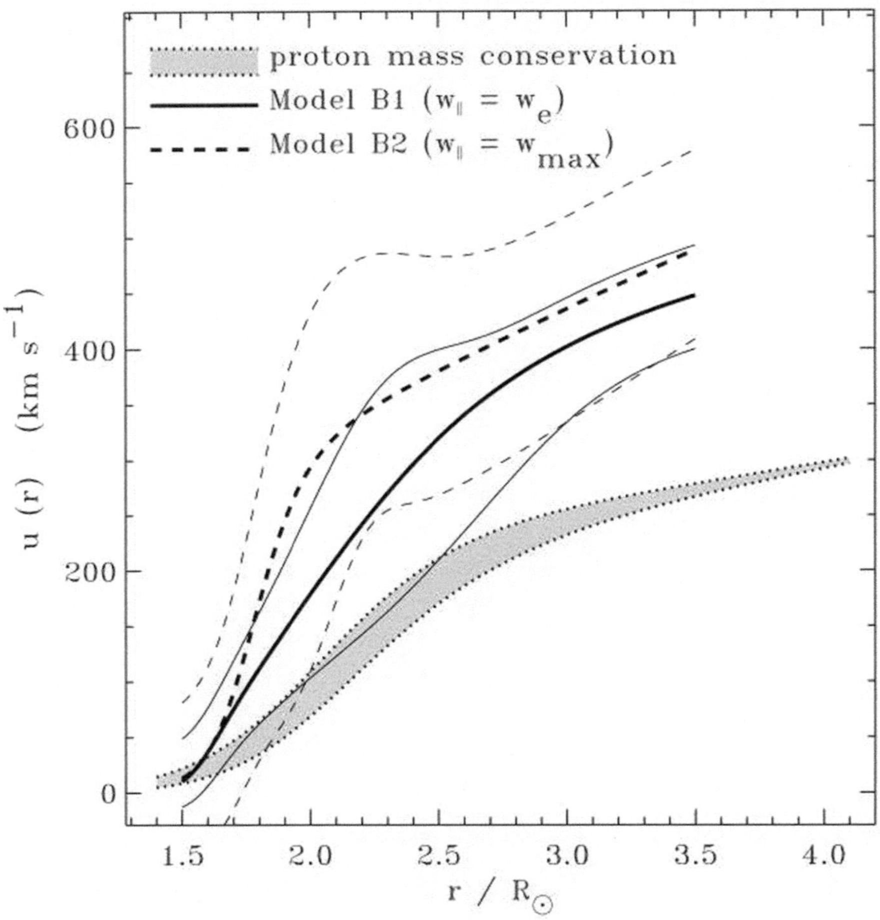

Fig. 9.10 Solar outflow speeds from polar coronal holes inferred from SOHO/UVCS Doppler dimming observations (*solid and dashed curves*) and from density measurements (*grey area*) (Cranmer et al. 1999)

a low-speed wind (see Fig. 9.5). Hence an important mechanism in the original solar wind model, namely that the acceleration was driven by electron heating and electron heat conduction (maintaining the constant temperature), does not seem to be operating in the high-speed wind originating in polar coronal holes. On the contrary, the observed low T_e and rapid acceleration indicates that electrons play a nearly "passive" role (maintaining charge neutrality) in the high-speed wind, which must then be driven by proton heating instead. A low electron temperature means that both the electric field and the electron heat flux must be quite small. The high proton temperatures observed in the corona (Fig. 9.8) supports this new picture of the solar wind acceleration. However, as we discuss in Section 9.4, even a proton temperature of $2 - 3 \times 10^6$ K at $2 - 3\ R_{\odot}$ is too low to drive a high-speed wind. This implies that a substantial fraction of the energy needed to drive the high-speed wind has to be supplied well beyond the corona, possibly in the form of pushing by Alfvén waves beyond the critical point.

The very high heavy ion temperatures in the corona also indicates that protons and ions, rather than electrons, are heated in the corona. However, as discussed in Section 9.4, high ion temperatures are expected even without preferential heating of ions, so this in itself is by no means a proof of strong ion heating. The temperature anisotropies, on the other hand, do provide evidence that wave-particle interactions are affecting the ion velocity distributions in the corona, and these processes may of course also heat the ions.

The most important implications that one can derive from these observations can be summarized as follows:

- Rapid acceleration requires heating close to the Sun (within a few $R_{\odot}$).
- Hence coronal and solar wind heating are inseparable.

- $T_p > T_e$ in the corona suggests that most of the heating goes into protons.
- $T_e \leq 1 \times 10^6$ K in coronal holes (Wilhelm et al. 1998) means
  - the electric field is weak and electron heat conduction cannot be large since it is proportional to $T_e^{5/2}$
  - hence electrons play a small role in the solar wind acceleration
- $T_\perp > T_\parallel$ (protons and ions) suggests heating by cyclotron waves.
- $T(\mathrm{O}) \sim 10^8$ K is not by itself a proof that heavy ions are preferentially heated.

9.4 Energy Budget of the Corona and Solar Wind

Regardless of which model one chooses to describe the expansion of the solar wind, fluid or kinetic, 1 or 3 dimensional, time dependent or static, there are certain aspects, mostly connected to the conservation of energy, that apply to all of them, in addition to the constraints placed by the observations. In the following we will discuss the implications derived from energy conservation in more detail.

As will be demonstrated below, the energy budget of the solar wind cannot be separated from the energy budget of the upper solar atmosphere, which here means the upper chromosphere, transition region, and corona. The upper atmosphere has three possible energy loss mechanisms: the solar wind energy loss, heat conduction downwards from the corona to the transition region and chromosphere, and radiation. Radiative energy loss is proportional to density squared, and therefore rapidly decreases with increasing altitude. The radiative loss also rapidly decreases as hydrogen becomes fully ionized in the low transition region, and it is negligible in coronal holes.

The temperature of the corona is thus set by the balance between the heating and energy loss through downward heat conduction and the solar wind energy loss, the latter only applicable to a magnetically open corona such as polar coronal holes. The nature of these two loss mechanisms is discussed below.

9.4.1 The Solar Wind Energy Flux

To keep the arguments simple we assume a spherically symmetric geometry. In a spherically symmetric outflow the energy flux carried by the wind is

$$F_E = 4\pi r^2 nu\left[\frac{1}{2}m_p(u^2 + v_{\mathrm{esc}}^2) + \frac{5}{2}k(T_e + T_p)\right] + 4\pi r^2(q_e + q_p) \quad (9.7)$$

where u is the flow speed, $v_{\mathrm{esc}} = \sqrt{2G_N M_\odot / r}$ is the gravitational escape speed (the v_{esc} term accounts for the gravitational potential energy flux), and q_e and q_p are the electron and proton heat flux densities, respectively. In a steady state flow with no heating between the Sun and Earth, F_E is constant. Making use of the observed fact that the wind is supersonic at Earth, and that both enthalpy flux (proportional to T_e and T_p) and heat fluxes are negligible at Earth compared to the kinetic and potential energy fluxes, the required energy flux *density* in the corona becomes

$$\frac{F_E}{4\pi R_\odot^2} \approx \left(\frac{1\ \mathrm{AU}}{R_\odot}\right)^2 n_E u_E \frac{1}{2} m_p(u_E^2 + v_{\mathrm{esc}}^2) \approx 50 \text{ to } 80\ \mathrm{W/m^2}, \quad (9.8)$$

where n_E and u_E are the observed density and flow speed at 1 AU. In a funnel geometry where the solar wind emerges from only a fraction $1/f_{\max}$ of the Sun's surface area, the required energy flux density which has to be supplied at the base of the funnel would be a factor $f_{\max}$ larger than this estimate. Hence with Dowdy's (Dowdy et al. 1986) estimate for $f_{\max}$ the required energy flux density scaled to the funnel base would be of order 10^4 W/m^2. (The corresponding Poynting flux from visible radiation is ~ 60 MW/m^2, showing that the solar wind is of no importance for the overall energy budget of the Sun.) This estimate tells us how much energy is needed to drive the observed solar wind. It does not necessarily imply that depositing ~ 100 W/m^2 in a spherically expanding corona/wind will lead to a wind with the correct energy flux density, as most of the energy could possibly be lost as downward heat conduction.

From (9.7) we can demonstrate that a hot corona is required in order to have a solar wind. With no energy supply to the corona/wind beyond a radial distance r_0, energy flux conservation between Sun and Earth can be written:

$$4\pi r_0^2 n_0 u_0 \left[\frac{1}{2} m_p u_0^2 + 5kT_0 + \frac{q_0}{n_0 u_0}\right] = \tag{9.9}$$

$$4\pi (1\ \mathrm{AU})^2 n_E u_E \left[\frac{1}{2} m_p (u_E^2 + v_{\mathrm{esc}}^2) + 5kT_E + \frac{q_E}{n_E u_E}\right], \tag{9.10}$$

where subscript "0" denotes values at r_0 and subscript "E" values at Earth orbit. Making use of particle flux conservation, $r_0^2 n_0 u_0 = (1\ \mathrm{AU})^2 n_E u_E$, assuming a purely adiabatic expansion, $q_0 = 0$, and omitting the terms proportional to u_0, T_E and q_E (which, from observations, are negligible terms) we get

$$T_0 = \frac{m_p}{10k}\left(u_E^2 + v_{\mathrm{esc}}^2\right)$$
$$= 4.6 \times 10^6\ \mathrm{K}\left[\frac{R_\odot}{r_0} + \left(\frac{u_E}{620\ \mathrm{km/s}}\right)^2\right]. \tag{9.11}$$

Choosing, e.g., $r_0 = 2\,R_\odot$, implying that all the energy needed to drive the wind has been supplied by this distance, we find $T_0(u_E = 0) \approx 2 \times 10^6$ K. In other words, even if we attempted to generate a very low-speed wind, we still would need a million-degree corona. And if we want to drive a high-speed wind we find $T_0(u_E = 700\ \mathrm{km/s}) \approx 10^7$ K. Even if we heated the corona out to $r_0 = 4\,R_\odot$, we still require a coronal proton temperature of order 7×10^6 K in order to drive a high-speed wind.

Note also from (9.11) that unless $r_0 \gg R_\odot$ (which would imply extended heating far from the Sun), the solar wind basically "shuts off" completely at a coronal temperature of order 10^6 K as there is then not sufficient thermal energy in the corona to drive even a wind with zero terminal flow speed.

The above estimate is of course crude, in the sense that we have assumed adiabatic expansion beyond the distance r_0. However, even with a lot more "sophisticated" models that attempt to provide a better description of heat conduction (e.g., Lie-Svendsen et al. 2001) we find that the estimate above is basically correct: Protons are not good conductors of heat, both because they are much heavier than electrons, and because they quickly become collisionless as they are heated. And the observations of low coronal electron temperatures (Fig. 9.8) imply that the electron heat conduction must also be small. Without explicit energy addition throughout the solar wind acceleration region, a very high coronal proton temperature is necessary for a high-speed wind.

Even the original isothermal Parker model of the solar wind, which has implicitly very efficient heat conduction, shows basically the same behaviour as the adiabatic prediction (9.11). With heat conduction, the solar wind does no longer suddenly switch off if the coronal temperature becomes too low. However, as the temperature of the Parker model is decreased, the $\exp(-T_{\mathrm{esc}}/T)$ term in (9.6) leads to a solar wind flux that decreases drastically with decreasing T, as illustrated in Fig. 9.6. Hence even this model requires a million-degree corona to produce a solar wind mass flux that is not many orders of magnitude smaller than what is observed.

Comparing the results from (9.11) with the maximum proton temperature in Fig. 9.8, 4×10^6 K at $r \approx 3\,R_\odot$, shows that the observed temperature is too low to drive a high-speed wind without substantial energy addition beyond $3\,R_\odot$, either in the form of thermal heating or momentum addition beyond the critical point by Alfvén waves.

The above reasoning may also be applied to minor, heavy elements. For these, heat conduction is unlikely to contribute at all (due to their large mass). If we furthermore neglect energy loss or gain caused by collisions with protons, and choosing oxygen as the example ($m = 16\ m_p$), the coronal oxygen temperature needed to bring oxygen ions out of the gravitational potential and accelerate them to the terminal wind speed is:

$$T_i \approx \frac{16 m_p}{5k} v_{\mathrm{esc}}^2 \left[\frac{R_\odot}{r_0} + \left(\frac{u_E}{620\ \mathrm{km/s}}\right)^2\right]$$
$$= 1.5 \times 10^8\ \mathrm{K}\left[\frac{R_\odot}{r_0} + \left(\frac{u_E}{620\ \mathrm{km/s}}\right)^2\right]. \tag{9.12}$$

Hence of order 10^8 K is needed merely to bring oxygen out of the gravitational potential. This is exactly what can be seen in Fig. 9.8, which shows an ion temperature of that order. For this reason the observed minor ion temperatures are not proof of "preferential heating" of minor elements in the corona. The heating of oxygen ions in the corona is sufficient to bring these ions out of the corona and no extended heating is needed.

Finally we note from (9.11) that if the solar wind were the only energy loss available to the corona, the coronal temperature *must* be of order a million degree or more, irrespective of the amount of coronal heating, since it cannot lose energy otherwise.

9.4.2 Downward Heat Conduction

The downward conduction of heat from the corona to the transition region has two effects on the corona and solar wind: It represents a possibly important energy loss for the corona, energy that would otherwise be available to drive the solar wind. For a magnetically closed corona it is the only energy loss mechanism. Secondly, the downward heat conduction basically sets the pressure of the corona, and hence strongly influences how the solar wind energy is divided between potential energy (or mass flux) and kinetic energy (flow speed).

In a collision-dominated, fully ionized gas the elctron heat flux density is given as

$$q_e = -\frac{\kappa_e}{\ln \Lambda} T^{5/2} \nabla T \qquad (9.13)$$

where $\ln \Lambda$ is the Coulomb logarithm accounting for Debye screening of the electric field and $\kappa_e \approx 2 \times 10^{-10}$ W m^{-1} K$^{-7/2}$ (Braginskii 1965). The $T^{5/2}$-dependence is a consequence of the energy dependence of the Coulomb cross section and implies that heat conduction in a fully ionized plasma is very sensitive to temperature.

Let us first assume that there is no outflow of the plasma, and that the coronal magnetic field expands radially. Since there is (essentially) no radiation loss in the corona, the total heat flux (integrated over the cross section of a magnetic flux tube) has to remain constant with altitude, $r^2 q_e =$ constant. Integrating this latter equation from the coronal temperature maximum at $r = r_1$ and downwards, the maximum temperature can be written

$$\max(T) \approx \left[\frac{7}{2}\frac{|q_0| R_\odot}{\kappa_e}\left(1 - \frac{R_\odot}{r_1}\right)\right]^{2/7}, \qquad (9.14)$$

where q_0 is the heat flux density at $r = R_\odot$. Choosing, e.g., $r_1 = 2\ R_\odot$ (the 2/7 power makes results insensitive to the choice for r_1) we find for different choices for q_0:

$$q_0 = 0.1\ \text{W/m}^2 \rightarrow T_1 = 2.8 \times 10^5\ \text{K}$$
$$q_0 = 10\ \text{W/m}^2 \rightarrow T_1 = 1.1 \times 10^6\ \text{K}$$
$$q_0 = 100\ \text{W/m}^2 \rightarrow T_1 = 2.0 \times 10^6\ \text{K}$$
$$q_0 = 50\ \text{MW/m}^2(!) \rightarrow T_1 = 8.7 \times 10^7\ \text{K}.$$

This shows that, for a very wide range of heat fluxes, we obtain a coronal temperature of order a million degrees. In other words, if heat conduction were the only energy loss of the corona, the coronal temperature would be of order 10^6 K almost irrespective of the amount of coronal heating. Hence heat conduction works largely like a thermostat for the corona.

The second effect of heat conduction, setting the coronal pressure, can also be demonstrated straightforwardly in the case of no outflow. In the transition region we neglect the magnetic flux tube expansion. The downward heat flux will eventually be lost as radiation in the transition region, and the balance between radiative loss and the heat flux divergence can be written

$$P^2 \frac{L(T(r))}{4kT(r)^2} = -\frac{dq_e}{dr}, \qquad (9.15)$$

where $P = 2n_e kT$ is the (nearly constant) pressure of the transition region, and $L(T)$ is the radiative loss function. Using (9.13) this can be recast as

$$\frac{d(q_e^2)}{dT} = \frac{\kappa_e P^2}{2k^2} \sqrt{T} L(T), \qquad (9.16)$$

and integrated to yield (Landini 1975)

$$P \approx \sqrt{\frac{2}{\kappa_e K(T_2)}} k |q_2|, \qquad (9.17)$$

where

$$K(T_2) = \int^{T_2} \sqrt{T} L(T) dT \qquad (9.18)$$

and T_2 and q_2 are the temperature and heat flux density at the top of the transition region (i.e., in the corona). The loss function $L(T)$ becomes small at coronal temperatures so that $K(T)$ is nearly a constant for $T \sim 10^6$ K. Hence we have shown that the transition region pressure is directly proportional to the downward heat flux from the corona. Since the transition region is very thin, this is also the pressure of the low corona, and we have thus shown that the coronal density increases as the downward heat flux increases.

Hence the density of the corona is not a "free" parameter that may be set at will in a model. Rather, it is determined by the energy balance of the corona, or more precisely by the radiative loss in the transition region caused by the downward heat conduction.

In the case of a solar wind, the upflow through the transition region may modify the energy balance somewhat, as some of the heat conducted downwards is used to heat the upstreaming gas. Hence the pressure can be somewhat lower than predicted by (9.17) since less energy is converted into radiative loss.

9.4.3 The Coronal Energy Balance

Since both the solar wind energy loss and the heat conduction energy loss require coronal temperatures of order 10^6 K or more before they become large, the existence of a million degree corona is nearly unavoidable for our Sun, no matter how much the corona is heated. (Other stars, with a more extended atmosphere and hence a much lower gravitational escape speed, may not have coronae.) The coronal temperature is thus rather insensitive to the amount of heating, while the density is expected to be much more sensitive.

The energy balance considerations show that neither the coronal density nor the temperature are independent parameters that may be set at will, as they were, e.g., in the original Parker solar wind model. Both parameters result from the balance between heating and energy loss of the corona.

Since downward and outward energy loss both become important at approximately the same temperature, it is not obvious from the preceding arguments which process will dominate. To answer this one has to resort to numerical radiative energy balance models that span from the upper chromosphere and into the supersonic solar wind, and that can account for the balance between downward heat conduction and radiation. Such model have, despite their complexity, essentially only two input parameters that can and must be specified: The form, location, and amount of (coronal) heating, and the flow geometry (Withbroe 1988).

The modelling that has been carried out to date, assuming a Munro-Jackson type geometry (Munro and Jackson 1977), shows unequivocally that in a magnetically open corona the solar wind energy loss dominates, and that typically 80–90% of the energy deposited in the corona will end up in the solar wind (Hansteen and Leer 1995).

Perhaps the most important conclusion to be drawn from this result is that the solar wind is not an "appendix" to the corona, and that the corona and solar wind should be seen as one coupled system. It also implies that the solar wind energy flux measured at Earth orbit is a measure of the amount of energy deposited in the corona, in other words that the polar coronal holes are heated by an energy flux, averaged over the coronal hole area, of approximately 500 W/m^2, making use of (9.8) and assuming a flux tube expansion factor of 5–7 (Munro and Jackson 1977).

When discussing the Parker model, we noted that (9.6) leads to a mass flux problem: Small variations in the coronal temperature should lead to large variations in the solar wind mass flux, so why is then the observed fast solar wind mass flux so constant (see Fig. 9.7)? This is not only a problem with isothermal solar wind models, but with all models that fix the coronal density and temperature. With the new picture of the solar wind dominating the coronal energy loss we have at least a partial answer: The mass flux is limited by the amount of energy available in the corona, and the constant mass flux merely reflects that the coronal heating rate does not vary much in time (although rapid, hour-scale or shorter, fluctuations cannot be ruled out).

Finally, we note that if only of order 10% of the energy is conducted downwards in an open corona, the downward heat flux would be expected to be 10 times larger in a closed coronal loop where all the deposited energy must be conducted down, provided the heating rate is the same. From (9.17) we would then expect much higher coronal densities in loops than in coronal holes.

9.5 Applying Funnel Geometry to Radiative Energy Balance Models

In the above section it was stated that the radiative energy balance models have only two input parameters, the geometry and heating. We summarize in the following the results of several model studies where the funnel type geometry presented by Dowdy is used.

The heating parameters in the model are adjusted such that the model results are close to observed solar corona and wind parameters, such as presented in Section 9.3.

Using improved fluid transport equations (Killie et al. 2004, Janse et al. 2005) the effect of a funnel shape geometry on the solar wind properties was investigated by Esser et al. (2005). The heating parameters in the model were adjusted to produce the observed electron and proton temperatures described in Fig. 9.8. The observations indicate that electrons play a minor role and proton temperatures have to be of order 10^7 K to drive a solar wind (see above). On the other hand observations also indicate that the proton temperature is of order 3×10^6 K such that a significant fraction of the energy flux (about 1/2) has to be supplied in the supersonic solar wind. In the model by Esser et al. this energy was supplied by Alfvén waves. The applied geometry is shown in Fig. 9.11 (bottom panel) together with a Munro and Jackson (1977) (M&J) geometry with maximum expansion factor of 5. In the M&J geometry most of the heat conducted downward from the corona (about 60 W) is radiated away and only a negligible fraction of the energy is used to heat the upstreaming plasma. In the funnel geometry most of the heat conducted downward from the corona (about 2,000 W) goes into heating the upstreaming plasma and a small fraction goes into radiation. The model results for the outflow speeds, electron densities and temperatures are shown in the upper three panels of the figure. Very close to the sun the funnel shape geometry seems to reproduce the observations slightly better. However, taken into account the uncertainties in the model calculations and observations this difference is not significant. The main result is that it was established, for the first time, that the extreme funnel shape geometry suggested by the studies of the lower solar atmosphere, can lead to a realistic solar wind. Sufficient heat can be conducted into the transition region to supply the enthalpy flux necessary for reproducing the observed mass flux while maintaining a pressure small enough to agree with observed coronal pressures.

The main difference between the two geometries lies in the Lyα intensities predicted by the models. For the M&J geometry, the predicted Lyα intensity is only about 1/10 of the observed intensity, unless the pressure in the transition region is significantly increased in the model. However, since the solar wind mass flux is proportional to the density at the critical point, increasing the pressure sufficiently will lead to a mass flux far higher than observed. In the funnel geometry, on the other hand, the flow in the transition region is fast enough to bring hydrogen far out of ionisation equilibrium such that the Lyα radiation is mostly produced at a temperature of 5×10^4 K where it is about a factor of 10 more efficient. The radiation losses integrated along a vertical cylinder (accumulative radiation loss) are shown for the two geometries in Fig. 9.12.

The funnel geometry thus seems to solve a long-standing discrepancy between models of the lower solar atmosphere (e.g., Vernazza et al. 1981) and solar wind models. Models of the lower solar atmosphere produce the right amount of Lyα but have a pressure far higher than what is acceptable for solar wind models due to the relation between pressure and mass flux. Solar wind models with a M&J type geometry, on the other hand, produce the right mass flux but have a pressure far lower than what is necessary to produce the right amount of Lyα radiation. The above described calculations indicate that the solution lies in the high transition region outflow speeds that bring hydrogen sufficiently out of ionisation equilibrium.

Other interesting features of the funnel type solar wind were discovered by Janse et al. (2007) who added helium to the model described above. Firstly, helium is readily pulled out of the chromosphere in this geometry instead of settling in the chromosphere which it does in models with much smaller expansion factors. The frictional force on the neutral helium that is exerted by the neutral hydrogen is large enough in the funnel to increase the helium scale height to that of hydrogen. Secondly, in the presence of both helium and the funnel geometry the solar wind has two solutions for a rather large range of heating parameters. This was the first time that it was discovered that heating parameters and geometry alone did not uniquely determine the solar wind solution. For the same heating parameters one can get a slow and a fast wind depending on the starting solutions of the time dependent model. An example of the electron density for the slow and fast solar wind solution is shown in Fig. 9.13. Heating parameters and geometry are exactly the same for the two cases. The two solutions differ in the energy budget in the transition region. In the slow solar wind the energy conducted downward is relatively high (about 110 W) which is enough to sustain a large enthalpy flux and thus a high solar wind mass flux. Only about 36 W is needed to

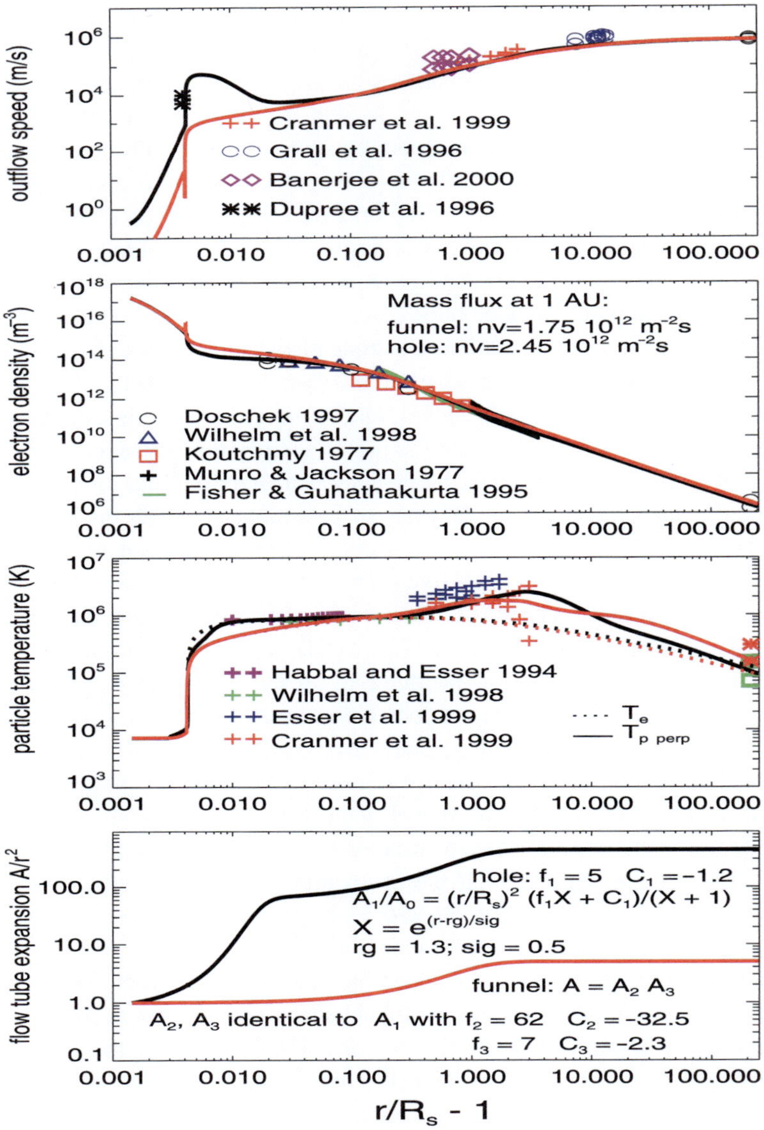

Fig. 9.11 The funnel model solution by Esser et al. (2005). *Red curves* represents the Munro and Jackson (1977) flow geometry while *black curves* represent the funnel geometry. The *bottom panel* shows the assumed flux tube area divided by r^2. Symbols denote observations

maintain the large enthalpy flux. In the fast solution only about 16 W are conducted downward and only 12 W are used to heat the upcoming plasma, the rest is needed to balance the radiation loss (2 W) and gravitational potential energy (4 W). The radiation loss in the slow wind solution is 64 W (due to the higher density). These two solutions are rather stable; in order for the wind to switch from one solution to the other, a large disturbance of either the electron or the helium heating has to occur (see Janse et al. 2007 for details).

Studies of heavy ions in the funnel geometry by Byhring et al. (2008) showed that observed blue shifts of transition region spectral lines (7–10 km/s for C V and O VI and 15–20 km/s for Ne VIII) place a constraint on the expansion of the flux tube in and above the transition region. The upper limit was found to be about a factor of 4 times the M&J expansion factor. However, additional expansion may take place in the chromosphere. Whether this expansion factor, which is much smaller than the factor suggested by Dowdy et al. (1986), is enough to account for the Lyα line intensities, pull helium out of the chromosphere, and allow for two solution for the same input parameters has not yet been investigated.

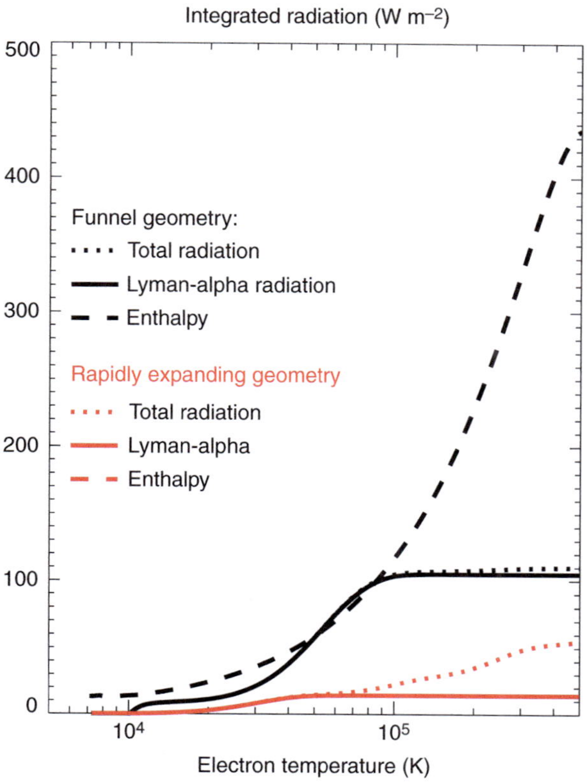

Fig. 9.12 Radiative loss integrated along a vertical cylinder and enthalpy flux density for the funnel (*black curves*) and a coronal hole geometry, for the model by Esser et al. (2005)

A completely different scenario for the origin of the solar wind was suggested by He et al. (2008). Their ideas are based on observational and model studies by Tu et al. (2005) and are similar to the ideas developed by Fisk and co-workers (e.g. Fisk 2003). A sketch of the model geometry is shown in Fig. 9.14. Closed loops surrounding a coronal funnel are pushed around by supergranulation convection. On contact between a loop and the funnel, reconnection occurs and the loop injects its mass and energy into the previously empty funnel. This injection is supposed to happen at 5 Mm and as indicated in the figure, the vertical flow speed is supposed to be 0 at this height. One-fluid equations of mass, momentum and energy flow are then used to model the upflowing plasma in region 2 and downflowing plasma in region 1. Assuming that heavy ions flow with the same speed as the protons, the Doppler shifts of selected spectral lines (from Si II, C IV and Ne VIII) are calculated from the model. It is shown that the calculated blue- and redshifts are in agreement with observations.

The mass and energy supplied to the solar wind thus originates in the coronal loops which reconnect with the empty funnel. At the time when the reconnection happens the solar wind flow tube thus consists of part of the loop and the upper part of the funnel, in other words a bent tube with constant cross section in the lower part and a cross section expanding with height in the upper part. At this point in time all the arguments put forward in Section 9.4 still apply, in particular the energy balance between the chromosphere, transition region (now in the loop) and corona. The funnel neck has now only to do with the downflow of the plasma.

This model approach is very interesting and the results reproduce most of the observed coronal and solar wind properties. However, several crucial questions have not yet been addressed. For example, what prevents the plasma in the transition region from streaming up into the funnel where the magnetic field is open at all times? Future models should also include the model of the loop and the expansion of the heavy ions which are used to calculate the spectral lines. In the present model it is inherently assumed that the loop contains the right amount of heavy ions. However, for the ions gravitational settling is a serious problem in loops. In addition it is not clear whether the ions would flow with the proton speed. Another question namely whether this time dependent mass and energy injection would lead to a steady solar wind should be relatively easy to investigate with this model. More difficult is supposedly to find a mechanism that determines how much of the mass flows up into the funnel to form the solar wind and how much flows down the neck of the funnel.

9.6 Summary

In the above presentation we have given a short review of how the idea of funnel type geometry developed in studies of the lower solar atmosphere. We have then given an overview of the type of observations that place constraints on solar wind expansion and have repeated some of the theoretical arguments that establish the connection between heating and energy loss from the corona. The main result from these arguments is that density and temperature in the corona are set by the radiative loss in the transition region caused by the downward heat conduction, increased downward heat

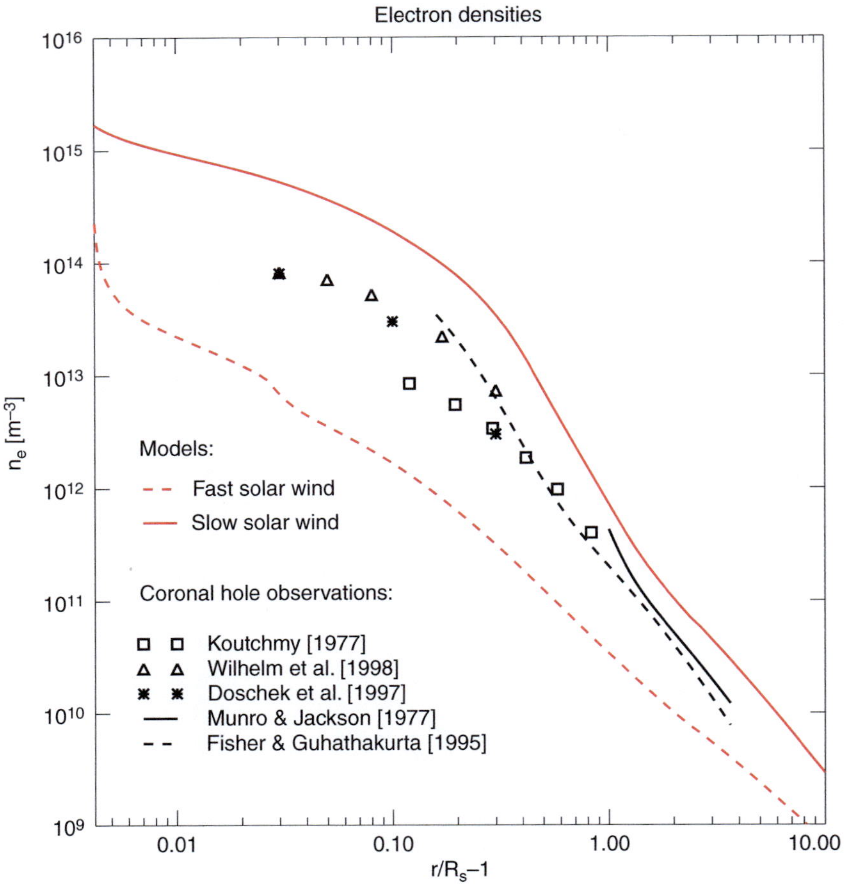

Fig. 9.13 Electron densities in a funnel model with helium included (Janse et al. 2007)

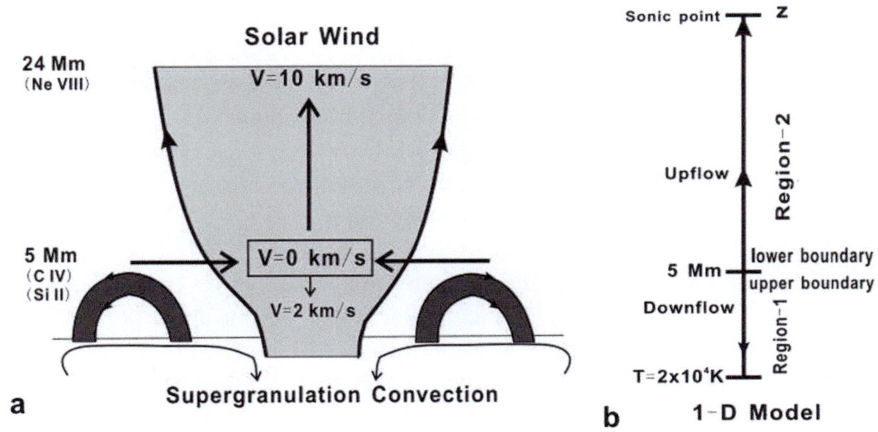

Fig. 9.14 Flow geometry suggested by He et al. (2008) based on Tu et al. (2005) (see also Fisk 2003). The funnel *grey area* is originally empty and is supplied with mass and energy during reconnection events with the surrounding loops. 1/2 of the loop mass flows down towards the transition region and 1/2 flows up to form the solar wind

flux increases the coronal density. The transition region pressure is proportional to the downward heat flux.

Incorporating funnel type geometries into solar wind expansion models that include the energy balance with the transition region and chromosphere shows that also in this type of geometry enough heat can be conducted down the transition region to produce a solar wind with reasonable mass flux and that the fast flow of hydrogen in this geometry leads to a predicted Lyα intensity in agreement with observations. Including helium in the funnel solar wind model reveals that two solar wind solutions can exist, one slow and one fast for the same heating parameters and geometry. The difference between them is the energy budget in the transition region. Removing helium or the funnel expansion from the model, removes the ambiguity.

Also time dependent reconnection type models are in agreement with observations, especially they provide a possible explanation for the red- and blueshifts of spectral lines that appear to originate in the same regions (flow tubes) on the sun. In these models small scale loops and open funnels are assumed to reconnect often enough to result in a continuous solar wind. At the time of the reconnection the solar wind geometry is determined by the open loop and upper part of a funnel. The arguments made in the present chapter about the importance of the energy balance between corona, transition region and chromosphere still applies, in the reconnection type models this connection happens primarily in the footpoints of the loop and should be included in future model studies of this kind.

References

Antonucci E, Giordano S, Dodero MA (2000) Adv Space Res 25(9):1923

Athay RG (1982) Astrophys J 263:982

Banerjee D, Teriaca L, Doyle JG, Lemaire P (2000) Sol Phys 194:43

Braginskii SI (1965) In: Leontovich MA (ed) Reviews of plasma physics, vol 1. Consultants Bureau, New York, NY, pp 205–311

Byhring HS, Esser R, Lie-Svendsen Ø (2008) Astrophys J Lett 673:L91. doi:10.1086/527518

Cranmer SR et al (1999) Astrophys J 511:481

Cranmer SR, Panasyuk AV, Kohl JL (2008) Astrophys J 678:1480. doi:10.1086/586890

Doschek GA, Warren HP, Laming JM, Mariska JT, Wilhelm K, Lemaire P, Schühle U, Moran TG (1997) Astrophys J 482:L109

Dowdy JF Jr, Rabin D, Moore RL (1986) Sol Phys 105:35

Dupree AK, Penn MJ, Jones HP (1996) Astrophys J Lett 467:L121

Esser R, Fineschi S, Dobrzycka D, Habbal SR, Edgar RJ, Raymond JC, Kohl JL, Guhathakurta M (1999) Astrophys J Lett 510:L63

Esser R, Habbal SR, Withbroe GL, Leer E (1986) J Geophys Res 91:2950. doi:10.1029/JA091iA03p02950

Esser R, Lie-Svendsen Ø, Janse ÅM, Killie MA (2005) Astrophys J Lett 629:L61

Fisher RR, Guhathakurta M (1995) Astrophys J 447:L139

Fisk LA (2003) J Geophys Res A108:1157. doi:10.1029/2002JA009284

Gabriel AH (1976) Proc R Soc London Ser A 281:339

Grall RR, Coles WA, Klinglesmith MT, Breen AR, Williams PJS, Markkanen J, Esser R (1996) Nature 379:429

Habbal SR, Esser R (1994) Astrophys J Lett 421:L59

Hammer R (1982a) Astrophys J 259:767

Hammer R (1982b) Astrophys J 259:779

Hansteen VH, Leer E (1995) J Geophys Res 100(A11):21577

Hassler DM, Rottman GJ, Shoub EC, Holzer TE (1990) Astrophys J Lett 348:L77

He JS, Tu CY, Marsch E (2008) Sol Phys 250:147

Janse ÅM, Lie-Svendsen Ø, Leer E (2005) J Plasma Phys 71:611C

Janse ÅM, Lie-Svendsen Ø, Leer E (2007) Astron Astrophys 474:997

Killie MA, Janse ÅM, Lie-Svendsen Ø, Leer E (2004) Astrophys J 604:842

Ko Y, Fisk LA, Geiss J, Gloeckler G, Guhathakurta M (1997) Sol Phys 171:345

Kohl JL, Esser R, Cranmer SR, Fineschi S, Gardner LD, Panasyuk AV, Strachan L, Suleiman RM, Frazin RA, Noci G (1999) Astrophys J Lett 510:L59

Koutchmy S (1977) Sol Phys 51:399

Landini M, Monsignori Fossi BC (1975) Astron Astrophys 42:213

Lie-Svendsen Ø, Leer E, Hansteen VH (2001) J Geophys Res 106:8217

McComas DJ et al (2000) J Geophys Res 105(A5):10419

Munro RH, Jackson BV (1977) Astrophys J 213:874

Parker EN (1958) Astrophys J 128:664

Tu C, Marsch E, Wilhelm K, Curdt W (1998) Astrophys J 503:475. doi:10.1086/305982

Tu CY et al (2005) Science 308:519

Vernazza JE, Avrett EH, Loeser R (1981) Astrophys J Suppl Ser 45:635

Wilhelm K, Marsch E, Dwivedi BN, Hassler DM, Lemaire P, Gabriel AH, Huber MCE (1998) Astrophys J 500:1023

Withbroe GL (1988) Astrophys J 325:442

Chapter 10

MHD Simulations of the Global Solar Corona and the Solar Wind

Roberto Lionello, Jon A. Linker, Zoran Mikić, Pete Riley, and Viacheslav S. Titov

Abstract We describe the latest applications of our global three-dimensional magnetohydrodynamic (MHD) model of the solar corona and the solar wind. The model uses boundary conditions based on observed photospheric magnetic fields. It has been used in the simplified, "polytropic" approximation of the energy equation to study the geometrical and topological properties of the magnetic field (e.g., the location and evolution of corona holes, the reproduction of streamer structure, the location of the heliospheric current sheet). However, this approximation does not reproduce the density and temperature contrasts between open and closed-field regions and does not address data from EUV and X-ray emission. Our improved MHD model that includes energy transport (radiative losses, anisotropic thermal conduction, and coronal heating) in the transition region and solar corona is capable of reproducing many emission properties as observed by SoHO and Hinode.

10.1 Introduction

Today there are still many key unsolved questions in solar physics: How is the corona heated and the solar wind accelerated? What is the origin of the slow solar wind? Why do certain regions erupt to produce coronal mass ejections (CMEs)? In the eruptive process, how are geo-effective ("southward") magnetic fields produced?

To investigate these questions we must develop an understanding of the connections between the solar surface, the corona, and the solar wind. Observations from present and future missions sampling diverse regions, both in parameter space and real space, are crucial to reach this goal. However, coronal and solar wind models are required to synthesize these measurements into a coherent picture. Among the desired capabilities of coronal models we must include predicting magnetic structure and topology, predicting the solar wind structure (e.g. fast wind streams), understanding the background through which CMEs propagate, providing magnetic connection of solar energetic particles (SEPs) to the Sun, and providing the Alfvén speed in coronal holes and active regions.

MHD models of the global solar corona can be broadly divided into two categories. The so called "polytropic" models avoid the complicated physics of the transition region by setting the ratio of specific heats γ to a reduced value (a spatially varying γ is sometimes used; Usmanov 1993, 1995; Mikić and Linker 1996; Linker et al. 1999; Mikić et al. 1999; Roussev et al. 2003; Riley et al. 2006; Cohen et al. 2007). Although polytropic MHD models can address many aspects of coronal physics, they do not correctly calculate the plasma density and temperature contrasts between open and closed-field regions, and between active regions and quiet Sun. Our polytropic model solves the 3D MHD equations on a grid in spherical coordinates:

$$\nabla \times \mathbf{B} = \frac{4\pi}{c}\mathbf{J}, \tag{10.1}$$

$$\nabla \times \mathbf{E} = -\frac{1}{c}\frac{\partial \mathbf{B}}{\partial t}, \tag{10.2}$$

R. Lionello (✉)
Predictive Science, Inc., San Diego, CA 92121-2910, USA
e-mail: lionel@predsci.com

M.P. Miralles, J. Sánchez Almeida (eds.), *The Sun, the Solar Wind, and the Heliosphere*, IAGA Special Sopron Book Series 4, DOI 10.1007/978-90-481-9787-3_10, © Springer Science+Business Media B.V. 2011

$$\mathbf{E} + \frac{\mathbf{v} \times \mathbf{B}}{c} = \eta \mathbf{J}, \qquad (10.3)$$

$$\frac{\partial \rho}{\partial t} + \nabla \cdot (\rho \mathbf{v}) = 0, \qquad (10.4)$$

$$\frac{1}{\gamma - 1}\left(\frac{\partial T}{\partial t} + \mathbf{v} \cdot \nabla T\right) = -T \nabla \cdot \mathbf{v} + \frac{m}{k\rho} S \qquad (10.5)$$

$$\rho\left(\frac{\partial \mathbf{v}}{\partial t} + \mathbf{v} \cdot \nabla \mathbf{v}\right) = \frac{1}{c}\mathbf{J} \times \mathbf{B} - \nabla(p + p_w) + \rho \mathbf{g} + \nabla \cdot (\nu \rho \nabla \mathbf{v}). \qquad (10.6)$$

S and p_w are not used in the polytropic approximation. $\mathbf{B}$ is the magnetic field, $\mathbf{J}$ is the electric current density, $\mathbf{E}$ is the electric field, ρ, $\mathbf{v}$, p, and T are the plasma mass density, velocity, pressure, and temperature, $\mathbf{g} = -g_0 R_\odot^2 \hat{\mathbf{r}}/r^2$ is the gravitational acceleration, η the resistivity, and ν is the kinematic viscosity. The ratio of specific heats is set to $\gamma = 1.05$.

Past (Skylab, Yohkoh), present (SoHO, STEREO, Hinode) and future (SDO) missions provide full disk emission images in EUV and X-rays that can put strong constraints on global coronal structure. Emission measurements have been used by comparison with 1D loop models to constrain coronal heating models (Klimchuk 2006). Sometimes authors model the corona as an ensemble of loops and solve 1D fluid and energy transport equations for geometry of selected field lines (Lundquist et al. 2004; Warren and Winebarger 2006). Our thermodynamic MHD model (Lionello et al. 2001) includes the following term in Eq. 10.5:

$$S = (-\nabla \cdot \mathbf{q} - n_e n_p Q(T) + H_{\mathrm{ch}}). \qquad (10.7)$$

S represents the contribution of thermal conduction along magnetic field lines ($\nabla \cdot \mathbf{q}$), radiative losses ($n_e n_p Q(T)$), and the specification of a heating function (H_{ch}). In the thermodynamic model we add the wave pressure term p_w in Eq. (10.6), which represents the contribution due to Alfvén waves (Jacques 1977) to the acceleration of the solar wind and is evolved using the WKB approximation. This model can reproduce the emission properties of the corona (Lionello et al. 2009).

Here we show first an application of our polytropic model to the study of reconnection in coronal holes (Section 10.2). Then we show how the thermodynamic MHD model is applied to study the May 12, 1997 CME (Section 10.3) and to predict the aspect of the corona during the July 19, 2009 solar eclipse (Section 10.4).

10.2 Interaction of Small Bipoles with Coronal Holes

Coronal holes play an important role in determining the magnetic flux balance in the heliosphere and the formation of the slow solar wind. Fisk and coworkers (1999), who argue that open magnetic flux is constant in the heliosphere, envision the transport of open flux out of coronal holes into the closed-field region. In their model, this is necessary in the context of flux reversal during the solar cycle and is accomplished solely through interchange reconnection (Crooker et al. 2002a) with small loops associated with parasitic polarities. The slow wind is released during this continuous reconnection process. This scenario is supported by older in-situ observations (Crooker et al. 2002b, 2003), although newer observations are more ambivalent (Owens and Crooker 2007). However, the transport of open flux through closed-field regions is difficult to reconcile with the result of Antiochos, who has claimed that every unipolar region on the photosphere can contain at most one coronal hole and that coronal holes of nested polarity regions must themselves be nested (Antiochos et al. 2007). We have investigated this problem by modeling the interaction of the magnetic field of two bipoles with a coronal hole using our 3D MHD algorithm in spherical coordinates. Two bipoles are introduced into a realistic background flux distribution obtained from a smoothed magnetogram for Carrington Rotation 1913 (Whole Sun Month, late August 1996). After calculating the corresponding potential magnetic field extrapolation and prescribing the initial velocity, density, and temperature in the corona from a Parker's solar wind solution, we have advanced the MHD equations until a steady state with a solar wind and a heliospheric current sheet has been reached. A surface flow $v_\Phi \lesssim 1$ km/s has been applied to advect the bipoles at localized latitudes. Figure 10.1 shows how open flux associated with the positive pole of a bipole eventually closes down as the bipole is advected into the closed-field region.

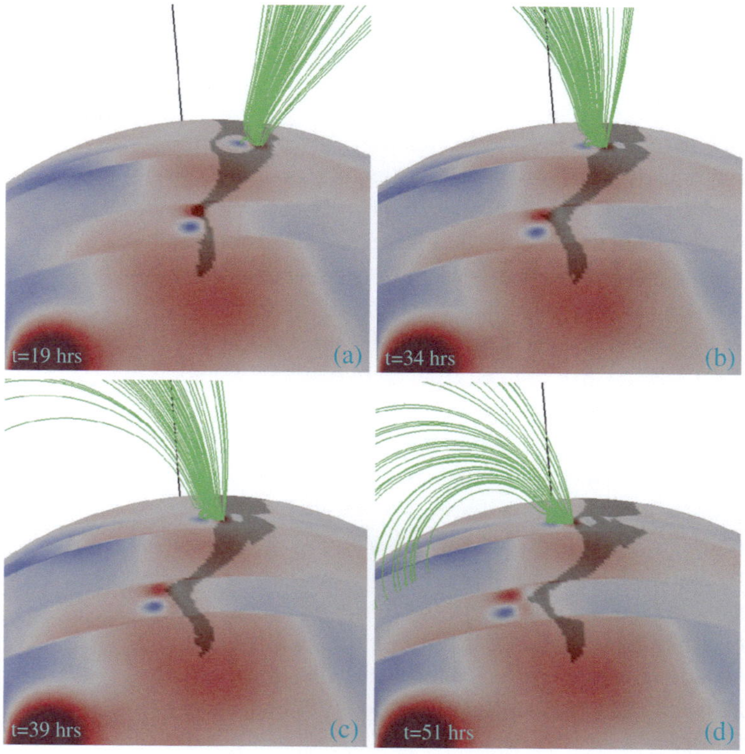

Fig. 10.1 Closing down of open flux associated with a positive polarity in a bipole as it is advected by surface flow in the simulation of Section 10.2. The magnetic flux map is superimposed to the coronal hole map (*shaded areas* indicate open flux). Eventually all the flux associated with the positive polarity reconnects and closes down

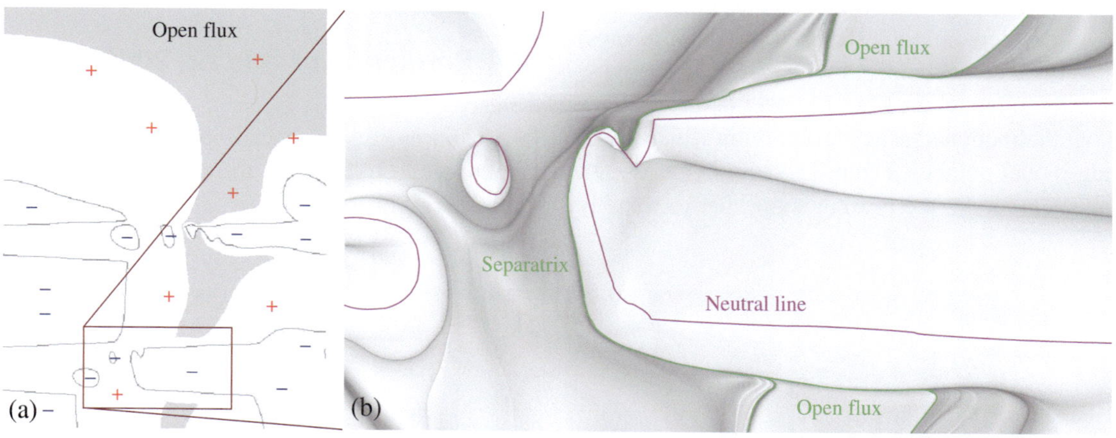

Fig. 10.2 Two maps showing the magnetic field topology at the end of the simulation of Section 10.2. **a** A coronal hole map (*shaded areas* indicate open field) with neutral lines. Within a single same (positive) polarity region, there seems to be two separated coronal holes, which have originated from a single coronal hole present at the beginning of the simulation. **b** An enlargement showing the squashing factor (Titov 2007) in the area in the *brown* box. Neutral lines are in *magenta*. The two coronal holes are actually linked through a zero-width separatrix line

Figure 10.2 shows that open flux areas may apparently be detached from the main coronal hole during this process. However, an analysis of the squashing factor (Titov 2007), which measures the elliptical deformation of the cross section of an infinitesimal flux tube and whose high values indicate topological features, reveals the presences of a separatrix line. This is virtually a "zero-width coronal hole", connecting the detached patch with the main coronal hole.

10.3 Thermodynamic MHD Modeling of the May 12, 1997 CME

Investigating CMEs is important not only because the physical mechanism that causes their release is still not understood, but also for their impact on our environment and life: when an Interplanatery CME reaches the magnetosphere it can trigger a geomagnetic storm affecting communication and navigation systems, electric power grids, and astronauts in orbit. On May 12, 1997 a halo CME erupted from NOAA Active Region 8038, accompanied by a C-class flare. Observed features include a double-dimming signature on the solar surface, a magnetic cloud in space, and a moderate geomagnetic storm on Earth. On the disk there was only AR 8038, to which two other limb-event CMEs were associated. Since the primary focus is on understanding how these eruptions are triggered by magnetic energy release, most CME models use simple energy equations. However, to develop a realistic model of an active region in the context of the global corona, energy transport processes cannot be ignored. In fact, present and future missions provide a wealth of data in emission lines. Therefore models that include energy transport can develop emission diagnostics (EUV, X-rays) that can be directly compared with these observations.

In order to model the May 12, 1997 CME, we have used our thermodynamic MHD algorithm with coronal heating model number 3 from Lionello et al. (2009). We have "tailored" the heating somewhat to roughly match the emission on May 11, 1997. First, we have extrapolated a potential magnetic field from a magnetic flux distribution obtained from an MDI magnetogram for May 11, 1997. Resolution was about 3.8″ in the active region, where the field strength reaches 115 G. Then we have advanced the MHD equations for approximately 3 days to develop a solar wind solution. Since there were no vector magnetograms of sufficient quality available, the active region in our calculation was energized via shearing flows, emerging of transverse magnetic field into the corona from below the photosphere, and flux cancellation. In the corona it is expected that all of these processes will occur simultaneously, in different proportions during individual events. From our results we have prepared synthetic emission images, which are shown in Fig. 10.3 with the corresponding magnetic field configuration. Although there are still too many free parameters, the model shows features that compare well with observations, such as the formation of a prominence, dimming regions visible in X-rays and in EUV, post-flare loops, and EIT waves. We hope that more detailed comparisons with vector magnetograms and emission images will help us to narrow the parameters down.

10.4 Prediction of July 22, 2009 Eclipse

Solar eclipses give us special opportunities to test our predicting capabilities and to refine our MHD model by comparing the spectacular white-light images that

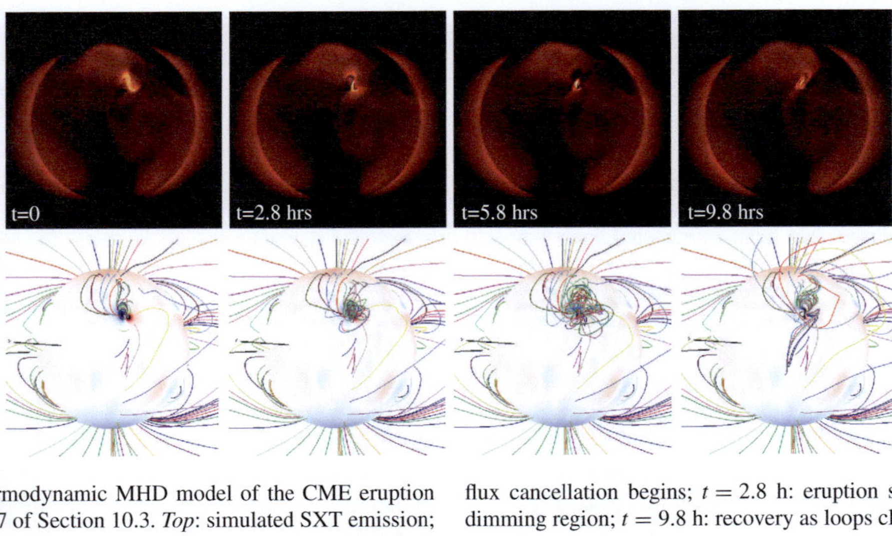

Fig. 10.3 Thermodynamic MHD model of the CME eruption of May 12, 1997 of Section 10.3. *Top*: simulated SXT emission; *bottom*: magnetic field evolution. $t = 0$: sheared configuration, flux cancellation begins; $t = 2.8$ h: eruption starts; $t = 5.8$ h: dimming region; $t = 9.8$ h: recovery as loops close

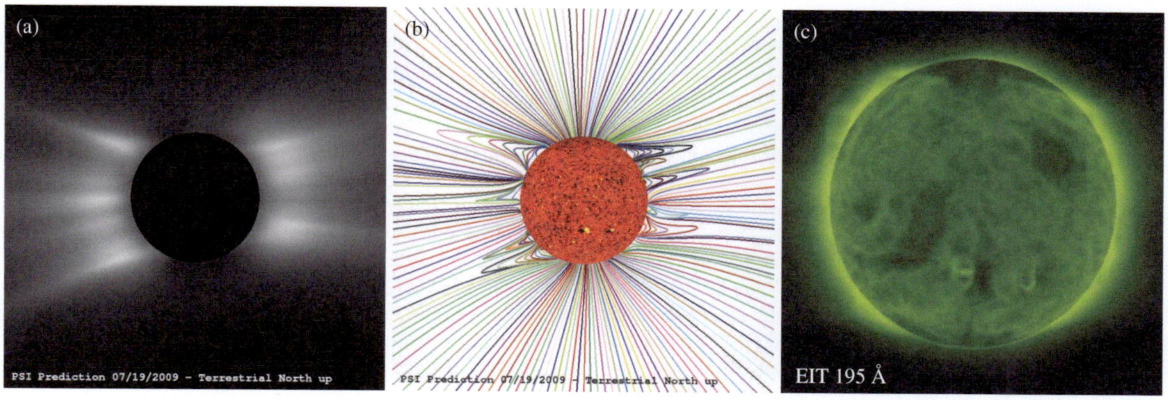

Fig. 10.4 Simulation of eclipse of July 22, 2009 using the thermodynamic MHD model (Section 10.4). **a** Predicted polarization brightness. The pB signal is produced by white light scattered off electrons in the coronal plasma. **b** Traces of the magnetic field lines in the solar corona. **c** Synthetic emission in EIT 195 Å

are captured around the maximum with those produced out of our calculations. We have used our thermodynamic MHD model to predict the aspect of the corona during the solar eclipse of July 22, 2009, which was visible in the equatorial regions spanning both hemispheres. This has been our highest resolution calculation ever, using approximately 21 million cells, running for 7 days on 4,992 processors on Ranger, the supercomputer at the Texas Advanced Computing Center. SoHO MDI data have been used to specify the magnetic field. As in the previous case, coronal heating model number 3 from Lionello et al. (2009) has been specified. In Fig. 10.4 we present our prediction of the corona, showing a synthetic polarization brightness image, magnetic field line traces, and synthetic emission in the EIT 195 Å band.[1] The Sun at the time was at solar minimum and had a rather weak polar magnetic field. We predicted the presence of several pseudostreamers surrounding the streamer formed under the current sheet. These appear bright in the white light image (Fig. 10.4a), since the plasma is trapped in closed-field regions (Fig. 10.4b). The synthetic EIT emission image (Fig. 10.4c) does not show any conspicuous active region.

[1] More images and animations are available from our web site: http://www.predsci.com/corona/jul09eclipse/

10.5 Conclusions

Global MHD models of the corona with simple energy treatments (e.g. polytropic) are mature and can be used to describe many coronal properties qualitatively, although they are not accurate quantitatively. In fact, they cannot predict emission, which is a major component of SoHO, TRACE, Hinode, STEREO, and SDO missions. However, they can be used to study topology and reconnection. We have developed a promising technique for modeling the global corona with realistic energy transport (thermal conduction, radiation, coronal heating). This kind of technique was previously confined to computing 1D loops or 2D configurations, but now we can expect realistic simulations of CMEs and high-resolution models of the global corona.

References

Antiochos SK, DeVore CR, Karpen JT, Mikić Z (2007) Structure and dynamics of the Sun's open magnetic field. ApJ 671:936–946. doi:10.1086/522489

Cohen O, Sokolov IV, Roussev II, Arge CN, Manchester WB, Gombosi TI, Frazin RA, Park H, Butala MD, Kamalabadi F, Velli M (2007) A semiempirical magnetohydrodynamical model of the solar wind. ApJ 654:L163–L166. doi:10.1086/511154

Crooker NU, Gosling JT, Kahler SW (2002a) Reducing heliospheric magnetic flux from coronal mass ejections without disconnection. J Geophys Res (Space Phys) 107:3–1

Crooker NU, Gosling JT, Kahler SW (2002b) Reducing heliospheric magnetic flux from coronal mass ejections without disconnection. J Geophys Res (Space Phys) 107:1028–+. doi:10.1029/2001JA000236

Crooker NU, Larson DE, Kahler SW, Lamassa SM, Spence HE (2003) Suprathermal electron isotropy in high-beta solar wind and its role in heat flux dropouts. Geophys. Res. Lett.**30**(12):21-1 to 21-4 1619. doi:10.1029/2003GL017036

Fisk LA, Zurbuchen TH, Schwadron NA (1999) On the coronal magnetic field: consequences of large-scalemotions. ApJ 521:868–877

Jacques SA (1977) Momentum and energy transport bywaves in the solar atmosphere and solar wind. ApJ 215:942–951

Klimchuk JA (2006) On solving the coronal heating problem. Sol Phys 234:41–77. doi:10.1007/s11207-006-0055-z

Linker JA, Mikić Z, Biesecker DA, Forsyth RJ, Gibson SE, Lazarus AJ, Lecinski A, Riley P, Szabo A, Thompson BJ (1999) Magnetohydrodynamicmodeling of the solar corona during Whole Sun Month. J Geophys Res 104:9809–9830

Lionello R, Linker JA, Mikić Z (2001) Including the transition region in models of the large-scale solar corona. ApJ 546:542–551

Lionello R, Linker JA, Mikić Z (2009) Multispectral emission of the Sun during the first Whole Sun Month: magnetohydrodynamic simulations. ApJ 690:902–912. doi:10.1088/0004-637X/690/1/902

Lundquist LL, Fisher GH, McTiernan JM, Régnier S (2004) In: Walsh RW, Ireland J, Danesy D, Fleck B (eds) Using synthetic emission images to constrain heating parameters. ESA SP-575: SOHO 15 coronal heating.ESA, Noordwijk, pp 306–+

Mikić Z, Linker JA (1996) The large-scale structure of the solar corona and inner heliosphere. In: Winterhalter D, Gosling JT, Habbal SR, Kurth WS, Neugebauer M (eds) Solarwind eight, Proceedings of the eighth international solarwind conference, American Institute of Physics conference proceedings 382. American Institute of Physics, Woodbury, New York, 1996, pp 104–107

Mikić Z, Linker JA, Schnack DD, Lionello R, Tarditi A (1999) Magnetohydrodynamic modeling of the global solar corona. Phys Plasmas 6:2217–2224

Owens MJ, Crooker NU (2007) Reconciling the electron counterstreaming and dropout occurrence rates with the heliospheric flux budget. J Geophys Res (Space Phys) 112:6106–+. doi:10.1029/2006JA012159

Riley P, Linker JA, Mikić Z, Lionello R, Ledvina SA, Luhmann JG (2006) A comparison between global solar magnetohydrodynamic and potential field source surfacemodel results. ApJ 653:1510–1516. doi:10.1086/508565

Roussev II, Gombosi TI, Sokolov IV, Velli M, Manchester W IV, DeZeeuw DL, Liewer P, Tóth G, Luhmann J (2003) A Three-dimensional model of the solar wind incorporating solar magnetogram observations. ApJ 595:L57–L61. doi:10.1086/378878

Titov VS (2007) Generalized squashing factors for covariant description of magnetic connectivity in the solar corona. ApJ 660:863–873. doi:10.1086/512671

Usmanov AV (1993) A global numerical 3-D MHD model of the solar wind. Sol Phys 146:377–396

Usmanov AV (1995) A global 3-DMHD model of the solar wind with Alfven waves. In: Solar wind conference, pp 65–+

Warren HP,Winebarger AR (2006) Hydrostatic modeling of the integrated soft X-Ray and extreme ultraviolet emission in solar active regions. ApJ 645:711–719. doi:10.1086/504075

Part IV
The Heliosphere

Chapter 11

Solar Wind Observations from the STEREO Perspective (2007–2009)

Antoinette B. Galvin

Abstract The STEREO two-spacecraft mission was launched in October 2006 and has been fully operational since January 2007. The scientific payload provides both in-situ and remote observations at 1 AU with (as of December 2009) up to 130° heliocentric longitudinal separation between the two spacecraft. The longitudinal and latitudinal vantage points available amongst the STEREO and L1 remote (SOHO) and in situ (Wind, ACE, SOHO) assets have provided unique opportunities for studying spatial and temporal variations in the solar wind at 1 AU. The longitudinal separation of the STEREO also provides a serendipitous proving ground for solar wind propagation and space weather predictive techniques and models. The extended period of solar quiet during the mission through 2009 has provided an opportunity for studying in-situ signatures of solar wind slow-high-slow speed stream interfaces and solar minimum ICMEs, yielding some new perspectives on their origins and propagation. In addition, with the incorporation of remote imaging that now extends from the solar disk to 1 AU, some solar wind features have been tracked directly and comprehensively from their solar source to the local measurements at 1 AU. In this chapter the solar wind observations in the STEREO era are presented in the context of the above and in relation to the prevalent solar cycle conditions.

A.B. Galvin (✉)
University of New Hampshire, Durham, NH 03824 USA
e-mail: toni.galvin@unh.edu

11.1 Introduction: The STEREO Mission and Its Place in Space

The Solar Terrestrial Relations Observatory (STEREO) mission is part of NASA's Solar Terrestrial Probes mission line with the primary science objectives of: understanding the causes and mechanisms of the initiation of coronal mass ejections (CMEs); characterizing the propagation of CMEs through the inner heliosphere to 1 AU; discovering the mechanisms and locations of solar energetic particle acceleration; and developing a three-dimensional, time-dependent model of the magnetic topology, kinetic properties and structure of the ambient solar wind plasma (Kaiser et al. 2007). The mission consists of two nearly identical spacecraft with similar payloads that have been placed into heliocentric orbits near 1 AU. The STEREO A spacecraft drifts ahead of the Sun-Earth line by approximately 22.5 ° per year, while the STEREO B spacecraft drifts behind the Sun-Earth line by about −22.5 ° per year (see Fig. 11.1, top). By December 2009, the two spacecraft had achieved a 130 ° separation in longitude from each other. The spacecraft are three-axis stabilized, Sun-pointing. The payload on each observatory consists of four packages: the Sun-Earth Connection Coronal and Heliospheric Investigation – or SECCHI, which includes coronagraphs, an extreme ultraviolet imager, and white-light heliospheric imagers (Howard et al. 2007); the In situ Measurements of PArticles and CME Transients – or IMPACT, which includes a magnetometer, thermal and suprathermal electron sensors, and energetic particle sensors (Luhmann et al. 2007); the PLAsma and SupraThermal Ion Composition investigation – or PLASTIC, which measures solar

M.P. Miralles, J. Sánchez Almeida (eds.), *The Sun, the Solar Wind, and the Heliosphere*, IAGA Special Sopron Book Series 4, DOI 10.1007/978-90-481-9787-3_11,

wind ions and their composition (Galvin et al. 2007); and the STEREO/WAVES – or S/WAVES, which observes traveling radio disturbances (Bougeret et al. 2007).

The STEREO spacecraft were launched in October 2006 and finished commissioning in January 2007. As seen in the bottom of Fig. 11.1, the early mission was under solar minimum conditions. The start of Cycle 24 was delayed longer than originally predicted, and – although well within historical variability – the minimum was longer than for other cycles in which in-situ solar wind observations are available. (For a review on solar cycle variability, the reader is referred to Richards et al. 2009.)

Solar wind and magnetic field observations have been taken by in-situ spacecraft in both the ecliptic and at high heliospheric latitudes during the approach to this "peculiar" solar minimum. Smith and Balogh (2008) reported a marked decrease in the open magnetic flux ($R^2\ Br$) measured during the

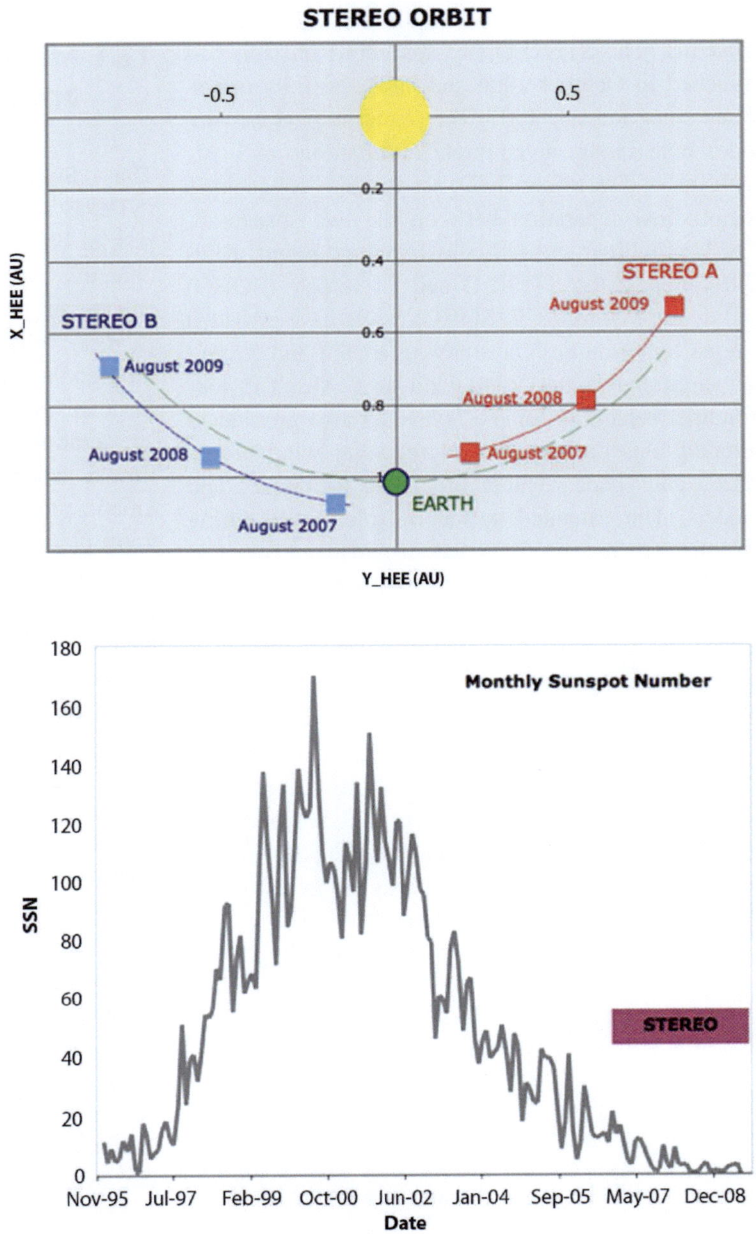

Fig. 11.1 *Top*: Orbits of the two STEREO spacecraft relative to the Sun-Earth line. Each spacecraft drifts about 22° from the Sun-Earth line per year. STEREO A is slightly closer to the Sun, at about 0.95 AU, while STEREO B is slightly further from the Sun, at about 1.08 AU. *Bottom*: Monthly sunspot numbers (courtesy NOAA National Geophysical Data Center). Solar Cycle 23 began in May 1996. The STEREO spacecraft was launched in October 2006, during the approach to solar minimum. The current prediction for the start of Cycle 24 is December 2008 (http://www.swpc.noaa.gov/SolarCycle/SC24/index.html)

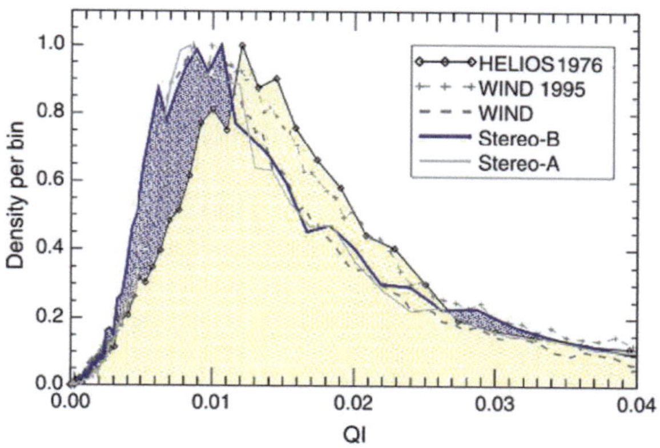

Fig. 11.2 From the heliospheric index, the quasi-invariant, it can be inferred that the recent solar minimum was weaker than the previous two minima. In this figure (adapted from Leitner et al. 2009), the QI distribution derived for STEREO A, B and Wind in 2007 are compared to that of Wind in 1995 and Helios in 1976. The STEREO B distribution is shaded *blue* and the Helios shaded *yellow*, to emphasize the shift toward lower QI for the current minimum

recent Ulysses latitudinal scan (time interval 2006.1–2007.4) compared to the measurements taken during a comparable period in the previous solar cycle (time interval 1993.5–1995.0); and McComas et al. (2008) reported the solar wind emanating from the polar coronal holes in this latest (third) Ulysses orbit was slightly slower, less dense, cooler, and contained less mass and momentum flux than seen in the previous solar minimum (first Ulysses orbit).

Leitner et al. (2009), utilizing the solar wind quasi-invariant as an in-situ proxy for solar activity, have inferred that solar activity was weaker in this solar minimum than in other recent minima periods (see Fig. 11.2). The quasi-invariant, or "QI", is defined as the ratio of the solar wind magnetic energy density to the plasma kinetic energy density, that is, the inverse square of the Alfven Mach number. Their study interval covered 30 years, using solar wind data from Helios, Wind, and STEREO A and B.

Using Wind as a baseline (Fig. 11.3), a comparison of solar wind speed distributions at the approach to the last solar minimum (1995–1996) to the recent solar minimum (2007–2008) indicates more high-speed structures were present during this latest solar minimum. The previous (1996) minimum was dominated by recurrent high speed streams emanating from a single extension of the north polar hole (Galvin and Kohl 1999), while the recent minimum was characterized by near-equatorial isolated coronal holes.

The primary science goals of the STEREO mission are strongly related to solar activity (generation and propagation of coronal mass ejections, solar energetic particle acceleration). During the two-year primary mission phase (January 2007–January 2009) and first year of extended operations, the Sun was in solar minimum conditions. In this chapter, emphasis is given to the STEREO accomplishments in this rather quiet interim period, prior to the Sun waking up and becoming more active.

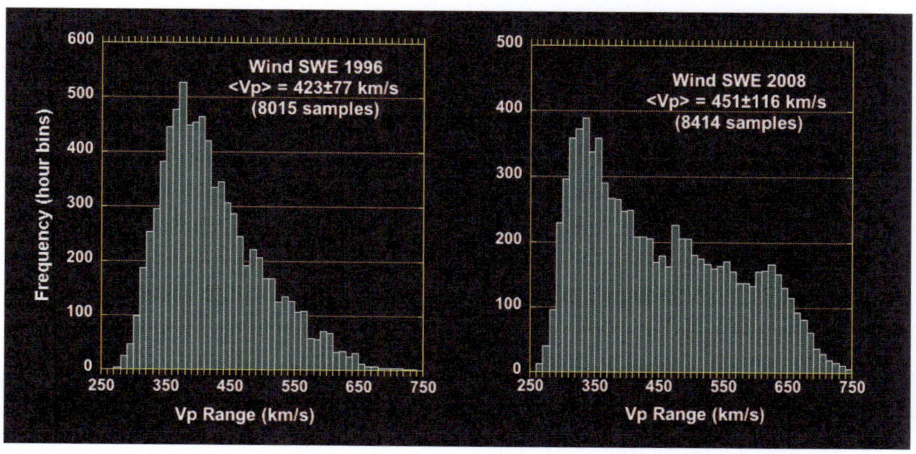

Fig. 11.3 Solar wind proton bulk speed distributions taken from the Wind Solar Wind Experiment (SWE) for the last solar minimum in 1996 (*left*) and the recent solar minimum in 2008 (*right*). SWE data courtesy K. W. Ogilvie and A. J. Lazarus

11.2 Solar and Solar Wind Conditions Near Solar Minimum: The View from STEREO

The solar conditions at the beginning of the STEREO mission included polar and near-equatorial isolated coronal holes, active regions, and a wavy and highly tilted neutral line (top synoptic coronal hole map, Fig. 11.4). One sees an evolution in the neutral line (and by extension, the equatorial streamer belt) from a sinusoidal pattern within a 40° band at the start of the mission, toward a flattened configuration. Near-solar minimum conditions are apparent in December 2008 (middle coronal hole map, Fig. 11.4). More recently there has begun a return of sun spots and active regions.

Given the low activity levels and sustained solar minima conditions, the solar wind observed by

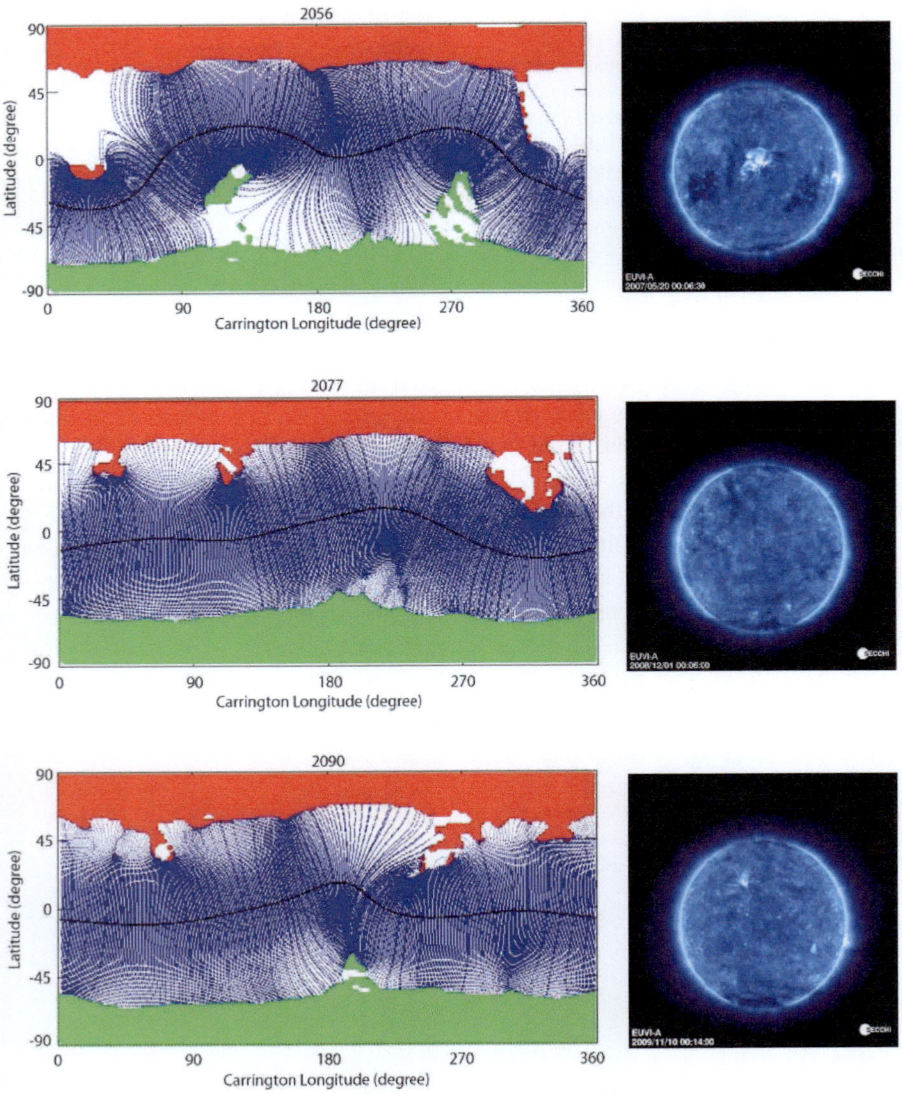

Fig. 11.4 *Left*: Synoptic coronal-hole model maps, courtesy NSO/GONG during three periods of the mission. Maps are provided in solar Latitude versus Carrington Longitude. Patches of *green* and *red* color indicate the location of the coronal holes and their magnetic polarity (*green* denotes positive polarity, *red* negative polarity). The neutral line, which may be identified with the equatorial streamer belt, is drawn in *black*. *Right*: Corresponding snapshots of the solar disk in extreme ultraviolet taken by STEREO. Coronal holes are characterized by less emission in EUV and therefore appear *dark*, while active regions appear bright. Extreme ultraviolet full solar disk images courtesy STEREO SECCHI, R. Howard, PI

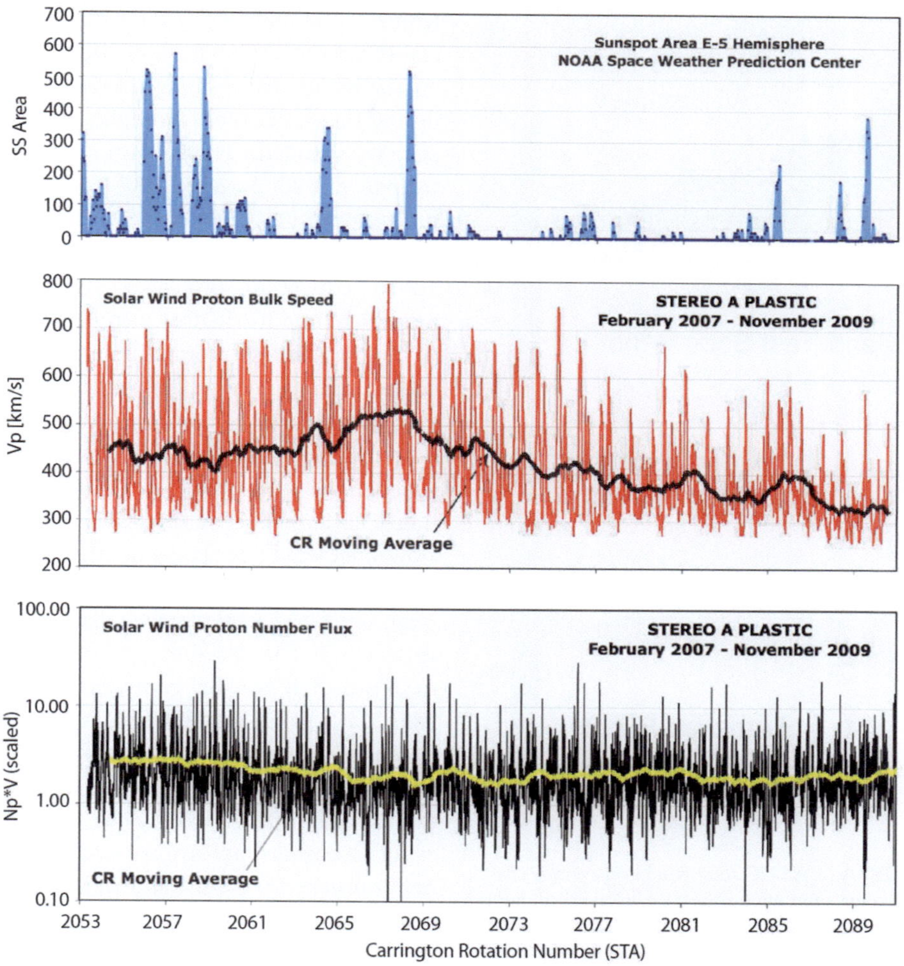

Fig. 11.5 *Top*: Monthly sunspot area (courtesy NOAA) indicating decreasing solar activity during the approach to solar minimum. *Middle*: Solar wind speed profile, as observed by STEREO A. *Bottom*: Solar wind proton number flux, scaled by R^2. In-situ data are hourly averages. Superimposed on the in-situ data are 27-day (i.e., a Carrington Rotation) running averages

STEREO and other spacecraft in the 2007–2009 time interval was predominately characterized by an alternating series of high speed solar wind (from coronal holes) and slow (interstream) wind, with an occasional transient event. In Fig. 11.5 is an overview of the solar wind proton speed profile and number flux plotted as a time series in Carrington Rotation Number (CRN 2053–2090) as observed by STEREO A from February 2007 to December 2009. The top panel gives the sunspot area on the solar disk, indicating the decline in solar activity from the start of the mission (sunspot data provided by NOAA). In the solar wind, one sees a general decline in the solar wind speed, and in its statistical variation, from early 2008 to the end of 2009. The average Vp for CRN 2054 was 451 ± 114 km/s, while for CRN 2089 the average was 334 ±59 km/s. The speed distribution peaked in CRN 2067, with 526 ±119 km/s, during a period prevalent with coronal hole high speed streams.

11.2.1 Solar Wind Flow Types in Solar Minima Conditions

The primary types of solar wind observed have been coronal hole solar wind (aka "fast wind"), a few transient events, and slow solar wind of likely multiple sources.

In Fig. 11.6 are contour plots of the solar wind iron average charge state plotted against the proton

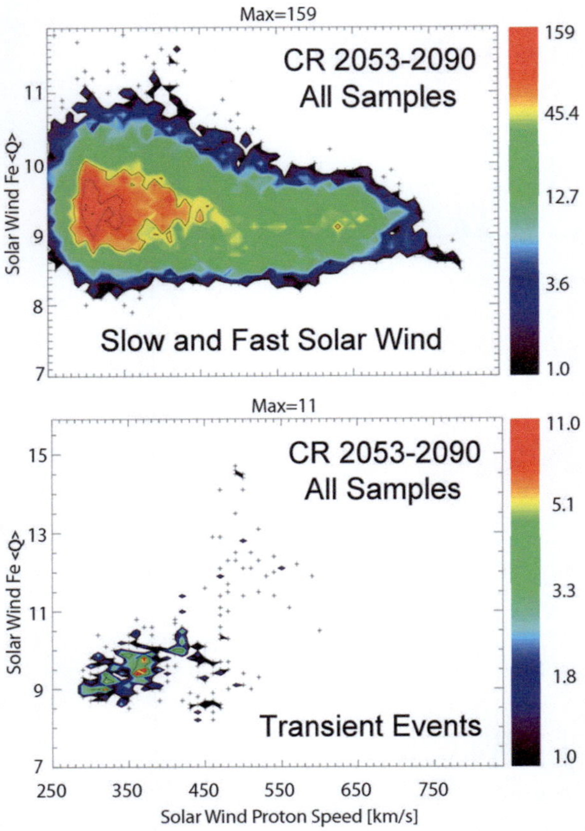

Fig. 11.6 STEREO A PLASTIC solar wind iron average charge state compared against solar wind speed. Included are all hourly samples from Carrington Rotation Numbers 2053–2090. *Top*: Slow and fast solar wind. *Bottom*: Transient events

bulk speed, measured by STEREO PLASTIC. The top panel data set has removed all identified transient (ICME) periods. The remaining data constitutes 23,460 one-hour samples of coronal hole-associated and slow solar wind, as well as their stream interfaces. The majority of these samples are low speed solar wind (67% of the samples are below 450 km/s). Within this speed group is included a number of slow-fast stream interfaces. Only 10% of the non-transient samples have speeds above 600 km/s. Over all non-transient samples, the average speed was 420 km/s (±110 km/s) and the average iron charge state was 9.38 (±0.45). The range of the average charge state over all (non-transient) samples was 8–12. This range is fully represented within the slow solar wind, however for the highest speeds (above 600 km/s), which are associated with solar wind emanating from coronal holes, the statistical variation in the observed charge state is significantly narrower (Galvin et al. 2009a, b).

The bottom panel of Fig. 11.6 are the samples taken during transient events (note the scale for iron charge state is expanded). These 441 transient hourly samples include the 22–24 May 2007 interplanetary coronal mass ejection (ICME) events, which were associated with the central meridian passage of solar active region 10956. This region produced several small flares and more than one ICME that were seen in situ by Wind, ACE and STEREO (for a full description of this event and its solar origins, see Kilpua et al. 2009a). The May events exhibit the highest iron charge states in this plot. Also included in the plot are the small-scale transients as identified by Kilpua et al. (2009b). The small-scale transients are magnetic structures embedded within the slow solar wind that appear to originate from the vicinity of sector boundaries on the Sun. In previous studies, these small transients were not distinguished from slow solar wind. The remaining ICMEs were taken from the Level 3 ICME list compiled by Jian (2009) and from the ICMEs identified by Kilpua et al. (2009c). It can be seen here that the majority of the transient time periods associated with this solar minimum, with the exception of the May 2007 samples, have modest charge states, not readily distinguishable from the non-transient solar wind shown in the top panel. This feature has been discussed by Kilpua et al. (2009b, c) and Galvin et al. (2009b). The non-exceptional charge states may be due to the specific solar origins of these ICMEs. Kilpua et al. (2009c) report that the ICMEs in the latter part of the mission are linked to high latitude active regions and polar crown filament eruptions, whereas the May 2007 events were associated with a flare-producing low-latitude active region (Kilpua et al. 2009a).

In Fig. 11.7, two different sets of samples representing the non-transient slow and fast solar wind are emphasized. In the top panel, non-transient solar wind from two consecutive Carrington Rotations, 2072–2073, have been evaluated separately from the full sampling distribution. This interval had the lowest solar activity in the past three years, when based upon the sun spot number, with an average daily sun spot number of 0.62 (average daily hemisphere area of 0.73 E-5). In contrast, the bottom contour plot contains samples from the two Carrington Rotations 2055.5–2057.5. During this second interval, which occurred early in the mission, the average daily sun spot num-

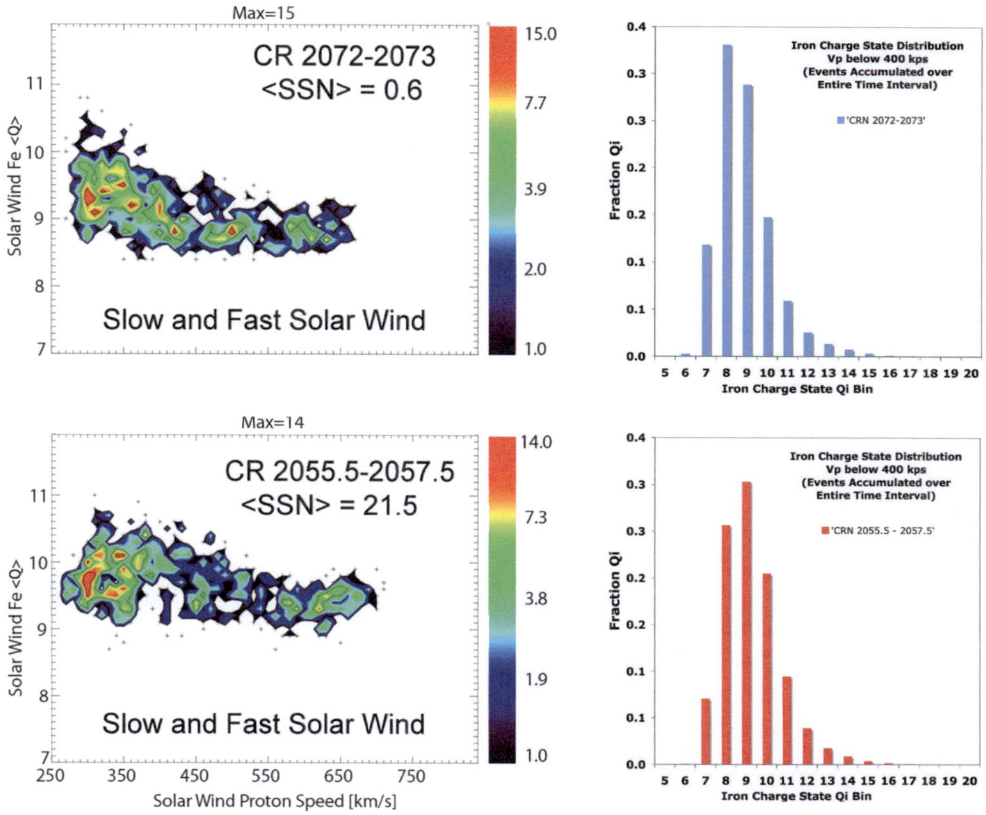

Fig. 11.7 STEREO A PLASTIC solar wind iron average charge state compared against solar wind speed. Subset of the fast and slow (non-transient) solar wind samples from Fig. 11.6. The two plots represent non-transient solar wind during intervals of relatively low and high sunspot numbers, respectively. *Left*: contour plots in same format as Fig. 11.6. *Right*: Fe charge state distributions for slow solar wind (proton speeds less than 400 km/s). Iron events were accumulated over the entire interval. (Data for Qi bin distributions courtesy M. Popecki.)

ber was 21.5 (hemisphere area of 192.4 E-5). The higher sunspot interval had transient events (including the May 2007 event), and these event periods have been removed for this sampling. As seen in the proton velocity profiles in Fig. 11.8, each interval contained fast and low speed solar wind. In Fig. 11.9, these two speed profiles are shown ballistically back-mapped to their estimated solar origins, illustrated by the corresponding SECCHI EUVI synoptic maps. The EUVI maps for both intervals (CRN 2072–2073 and CRN 2055.5–2057.5) show polar and isolated coronal holes. The second interval (CRN 2055.5–2057.5) is distinguished by the presence of many low latitude active regions.

The average proton speed for the low sunspot number contour was 419 $\pm$ 104 km/s, with an average iron charge state of 9.2 $\pm$ 0.4 charge units (range 8.4–10.9 units). For the more active period, the average speed (non-transient) was higher and slightly more variable at 434 $\pm$ 120 km/s, with an average iron charge state also higher at 9.8 $\pm$ 0.4 charge units (range of 8.7–11.0 units). The shift upwards in range of the iron charge state, while modest, is noticeable in the contour for the high sunspot number interval compared to the lower sunspot number interval. This is also illustrated for the slow solar wind by the charge state distribution for iron, provided to the right of each contour. The Fe Qi distribution is derived from the measured iron events accumulated over the entire sampling interval, limited to those events where the proton bulk speed was less than 400 km/s.

Coronal hole-associated solar wind can have a range of solar wind speeds, as can transients. Distinct from these is a general category of so-called slow solar wind. The slow solar wind is likely of multiple origins. Schwenn (2006) distinguishes two types of slow

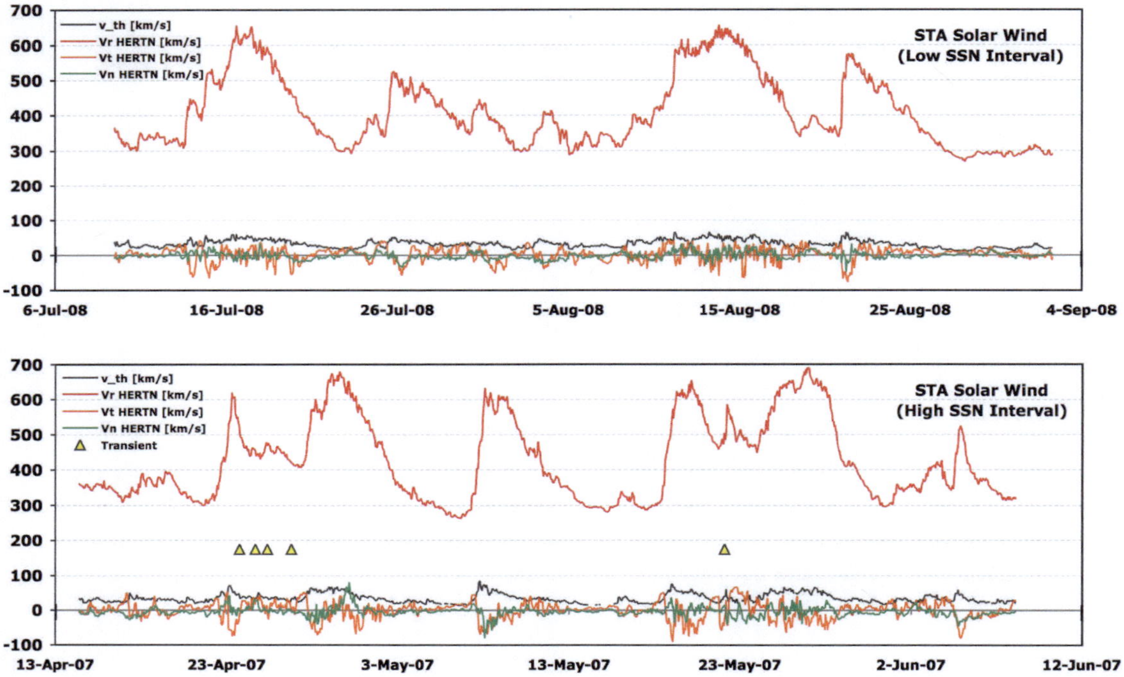

Fig. 11.8 STEREO A PLASTIC time profile of solar wind proton velocity components and thermal speed for the two contour plots shown in Fig. 11.7, representing periods with low and relatively high (within STEREO mission timeline standards) sun spot numbers. Transient events that were deleted from the corresponding slow and fast solar wind contours are marked

solar wind: that originating near the heliospheric current sheet (HCS), and that originating from above active regions. The early period of the STEREO mission characterized by higher solar activity likely contained both types of slow wind. As seen in the bottom panels of Fig. 11.9, many of the lower speed solar wind flows in the CRN 2055.5–2057.5 interval map back to the same longitudes as low latitude active regions. As the mission progressed, active regions became sparse (upper panels in Fig. 11.9) and the slow wind was likely associated with other sources. However, it must be emphasized that the sources of the slow solar wind are still a matter of active inquiry, and it is hoped that STEREO will contribute to resolving some of these issues.

11.2.2 Solar Wind Interfaces and Evolution

Having 1-AU spacecraft at multiple longitudes has greatly contributed to recent studies of solar wind correlation lengths (Podesta et al. 2008; Opitz et al. 2009) and the multipoint study of the evolution of the stream interfaces between slow and fast solar wind.

The transition from slow to fast solar wind, known as the co-rotation interaction region (CIR) with its characteristic compression region, is associated with the acceleration of particles and recurrent geomagnetic storms and as such is of broad interest. Jian (2009) and Simunac et al. (2009a) utilized two different techniques for identifying the interface between slow and fast solar wind at co-rotating interaction regions. By observing the same stream interface in succession at multiple spacecraft located at different longitudes (STEREO B, Wind, ACE, and STEREO A), the evolution of the interface and the shocks created by the stream interactions can be determined. Jian (2009) found that the CIR shocks that form by 1 AU are relatively weak and can vary significantly from spacecraft to spacecraft. Simunac et al. (2009a), concentrating on the predictability of stream co-rotation, found good agreement (within 5°) for spacecraft separated by no more than 20° in longitude. Beyond 20°, equivalent to about a day and half of solar rotation, the affects of different spacecraft latitudes and temporal evolution of the solar source (seen by SECCHI) become evident.

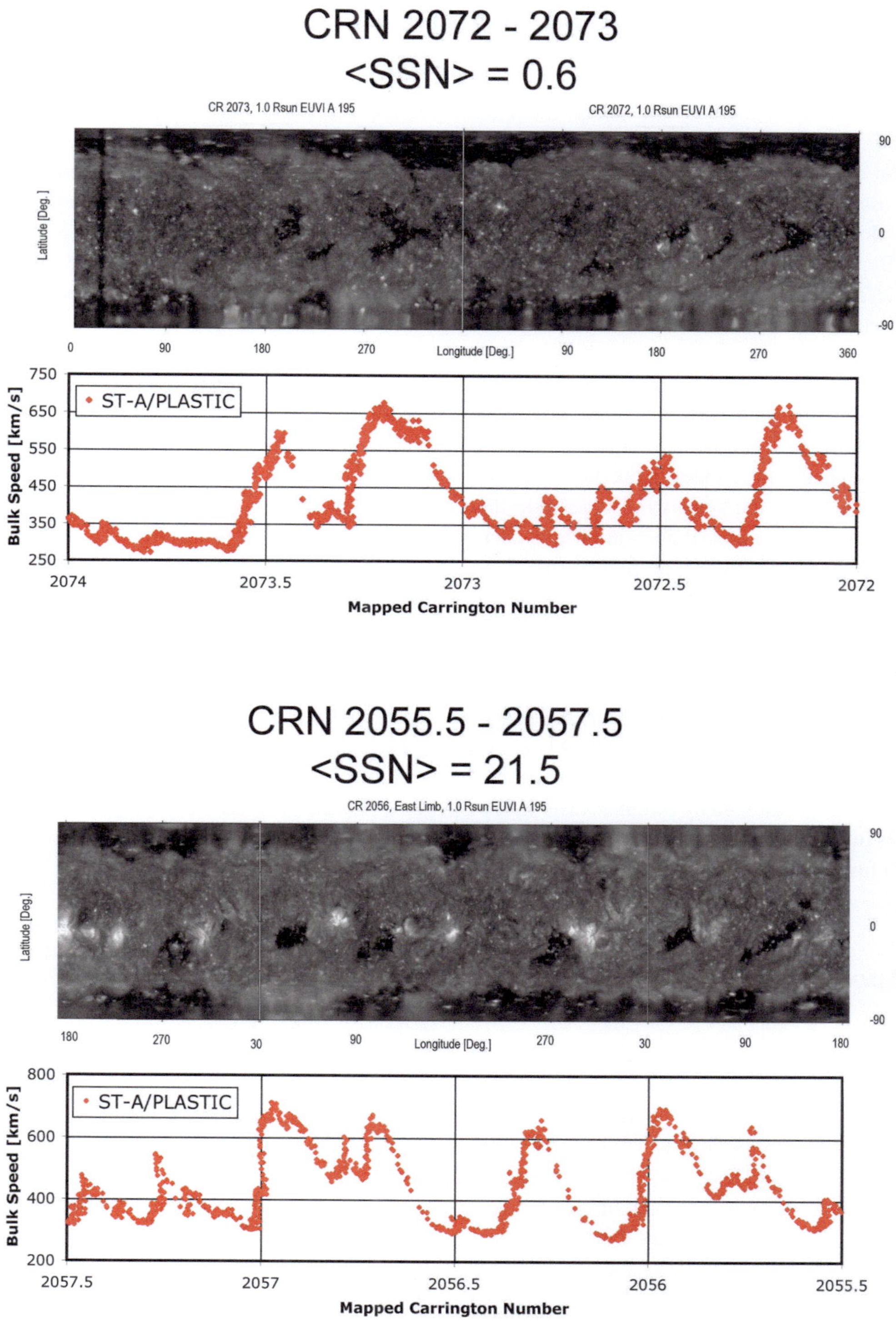

Fig. 11.9 Mapping solar wind to solar origins. STEREO A (STA) PLASTIC proton speeds for the periods provided in Fig. 11.8, back-mapped to their originating Carrington Rotation Number (CRN), or equivalently, Carrington Longitude. Above each speed profile is the corresponding solar synoptic maps from STEREO SECCHI EUVI. Note that time runs right to left in this format. (SECCHI maps courtesy R. Howard, SECCHI PI.)

The transition from fast to slow solar wind, associated with rarefaction regions, has not received as much attention. This region is important however, because (as noted by Burton et al. 1999) the solar wind speed profile is not distorted at this interface as it is by the acceleration and deceleration processes that occur in the compression region (an important factor in mapping the solar wind boundaries back to solar source regions). Burton et al. (1999) found the fast-to-slow stream interface could be identified by an abrupt drop in specific entropy, with coincident but more gradual changes in solar wind minor ion composition. Their Ulysses results were obtained at distances > 4.5 AU. Using solar wind plasma data from STEREO, Simunac et al. (2010) have given particular emphasis on possible correlations between changes in the proton specific entropy argument and the solar wind in-ecliptic flow direction for the fast-to-slow transition. The expectation was to find a discontinuous drop in entropy associated with the change in flow angle that was used to identify the point of transition. In many cases, however, they found that the change in flow angle took place while there was a plateau in the proton specific entropy argument. These results suggest that the sharp entropy drops observed in the outer heliosphere in the Ulysses study may sometimes develop outside of 1 AU. In all 20 cases examined by Simunac et al. (2010), the changes in the flow angle suggested a gradual transition from fast to slow solar wind near 1 AU.

11.2.3 Space Weather Implications

The age of STEREO is showing a maturity in advancing remote sensing and the joined analysis between remote observations and in situ solar wind measurements. Using the stereoscopic capabilities of SECCHI HI, Rouillard et al. (2009) were able to map the radial and latitudinal evolution of compressed solar wind inside a CIR. STEREO B was able to track such an event from the inner heliosphere to STEREO A at 1 AU. At STEREO A, the in situ instruments took over – the solar wind structure passage was confirmed by PLASTIC and IMPACT.

Co-rotating interaction regions (CIRs) are the cause of recurrent magnetic storms. In fact, although solar activity had been minimal during the STEREO mission thus far, certain types of geomagnetic activity were at high levels in 2008. As noted by Gibson et al. (2009) the relativistic electrons in the outer radiation belt were significantly increased (over a factor of 3) in 2008 over the 1996 solar minimum due to the larger number of high speed streams (Figs. 11.3 and 11.5) emanating from long-lived low latitude coronal holes in the recent minimum. The L5 Lagrange point is a notational location for a future solar wind monitor, in particular for detecting CIRs prior to their encounter with Earth. One of the serendipitous aspects of the steadily increasing longitudinal separation of the STEREO spacecraft is that they provide a platform for testing the usefulness of having a spacecraft parked at 60° off-longitude from Earth. This was tested by Simunac et al. (2009b), who obtained predictive results within 10% of the total co-rotation time between the observatories.

As seen in Fig. 11.5, the typical solar wind proton speed peaked in CR 2067 (March 2008) and has since been declining. This decline in solar wind speed may present new implications for the solar wind effects on geospace, from those reported already by Gibson et al. (2009).

In conclusion, although the Sun has been slumbering through most of the STEREO mission thus far, the STEREO community has been wide awake. The solar minimum has presented unusual conditions for which the STEREO and L1 vantage points and STEREO advanced instrumentation have been able to take full advantage.

Acknowledgements This work was funded at UNH under NASA contract NAS5-00132. The author thanks the Project Office at NASA GSFC, JHU/APL, and PLASTIC team members at UNH, CAU, MPE, GSFC, CESR, and U. Bern. Particular thanks to PLASTIC team members L. Ellis (programming), K. Simunac (PLASTIC Level 2 proton data), M. Popecki (PLASTIC Level 2 Fe Qi distributions), and K. Donahue (illustrations). We thank K. Ogilvie and A. Lazarus (Wind SWE), R. Howard (STEREO SECCHI), J. Luhmann (STEREO IMPACT), and J.L. Bougeret (STEREO SWAVES). CH models are from the Global Oscillation Network Group (GONG) program, managed by the National Solar Observatory, which is operated by AURA, Inc., under a cooperative agreement with the NSF. Solar indices are from the NOAA National Geophysical Data Center.

References

Bougeret JL, Goetz K, Kaiser ML, Bale SD, Kellogg PJ, Maksimovic M, Monge N, Monson SJ, Astier PL, Davy S et al (2007) S/WAVES: The radio and plasmawave investigation on the STEREO mission. Space Sci Rev 136:487–528

Burton ME, Neugebauer M, Crooker NU, von Steiger R, Smith EJ (1999) Identification of trailing edge solar wind stream interfaces: a comparison of Ulysses plasma and compositionmeasurements. J Geophys Res 104: 9925–9932

Galvin AB, Kohl JL (1999) Whole Sun Month at solar minimum: an introduction. J Geophys Res 104:9673–9678

Galvin AB, Kistler LM, Popecki MA, Farrugia CJ, Simunac KDC, Ellis L, Moebius E, Lee MA, Boehm M, Carroll J, et al (2007)The plasma and suprathermal ion composition (PLASTIC) investigation on the STEREO observatories. Space Sci Rev 136:437–486

Galvin A, Popecki M, Simunac K, Kistler L, Ellis L, Farrugia C (2009a) Solar wind characteristics during the current solar minimum (2007–2009) - STEREO PLASTIC observations, SOHO-23: Understanding a peculiar solar minimum. ASP Conference Series, ed. Steven R. Cranmer, J. Todd Hoeksema, and John L. Kohl, Vol. 438, 275–278, 2010

Galvin AB, Popecki MA, Simunac KDC, Kistler LM, Ellis L, Barry J, Berger L, Blush LM, Bochsler P, Farrugia CJ et al (2009b) Solar wind trends and signatures: STEREO PLASTIC observations approaching solar minimum. Ann Geophys 27:3909–3922

Gibson SE, Kozyra JU, de Toma G, Emery BA, Onsager T, Thompson BJ (2009) If the Sun is so quiet, why is the Earth ringing? A comparison of two solar minimum intervals. J Geophys Res 114:A09105. doi:10.1029/2009JA014342

Howard, R.A., Moses, J.D., Vourlidas, A., Newmark, J.S., Socker, D.G., Plunkett, S.P., Korendyke, C.M., Cook, J.W., Hurley, A., Davila, J.M., et al (2007) Sun Earth connection coronal and heliospheric investigation (SECCHI). Space Sci Rev 136:67–116

Jian L (2009) Level 3 Results of STEREO IMPACT/PLASTIC, http://www-ssc.igpp.ucla.edu/forms/stereo/stereo_level_3.html Cited 20 October 2009

Jian LK, Russell CT, Luhmann JG, Galvin AB, MacNeice PJ (2009) Multispacecraft observations: stream interactions and associated structures. Solar Phys 259:345–360

Kaiser ML, Kucera TA, Davila JM, St Cyr OC, Guhathakurta M, Christian E (2007) The STEREO mission: an introduction. Space Sci Rev 136:1–16

Kilpua EKJ, Liewer P, Farrugia C, Luhmann J,Mstl C, Li Y, Liu Y, Lynch B, Vourlidas A, Acuna M et al (2009a) Multispacecraft observations of magnetic clouds and their solar origins between 19 and 23 May 2007. Solar Phys 254:325–344. doi:10.1007/s11207-008-9300-y

Kilpua EKJ, Luhmann JG, Gosling J, Li Y, Elliott H, Russell CT, Jian L, Galvin AB, Larson D, Schroeder P, Simunac K, Petrie G (2009b) Small solar wind transients and their connection to the large-scale coronal structure. Solar Phys 256:327–344. doi:10.1007/s11207-009-9366-1

Kilpua EKJ, Pomoell J, Vainio R, Luhmann J, Li Y, Vourlidas A, Schroeder P, Galvin AB, Simunac K (2009c) STEREO observations of interplanetary coronal mass ejections and prominence dection during solar minimum period. Ann Geophys 27:4491–4503

Leitner M, Farrugia CF, Galvin A, Simunac KDC, Biernat HK, Osherovich VA (2009) The solar wind quasi-invariant observed by STEREO A and B at solar minimum 2007 and comparison with two other minima. Solar Phys 259:381–388

Luhmann JG, Curtis DW, Schroeder P, McCauley J, Lin RP, Larson DE, Bale SD, Sauvaud J-A, Aoustin C, Mewaldt RA et al (2007) STEREO IMPACT investigation goals, measurements, and data products overview. Space Sci Rev 136:117–184

McComas DJ, Ebert RW, Elliott HA, Goldstein BE, Gosling JT, Schwadron NA, Skoug RM (2008) Weaker solar wind from the polar coronal holes and the whole Sun. Geophys Res Lett 35:L18103. doi:10.1029/2008GL034896

Opitz A, Karrer R, Wurz P, Galvin AB, Bochsler P, Blush LM, Daoudi H, Ellis L, Farrugia CJ, Giammanco C et al (2009) Temporal evolution of the solar wind bulk velocity at solar minimum by cross-correlating the STEREO A and B measurements. Solar Phys 256:365–377. doi:10.1007/s11207-008-9304-7

Podesta JJ, Galvin AB, Farrugia CF (2008) Correlation length of solar wind fluctuations measured tangent to the earth's orbit: first results from STEREO. J Geophys Res 113(A09):104. doi:10.1029/2007JA012865

Richards MT, Rogers ML, Richards DStP (2009) Long-term variability in the length of the solar cycle. Pub Astron Soc Pac 121:797–809

Rouillard AP, Savani NP, Davies JA, Lavraud B, Forsyth RJ, Morley SK, Optiz A, Sheeley NR, Burlaga LF, Sauvaud J-A et al (2009) A multispacecraft analysis of a small-scale transient entrained by solar wind streams. Solar Phys 256:307–326. doi:10.1007/s11207-009-9329-6

Schwenn R (2006) Solar wind sources and their variations over the solar cycle. Space Sci Rev 124:51–76. doi:10.1007/s11214-006-9099-5

Simunac KDC, Kistler LM, Galvin AB, Lee MA, Popecki MA, Farrugia CJ, Mbius E, Blush LM, Bochsler P, Wurz P et al (2009a) In situ observations of solar wind stream interface evolution. Solar Phys 259:323–244. doi:10.1007/s11207-009-9393-y

Simunac KDC, Kistler LM, Galvin AB, Popecki MA, Farrugia CJ (2009b) In situ observations from STEREO/PLASTIC: a test for L5 space weather monitors. Ann Geophys 27:3805–3809

Simunac KDC, Galvin AB, Barry J, Farrugia C, Kistler LM, Kucharek H, Lee MA, Liu YC-M, Moebius E, Popecki MA (2010) Identifying the ends of high-speed streams near 1 AU with in situ data from STEREO/PLASTIC. In: Maksimovic M, Issautier K, Meyer-Vernet N, Moncuquet M, Pantellini F (eds) AIP Proceedings for the twelfth international solar wind conference, Saint-Malo, France, pp 351–354

Smith EJ, Balogh A (2008) Decrease in heliospheric magnetic flux in this solar minimum: recent Ulysses magnetic field observations. Geophys Res Lett 35:L22103. doi:10.1029/2008GL035345

Chapter 12

Shocks in the Heliosphere

Toshio Terasawa

Abstract Most of astrophysical shocks are collisionless, where the collective electromagnetic interaction plays the essential role in the dissipation process and results in a large deviation of particle phase-space distribution from the thermal equilibrium. An ideal laboratory for collisionless shocks is the heliosphere, in which shocks are formed ahead of coronal mass ejections (CME), ahead of planetary/cometary magnetospheres/ionospheres, around corotating interaction regions (CIR), and ahead of the heliopause. This review gives three topics about the nonthermal particle acceleration at these heliospheric shocks, *nonlinear reaction of shock acceleration, the maximum energy of shock-accelerated particles*, and the problem of *shock acceleration of electrons*.

12.1 Introduction

Shock waves are ubiquitous in the universe, ranging from mini-bow shocks ahead of lunar magnetic anomalies (Lin et al. 2007), to largest-scale shocks generated at the cosmological structure formation (Voit 2005). In these astrophysical shocks the two-body Coulomb collisions are relatively unimportant and collective interactions through long-range electromagnetic forces become dominant. The physical status of astrophysical plasmas, therefore, can be far from the thermodynamic equilibrium, producing 'nonthermal' particles, such as cosmic ray particles. In 1977–1978 several authors (Axford et al. 1977; Bell 1978; Blandford and Ostriker 1978; Krymsky 1977) published almost simultaneously their ideas on the stochastic shock acceleration of cosmic rays, now called 'diffusive shock acceleration' (DSA), a physical mechanism which is reviewed by Drury (1983); Jones and Ellison (1991); Toptyghin (1980).

It is generally believed that the cosmic ray particles (Fig. 12.1) up to $\sim 10^{15}$ eV are diffusively accelerated at supernova shocks within our galaxy (Schlickeiser 2002). On the other hand, ultra-high-energy cosmic rays (UHECR), cosmic ray particles above $\sim 10^{19}$ eV, are certainly of extragalactic origin (Hillas 1984; Nagano and Watson 2000; Norman et al. 1995). Candidate sites of UHECR acceleration are shocks formed in gamma ray bursts (GRB; Meszaros 2002; Vietri 1995), active galactic nuclei (AGN; Dermer et al. 2009; Rachen and Biermann 1993), and galaxy clusters (GC; Inoue et al. 2005). While these shocks in GRB/AGN/GC largely differ from heliospheric shocks in their relativistic speed (GRB and AGN), and in their cosmological time/spatial scale (GC), there are certain commonalities in basic physical mechanisms, where our understanding of heliospheric shocks could lead to new understanding shock physics in general.

Table 12.1 gives the typical parameters found in the heliosphere at (1) CME shocks, (2) the earth's bow shock, (3) CIR shocks, and (4) the solar wind termination shock. We look through these shocks in what follows.

(1) *CME shocks*: When some amount of magnetic energy is suddenly released on the solar surface, coronal mass ejection (CME) occurs. If the CME speed exceeds the local fast MHD wave speed in the solar corona, a fast MHD shock is created

T. Terasawa (✉)
Institute for Cosmic Ray Research, University of Tokyo, 5-1-5 Kashiwa-no-Ha, Kashiwa City Chiba, 277-8582, Japan
e-mail: terasawa@icrr.u-tokyo.ac.jp

M.P. Miralles, J. Sánchez Almeida (eds.), *The Sun, the Solar Wind, and the Heliosphere*, IAGA Special Sopron Book Series 4, DOI 10.1007/978-90-481-9787-3_12, © Springer Science+Business Media B.V. 2011

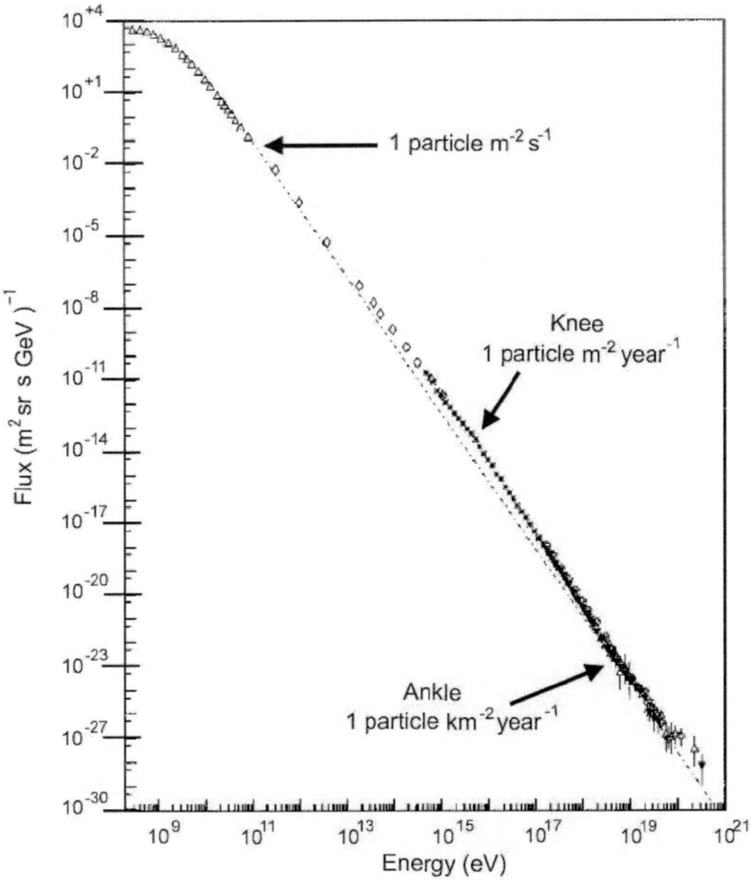

Fig. 12.1 The energy spectrum of cosmic rays for 10^8–10^{21} eV. For 10^{10}–10^{15} eV the energy spectrum has the energy dependence $E^{-2.7}$, while it has the slightly softer dependence ($\sim E^{-3.1}$) above 10^{15} eV. The highest energy edge of cosmic rays is found at $\sim 10^{20}$ eV. Below $\sim 10^{10}$ eV the solar wind modulates the energy spectrum significantly

Table 12.1 Examples of parameters at heliospheric shocks

Classification	Location/distance from the sun	Spatial scale	Shock velocity[a]	E^{b}_{acc}
CME shocks	Lower corona $\sim$ several AU	10^7–10^{14} cm	several tens of km s^{-1} $\sim$ 4,000km s^{-1}	10^6–10^{10} eV
Earth's bow shock	Ahead of the earth's magnetoshere	10^{10} cm	V^{c}_{sw}	10^4–10^5 eV
CIR shocks	Several AU	10^{13}–10^{14} cm	$\sim \Delta V_{sw}$	10^6–10^7 eV
Termination shock	$\sim$ 100 AU	10^{15} cm	V^{c}_{sw}	10^5eV (-10^8 eV?)[d]

[a]Here the shock velocities in the upstream plasma rest frames are shown. The minimum shock velocity is determined by the local fast MHD wave speed, which is $\sim$ several tens of km s^{-1}.
[b]Typical energies of protons diffusively accelerated at corresponding shocks.
[c]The solar wind velocity V_{sw} is in the range 200–800 km s^{-1}. Since both the earth's bow shock and the termination shock are standing in the solar wind flow, their shock velocities are given by V_{sw}.
[d]10^5eV is the energy of shock-accelerated pickup interstellar protons (Section 12.2.1, Fig.12.6a), while 10^8 eV is the energy of the anomalous component of cosmic rays. The latter has been believed to originate at the termination shock. However, recent Voyager observations have cast doubt on this belief (see text).

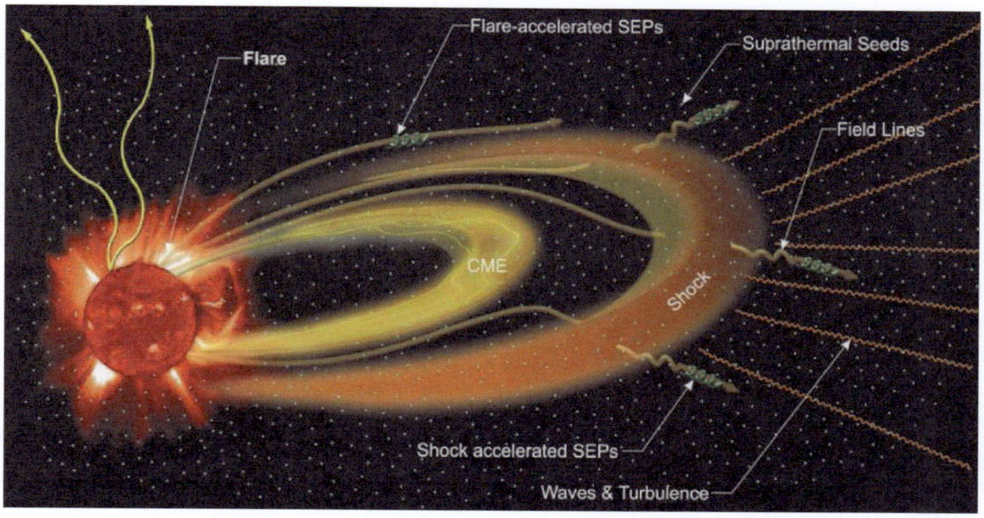

Fig. 12.2 An artist's view of a CME and the related interplanetary shock. (Reproduced from Fig. 6 of Desai and Burgess 2008, by permission of American Geophysical Union.)

ahead of the CME (Fig. 12.2). It is believed that solar energetic particles (SEPs) in gradual events are accelerated by these CME shocks along their propagation from the solar corona to the earth orbit and beyond (Lee 2005; Klecker et al. 2006; Reams 1999). I will discuss the nonlinear aspect of the acceleration process at CME shocks (Section 12.2.1).

The fastest CME shock velocity tabulated in Table 12.1 ($\sim$ 4,000 km s^{-1}) is from the the historical August 1972 event (Smart and Shea 1985). The energy $E_{\rm acc}$ of the SEPs in Table 12.1 ranges widely from 10^6 eV to 10^{10} eV: For usual CME shocks $E_{\rm acc}$ is 10^{6-8} eV, but at strongest CME shocks occurring a few times in one solar cycle $E_{\rm acc}$ reaches to $\sim 10^{9-10}$ eV or higher (Section 12.2.2).

(2) *The earth's bow shock*: Ahead of the objects standing in the supersonic solar wind, such as planetary/cometary magnetospheres/ionospheres, bow shocks (BSs) are formed (Treumann and Jaroschek 2008). In the earth's bow shock the stochastic acceleration of energetic ions (so-called diffuse ions. See Eastwood et al. 2005; Paschmann et al. 1981; Trattner et al. 1994) occurs in the noon-dawn foreshock region, while the shock-drift acceleration (e.g., Webb et al. 1983) of electrons occurs on the dusk side of the BS where the interplanetary magnetic field is in a quasi-perpendicular shock geometry.

Recent progress in the study of the earth's bow shock has been the realization of three-dimensional multipoint observations by the fleet of the Cluster spacecraft (e.g., Kis et al. 2004, 2007; Kronberg et al. 2009; Lucek et al. 2008). Through these observations one can differentiate spatial and temporal variations of physical quantities without assumptions. Such separations are essential in deriving the spatial diffusion coefficients K for diffuse ions (Fig. 12.3), where

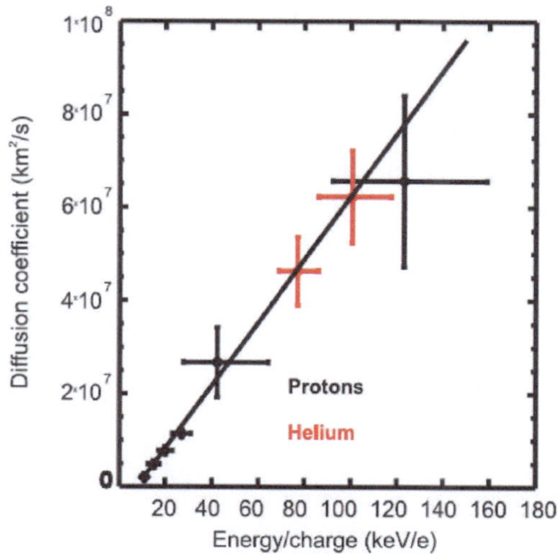

Fig. 12.3 Energy dependence of spatial diffusion coefficients, K, for the proton and helium components of diffuse ions derived from the Cluster spacecraft observation. For the horizontal axis, the energy-per-charge unit is used. (Reproduced from Fig. 5 of Kronberg et al. 2009, by permission of American Geophysical Union.)

a systematic energy dependence of K is seen. However, further studies are needed to understand the discrepancy between the observed energy dependence of K and the theoretical expectation based on the quasi-linear treatment of the foreshock turbulence (Kronberg et al. 2009).

(3) *CIR shocks*: A corotating interaction region (CIR) is formed by the solar rotation if the spatial inhomogeneity on the source surface on the sun causes a significant radial variation of the solar wind speed, ΔV_{SW}, which ranges from several tens of km s^{-1} to $\sim$500 km s^{-1}. At the interface between the slow and fast solar wind streams, the plasma and magnetic field are compressed, which eventually evolve to a forward-reverse shock pair, known as a CIR shock pair. It has been known that nucleons are accelerated up to several MeV/n at such CIR shocks (Barnes and Simpson 1976). As predicted, there is a strong association between long-lived coronal holes and CIR particle events (Fig. 12.4). However, even if the source coronal hole remains stable, transient variations in a CIR can be generated by the interaction with CMEs propagating through the interplanetary medium (Gómez-Herrero et al. 2009).

(4) *The solar wind termination shock*: Recently Voyager 1 and 2 spacecraft successively crossed the solar wind termination shock (TS) and provided detailed information on the nature of TS (Florinski et al. 2009a). It was a big surprise that the anomalous component of cosmic rays (ACR), which had long been attributed to the acceleration process at TS, does not stop increasing at TS but shows a continuous increase in the heliosheath region. The interpretation of this behavior of ACR is still an open question. Alternative acceleration processes for ACR being proposed are stochastic acceleration in the heliosheath turbulence (Fisk and Gloeckler 2009) and acceleration accompanying with magnetic reconnection process (Drake et al. 2010; Lazarian and Opher 2009).

12.2 Selected Topics

While the basic concept of shock acceleration process has been established, there are wide research areas in related phenomena which need further investigation. In this section I will review some of the selected topics. Further reviews about heliospheric shocks in general can be found in semi-classical monographs (Stone and Tsurutani 1985; Tsurutani and Stone 1985) and in a recent article (Schwartz 2006).

12.2.1 Nonlinear Reaction of Shock Acceleration

Acceleration at astrophysical shocks (those in supernova remnants, gamma ray bursts, etc.) is expected to be so efficient that the energy densities of accelerated particles can become comparable (or even exceed?) the thermal/magnetic energy densities of the background plasma. In such circumstance, the back-reaction of accelerated particles modifies the shock structure itself. This modification has been discussed in terms of "cosmic-ray-modified" or "cosmic-ray-

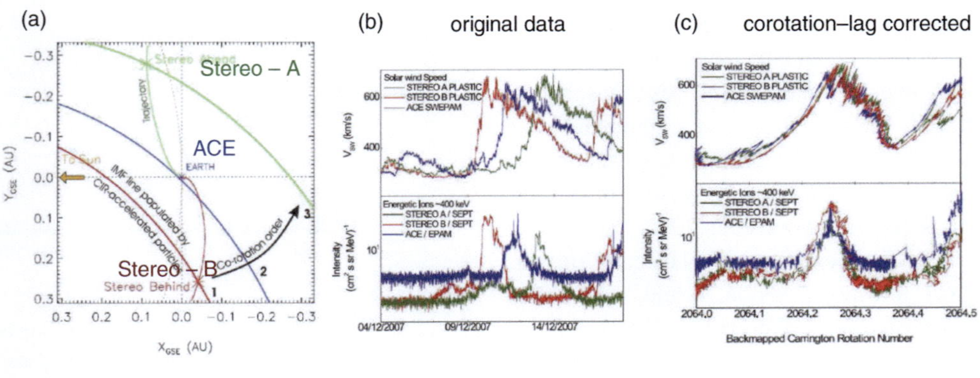

Fig. 12.4 **a** Geometry of three-spacecraft observation. **b** The solar wind speed (*upper panel*) and the energetic ion flux (*lower panel*) observed at three positions (STEREO-A:*green*, STEREO-B:*brown*, and ACE:*blue*). **c** The same as **b** but after the correction of corotation lag. (Reproduced from Figs. 2 and 6 of Gómez-Herrero et al. 2009, by permission of American Geophysical Union.)

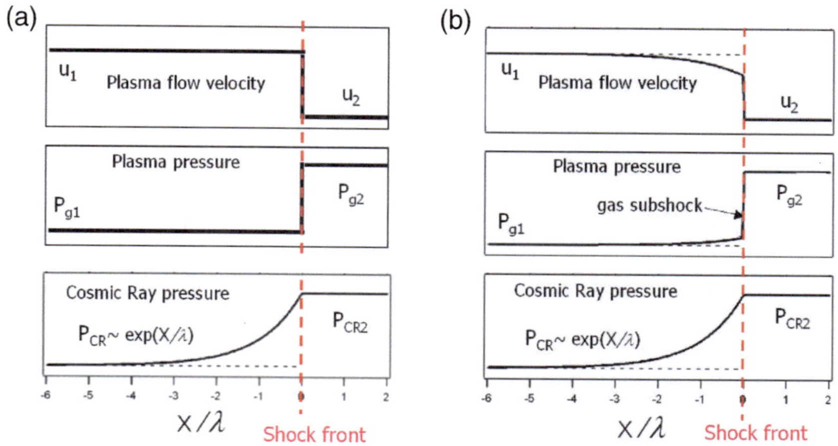

Fig. 12.5 **a** Structure of usual (non-cosmic-ray-mediated) shocks, where the upstream plasma velocity (*top panel*) and pressure (*middle panel*), u_1 and P_{g1}, jump to the downstream values, u_2 and P_{g2}, at the shock front ($X = 0$). The cosmic ray partial pressure (*bottom panel*) in the upstream region, on the other hand, shows an exponential increase toward the shock front with a spatial scale λ. **b** Structure of cosmic-ray-mediated shocks, where plasma velocity and pressure show a gradual change in the upstream region according to the nonlinear effect of the cosmic ray pressure gradient

mediated" shocks (CRMS) (Fig. 12.5, see Drury and Völd 1981; Malkov et al. 2001).

It is noted that studies of basic physics of CRMS can be made observationally at some of strong heliospheric shocks.[1] Fig. 12.6a shows the structure of the solar wind TS. The gradual decrease of the solar wind velocity from ~0.4 AU before the crossing of TS as well as the gradual increase of shock-accelerated interstellar pickup protons is what is expected at the CRMS.

Figure 12.6b shows the deceleration of solar wind speed in the BS foreshock region, where a positive correlation is seen between the amount of deceleration (vertical axis) and the normalized wave amplitude (horizontal axis: the proxy of the partial pressure of diffuse ions). This correlation substantiates that the nonlinear effect is important for the formation process of diffuse ions (Bonifazi et al. 1983; Terasawa 2005; Zhang et al. 1995).

Interplanetary CME shocks also show the CRMS property occasionally (Fig. 12.7). The energy densities of SEPs integrated over 60–230 keV reached 15% (Fig. 12.7a: 21 February 1994 case) or 6% (Fig. 12.7b: 29 October 2003 case) of the upstream solar wind kinetic energy density, where the solar wind velocity changed correspondingly (increased in the observers' frame, or decreased in the shock rest frame). Nonlinearity of acceleration at interplanetary CME shocks not only appears on the local plasma structure but also on the property of SEPs. It is known that there are energy-dependent limits of the intensity of SEPs which are interpreted to be the result of the nonlinear self-regulation of the shock acceleration process of SEPs (Reams and Ng 1998).

12.2.2 The Maximum Energy of Shock-Accelerated Particles

In most of gradual SEP events the highest energies of SEP, E_{max}, are found at several tens-100 MeV. However, a few times in a solar cycle, E_{max} reaches or even exceeds 1–10 GeV, so that the SEP of the highest energies can be detected as "ground level enhancements" (GLEs) by ground-based neutron monitor networks, which are originally designed to measure the galactic cosmic ray variation. Figure 12.8 shows a GLE event on 20 January 2005. At the bottom panel of Fig. 12.8, relativistic protons showed a rapid rise (<hour) and prompt decay (~3–4 h), while the SEPs of energy <~300 MeV showed a similar rapid rise but gradual decay. How to explain the rapid acceleration of GeV protons in such GLE events is a challenging problem. Based on the anisotropy observation McCracken et al. (2008) presented a two-pulse

[1] At the heliospheric shocks, the role of "cosmic rays" in CRMS is played by various components of energetic particles (interstellar pickup protons at TS, diffuse ions at BS, and SEPs at CME shocks).

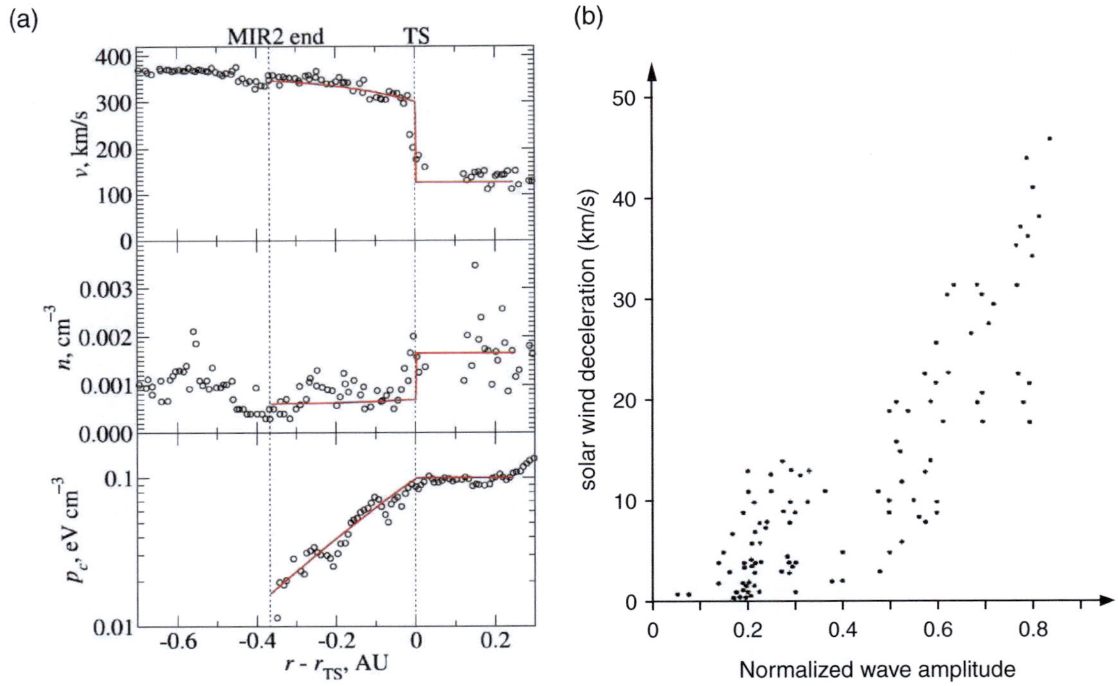

Fig. 12.6 **a** Solar wind speed (*top*), density (*middle*), and partial pressure of energetic ions (*bottom*) near the 2007 Voyager 2 TS (termination shock) encounter, where the horizontal axis shows the distance from the shock front. **b** Solar wind deceleration in the earth's BS (bow shock) foreshock region is plotted against the normalized wave amplitude, which is the proxy of the partial pressure of the diffuse ions. (Reproduced from Fig. 2 of Florinski et al. 2009b and Fig. 9 of Bonifazi et al. 1983 by permission of American Geophysical Union.)

acceleration model, where the first pulse was produced in the low corona near the flare loop where the initial energy release occurred (likely reconnection-related acceleration), and the second pulse was related to acceleration by the CME shock. Understanding the relative contributions of reconnection-related acceleration and shock acceleration to GLE events is an important unsolved issue. For the contribution of shock acceleration, there have been several theoretical efforts (Ng and Reams 2008; Vainio and Laitinen 2007; Zank et al. 2000). It seems necessary to develop a comprehensive acceleration model in which both the initial reconnection-related injection process and the follow-up shock acceleration process are coupled.

12.2.3 Shock Acceleration of Electrons?

Electron acceleration at astrophysical shocks is very efficient. For example, acceleration of TeV electrons occur within a few years at a supernova remnant RXJ1713.7-3946 (Uchiyama et al. 2007). On the other hand, the electron behavior at heliospheric shocks is not well understood (Treumann and Terasawa 2001). In a statistical study of 4 years of ACE data of transient interplanetary shocks (Lario et al. 2003) well-defined evidence of diffusive shock acceleration was found in 26 events for ions but only in 4 events for electrons. It is noted that the energy coverage of the statistical study (Lario et al. 2003) was limited to electrons $>$ 38 keV. One of the rare clear electron events is shown in Fig. 12.9 (Shimada et al. 1999), where electron acceleration at an interplanetary CME was seen from the suprathermal energy ($\sim$several tens of eV) to $\sim$40 keV. Further statistical surveys of electron events with energy coverage extending down to suprathermal energy are highly desirable.

12.3 Concluding Remarks

In this chapter I have briefly reviewed several topics of heliospheric shocks, which relate to particle accel-

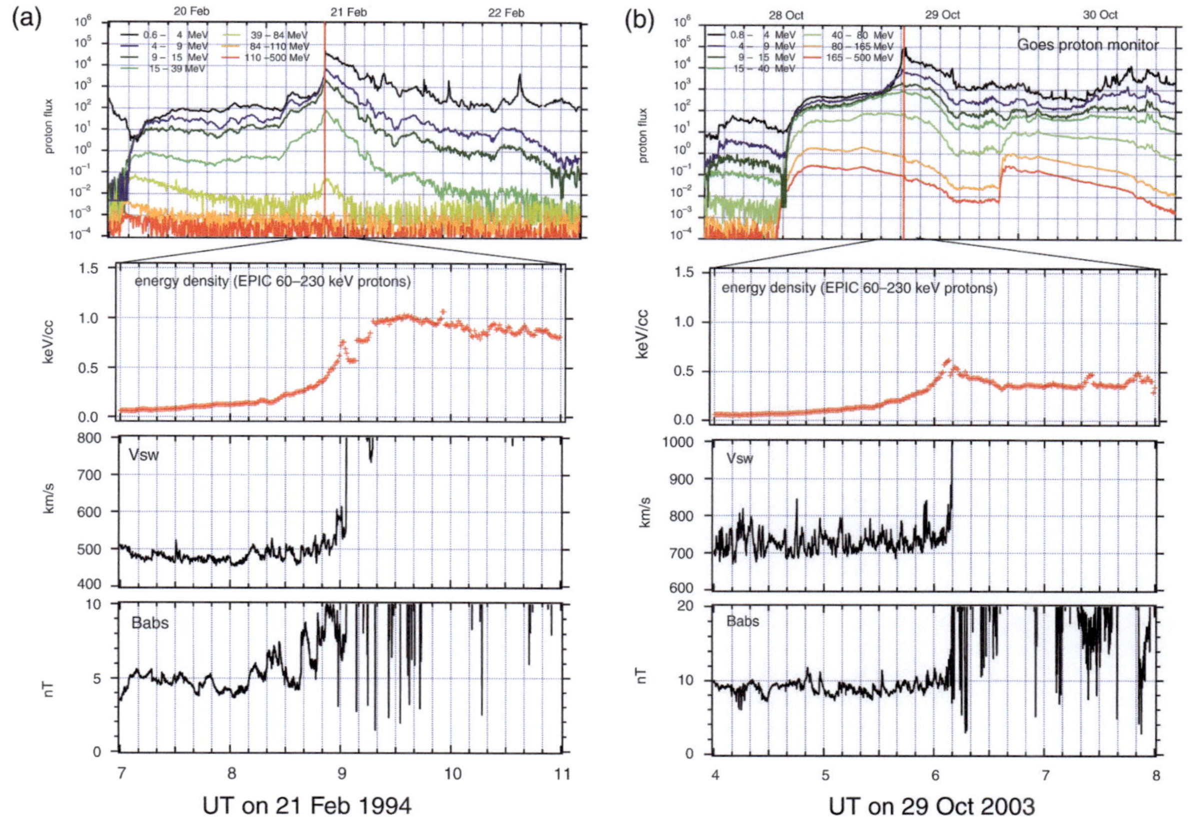

Fig. 12.7 **a** 1994 event between 19 and 23 February, during which a moderately strong interplanetary CME shock came on 21 February. From top, MeV proton fluxes obtained by the GOES particle monitor, energy density of energetic protons in the range of 60–230 keV, the solar wind velocity V_{SW}, and the magnitude of the magnetic field $|B|$. After the shock arrival V_{SW} and $|B|$ jumped up scaling out of the coverages of these panels. **b** 2003 event between 28 October and 1 November, where the format being the same as (**a**). (Reproduced from Figs. 1 and 2 of Terasawa et al. 2006 by permission of Committee on Space Research.)

eration phenomena. The first topic, the nonlinear reaction of shock acceleration, is one of the hottest topics in the field of astrophysical shocks, such as those in supernova and gamma ray bursts. Although the typical Mach numbers of heliospheric shocks are much smaller than astrophysical shocks, the study of the former shocks can give physical insight into the elementary physical processes governing the shock dynamics in the latter shocks. The second topic, the maximum energy of shock-accelerated particles, is another important topic in the study of astrophysical shocks. It is hoped that the study of heliospheric shocks leads to the understanding of the physical mechanisms which determine the maximum energies of galactic cosmic rays, and even those of extragalactic cosmic rays. Finally, the third topic, the electron acceleration by shocks, seems somewhat paradoxical: While electron acceleration at strong astrophysical shocks is obviously a common phenomenon, it is very rare at heliospheric shocks. One likely origin is the difference in Mach numbers as stated above. Further observational and theoretical study is needed to understand the behavior of electrons at these different shock environments.

Acknowledgements The author thanks Drs. R. Kataoka, K. Asano, K. Shibata, N. Nitta, R. Nakamura, W. Baumjohann, B. Klecker, and M. Scholer for valuable comments and discussions. This work is supported in part by the Grant-in-aid 21540259 for Scientific Research from the MEXT, Japan. Neutron monitors of the Bartol Research Institute are supported by NSF grant ATM-0527878.

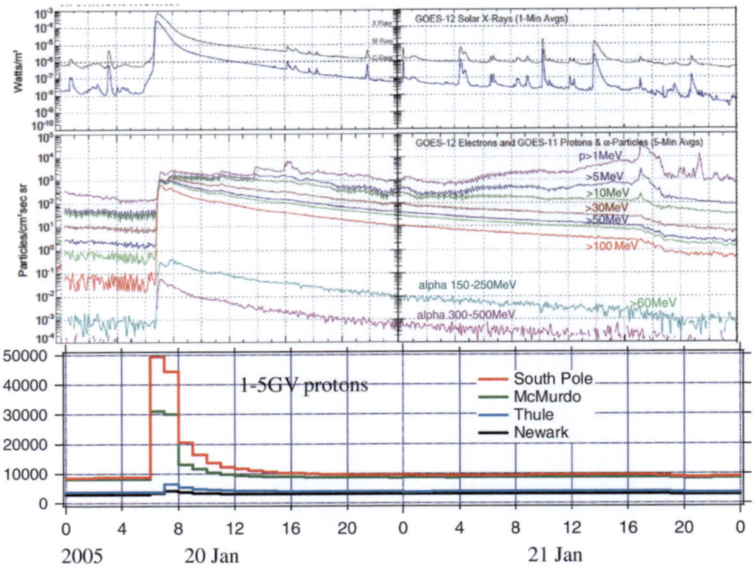

Fig. 12.8 The *top panel* shows the variation of solar soft X-rays. A strong X-ray flare of class X7.1 was seen peaking at 07:01 UT on 20 January 2005, whose Hα counterpart (2B) was identified at N12° W58°. The *middle panel* shows fluxes of protons and alpha particles below several hundred MeV (GOES data 2005). The *bottom panel* shows the hourly values of relativistic proton fluxes obtained by the ground neutron monitor network (Bartol 2005). It is noted that 1 h resolution in the *bottom panel* is not enough to resolve the initial increase of GeV protons, which occurred within ~10 min (Masson et al. 2009)

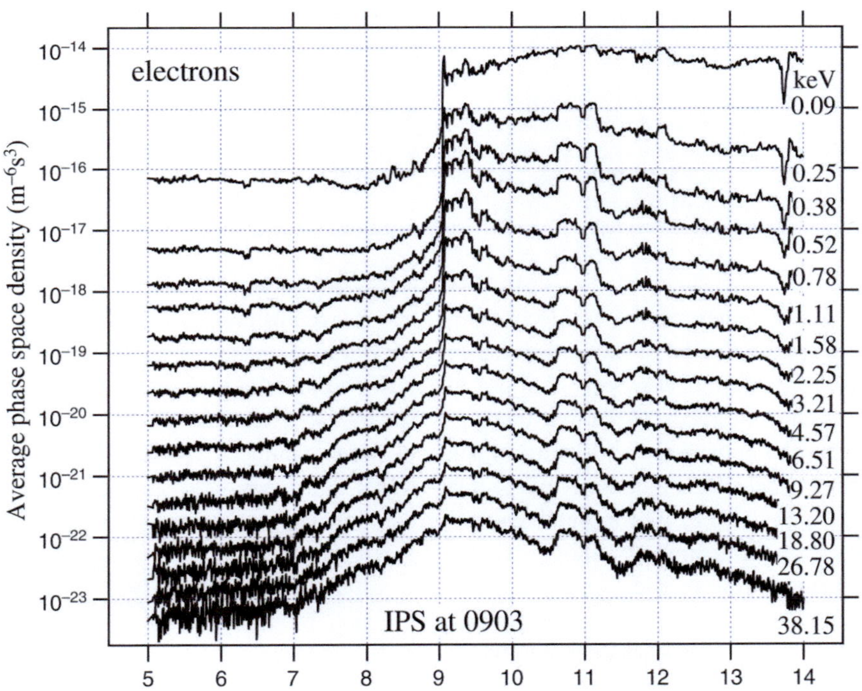

Fig. 12.9 The phase space densities of electrons averaged over the pitch angle during 05:00–14:00 UT on 21 February 1994, which is a subinterval of the event shown in Fig. 12.7a. (Reproduced from Fig. 2 of Shimada et al. 1999 by permission of Springer.)

References

Axford IA, Leer E, Skadron G (1977) The acceleration of cosmic rays by shock waves. Proc 15th Int Conf Cosmic Rays 11:132–137

Barnes CW, Simpson JA (1976) Evidence for interplanetary acceleration of nucleons in corotating interaction regions. Astrophys J 210:L91–L96

Bartol Research Institute Neutron monitor data (2005) Data list. From ftp://ftp.bartol.udel.edu/pyle/BRIData/BRI2005.txt

Bell AR (1978) The acceleration of cosmic rays in shock fronts, 1. Mon Not R Astron Soc 182:147–156

Blandford RD, Ostriker JP (1978) Particle acceleration by astrophysical shocks. Astrophys J 221:L29–L32

Bonifazi C, Moreno G, Russell CT et al (1983) Solar wind deceleration and MHD turbulence in the earth's foreshock region: ISEE 1 and 2 and IMP 8 observations. J Geophys Res 88:2029–2037

Dermer CD, Razzaque S, Finke JD et al (2009) Ultra-high-energy cosmic rays from black hole jets of radio galaxies. New J Phys 11:065016

Desai MI, Burgess D (2008) Particle acceleration at coronal mass ejection-driven interplanetary shocks and the earth's bow shock. J Geophys Res 113:A00B06

Drake JF, Opher M, Swisdak M et al (2010) A magnetic reconnection mechanism for the generation of anomalous cosmic rays. Astrophys J 709:963–974

Drury LOC (1983) An introduction to the theory of diffusive shock acceleration of energetic particles in tenuous plasmas. Rep Prog Phys 46:973–1027

Drury LOC, Völd HJ (1981) Hydromagnetic shock structure in the presence of cosmic rays. Astrophys J 248:344–351

Eastwood JP, Lucek EA, Mazelle C et al (2005) The foreshock. Space Sci Rev 118:41–94

Fisk LA, Gloeckler G (2009) The acceleration of anomalous cosmic rays by stochastic acceleration in the heliosheath. Adv Space Res 43:1471–1478

Florinski V, Balogh A, Jokipii JR et al (2009a) The dynamic heliosphere: outstanding issues. Space Sci Rev 143: 57–83

Florinski V, Decker RB, le Roux JA et al (2009b) An energetic-particle-mediated termination shock observed by Voyager 2. Geophys Res Lett 36:L12101

Geophysical Monograph 34 (1985) In: Stone RG, Tsurutani BT (eds) Collisionless shocks in the heliosphere: a tutorial review. American Geophysical Union, Washington, DC

Geophysical Monograph 35 (1985) In: Tsurutani BT, Stone RG (eds) Collisionless shocks in the heliosphere: Review of current research. American Geophysical Union, Washington, DC

GOES space environment monitor (2005) Data plot. From http://goes.ngdc.noaa.gov/data/plots/2005/GOES-200501.pdf

Gómez-Herrero R, Klassen A, Müler-Mellin R et al (2009) Recurrent CIR accelerated ions observed by STEREO SEPT. J Geophys Res 114:A05101

Hillas AM (1984) The origin of ultra-high-energy cosmic rays. Annu Rev Astron Astrophys 22:425–444

Inoue S, Aharonian FA, Sugiyama N (2005) Hard X-ray and gamma-ray emission induced by ultra-high-energy protons in cluster accretion shocks. Astrophys J 628:L9–L12

Jones FC, Ellison DC (1991) The plasma physics of shock acceleration. Space Sci Rev 58:259–346

Kis A, Scholer M, Klecker B et al (2004) Multi-spacecraft observations of diffuse ions upstream of earth's bow shock. Geophys Res Lett 31:L20801

Kis A, Scholer M, Klecker B et al (2007) Scattering of field-aligned beam ions upstream of earth's bow shock. Ann Geophys 25:785–799

Klecker B, Kunow H, Canve HV et al (2006) Energetic particle observations, Report of working group C. Space Sci Rev 123:217–250

Kronberg EA, Kis A, Klecker B et al (2009) Multipoint observations of ions in the 30–160 keV energy range upstream of the earth's bow shock. J Geophys Res 114:A03211

Krymsky GF (1977) A regular mechanism for accelerating charged particles at the shock front. Dokl Akad Nauk SSSR 234:1306–1308 (Soviet Phys Dokl 22:327–328)

Lario D, Ho GC, Decker RB et al (2003) ACE observations of energetic particles associated with transient interplanetary shocks. In: Velli M, Bruno R, Malara F (eds) Solar wind ten, AIP, Melville, NY, pp 640–643

Lazarian A, Opher M (2009) A model of acceleration of anomalous cosmic rays by reconnection in the heliosheath. Astrophys J 703:8–21

Lee MA (2005) Coupled hydromagnetic wave excitation and ion acceleration at an evolving coronal/interplanetary shock. Astrophys J Suppl Ser 158:38–67

Lin RP, Mitchell DL, Curtis DW et al (2007) Lunar surface magnetic fields and their interaction with the solar wind: Results from Lunar Prospector. Science 281:1480–1484

Lucek EA, Horbury TS, Dandouras I et al (2008) Cluster observations of the earth's quasi-parallel bow shock. J Geophys Res 113:A07S02

Malkov MA, Drury LOC (2001) Nonlinear theory of diffusive acceleration of particles by shock waves. Rep Prog Phys 64:429–481

Masson S, Klein KL, Bütikofer R et al (2009) Acceleration of relativistic protons during the 20 January 2005 flare and CME. Solar Phys 257:305–322

McCracken KG, Moraal H, Stoker PH (2008) Investigation of the multiplecomponent structure of the 20 January 2005 cosmic ray ground level enhancement. J Geophys Res 113: A12101

Meszaros P (2002) Theories of gamma-ray bursts. Annu Rev Astron Astrophys 40:137–169

Nagano M, Watson AA (2000) Observations and implications of the ultrahighenergy cosmic rays. Rev Mod Phys 72:689–732

Ng CK, Reams DV (2008) Shock acceleration of solar energetic protons: The first 10 minutes. Astrophys J 686:L123–L126

Norman CA, Melrose DB, Achterberg A (1995) The origin of cosmic rays above $10^{18.5}$ eV. Astrophys J 454:60–68

Paschmann G, Sckopke N, Papamastorakis I et al (1981) Characteristics of reflected and diffuse ions upstream from the earth's bow shock. J Geophys Res 86:4355–4364

Rachen JP, Biermann PL (1993) Extragalactic ultra-high energy cosmic rays I. Contribution from hot spots in FR-II radio galaxies. Astron Astrophys 272:161–175

Reams DV, Ng CK (1998) Streaming-limited intensities of solar energetic particles. Astrophys J 504:1002–1005

Reams DV (1999) Particle acceleration at the sun and in the heliosphere. Space Sci Rev 90:413–491

Schlickeiser R (2002) Cosmic ray astrophysics. Springer, Berlin
Schwartz SJ (2006) Shocks: commonalities in solar-terrestrial chains. Space Sci Rev 124:333–344
Shimada N, Terasawa T, Hoshino M et al (1999) Diffusive shock acceleration of electrons at an interplanetary shock observed on 21 Feb 1994. Astrophys Space Sci 264:481–488
Smart DF, Shea MA (1985) A simplified model for timing the arrival of solar-flare initiated shocks. J Geophys Res 90:183–190
Terasawa T (2005) GEOTAIL observation of solar wind and interplanetary phenomena. In: Hoshino M, Omura Y, Lanzerotti LJ (eds) Frontiers in magnetospheric plasma physics, COSPAR Colloquium Series 16:267–280
Terasawa T, Oka M, Nakata K et al (2006) 'Cosmic-ray-mediated' interplanetary shocks in 1994 and 2003. Adv Space Res 37:1408–1412
Toptyghin IN (1980) Acceleration of particles by shocks in a cosmic plasma. Space Sci Rev 26:157–213
Trattner KJ, Möbius E, Scholer M et al (1994) Statistical analysis of diffuse ion events upstream of the earth's bow shock. J Geophys Res 99:13389–13400
Treumann RA, Terasawa T (2001) Electron acceleration in the heliosphere. Space Sci Rev 99:135–150
Treumann RA, Jaroschek H (2008) Planetary bow shocks. arXiv:0808.1701v1 [astro-ph.EP]
Uchiyama Y, Aharonian FA, Tanaka T et al (2007) Extremely fast acceleration of cosmic rays in a supernova remnant. Nature 449:576–578
Vainio R, Laitinen T (2007) Monte Carlo simulations of coronal diffusive shock acceleration in self-generated turbulence. Astrophys J 658:622–630
Vietri M (1995) The acceleration of ultra-high-energy cosmic rays in gamma-ray bursts. Astrophys J 453:883–889
Voit GM (2005) Tracing cosmic evolution with clusters of galaxies. Rev Mod Phys 77:207–258
Webb GM, Axford WI, Terasawa T (1983) On the drift mechanism for energetic charged particles at shocks. Astrophys J 270:537–553
Zank GP, Rice WKM, Wu CC (2000) Particle acceleration and coronalmass ejection driven shocks: A theoretical model. J Geophys Res 105:25079–25095
Zhang TL, Schwingenschuh K, Russell CT (1995) A study of the solar wind deceleration in the earth's foreshock region. Adv Space Res 15:137–140

Chapter 13

Observations of the Termination Shock and Heliosheath

John D. Richardson

Abstract Voyagers 1 and 2 have both crossed the solar wind termination shock and are now in the heliosheath, the region of shocked solar wind plasma before the heliopause. This paper reviews observations of the termination foreshock, the termination shock (TS), and the heliosheath. The TS foreshock is characterized by anisotropic (streaming along the magnetic field) particles with energies of tens of keV to tens of MeV. The lower energy particles are accelerated at the TS. The particles are observed upstream of the TS when magnetic field lines connect the TS to the Voyager spacecraft. The TS is a weak shock, with a shock strength of about 2. The TS is the first observed particle mediated shock; the energetic particles cause the upstream solar wind to slow many days before the actual TS crossing. Three TS crossings were observed. Two of the TS crossings look like classic quasi-perpendicular supercritical shocks, but the third looked very different, with slowly varying N and V across the shock and two magnetic field ramps. These data have been interpreted as evidence that the TS is reforming upstream. Only about 20% of the flow energy at the TS was transferred to the thermal protons; the rest of this energy probably heats the pickup ions. The heliosheath is a highly variable region. Magnetic field and plasma parameters fluctuate by factors of up to five on scales of tens of minutes and the standard deviations of these parameters are large. The flows are away from the nose of the heliosphere, as expected, but the flow speeds differ at V1 and V2 and the speeds at both spacecraft differ from model predictions. Before the Voyager TS encounters, the TS was thought to be the source of the anomalous cosmic rays (ACRs), but at the Voyager crossing no evidence of the ACR source was observed, so these particles must be accelerated elsewhere on the TS or elsewhere in the heliosphere.

13.1 Introduction

This paper reviews the Voyager observations of the termination shock (TS) and heliosheath. Although lots of important theoretical and modeling work have been done, this paper focuses mainly on observations and so includes theory and model results only when they directly explain the observations. Soon after Parker proposed a supersonic solar wind, he and others realized a termination shock must exist [Parker, 1963]. The distance to the heliospheric boundaries are determined by pressure balance between the solar wind and the local interstellar cloud (LIC). Many estimates were made of the distance the TS boundary, hampered by the lack of knowledge of the pressure of the interstellar medium.

Models provided a picture of the heliospheric structure which was lent credence by Hubble Space telescope pictures of other astrospheres. Figure 13.1 shows a schematic diagram of the heliosphere. The colors in the top panel show the plasma temperature and the arrows show the flow streamlines. The LIC flows to the left at 26.4 km/s relative to the Sun. The solar wind moves radially outward from the Sun. Since these plasmas are magnetized, they cannot flow through each other, and a boundary forms between the solar wind and the LIC which is called the heliopause. The

J.D. Richardson (✉)
Kavli Center for Astrophysics and Space Science,
Massachusetts Institute of Technology, Cambridge, MA, USA
e-mail: jdr@space.mit.edu

M.P. Miralles, J. Sánchez Almeida (eds.), *The Sun, the Solar Wind, and the Heliosphere*, IAGA Special Sopron Book Series 4, DOI 10.1007/978-90-481-9787-3_13, © Springer Science+Business Media B.V. 2011

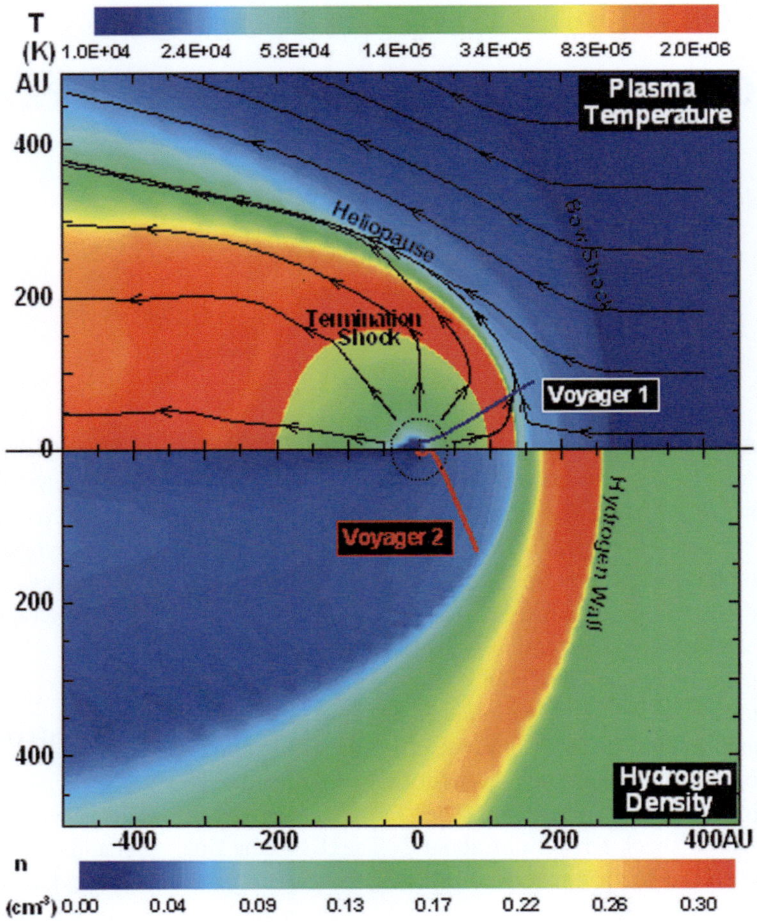

Fig. 13.1 A plot of the heliosphere from a plasma (*top*) and neutral (*bottom*) perspective. The figure shows the equatorial plane. The color bar on the *top panel* shows the plasma temperature. The *lines* show the plasma flow. The main boundaries, the termination shock, heliopause, and bow shock are labeled. The color bar on the *bottom panel* shows the H density; the hydrogen wall in front of the heliopause is labeled and the trajectories of the Voyager spacecraft are shown. Figure courtesy of H. Müller

heliopause (HP) is probably a rotational discontinuity analogous to a planetary magnetopause. The TS is where the solar wind becomes subsonic as it senses the LIC and starts to divert down the heliotail. The region of shocked solar wind between the TS and HP is the heliosheath; Voyager 1 (V1) crossed the TS in 2004 and Voyager 2 (V2) crossed it in 2007; both spacecraft remain in the heliosheath. The LIC may be supersonic with respect to the Sun; if it were, a bow shock would form in the LIC ahead of the HP.

The color bar in the bottom panel of Fig. 13.1 shows the neutral H density. Roughly 2/3 of the LIC is neutral and 1/3 plasma. These LIC neutrals are not bound by the magnetic field and move into the heliosphere, where most become ionized, are accelerated to the solar wind energy, and form a population of hot (1 keV) pickup ions. Before the Voyagers crossed the TS, the generally accepted paradigm was that the anomalous cosmic rays, singly ionized particles with energies 20–100 MeV, were pickup ions accelerated at the TS and then transported to the inner heliosphere.

The hydrogen wall labelled in Fig. 13.1 is near the nose of the heliosphere; the plasma slows down in this region as it approaches the HP. The ions and neutrals are coupled via charge exchange, so the neutrals slow as well, creating the denser hydrogen wall. This wall is observed in Lyman-alpha radiation both in our heliosphere and in other astrospheres [Linsky and Wood 1996; Wood et al. 1996]. The Voyager spacecraft were launched in 1977 and by mid-2009 were at 108 (V1) and 90 AU (V2). V1 is at about 35°N heliolatitude and V2 at 30°S. Both have a complement of scientific instruments to measure the magnetic field, plasmas (on V2, the V1 plasma detector no longer works), particles, and waves. This paper will first discuss the TS precursors, then the TS crossing, the energetic particles observations near the TS, and finally the data from the heliosheath.

13.2 The Termination Shock Foreshock

The HP and TS locations are determined by pressure balance between the solar wind and LIC plasmas. We measure the solar wind dynamic pressure, which ranges from 1–4 nP. However, we cannot measure the LIC densities or magnetic field and so the TS location was the subject of much speculation and prediction before V1 crossed it in 2004 at 94 AU. Figure 13.2 shows one of the early precursors to the TS crossing. In mid-2002, the particle fluxes on V1 increased by a factor of 20–30 over those observed by V2. Since V1 was 18 AU further out, a TS source was a natural hypothesis. The V1 fluxes remained enhanced for about 6 months, decreased in early 2003, then increased later in 2003. The particles were observed at energies from tens of keV to tens of MeV and streamed along the magnetic field, primarily in the direction outward from the Sun. This outward streaming led to the suggestion that the increase in particle flux signaled the TS crossing, with another crossing back into the solar wind in early 2003 (Krimigis et al. 2003). Counterarguments to this hypothesis were that the magnetic field strength did not increase when the particles increased at the first "crossing", as expected in the heliosheath, and that the

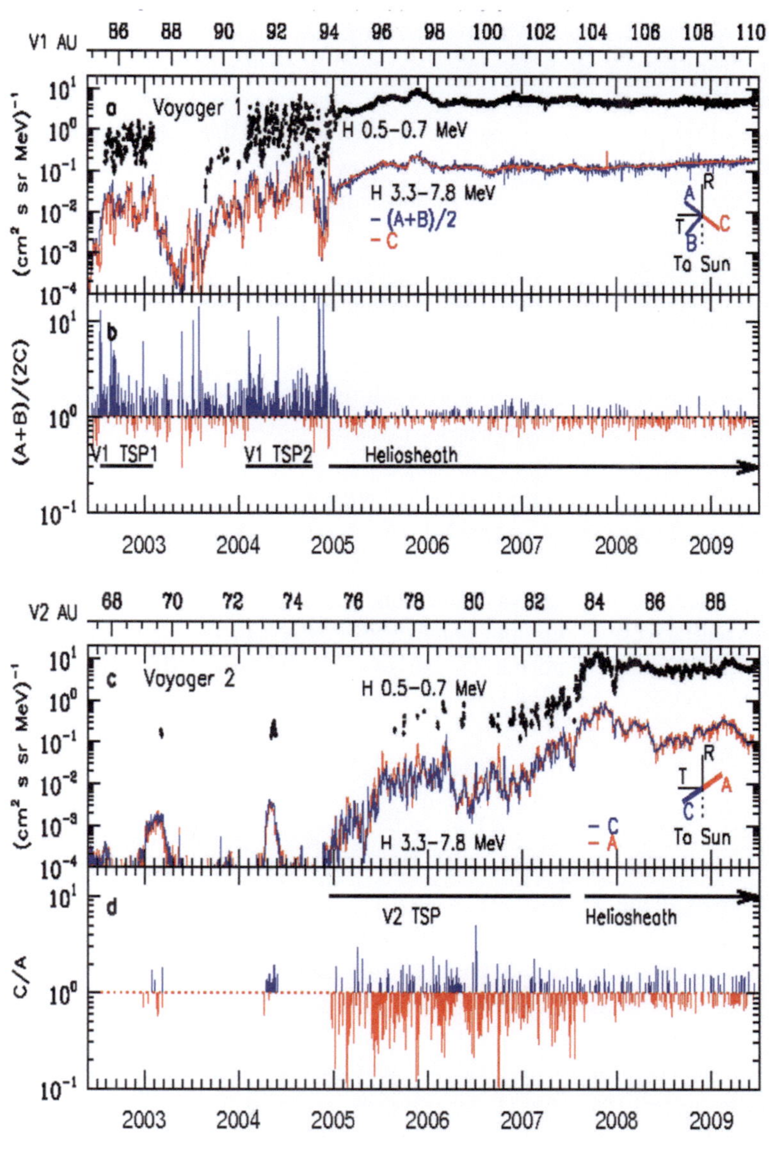

Fig. 13.2 Overview of the Cosmic Ray Subsystem (CRS) data near the V1 and V2 TS crossings. The *top panel* of each plot shows two ion energy ranges, 0.5–0.7 and 3.3–7.8 MeV. The 3.3–7.8 ions are split into fluxes in the roughly T and –T directions. The *bottom panels* show that upstream of the TS, the ions at V1 flow in the T direction and those at V2 in the –T direction, in both cases away from the nose of the heliosphere. The foreshock (TSP) regions are labeled for each spacecraft. In the heliosheath, downstream of the TS, the particles are nearly isotropic. Figure courtesy of A. Cummings and E. Stone

field increased when the particle flux decreased at the second "crossing", the opposite of what is expected when going from the heliosheath to the solar wind (Burlaga et al. 2003). Another counterargument was that the energy spectra of the ACR particles remained modulated, whereas it was expected that the spectra would unfold to a power law at their source (McDonald et al. 2003).

Soon it was realized that these streaming particles signaled that V1 was in a foreshock region, analogous to planetary foreshocks, where field lines are connected to the TS (Jokipii et al. 2004). Models predict that the TS is blunt (see Zank 1999), so that spherical magnetic field lines intersect the TS first toward the nose of the heliosphere. Since V1 and V2 are on opposite sides of the nose, V1 observes particles streaming from the TS along the magnetic field lines in the direction away from the Sun, while V2 on the other side of the nose observes particles streaming from the opposite direction, along field lines toward the Sun (Decker et al. 2006). Figure 13.2 shows particle fluxes and anisotropies observed by V1 and V2; when V2 entered the foreshock region in 2005 it did see particles streaming in the opposite direction as predicted (Decker et al. 2008; Stone et al. 2008).

Additional TS precursors were the electron plasma oscillations observed by V1 and V2 in the foreshock (Gurnett and Kurth 2005). These waves are generated when magnetic field lines are tangent to a shock, in this case the TS; electrons flowing along field lines from the TS excite the waves. V1 saw six episodes of these waves from 91–94 AU. The last episode, the day before the TS crossing, was coincident with streaming ions and electrons. Only one wave event was observed at V2, 30 days before the TS crossing (Gurnett and Kurth 2008).

13.3 The Termination Shock

V1 crossed the TS in 2004 at 94 AU, revealing for the first time the size of the heliosphere. At the TS, the fluxes of particles with energies from tens of keV to few MeV peaked, indicating that the TS was the source of these particles. The higher-energy ACR particles do not peak at the TS; the ACR fluxes at the shock are less than some fluxes observed in the foreshock (Decker et al. 2005). Contrary to expectations, the ACR spectra did not unroll to a power law, so the TS is not the ACR source, at least where crossed by V1 (Stone et al. 2005). The plasma instrument on V1 is not working, but the Low-Energy Charged Particle (LECP) instrument team uses the Compton-Getting effect to derive the radial and tangential solar wind speeds at V1 (Decker et al. 2005). The radial speed decreased at the TS and the tangential speed increased as expected. The particle fluxes are nearly isotropic in the heliosheath, not streaming along the field lines as in the foreshock. The magnetic field strength increased at the TS by roughly a factor of 3 and the standard deviation of the field also increased, as expected, in the heliosheath. The December 2004 data clearly showed a TS crossing, although the actual crossing occurred in a data gap.

At about the same time V1 crossed the TS at 94 AU, V2 entered the foreshock at 75 AU, 10 AU closer to the Sun than V1 first observed the foreshock. This difference implied an asymmetric heliosphere; either the foreshock region was much thicker in the V2 direction or the TS was closer in that direction (or both). Voyager 2 crossed the TS in August 2007 at 84 AU, confirming that the heliosphere is asymmetric (Burlaga et al. 2008; Decker et al. 2008, Richardson 2008; Stone et al. 2008). Figure 13.3 shows the V2 data near the TS. The speed began to decrease about 80 days before the TS, from 400 km/s to 300 km/s at the TS crossing, then decreased by a factor of 2 at the TS. The density increased by a factor of 2 at the TS and became highly variable in the heliosheath. The temperature increased by a factor of 10 from 10^4 to 10^5K. The flow deflected away from the nose of the heliosphere, as expected. The magnetic field went up by a factor of 2 at the TS and the standard deviation of the field increased (Burlaga et al. 2008).

The first crossing of the TS by V2 occurred in a data gap. But then the TS moved outward, past V2, giving the first direct observations of the TS; three TS crossings occurred while V2 was being tracked and at least two others in data gaps. Figure 13.4 shows the three TS crossings. The second crossing was a classic quasi-perpendicular super-critical shock (Burlaga et al. 2008; Richardson et al. 2008). Ahead of the TS was the foot region, formed by ions reflecting from the TS, in which the speed decreases and magnetic field increases. The main shock jump occurs a few minutes later at the ramp. The magnetic field has an enhanced magnitude region after the ramp called the

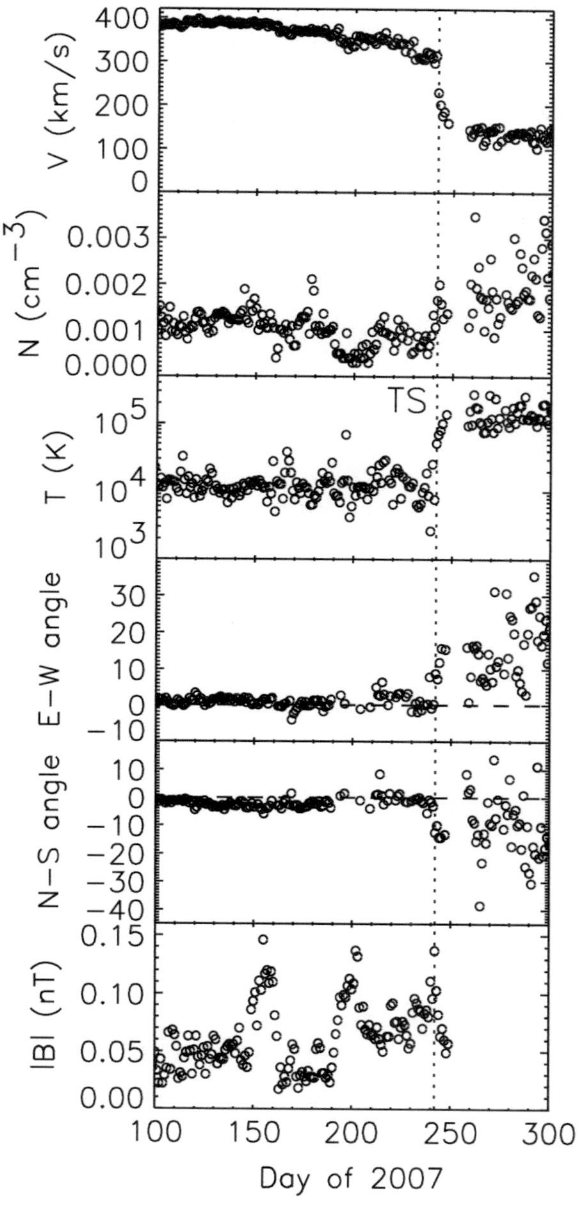

Fig. 13.3 Overview of the TS crossing (from Richardson et al. 2008). The speed decreases from 400 to 300 km/s before the shock. At the shock the speed decreases and N and T increase. The flows divert away from the nose of the heliosphere

overshoot, followed by an undershoot with low field, then oscillations in field strength. The TS at 243.82 was very different. The speed increased slowly over 20 min, the density decreased monotonically, and the magnetic field had two jumps that looked like ramp regions. These data were interpreted as a shock which was reforming (Burlaga et al. 2008); theoretical work suggests that quasi-perpendicular super-critical shocks reform a gyroradius upstream, and the TS seems to be the first shock at which this process is observed. The TS is clearly dynamic, changing in structure over time scales of a few hours.

Broadband electrostatic noise was predicted to occur at the TS crossing. Plasma waves were observed at tens of Hz coincident with the second two TS crossings (Gurnett and Kurth 2008). Emissions were also observed at 1,645 on day 243, when other instruments did not observe a shock. No electrostatic noise was observed at the TS crossing at 243.82, perhaps because this shock was reforming.

Figure 13.3 shows that the speed decreased before the TS; this decrease occurred in discrete steps. The last half of the decrease, from 350 to 310 km/s, is coincident with an increase in the tens of keV energy particle intensity (Florinski et al. 2009). The inward pressure gradient which results is sufficient to slow the solar wind as observed. Thus V2 observed for the first time a particle mediated shock, where the particles accelerated at the shock move upstream and change the shock structure. Shocks mediated by ACRs and other particles had been hypothesized, but never before observed.

The thermal plasma is heated to 10^5K at the TS; however, this heating accounts for only 15% of the flow energy dissipated at the TS. The ions gain so little energy that the heliosheath plasma is still supersonic with respect to the the thermal ions downstream of the TS (Richardson 2008). The flow energy instead heats the pickup ions, as predicted by Zank et al. (1996) for a plasma thermally dominated by pickup ions. The V1 LECP energy spectra, if extrapolated to pickup ion energies, require that 80% of the flow energy go into the pickup ions, consistent with the V2 plasma data (Gloeckler et al. 2005). The wave speed in the heliosheath is then determined by the heated pickup ions and the overall plasma is subsonic, as it must be downstream of the shock.

These observations show that the TS is asymmetric; one way to create such an asymmetry is if the LIC magnetic field is not aligned with the LIC flow. Fig. 13.5 shows qualitatively how the tilt in the LIC field leads to enhanced field strength in the south as the field drapes over the heliosphere, compressing the heliosphere in the south (Opher et al. 2007). The observed TS crossings were 10 AU apart in radial distance. The TS also moves due to changes in the solar wind pressure; models show the V1 TS distance would be 2–3 AU closer

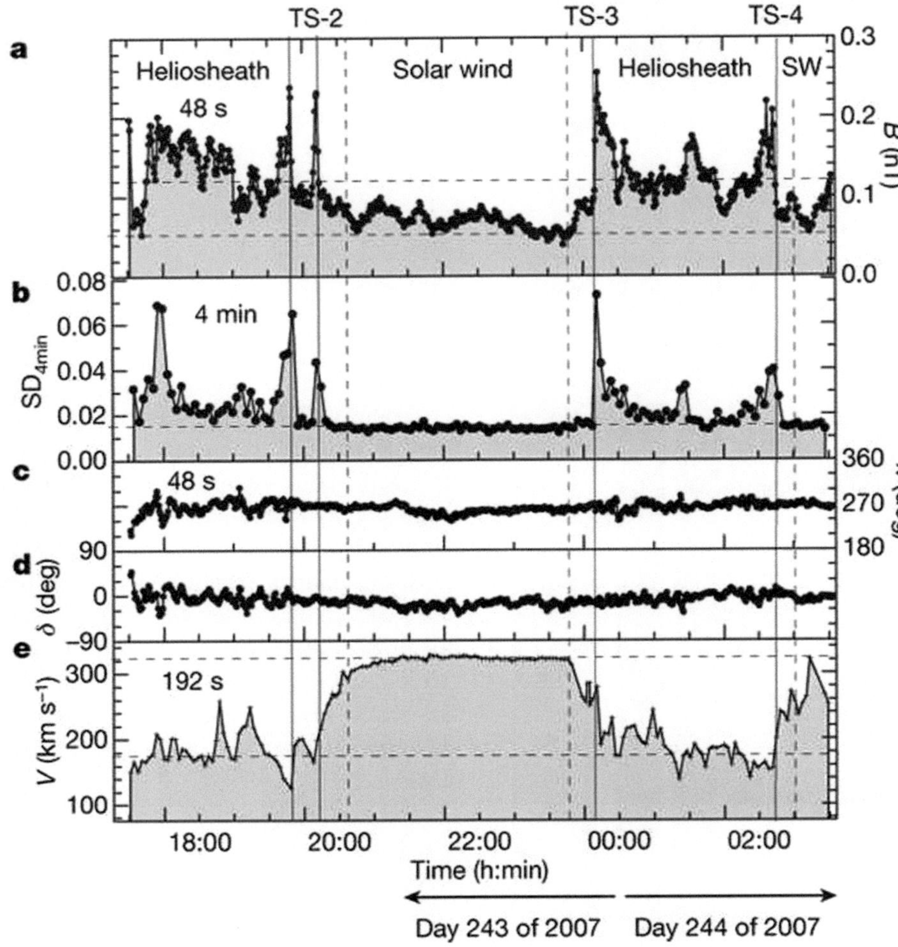

Fig. 13.4 A detailed view of the TS crossing showing the magnetic field magnitude, standard deviation, azimuthal and elevation angles, and the plasma speed (from Burlaga et al. 2008). The last two TS crossings look like standard quasi-perpendicular, supercritical shocks whereas the first shock crossing has a slowly varying speed and two magnetic field peaks. Burlaga et al. (2008) suggest that the TS is reforming upstream in this case

when V2 crossed the TS, giving an asymmetry of 7–8 AU (Richardson 2008). Models indicate that this large an asymmetry requires a large LIC field, of order 4 μG (Pogorelov et al. 2009; Opher et al. 2009). This large a LIC field would probably result in a high enough LIC Alfvén speed that the LIC flow would be subsonic, so no bow shock would form in the LIC flow upstream of the HP.

13.4 Energetic Particles

Figure 13.6 shows the three energetic particle populations observed in the heliosheath (Decker et al. 2008). The termination shock particles (TSP) protons, discovered in the foreshock, have energies less than 10 MeV, the ACRs have energies from 10–150 MeV, and the galactic cosmic rays (GCRs) are at higher energies. The figure shows the ratios of particle intensities at V2 compared to those at V1 after the V2 TS crossing, when V1 was deep in the heliosheath. The TSP intensities are larger at V2 than V1, consistent with the TS being the source of the TSPs and the intensities decreasing away from the source. The ACR intensities are higher at V1 than V2, so the ACR source is not near the V2 TS crossing. The GCRs and highest energy ACRs have equal fluxes at V1 and V2, indicating that these particles are not modulated in the heliosheath. The bottom panel shows that the He profile looks

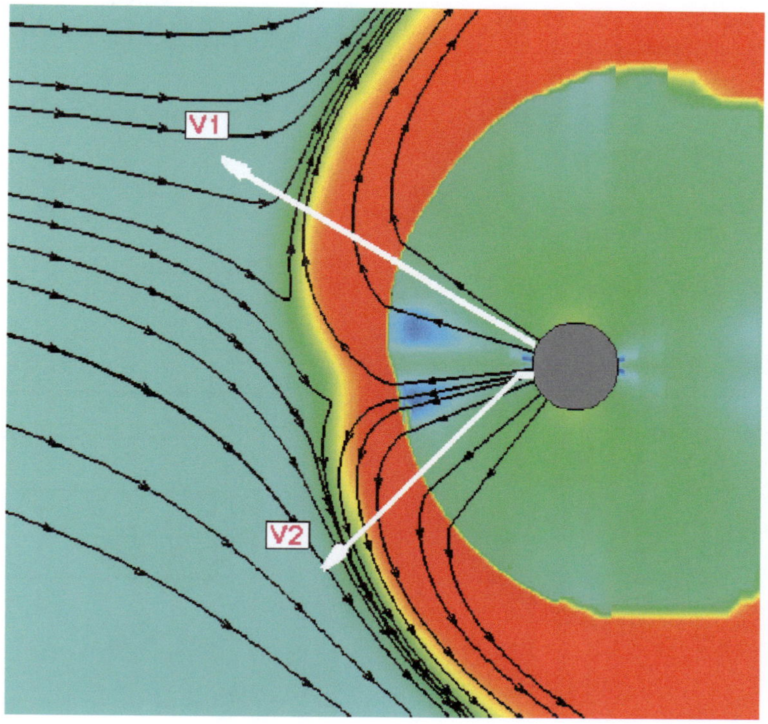

Fig. 13.5 A model of the heliospheric configuration with the LIC magnetic field tilted from the LIC flow direction. The colors show plasma temperature, with *green* cool and *red* hot, and the *arrows* the magnetic field direction. The field lines drape around the heliosphere in the south, compressing the heliosphere in the south compared to the north. The *white lines* show the Voyager trajectories. Figure courtesy of M. Opher

similar to the proton profile. The dotted lines in each panel show the He or H profiles shifted in energy to account for the mass difference. These shifted profiles are identical, so the modulation and sources for He and H are the same.

Figure 13.7 shows the He spectra observed at V1 and V2 as a function of time. The first two panels show the spectra just before and after the V1 TS crossing. The lowest energy TSP fluxes increase across the TS, but the ACR particle intensities do not change, contrary to pre-Voyager predictions (Stone et al. 2005, 2008). The dotted line shows the profile expected at the particle source. After the V1 TS crossing the V2 TSP flux begins to increase. The V2 TSP flux continues to increase and after the V2 TS crossing (panel 4) it is a factor of 3 above the peak V1 flux at the lowest energies. The ACR flux also did not change at the V2 TS crossing. The ACR fluxes have continued to increase with distance at both V1 and V2, and the power spectra have continued to unroll towards a power law. At the highest energies the ACRs are not modulated.

Before the Voyagers crossed the TS, the ACR source was thought to be the TS accelerating pickup ions formed from interstellar neutrals. With the falsification of this hypothesis by V1 and V2, other theories of ACR production have been proposed. McComas and Schwadron (2006) suggest that the acceleration occurs at the TS, not at the nose but in the flanks. Since the TS is blunt, the field lines are only briefly connected to the shock near the nose, whereas the field lines in the flanks and tail have much longer shock connection times which may allow these particles to be accelerated to higher energies; these ACRs then move toward the nose of the heliosphere, producing a rise in ACR intensity across the heliosheath. Fisk and Gloeckler (2007) suggest that the pile-up of the magnetic field ahead of the HP, the Axford-Cranfill effect, could produce a region of large magnetic variations that could accelerates the ACRs, which would also produce increasing ACR intensities with distance. As the Voyager's near the HP they should be able to differentiate between these two hypotheses.

13.5 The Heliosheath

The heliosheath is the largest region of the heliosphere and has highly variable magnetic fields and plasmas. The magnetic field increases by a factor of 2 at the TS, and the standard deviation of the field increases by a

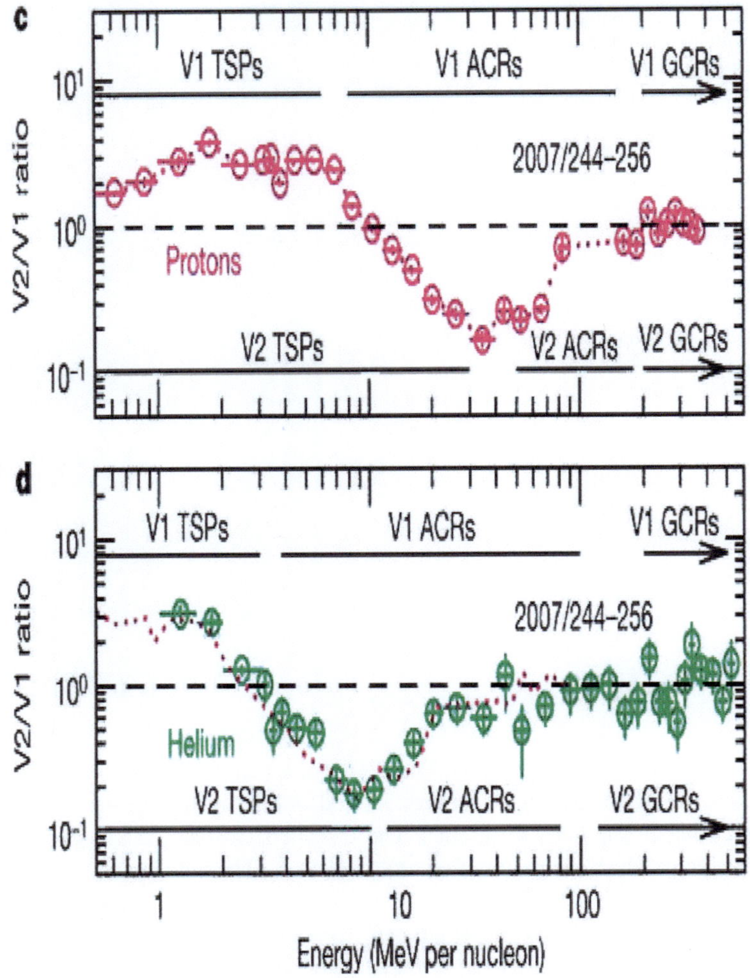

Fig. 13.6 The ratio of H and He intensities observed by LECP on V1 and V2 immediately after the V2 termination shock crossing. The particle populations are labeled. The *dashed lines* show the He (*top*) and H (*bottom*) adjusted for mass and overplotted on the H (*top*) and He (*bottom*) intensities

factor of about 3 (Burlaga et al. 2005, 2008). Magnetic field fluctuations of up to a factor of 5 are observed on times scales as small as 5–10 min (Burlaga et al. 2008). These fluctuations are generally in the magnetic field magnitude, while the direction remains fairly constant, so these variations are compressive (Burlaga et al. 2005, 2006a, 2009; Burlaga and Ness 2009). V1 observed magnetic humps and holes, regions of enhanced and diminished field strength, with periods of a few hours. These features are similar to the magnetic mirror mode structures observed in other sheaths but could also be soliton waves (Burlaga et al. 2006b).

Plasma variations are observed on similar time scales, often anti-correlated with the field changes (Richardson et al. 2010). Density changes of a factor of two are common and the temperature often changes by several orders of magnitude (Richardson 2008). Some spectra in the heliosheath are very cold, with temperatures consistent with the solar wind being heated only by compression as it enters the heliosheath, with no additional shock heating. These very cold spectra make up a few percent of the observations and persist for only a few minutes when they are observed. These may be ions which passed through the TS while it was reforming, and thus were not heated by the shock. Angular flows also vary on short time scales, suggesting some of the upstream flow energy powers turbulent flows downstream.

The folds in the magnetic field associated with the heliospheric current sheet (HCS) tilt are compressed in the heliosheath since the solar wind has slowed, so the oppositely directed fields are in close proximity. Reconnection was hypothesized to be common at the HCS and could provide heating of the heliosheath

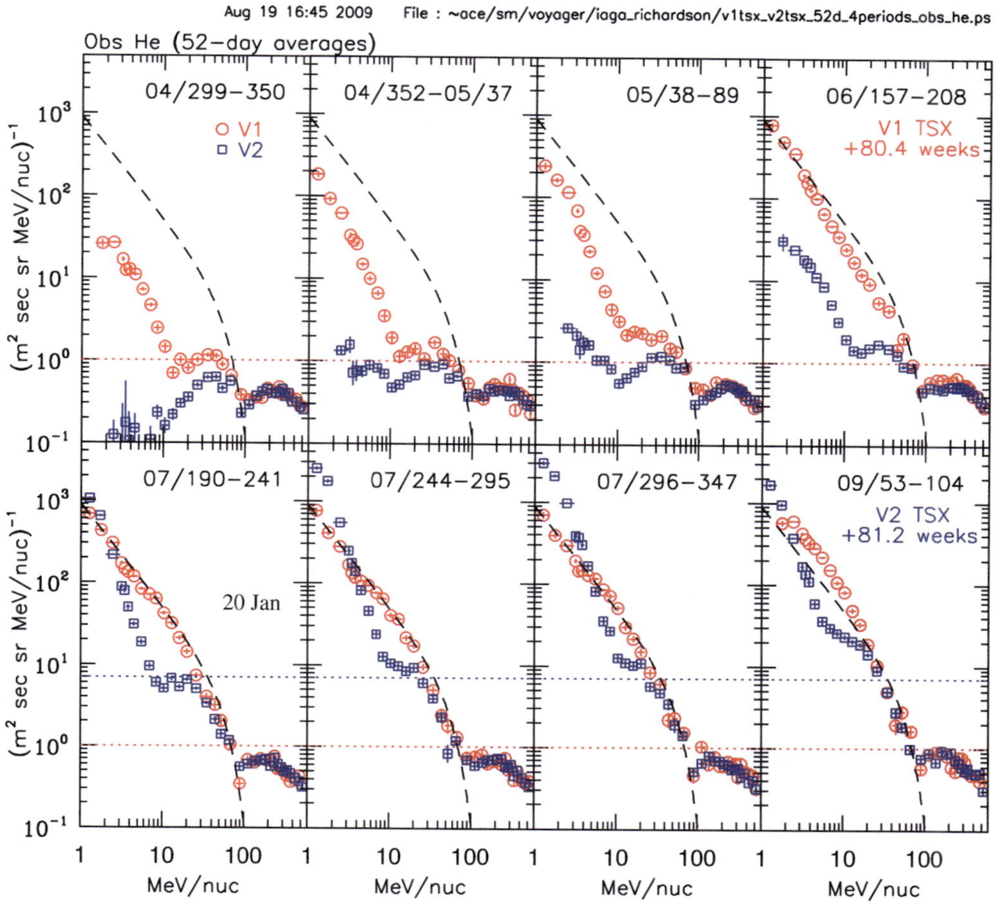

Fig. 13.7 Helium spectra from the CRS instrument from 2004 to 2009. The *red circles* show V1 observations and the *blue squares* V2 observations. The *dashed line* shows the power law spectra predicted at the TS before the Voyager observations showed otherwise. Figure courtesy of A. Cummings and E. Stone

plasma. The magnetic field data show one clear reconnection signature; on day 105 of 2005 a dropout of B was observed coincident with the HCS crossing (Burlaga et al. 2006b). However, this event was the only example of reconnection reported to date, so reconnection is likely not an important process in at least the inner heliosheath.

Figure 13.8 shows the plasma data from the heliosheath. The speed has remained on average constant with an average speed of 138 km/s, but with variations from 100 to almost 300 km/s. The typical time scale for speed changes is tens of days. The density decreased by factor of 2 as V2 moved across the heliosheath and has become much less variable. The temperature also decreased by about a factor of 2. We note that there was a large transient which lasted about 10 days at 2008.6, with increased speed, density, temperature, and different flow angles, whose source is not understood. The non-radial flows increase across the heliosheath as expected as the flow turns tailward.

Although the V1 plasma experiment is not working, the LECP team derives plasma flow speeds in the R and T directions using the Compton-Getting effect. They find that the speed was slow, 200–300 km/s, before V1 crossed the TS and was briefly inward after that crossing (Decker et al. 2005). The radial speed increased to 90 km/s after the TS effects ended and has slowly decreased to about 20 km/s in mid-2009 (Decker, private communication, 2009). The V_T component has remained constant at about −57 km/s. The difference between the speeds derived from the V1 and V2 data is an outstanding problem (Richardson et al. 2009). The LECP and plasma instruments on V2 observe very similar speeds, so the

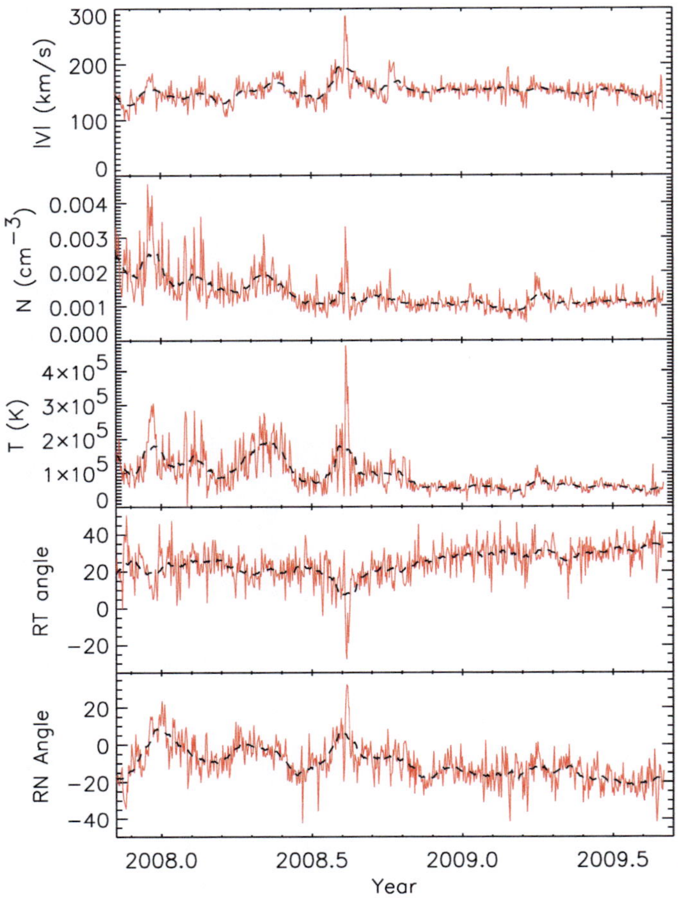

Fig. 13.8 Plasma data in the heliosheath. The panels show daily averages of the speed, density, temperature, and flow angles of the heliosheath plasma. The dashed lines show 25-day running averages

difference is probably not instrumental. Models predict that V1 and V2 should observe comparable radial speeds and that the observed speeds should decrease across the heliosheath (Pogorelov et al. 2009). The V1 observations match the predictions well, the V2 data do not.

The non-radial flows at V2 are more in the T than N directions. Figure 13.9 shows the orientation of the coordinate system; looking radially outward from the Sun along R, the T direction is parallel to the solar equatorial plane and in the direction of solar rotation and N completes a right-handed system (and is roughly northward). The locations of V1 and V2 are shown; flow away from the nose would be mainly in the N direction with lesser flow in the -T direction for V1 and equally in the T and -N directions for V2. The data show larger flow in the T than -N directions at V2 and flow in the -T direction at V1 (the N component cannot be measured by LECP). Thus these flows also differ from expectations. Another unexpected feature in the flow was a quasi-periodic oscillation in V_N with a period of 110 days observed for a year after the TS crossing (Richardson et al. 2009).

If the non-radial flows after the TS result from the angle between the TS normal and the flow direction, then the flow directions give estimates of the average TS orientation. Since the average heliosheath flows are larger in the T than N directions, the TS must be more blunt in the T than N directions, which implies that the TS is flattened at the poles compared to the sides (at least in the V2 direction) as shown in Fig. 13.9. The large-scale structure of the heliosphere should soon be revealed by the IBEX spacecraft.

The flows observed after V2 crossed the TS oscillated in the RN (north-south) plane with a period of about 110 days. The amplitude of these variations was about 17°S, with the TN angle varying between +3° and −31°. One possible explanation for the period is that the round-trip time for waves to travel from the TS to the HP is about 110 days. Another possible source

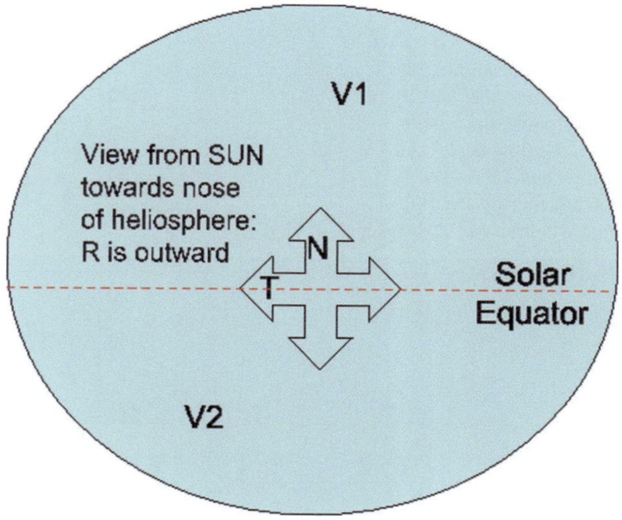

Fig. 13.9 Schematic diagram of the heliosphere looking outward from the Sun showing the directions of the R and T axes and the squashed shape of the TS

for these waves is that they are driven by interactions at the boundary between the fast and slow solar wind.

The flows in the heliosheath are ultimately determined by the interaction between the solar wind and the interstellar plasma, and this interaction depends on the LIC magnetic field. Given the TS crossing distances at the V1 and V2 locations, the direction of plasma flow in the non-radial direction, and the observations of the location of the foreshock region, the magnetic field strength and direction can be constrained. Opher et al. (2009) use a MHD model which includes pickup ions to show that the LIC magnetic field must be strong, near 4 μG, tilted 20–30° from the ecliptic plane and about 90° from the LIC flow.

13.6 The Future

The next milestone for the Voyagers will be the crossing of the HP. As with all new discoveries, the location will be known only when we reach this boundary. We expect the magnetic field to increase, the plasma density to increase but the flow to slow, and perhaps to observe increased particle fluxes. Model estimates place the HP location 30–60 AU beyond the TS, with the HP location closer in the V2 than V1 directions. As we get closer to the HP, the magnetic field strength is expected to increase as the field piles up against the HP, the density should decrease since the increased field pressure will cause it to move tailward along field lines, and the flow direction will become completely non-radial. Increases in the magnetic field and decreases in the plasma density are already being observed. The V1 LECP instrument has observed a substantial flow deflection, shown in Fig. 13.10. The flow in the RT plane is already 60°S from radial in mid-2009; the linear fit to the angle shown in the figures shows a slope of 10°S/year, giving a HP crossing in 3–4 years; this extrapolation would place the HP at about 128 AU, about 34 AU past the TS (neglecting the motions of both boundaries) in the V1 direction. Thus we may not have too long a wait before the Voyagers sample the LIC in situ.

13.7 Summary

The Voyager spacecraft found many surprises as they approached, crossed, and moved beyond the TS. An early indication of the approaching TS was an increase in particle fluxes flowing along the magnetic field. These particles form the TS foreshock region, where field lines in the solar wind are connected to the blunt shock in the nose direction and particles energized at the TS stream along these field lines into the heliosphere. V1 crossed the TS at 94 AU and observed the predicted increase in the magnetic field magnitude and magnetic turbulence, revealing the size of the heliosphere. Voyager 2 crossed the TS at 84 AU, revealing that the heliosphere was asymmetric. This

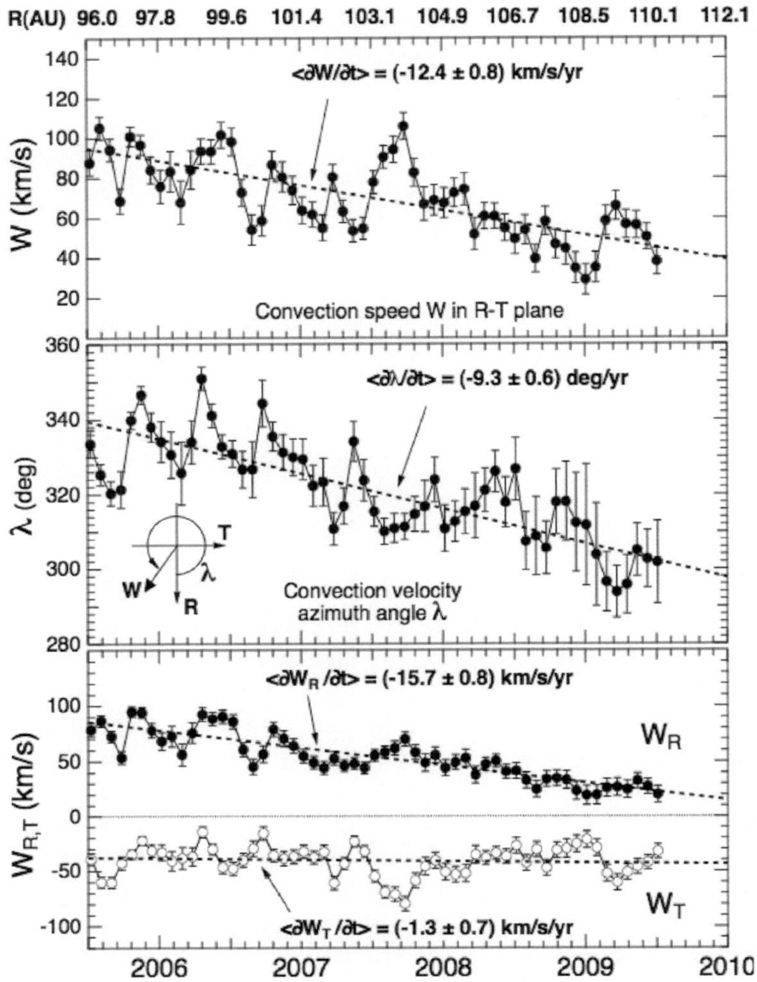

Fig. 13.10 Speeds in the heliosheath observed by V1. The *top panel* shows the magnitude of the velocity in the RT plane, the *middle panel* the flow angle, and *bottom panel* the velocity components V_R and V_T. Figure courtesy of R. Decker

asymmetry probably results from a strong LIC magnetic field which is tilted with respect to the LIC flow direction. The solar wind speed starts to decrease about 80 days before the TS; at least some of this decrease is attributed to a gradient in the particle pressure ahead of the TS indicating that the TS is a particle-mediated shock. The TS was crossed at least 5 times, showing that it moves on small time scales. One crossing had the structure of an ideal quasi-perpendicular super critical shock. Another had only slow changes in plasma parameters and two ramps in the field, which suggest it was is the process of reforming upstream. The energy transfer at the TS is different from planetary bow shocks in that much less of the flow energy (20%) goes into the thermal ions. The difference is that pickup ions comprise about 18% of the solar wind upstream of the TS and this more energetic plasma component gains most of the flow energy. As expected, the heliosheath is a highly variable region. The magnetic field and plasma parameters vary on scales of tens of minutes. The magnetic field fluctuations are mainly compressive, meaning that the field strength changes but the direction does not. The plasma temperature has a small number of spectra which appear not to have been heated at the TS, perhaps the plasma which passed the TS as it was reforming. The V2 radial flows are much larger than the V1 flows and do not decrease across the first 6 AU of the heliosheath as does the V1 radial speed (and as models predict). These radial flows and the relative flows in the T and N directions are not understood. The long-held hypothesis that the ACR source was the TS was not verified. No increase in the ACR component was observed across the TS at either spacecraft. Other regions of the heliosphere, such as the flanks or tail of the heliosphere of the nose of the HP, have been suggested as alternates source regions. The Voyagers found a new particle population, the TSP particles, with energies of tens of keV to a few MeV. These particles were first observed in the TS foreshock and have peak intensities at the TS, indicating the TS accelerates these particles. The Voyagers continue onward toward the HP and interstellar medium, which is estimated to be at 120–150 AU. The spacecraft have power to last through roughly 2022, when they will be at 130 (V2) and 152 (V1) AU, so there is a good chance they will make the first measurements of the environment outside our heliosphere.

Acknowledgements This work was supported under NASA contract 959203 from the Jet Propulsion Laboratory to the Massachusetts Institute of Technology and NASA grants NAG5-8947 and NNX08AE49G.

References

Burlaga LF, Ness NF, Stone EC, McDonald FB, Acuna MH, Lepping RP, Connerney JEP (2003) Search for the heliosheath with Voyager 1 magnetic field measurements. Geophys Res Lett 30:2072–2075

Burlaga LF et al (2005) Crossing the termination shock into the heliosheath: magnetic fields. Science 309:2027–2029. doi:10.1126/science.1117542

Burlaga LF, Ness NF, Acuna MH (2006a) Magnetic fields in the heliosheath: Voyager 1 observations. Astrophys J 642:584–592

Burlaga LF, Ness NF, Acuna MH (2006b) Trains of magnetic holes and magnetic humps in the heliosheath. Geophys Res Lett 33:L21106

Burlaga LF, Ness NF, Acuna MH, Lepping RP, Connerney JEP, Richardson JD (2008) Observations of magnetic fields at the termination shock by Voyager 2. Nature 454:75–77

Burlaga LF, Ness NF (2009) Compressible "turbulence" observed in the heliosheath by Voyager 2. Astrophys J 703:311–324

Burlaga LF, Ness NF, Acuna MH, Richardson JD, Stone EC, McDonald FB (2009) Observations of the heliosheath and solarwind near the termination shock by Voyager 2. Astrophys J 692:1125–1130

Decker RB et al (2008) Shock that terminates the solar wind is mediated by non-thermal ions. Nature 454:67–70

Decker RB, Krimigis SM, Roelof EC, Hill ME (2006) Low-energy ions near the termination shock. In: Physics of the inner heliosheath: Voyager observations, theory, and future prospects, AIP conference proceedings 258, pp 73–78

Decker RB, Krimigis SM, Roelof EC, Hill ME, Armstrong TP, Gloeckler G, Hamilton DC, Lanzerotti LJ (2005) Voyager 1 in the foreshock, termination shock, and heliosheath. Science 309:2020–2024

Fisk LA, Gloeckler G (2007) Acceleration and composition of solar wind suprathermal tails. Space Sci Rev 130:153–160

Florinski V, Decker RB, le Roux JA, Zank GP (2009) An energetic-particle-mediated termination shock observed by Voyager 2. Geophys Res Lett 36:L12101

Gloeckler G, Fisk LA, Lanzerotti LJ (2005) Acceleration of solar wind and pickup ions by shocks. Proceedings of solar wind 10/SOHO 16 "Connecting Sun and Heliosphere", ESA SP-592:107–111

Gurnett DA, Kurth WS (2005) Electron plasma oscillations upstream of the solar wind termination shock. Science 309:2025–2027

Gurnett DA, Kurth WS (2008) Intense plasma waves at and near the solar wind termination shock. Nature 454:78–80

Jokipii JR, Giacalone J, Kota J (2004) Transverse streaming anisotropies of charged particles accelerated at the solar wind termination shock. Astrophys J 611:L141–L144

Krimigis SM et al (2003) Voyager 1 exited the solar wind at a distance of 85 AU from the Sun. Nature 426:45–48

Linsky JL, Wood BE (1996) The alpha centauri line of sight: D/H ratio, physical properties of local interstellar gas, and measurement of heated hydrogen (the 'hydrogen wall') near the heliopause. Astrophys J 463:254–270

McComas DJ, Schwadron NA (2006) An explanation of the Voyager paradox: particle acceleration at a blunt termination shock. Geophys Res Lett 33:L04102

McDonald FB et al (2003) Enhancements of energetic particles near the heliospheric termination shock. Nature 426:48–51

Opher M, Stone EC, Gombosi TI (2007) The orientation of the local interstellar magnetic field. Science 316:875–878

Opher M, Alouani Bibi F, Toth G, Richardson JD, Izmodenov VV, Gombosi TI (2009) A strong, highly-tilted interstellar magnetic field near the Solar System. Nature 462:1036–1038

Parker EN (1963) Interplanetary dynamical processes. Interscience Publishers, New York, NY

Pogorelov NV, Heerikhuisen J, Mitchell JJ, Cairns IH, Zank GP (2009) Heliospheric asymmetries and 2–3 kHz radio emission under strong interstellar magnetic field conditions Astrophys. J Lett 695:L31–L34

Richardson JD (2008) Plasma temperature distributions in the heliosheath. Geophys Res Lett 35:L23104–L23108

Richardson JD, Kasper JC, Wang C, Belcher JW, Lazarus AJ (2008) Cool heliosheath plasma and deceleration of the upstream solar wind at the termination shock. Nature 464:63–66

Richardson JD, Stone EC, Kasper JC, Belcher JW, Decker RB (2009) Plasma flows in the heliosheath. Geophys Res Lett 36:L10102. doi:10.1029/2009GL038421

Richardson JD (2010) The plasma in the heliosheath. In: Maksimovic M et al (eds) Twelfth international solar wind conference, AIP conference proceedings 1216, Saint-Malo, France, p 551

Stone EC et al (2008) Voyager 2 finds an asymmetric termination shock and explores the heliosheath beyond. Nature 454:71–74

Stone EC, Cummings AC, McDonald FB, Heikkila B, Lal N, Webber WR (2005) Voyager 1 explores the termination shock region and the heliosheath beyond. Science 309:2017–2020

Wood BE, Alexander WR, Linsky JL (1996) The properties of the local interstellar medium and the interaction of the stellar winds of Epsilon Indi and Lambda Andromedae

with the interstellar environment. Astrophys J 470: 1157–1171
Zank GP (1999) Interaction of the solar wind with the local interstellar medium: a theoretical perspective. Space Sci Rev 89:413–687
Zank G, Pauls H, Cairns I, Webb G (1996) Interstellar pickup ions and quasiperpendicular shocks: Implications for the termination shock and interplanetary shocks. J Geophys Res 101:457–466

Part V
Heliophysical Processes

Chapter 14

Three Dimensional Magnetic Reconnection at Null Points and Separators

Clare E. Parnell, Andrew L. Haynes, and Rhona C. Maclean

Abstract Three-dimensional reconnection is much more diverse than two-dimensional reconnection. In deed, the characteristics of these two types of reconnection are very different. For instance, three-dimensional reconnection can occur both in the vicinity of null points, as well as in the absence of null points, whereas two-dimensional reconnection must occur at null points. Since, in two-dimensions reconnection only occurs at a single point, at most one pair of field lines can be reconnected to form a new pair of field lines at any instant in time. In three dimensions, however, there is a finite diffusion volume in which many field lines can be processed simultaneously. For the entire time that a portion of a field line is in the diffusion volume it will reconnect continuously and continually with all the other field lines it meets. This means that in three-dimensions field lines do not reconnect in pairs of lines making understanding three-dimensional reconnection difficult. In this chapter, we review the characteristics of two-dimensional and three-dimensional reconnection in more detail and also discuss some of the consequences of these characteristics. It is well know that the magnetic fields that thread the Sun's surface cover a wide range of scales from tiny intranetwork features of just 10^{16} Mx up to sunspots with 10^{23} Mx of flux. There is an extensive mix of features of opposite polarities and scales which carpet the Sun. However, unlike a carpet the pattern is not static, and these features are highly dynamic. This inevitably leads to a complex and highly structured magnetic field in the solar atmosphere. It has been thought that null points rarely occurred in the solar atmosphere's complex magnetic field, however, we show here that null points can readily occur. Furthermore, by studying the magnetic topology of a flux emergence experiment we reveal that reconnection occurring at separators linking two clusters of nulls is the mechanism by which the newly emerged flux interacts with the pre-existing flux in the solar atmosphere. Furthermore, we show that separator reconnection occurs along the length of the separator and not at the nulls at the ends of the separator. Thus, if a separator reaches up into the corona, even if its null points reside below the corona, it can be an important site for reconnection and thus a site of coronal heating.

14.1 Introduction

The idea that tangled and stressed magnetic fields can restructure and reduce to a lower energy state via a process known as magnetic reconnection has been around for many years (Giovanelli 1946; Parker 1957; Sweet 1958; Dungey 1961; Petschek 1964). Tangled magnetic fields can store energy in the form of currents. Magnetic reconnection is the mechanism by which this excess magnetic energy can be converted into thermal energy, kinetic energy and the acceleration of particles. Although reconnection has been known about for many years there are many aspects of reconnection, especially in three dimensions, that are currently not well understood.

Work on magnetic reconnection has generally been motivated by events on the Sun, in particular, solar flares and the heating of the solar corona and by the

C.E. Parnell (✉)
School of Mathematics and Statistics, University of St Andrews, St Andrews, Fife, KY16 9SS, Scotland, UK
e-mail: clare@mcs.st-and.ac.uk

M.P. Miralles, J. Sánchez Almeida (eds.), *The Sun, the Solar Wind, and the Heliosphere*, IAGA Special Sopron Book Series 4, DOI 10.1007/978-90-481-9787-3_14,

interaction of the solar wind with the Earth's magnetic field. There has also been interest from laboratory plasma physicists who have studied reconnection in fusion devices. In all of these scenarios one major question is, where does the reconnection occur? A second is, how will it occur? Using three-dimensional magnetohydrodynamic models we have been investigating the answers to these questions. First, however, before we discuss our experiments and our findings, we give a brief review of the main attributes of two-dimensional (2D) and three-dimensional (3D) reconnection.

Naturally, the original reconnection scenarios were 2D and a considerable body of work has been built up over several decades covering many aspects of 2D reconnection. Extensive reviews of this work can be found in Priest and Forbes (2000) and Biskamp (2000).

In 2D, magnetic null points – points at which all components of the magnetic field vanish – are essential for reconnection. 2D magnetic nulls may be either X-type or O-type, but for reconnection X-type nulls are required. From such a null four field lines, called separatrices, extend out with two ending in sources and the other two ending in sinks. The separatrices divide the magnetic field up into four flux domains which contain topologically distinct field lines with different connectivities.

Pairs of field lines from flux domains on opposite sides of the null with connectivities $A \to A'$ and $B \to B'$ say, may reconnect at the null to form two new pairs of field lines in the other two domains. The field lines in these other domains will have connectivities $A \to B'$ and $B \to A'$. Reconnection in 2D therefore involves a transfer of flux from one pair of domains into another pair of flux domains and it results in a discontinuous mapping of field lines at the instant reconnection takes place at the null point. In order for reconnection to occur at the null point there must be an enhanced current (a current sheet), but there must also be an appropriate flow such as an X-type stagnation flow to facilitate reconnection.

In 3D, however, the magnetic field structures and the process of magnetic reconnection are not quite so simple, but magnetic null points do also exist in 3D (Parnell et al. 1996). The equivalent of the 2D separatrices are a pair of field lines called spines that extend out of (into) the null and a surface of field lines that extends into (out of) the null known as a separatrix surface. Furthermore, in 3D, nulls can be either positive or negative depending on whether the field lines in the separatrix surface extend away from or into the null, respectively. Separatrix surfaces from null points divide up the magnetic field into topologically distinct flux domains. If the separatrix surfaces from a positive and a negative null intersect then there must exist a special field line called a separator that connects the two nulls and lies along the intersection of the two separatrix surfaces (e.g. Lau and Finn 1990).

Many different definitions have been proposed for reconnection in 3D. Some require the existence of features associated with null points. For example, Vasyliunas (1975) suggested that reconnection occurred when there was plasma flow across separatrices (i.e. across the field lines lying in a separatrix surface), whereas Sonnerup (1979) stated that for reconnection to occur in 3D the presence of an electric-field along a separator was required. Others have prosed more general definitions such as Axford (1984) who suggested that reconnection was the transfer of plasma-elements from one field line to another (i.e. a break down of the frozen-in flux theorem) and Greene (1993) defined it as any magnetic field evolution that is not flux preserving. Then, in the late eighties, a theory of generalised magnetic reconnection was proposed which encompassed the reconnection definitions previously suggested (Schindler et al. 1988; Hesse and Schindler 1988). In their general theory, Schindler et al. (1988) and Hesse and Schindler (1988) established that 3D reconnection does not require nulls or structures associated with nulls, but can readily occur away from these features. In particular, 3D magnetic reconnection can occur not only at 3D nulls (e.g., Craig et al. 1995; Craig and Fabling 1996; Priest and Titov 1996; Pontin and Craig 2005; Pontin and Galsgaard 2007; Pontin et al. 2007a, b), but more commonly it can occur in a null-less regions such as hyperbolic flux tubes (Priest and Démoulin 1995; Démoulin et al. 1996; Titov et al. 2003; Galsgaard et al. 2003; Linton and Priest 2003; Aulanier et al. 2005; Pontin et al. 2005a; Aulanier et al. 2006; De Moortel and Galsgaard 2006a, b; Wilmot-Smith and De Moortel 2007) or at mode-rational surfaces (Browning et al. 2008; Hood et al. 2009).

Some of the above situations depend on the fact that field lines from flux domains with two different connectivities, say $A \to A'$ and $B \to B'$, are reconnected to form field lines in two new flux domains with connectivities $A \to B'$ and $B \to A'$, exactly as in the 2D case. However, not all 3D reconnection scenarios require a transfer of flux between different topologically distinct domains. For instance, the tangled flux within a flux tube may reconnect within itself in order to untangle and relax to a lower energy state (e.g. Galsgaard and Nordlund 1996; Wilmot-Smith et al. 2009). On a global scale, the magnetic field lines have the same connectivity, however, locally the field may be sufficiently tangled that groups of field lines can only access a more energetically favourable path by reconnecting.

Furthermore, unlike in 2D, it is generally not possible to identify pairs of field lines that reconnect to form new pairs of field lines in 3D reconnection (e.g. Hornig and Priest 2003; Pontin et al. 2005b). This is a result of the fact that 3D reconnection occurs continually and continuously throughout a finite diffusion volume as opposed to at a single point. Another consequence of the reconnection being continual and continuous throughout the diffusion volume is that in 3D the field line mapping is continuous between pre- and post-reconnected field lines as opposed to discontinuous. Hornig and Priest (2003) explain the theory behind this behaviour using a kinematic model whilst Pontin et al. (2005a) and Aulanier et al. (2006) provide nice illustrations using numerical experiments.

Finally, we note that the flow pattern required for 3D reconnection is also different from that required for 2D reconnection. Both Hesse (1995) and Hornig and Priest (2003) indicate that counter rotational flows are an essential ingredient of 3D reconnection. This idea is confirmed by Parnell et al. (2010) who have found that counter rotating flows are important for reconnection about separators.

As mentioned above there are a number of differences between 2D and 3D reconnection, however there are also some similarities too. Reconnection in 2D and 3D both result in fast outflow jets away from the reconnection site. These jets do not have to be symmetric and neither do they have to be Alfvénic. Also, reconnection in both 2D and 3D occurs at current sheets and lead to the dissipation of currents, thus enabling the conversion of magnetic energy into other forms of energy. Finally, reconnection in 2D and 3D is the only mechanism by which flux can be transferred between topologically distinct flux domains making it an extremely important and fundamental plasma physics process.

14.2 Magnetic Fields and Null Points in the Solar Atmosphere

The Sun's surface is threaded by a complex seething patchwork of magnetic field lines that are directed either into or out-of the Sun (Fig. 14.1). During solar maximum the magnetic field in the Sun's atmosphere is dominated by active regions containing large paired regions of opposite-polarity flux. Whilst at solar minimum magnetic features are much smaller and the surface is carpeted in a speckled mass of small-scale features. All these magnetic features are highly dynamic, especially the small-scale ones which have lifetimes of just a few minutes. The large features, such as sunspots, are much more stable and can have lifetimes of days to weeks. Estimates have been made to determine the time it takes to completely replace all the flux in the solar surface during solar minimum. They suggest a turnover time of just 14 hours (Hagenaar 2001).

The magnetic fields themselves that extend out from these surface magnetic flux patches have an even more rapid turn over time. Close et al. (2004) showed that on the quiet-Sun the time to change all the magnetic connections from magnetic features observed by MDI is just 1.4 h. Higher-resolution observations are likely to reveal an even more rapid recycling of the surface magnetic features and the atmospheric magnetic fields. The process by which all the magnetic connections can change is magnetic reconnection, thus this result shows that there are considerable amounts of reconnection occurring continually and continuously all over the solar surface. This naturally begs a number of questions. Where exactly does this reconnection occur? How does it occur? And how much energy does it release?

In 3D, as already discussed, 3D reconnection can occur in many different places. It is thought by some that null points rarely exist in the solar atmosphere and so null point and separator reconnection are unlikely to be important for coronal heating. To establish if these features are likely to be important locations of reconnection we need to determine if there are many magnetic null points in the solar atmosphere.

Estimates have been made as to the numbers of nulls in the atmosphere above the quite-Sun by Longcope and Parnell (2009). They used the tri-linear null finding method of Haynes and Parnell (2007) to locate nulls in potential field extrapolations from observed MDI

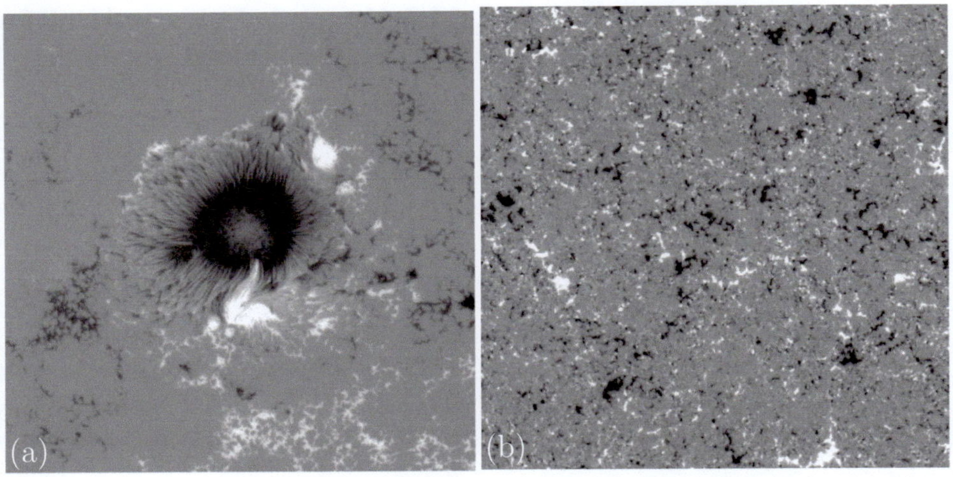

Fig. 14.1 Solar photospheric magnetograms taken by Hinode/SOT showing (**a**) large-scale magnetic features in an active region and (**b**) showing small-scale magnetic features in the quiet-Sun. *White* indicates flux directed out from the Sun, *black* indicates flux directed into the Sun. Taken from Parnell et al. (2009)

quiet-Sun magnetograms. On the base of the numerical box the complete magnetic field was used providing continuous field through the base. Thus, all the nulls found were located above this in what would be considered as the solar atmosphere. Figure 14.2 shows a single MDI magnetogram with the numerous nulls

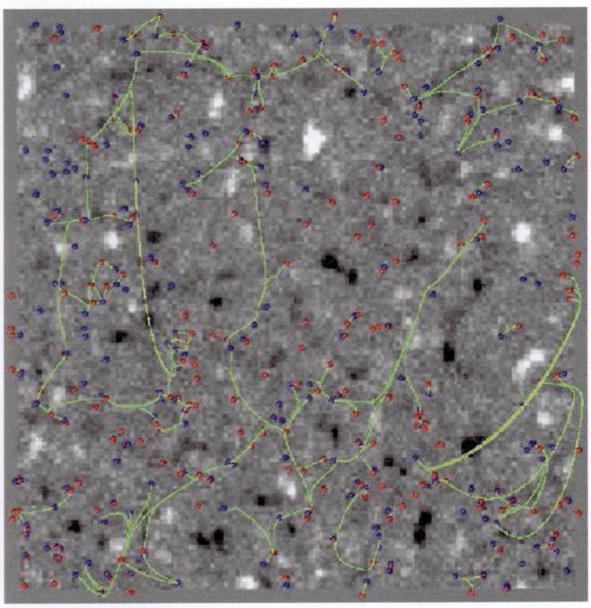

Fig. 14.2 Solar photospheric magnetograms taken by SOHO/MDI showing the positive (*red dots*) and negative (*blue dots*) nulls and separators (*green lines*) in the atmosphere found in the potential magnetic field extrapolated from the continuous field of this magnetogram

(red and blues dots) found in the extrapolated potential magnetic field. For nulls at heights of more than 1.5 Mm above the photosphere there is typically only one null every 322 Mm^2, however, there are considerably more nulls below this height. Accurate estimates of the exact numbers of nulls below 1.5 Mm are difficult to make with MDI data, however, such estimates should be possible using data from the Helioseismic and Magnetic Imager on the Solar Dynamics Observatory.

With only 1 null every 322 Mm^2 in the region higher than 1.5 Mm above the solar surface it would seem that there are relatively a few nulls in the solar corona. However, there is new evidence to suggest that the number of nulls found in potential field extrapolations are an underestimate of the true number of nulls. Maclean et al. (2009) investigated the magnetic topology of a flux emergence experiment where a twisted flux rope situated in the convection zone rises up through the photosphere into a region of uniform overlying field representing the background magnetic field in the corona. Various aspects of this flux emergence experiment are described in detail by Galsgaard et al. (2005); Archontis et al. (2005) and Galsgaard et al. (2007), however, none of these papers considered the topology of the magnetic field. The reason for this was because they did not have the tools required to determine it. From a potential field extrapolation of the photospheric field it is unlikely that any nulls would be found, as the magnetic configuration is very simple. However, when the magnetic field time series from the resistive MHD experiment was analysed using the null finding method of Haynes and Parnell (2007) there were more than 300 different nulls found that lasted for various lengths of time. These nulls did not appear at random locations, but rather they were found in two loose clusters on either side of the emerging flux region (Fig. 14.3). The overall topological degree of the numerical volume remained at zero, as did the topological degree of the individual clusters indicating that the nulls were created and destroyed in opposite-polarity pairs within each cluster. Tests were made to show that these nulls were genuine and were not simply due to numerical oscillations.

It is likely that nulls cluster because they tend to occur in regions of weak magnetic field (Albright, 1999) and the highly dynamic behaviour of the magnetic field can lead to the creation of more nulls in this weak field region. However, this dynamic behaviour inevitably makes it not only easy to create new nulls, but, by the same argument, it means it is easy to destroy them too. Thus, unsurprisingly, Maclean et al. (2009) found that many of the nulls in her experiment lasted just minutes, however, a few nulls lasted considerably longer with lifetimes equivalent to the duration of the flux rope interaction which lasted more than an hour.

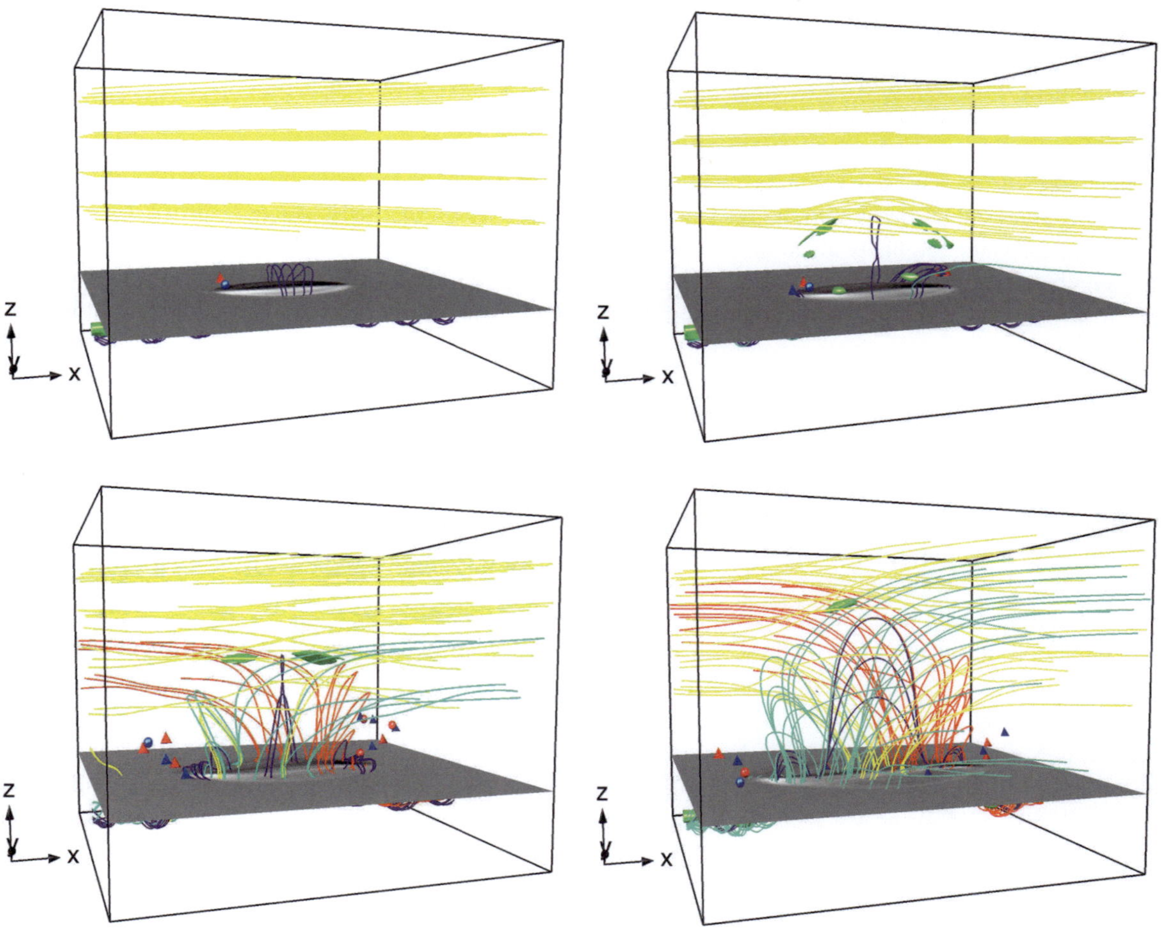

Fig. 14.3 Three-dimensional snapshots during the emergence of a flux tube into an overlying horizontal magnetic field orientated at 135° to the axis of the flux tube. Sample field lines extending the full length of the flux tube (*purple*), extending from the corona into the flux tube (*orange*), from the flux tube into the corona (*blue*) and lying purely in the corona (*yellow*) are drawn. The photosphere is indicated by the grey plane with contours of the vertical magnetic field shown. The *red* and *blue* symbols above this plane represent the positive and negative nulls, respectively. The *green* isosurfaces indicate regions of strong parallel electric field. Figure courtesy of Maclean et al. (2009)

Magnetic fields that evolve via resistive MHD are likely to be much more complex than those that evolve though a series of equi-potential states (Haynes et al. 2007). This is because in an evolution through a series of equi-potential states reconnection occurs instantaneously and perfectly, but during a resistive MHD evolution reconnection is held off allowing the magnetic field to become more tangled and stressed. In these situations magnetic null points are more likely to arise. Thus the numbers of nulls estimated by Longcope and Parnell (2009) is likely to be a lower bound on the true number of nulls.

14.3 Separators

A magnetic field that has been stressed will hold its excess energy in the form of electric currents. These currents may be distributed over a large volume and be aligned with the magnetic field such that there are no magnetic forces, i.e., the magnetic field may be force-free. Also, the currents may be held in narrow current sheets which are likely locations for reconnection. For reconnection to occur it is not sufficient to have a favourable magnetic field configuration, but it is also essential that there is an appropriate plasma flow. Furthermore, reconnection will not activate until the gradients in the magnetic field have reached appropriately small-scales relative to the magnetic resistivity. Exactly how small these scales are depends on the characteristics of the plasma and requires knowledge of the kinetic behaviour of the plasma.

Numerical experiments investigating reconnection are often designed with initial magnetic configurations that contain particular magnetic features, such as null points or QSLs (e.g. Aulanier et al. 2006). Then a specific driver is imposed on the boundary to ensure that reconnection occurs at these required locations (e.g. Antiochos et al. 2002; Masson et al. 2009).

Maclean et al. (2009) and Maclean et al. (2010) considered a different approach and deliberately started with a magnetic configuration that did not contain any particularly favourable sites for reconnection. Their flux emergence experiment was initiated by reducing the density in the central part of the submerged flux tube such that a magnetic buoyancy instability lead to the tube rising and interacting with the overlying magnetic field causing reconnection to occur. For the reconnection to take place the magnetic field and plasma had to create favourable reconnection sites. What was the nature of the sites it created?

In 3D, an electric field component parallel to the magnetic field is a necessary requirement for reconnection (e.g. Sonnerup 1979; Schindler et al. 1988; Hesse and Schindler 1988; Hornig and Priest 2003; Parnell et al. 2008). Figure 14.3 shows isosurfaces of the parallel electric field. There is from the outset a tube of strong negative parallel electric field that lies along the axis of the flux rope, however, this only results in a negligible diffusion of the magnetic field within the flux rope and is not associated with the main reconnection between the flux rope and the overlying coronal field. Instead, the reconnection between these two different magnetic fields is associated with regions of strong positive parallel electric field that develops between them just before the onset of reconnection. They are situated in the central region of the box in the corona and lie in the regions between the four different connectivities of flux. The null points are clearly not associated with these regions of strong positive parallel electric field. Instead, it was found that these regions were threaded by separators that linked the two null clusters Fig. 14.4 (Maclean et al. 2010).

Maclean et al. (2010) found that in the flux emergence experiment there were a large number of separators created. Within each null cluster most of the nulls were linked by short separators that linked the nulls to form a chain (e.g. Fig. 14.4). These short separators did not tread high parallel electric field regions and thus little reconnection can be occurring along them. However, there were also a large number of long separators that connect the two null clusters. These separators thread the high positive parallel electric field regions and thus they are important locations for reconnection. These separators lie along the surface at the top of the emerging flux domain where it pushes up against the overlying field and are associated with the intersection of the four flux connectivities as they must be if they are involved in the reconnection of these flux regions.

Here we note a few interesting generic properties of the separators found in the flux emergence experiment of Maclean et al. (2010). The separators linking the two null clusters start in one cluster at just one or two nulls and end in the other cluster at either one or two nulls. These nulls are the long lived nulls. The magnetic field in the separators all go the same way and so on the left-hand side they originate from negative (blue) nulls and end in positive (red) nulls. In anyone snapshot many tens of separators connect a single pair of nulls between the two null clusters. Multiply-connected null pairs easily arise in all magnetic fields even potential magnetic fields (Parnell 2007). The separators linking

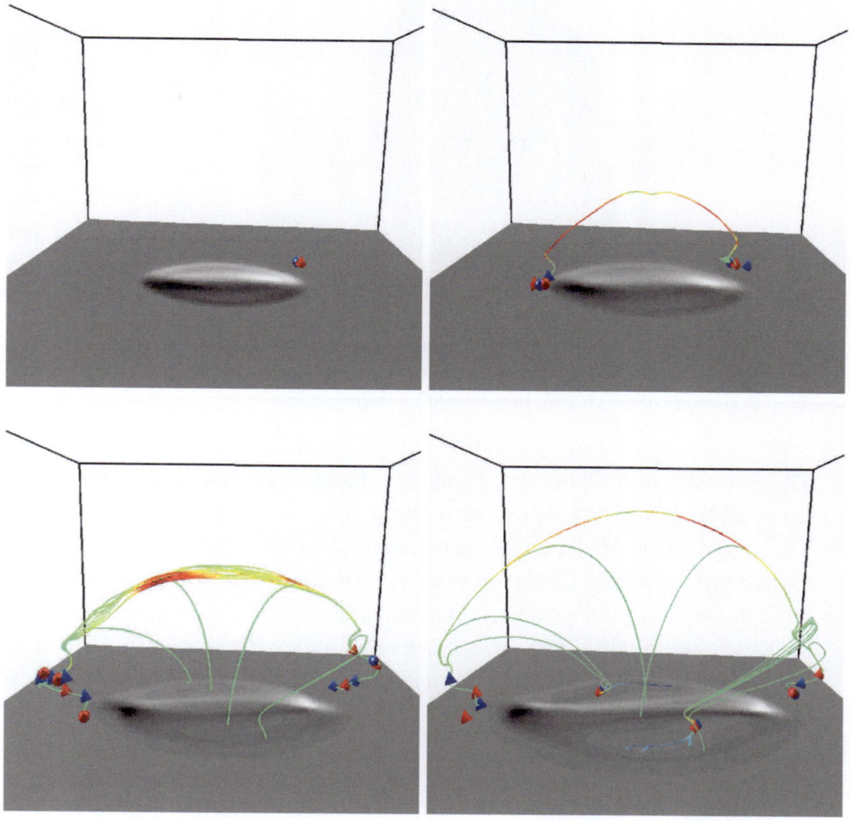

Fig. 14.4 Three-dimensional snapshots taken at the same time as those in Fig. 14.3 showing the null points (*red and blue symbols*, as before) and magnetic separators in the solar atmosphere. Here, the separators have been colour coded according to the amount of parallel electric field along them. *Yellow* and *red* indicate large positive parallel electric field, *green* represents zero parallel electric field whereas *blues* indicate negative parallel electric field

the two null clusters mostly reside at the outer surface of the emerging flux rope in the corona, but some run along the top of the emerging field region in the corona and then drop down and thread through the flux rope in the convection zone, before rising up into the corona again. The complex network of separators indicates that the magnetic field itself is very complex. Finally, we note that not all of the long separators show evidence of strong parallel electric field. As we have already said, for reconnection to occur not only must a favourable magnetic configuration be created, but also a favourable plasma flow must occur about the configuration for reconnection to occur. Clearly, not all of the separators created coincide with favourable plasma flows and thus they only have weak parallel electric fields along them.

14.4 Separators and Separator Reconnection

The structure of separators and the nature of separator reconnection has been considered by Parnell et al. (2010). They find that:

1. Reconnection occurs in "hot spots" of enhanced parallel electric field along separators and it does not appear to occur at the null points at the ends of the separator. The enhanced parallel electric field along separators changes both temporally and spatially during a period of reconnection (Fig. 14.5).

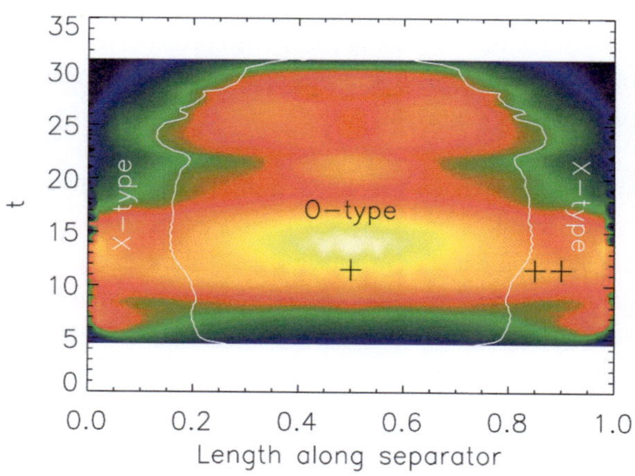

Fig. 14.5 Time evolution of the parallel electric field (location of reconnection) along the length of a separator. The high parallel electric field (*yellow*) region in the centre of the separator and low parallel electric field (*blue*) at the nulls demonstrate that separator reconnection occurs along the separator and not at the nulls. From Parnell et al. (2010)

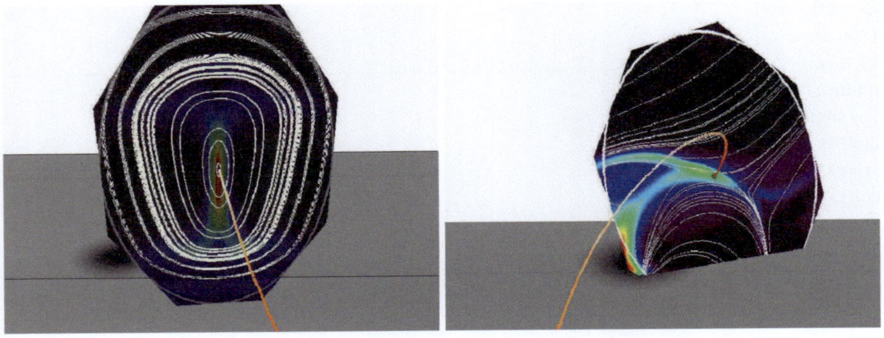

Fig. 14.6 2D planes perpendicular to the separator showing that the perpendicular projected field about separators may be either elliptic (*left*) or hyperbolic (*right*). In particular, a single separator may have hyperbolic magnetic fields near its nulls and elliptic field near its centre. From Parnell et al. (2010)

2. The reconnection "hot spots" along separators occur between regions of counter rotating flow.
3. The magnetic field about the separators may be elliptic or hyperbolic. That is in planes perpendicular to the separator the projected magnetic field may be X-type or spiral/O-type. This means that hyperbolic magnetic field configurations are not essential for reconnection (Fig. 14.6).

14.5 Implications and Concluding Remarks

These results indicate that even though many null points may reside below the solar corona, the separators linking these nulls can extend up into the corona. Reconnection can then occur in the corona about these magnetic topological features provided an appropriate plasma flow is present.

As we have seen many separators may exist each showing evidence of reconnection. This indicates that reconnection actually occurs over a wide region, and not at a single location. This means that the energy can be deposited along many newly reconnected field lines simultaneously leading to a potentially wide spread and rapid distribution of heat.

Finally, we note that multiply connected null pairs, as seen in the flux emergence experiment of Maclean et al. (2010), can result in recursive reconnection (Parnell et al. 2008), i.e., the reconnection of magnetic flux multiple times. This occurs, because the magnetic field is not clever. It does not know the quickest route to obtain a lower-energy state. Instead it simply reconnects wherever it can leading to a gradual edging closer to a lower energy state. This process of recursive reconnection, which occurs due to multiple separators, can explain very naturally how flares and other large solar events can have a long duration since the resulting energy release in these situations is typically exponential.

Observationally, the implications of these complex flux interaction events means that reconnection can easily lead to multiple, long duration heating events, that provide wide spread distributed heating. Thus making reconnection an ideal coronal heating mechanism.

Acknowledgements CEP gratefully acknowledges support from the Leverhulme Trust in the form of a Leverhulme Prize which she has used to fund ALH as a postdoc. RCM is grateful to STFC for providing financial support.

References

Albright, BJ (1999) The density and clustering of magnetic nulls in stochastic magnetic fields. Phys Plasmas 6:4222–4228

Antiochos SK, Karpen JT, DeVore CR (2002) Coronal magnetic field relaxation by null-point reconnection. Astrophys J 575:578–584. doi:10.1086/341193

Archontis V, Moreno-Insertis F, Galsgaard K, Hood AW (2005) The three dimensional interaction between emerging magnetic flux and a large-scale coronal field: reconnection, current sheets, and jets. Astrophys J 635:1299–1318. doi:10.1086/497533

Aulanier G, Pariat E, Démoulin P (2005) Current sheet formation in quasi-separatrix layers and hyperbolic flux

tubes. Astron Astrophys 444(3):961–976. doi:10.1051/0004-6361:20053600
Aulanier G, Pariat E, Démoulin P, Devore CR (2006) Slip-running reconnection in quasi-separatrix layers. Solar Phys 238(2):347–376. doi:10.1007/s11207-006-0230-2
Axford WI (1984) Magnetic field reconnection. In: Hones EW Jr (ed) Magnetic reconnection in space and laboratory plasmas, AGU Publishers, Washington, DC, pp 1–+
Biskamp D (2000) Magnetic reconnection in plasmas. CambridgeUniversity Press, Cambridge, UK
Browning PK, Gerrard C, Hood AW, Kevis R, van der Linden RAM (2008) Heating the corona by nanoflares: simulations of energy release triggered by a kink instability. Astron Astrophys 485:837–848. doi:10.1051/0004-6361:20079192
Close RM, Parnell CE, Longcope DW, Priest ER (2004) Recycling of the solar corona's magnetic field. Astrophys J Lett 612:L81–L84. doi:10.1086/424659
Craig IJD, Fabling RB (1996) Exact solutions for steady state, spine, and fan magnetic reconnection. Astrophys J 462: 969–+. doi:10.1086/177210
Craig IJD, Fabling RB, Henton SM, Rickard GJ (1995) An exact solution for steady state magnetic reconnection in three dimensions. Astrophys J Lett 455:L197+. doi:10.1086/309822
De Moortel I, Galsgaard K (2006a) Numerical modelling of 3D reconnection due to rotational footpoint motions. Astron Astrophys 451:1101–1115. doi:10.1051/0004-6361:20054587
De Moortel I, Galsgaard K (2006b) Numerical modelling of 3D reconnection. II. Comparison between rotational and spinning footpoint motions. Astron Astrophys 459:627–639. doi:10.1051/0004-6361:20065716
Démoulin P, Priest ER, Lonie DP (1996) Three-dimensional-magnetic reconnection without null points 2. Application to twisted flux tubes. J Geophys Res 101:7631–7646. doi:10.1029/95JA03558
Dungey JW (1961) Interplanetary magnetic field and the auroral zones. Phys Rev Lett 6:47–48. doi:10.1103/PhysRevLett.6.47
Galsgaard K, Nordlund Å (1996) Heating and activity of the solar corona 1. Boundary shearing of an initially homogeneous magnetic field. J Geophys Res 101:13445–13460. doi:10.1029/96JA00428
Galsgaard K, Titov VS, Neukirch T (2003) Magnetic pinching of hyperbolic flux tubes. II. Dynamic numerical model. Astrophys J 595:506–516. doi:10.1086/377258
Galsgaard K, Moreno Insertis F, Archontis V, Hood A (2005) A three-dimensional study of reconnection, current sheets, and jets resulting from magnetic flux emergence in the Sun. Astrophys J Lett 618:L153–L156. doi:10.1086/427872, arXiv:astro-ph/0410057
Galsgaard K, Archontis V, Moreno-Insertis F, Hood AW (2007) The effect of the relative orientation between the coronal field and new emerging flux. I. Global properties. Astrophys J 666:516–531. doi:10.1086/519756, 0705.1097
Giovanelli RG (1946) A theory of chromospheric flares. Nature 158:81–82. doi:10.1038/158081a0
Greene JM (1993) Reconnection of vorticity lines and magnetic lines. Phys Fluids B 5:2355–2362. doi:10.1063/1.860718
Hagenaar HJ (2001) Ephemeral regions on a sequence of full-disk Michelson Doppler Imager magnetograms. Astrophys J 555:448–461. doi:10.1086/321448
Haynes AL, Parnell CE (2007) A trilinear method for finding null points in a three-dimensional vector space. Phys Plasmas 14(8):082107–+. doi:10.1063/1.2756751, arXiv:0706.0521
Haynes AL, Parnell CE, Galsgaard K, Priest ER (2007) Magnetohydrodynamic evolution of magnetic skeletons. R Soc Lond Proc Ser A 463:1097–1115, arXiv:astro-ph/0702604
Hesse M (1995) Three-dimensional magnetic reconnection in space- and astrophysical plasmas and its consequences for particle acceleration. In: Klare G (ed) Reviews in modern astronomy, vol 8, pp 323–348 Astronomische Gelleschaft, Hamburg
Hesse M, Schindler K (1988) A theoretical foundation of general magnetic reconnection. J Geophys Res 93:5559–5567. doi:10.1029/JA093iA06p05559
Hood AW, Browning PK, van der Linden RAM (2009) Coronal heating by magnetic reconnection in loops with zero net current. Astron Astrophys 506:913–925
Hornig G, Priest ER (2003) Evolution of magnetic flux in an isolated reconnection process. Phys Plasma 10:2712–2721
Lau YT, Finn J (1990) Three-dimensional kinematic reconnection in the presence of field nulls and closed field lines. Astrophys J 350:672–691
Linton MG, Priest ER (2003) Three-dimensional reconnection of untwisted magnetic flux tubes. Astrophys J 595:1259–1276. doi:10.1086/377439
Longcope DW, Parnell CE (2009) The number of magnetic null points in the quiet sun corona. Solar Phys 254:51–75. doi:10.1007/s11207-008-9281-x, arXiv:0811.0097
Maclean RC, Parnell CE, Galsgaard K (2009) Is null-point reconnection important for solar flux emergence? Solar Phys 260:299–320. doi:10.1007/s11207-009-9458-y, arXiv:0910.0368
Maclean RC, Haynes AL, Parnell CE (2010) The detection of numerous magnetic separators in a 3D MHD model of solar emerging flux. Astrophys J Lett, submitted
Masson S, Pariat E, Aulanier G, Schrijver CJ (2009) The nature of flare ribbons in coronal null-point topology. Astrophys J 700:559–578. doi:10.1088/0004-637X/700/1/559
Parker EN (1957) Sweet's mechanism for merging magnetic fields in conducting fluids. J Geophys Res 62:509–520. doi:10.1029/JZ062i004p00509
Parnell CE (2007) Multiply connected source and null pairs. Solar Phys 242:21–41. doi:10.1007/s11207-007-0329-0
Parnell CE, Smith JM, Neukirch T, Priest ER (1996) The structure of threedimensional magnetic neutral points. Phys Plasmas 3:759–770. doi:10.1063/1.871810
Parnell CE, Haynes AL, Galsgaard K (2008) Recursive reconnection andmagnetic skeletons. Astrophys J 675:1656–1665. doi:10.1086/527532
Parnell CE, DeForest CE, Hagenaar HJ, Johnston BA, Lamb DA, Welsch BT (2009) A power-law distribution of solarmagnetic fields overmore than five decades in flux. Astrophys J 698:75–82. doi:10.1088/0004-637X/698/1/75
Parnell CE, Haynes AL, Galsgaard K (2010) Magnetic separators and separator reconnection. J Geophys Res 115(A14):2102–, doi:10.1029/2009JA014557

Petschek HE (1964) Magnetic field annihilation. NASA Spec Publ 50:425–+

Pontin DI, Craig IJD (2005) Current singularities at finitely compressible threedimensional magnetic null points. Physics of Plasmas 12:072112

Pontin DI, Galsgaard K (2007) Current amplification and magnetic reconnection at a three-dimensional null point: physical characteristics. J Geophys Res (Space Phys) 112(A11): 3103–+. doi:10.1029/2006JA011848, arXiv:astro-ph/0701555

Pontin DI, Galsgaard K, Hornig G, Priest ER (2005a) A fully magnetohydrodynamic simulation of three-dimensional non-null reconnection. Phys Plasmas 12:052307

Pontin DI, Hornig G, Priest ER (2005b) Kinematic reconnection at a magnetic null point: fan-aligned current. Geophys Astrophys Fluid Dyn 99:77–93

Pontin DI, Bhattacharjee A, Galsgaard K (2007a) Current sheet formation and nonideal behavior at three-dimensional magnetic null points. Phys Plasmas 14(5):052106–+. doi:10.1063/1.2722300, arXiv:astro-ph/0701462

Pontin DI, Bhattacharjee A, Galsgaard K (2007b) Current sheets at threedimensional magnetic nulls: Effect of compressibility. Phys Plasmas 14(5):052109–+. doi:10.1063/1.2734949, arXiv:physics/0701197

Priest ER, Démoulin P (1995) Three-dimensional magnetic reconnection without null points. 1. Basic theory of magnetic flipping. J Geophys Res 100:23443–23464. doi:10.1029/95JA02740

Priest ER, Forbes TG (2000) Magnetic reconnection. Cambridge University Press, Cambridge, UK

Priest ER, Titov VS (1996) Magnetic reconnection at three-dimensional null points. Phil Trans R Soc Lond 355: 2951–2992

Schindler K, Hesse M, Birn J (1988) General magnetic reconnection, parallel electric fields, and helicity. J Geophys Res 93:5547–5557. doi:10.1029/JA093iA06p05547

Sonnerup BUÖ (1979) Magnetic field reconnection. In: Space plasma physics: the study of solar-system plasmas, vol 2. National Academy of Sciences, Washington, DC, pp 879–+

Sweet PA (1958) The neutral point theory of solar flares. In: Lehnert B (ed) Electromagnetic phenomena in cosmical physics, IAU symposium, vol 6. Cambridge University Press, Cambridge, pp 123–+

Titov VS, Galsgaard K, Neukirch T (2003) Magnetic pinching of hyperbolic flux tubes. I. Basic estimations. Astrophys J 582:1172–1189. doi:10.1086/344799, arXiv:astro-ph/0208112

Vasyliunas VM (1975) Theoretical models of magnetic field line merging. I. Rev Geophys Space Phys 13:303–336

Wilmot-Smith AL, De Moortel I (2007) Magnetic reconnection in flux-tubes undergoing spinning footpoint motions. Astron Astrophys 473:615–623. doi:10.1051/0004-6361:20077455

Wilmot-Smith AL, Hornig G, Pontin DI (2009) Magnetic braiding and parallel electric fields. Astrophys J 696:1339–1347. doi:10.1088/0004-637X/696/2/1339, arXiv:0810.1415

Chapter 15

Current Sheets in the Solar Atmosphere

Giannina Poletto

Abstract Current sheets (CSs) appear in many solar events and are invoked to account for a variety of energy release processes. In this review, which focusses on observational aspects of CSs, I will initially describe a few typical phenomena where current sheets have been assumed to play a key role and I will analyze the (mostly indirect) evidence we have for their occurrence. As a consequence of recent advances in the observations of large-scale events like Coronal Mass Ejections (CMEs), we acquired an improved knowledge of CSs and of their physical parameters. Results about CSs temperature, density and resistivity, inferred from UV spectra and White Light (WL) data, as a function of time and heliocentric distance, are illustrated in some detail. The internal structure of CSs has been deemed to be unaccessible to observations, because of the small scales involved: however, recently, interesting ideas have been proposed to advance our understanding of this unexplored region. Open questions, and promising attempts to answer some of them, conclude the chapter.

15.1 Introduction

As fully illustrated in many textbooks (e.g. Priest 1982), the induction equation

$$\frac{\partial B}{\partial t} = \nabla \times (v \times B) + \eta \nabla^2 B \qquad (15.1)$$

shows that, in the highly conducting solar plasma, magnetic field diffusion can occur only in thin layers where the resistivity ($1/\sigma$, with σ conductivity) is dramatically enhanced. In Eq. 15.1 η is the magnetic diffusivity (which has the units of $l^2 t^{-1}$) and is given by $\eta = 1/\mu\sigma$, where μ is the permeability of free space. These thin layers where plasma can flow across the fieldlines and the magnetic field changes direction/magnitude, are dubbed Current Sheets (CSs).

The ratio between the first (convective) and the second (diffusive) term of the induction equation – the magnetic Reynolds number R_m – is for typical coronal parameters of the order of $10^8 - 10^{12}$. Hence the diffusion term is altogether negligible. On the contrary, if we want to be in the diffusive limit, we require $R_m << 1$ and we are forced to look for small-scale regions with high resistivity. In the Sweet-Parker reconnection model, for instance, that assumes a diffusion region whose length is much longer than its width, it can be easily shown that the CS thickness is on the order of 10^{-5} of its length: hence, a CS 10^8 cm long is only 10^3 cm wide, too thin to be directly observable. A length on the order of about 10^3 cm corresponds to the proton Larmor radius in the solar corona: a typical dimension assumed for CSs which, as a consequence, are expected to be undetectable. This widespread idea, however, is based on laboratory experiments and on quasi-static processes (e.g. Ono 1997; Wood and Newkirch 2005) and does not consider turbulent CSs, whose conditions may differ from those predicted by traditional theories.

In the following, I will give several examples of "observed" CSs, whose thickness appears to be much larger than expected from previous considerations and I will discuss what processes can lead to these sizes. Whether we are really observing CSs or, rather, plasma

G. Poletto (✉)
INAF - Arcetri Astrophysical Observatory, Largo Fermi, 5, 50125 Firenze, Italy
e-mail: poletto@arcetri.astro.it

M.P. Miralles, J. Sánchez Almeida (eds.), *The Sun, the Solar Wind, and the Heliosphere*, IAGA Special Sopron Book Series 4, DOI 10.1007/978-90-481-9787-3_15, © Springer Science+Business Media B.V. 2011

sheets within which CSs are embedded, is a matter of debate. In this chapter the term CS refers to the region where diffusion occurs (as in Lin et al. 2009), hence it may refer either to a plasma sheet, with its embedded CS, or to a turbulent, broadened, CS whose average properties only can be observed. Section 15.2 deals with the observational evidence of CSs, in different solar regions, while Section 15.3 describes recent results about the physical parameters of CSs. Recent ideas about CSs internal structure, open problems and promising attempts to answer some of them are illustrated, respectively, in Sections 15.4 and 15.5.

15.2 Current Sheets: Why and Where?

The very first question we need to answer is: Why are we interested in the occurrence and detection of CSs? We know that CSs are sites where magnetic field reconfiguration is occurring, in other terms CSs are sites where reconnection occurs and this process releases energy that can be converted into other energy forms. However, at the time the Sweet-Parker model was proposed (Sweet 1958; Parker 1963), it looked like CSs might be, after all, not that interesting, because the time over which energy dissipates was much too long to account for explosive phenomena in the solar atmosphere. The Petschek (Petschek 1964) modification of the previous model, by which a faster energy release is achieved, showed the relevance that CSs might have in accounting for solar energetic phenomena and promoted an interest in the theory and observations of CSs that continues nowadays. Time-dependent Petschek reconnection (e.g. Biernat et al. 1987), bursty reconnection (e.g. Riley et al. 2007; Bárta et al. 2008), fractal reconnection (e.g. Shibata and Tanuma 2001), collisionless reconnection (e.g. Cassak et al. 2006), 3-D reconnection (e.g. Linton and Longcope 2006) are examples of the variety of theoretical descriptions/developments that have been proposed through the years.

What about the observations, direct or indirect, of CSs? Where can we expect CSs to form? If we focus on small-scale CSs we may look at the outcome from the turbulent photospheric magnetic field; if we focus on large-scale CS we may look at flux rope/arcade eruption, or, going to faraway regions in the solar atmosphere realm, to the effect of the interaction of solar wind and the geomagnetic field. In the following, examples of these diverse CSs will be given.

15.2.1 Small-Scale Current Sheets

The most popular process where CSs are involved is the "nanoflare" coronal heating mechanism proposed by Parker (e.g. Parker 1987). According to Parker, small-scale CSs form because photospheric convection drags magnetic fieldlines around. Wandering fieldlines interlink with each other, eventually leading to the formation of CSs, where energy release, albeit tiny, may occur frequently enough to account for coronal heating. These CSs are not directly identifiable, although indirect evidence of their presence may be inferred from the observations of the flare frequency distribution (e.g. Veronig et al. 2002).

Small-scale CSs may form in the low photosphere – around the region of temperature minimum – when approaching magnetic features result in cancellation events. According to Litvinenko et al. (2007) reconnection jets associated with canceling magnetic fragments may provide for the filament mass. Once more, there is no direct observation of the CSs, whose size, according to the authors, is on the order of a few hundred meters.

These two examples show how CSs can be invoked in disparate phenomena. We cannot, however, really talk of "observation" of CSs, most likely easier in large scale events. Indeed, the first direct evidence for the presence of CSs in space plasma came from observations acquired in the magnetosphere, described in the following.

15.2.2 Large-Scale Current Sheets

Let us now look at the interaction of solar wind and the Earth's magnetosphere. Current sheets that form in this way are identified by the observation of accelerated/decelerated plasma flows, assumed to be signatures of Petschek-type exhausts: plasma brought into the CS sideways is expelled from the tips of the CS and observed as "reconnection exhausts". These are accompanied by further evidence that supports their interpretation in terms of the occurrence of CS: for instance, the profile of the magnetic field across the exhaust shows abrupt changes in the magnetic field component B_z at the edges of the exhaust and a B_z plateau within the exhaust (e.g. Gosling et al. 2006).

These CSs have been observed in situ by space experiments onboard Ulysses, ACE, Cluster and Wind: the successive observation of the same CS by different missions, led to the capability of estimating the length of the CS. Phan et al. (Phan et al. 2006) give a value on the order of 1/3 of the solar radii for the length of the CS observed by ACE, CLUSTER and WIND in February 2002. We will see in Section 15.5 how the near Earth space provides evidence also of altogether different, tiny CSs.

Another case where large-scale CSs can be expected is at the time of Coronal Mass Ejection (CMEs). Indeed, it has been the analysis of CME data that allowed us to evaluate previously unknown physical CS parameters. This is illustrated in the next Section.

15.3 Physical Parameters of Current Sheets

Large-scale catastrophic events like CMEs should offer the best opportunity for the observational detection of CSs, whose occurrence is predicted by CME models where the eruption of an unstable magnetic structure is accompanied by the formation of a large-scale CS (e.g. Antiochos et al. 1999; Lin and Forbes 2000). Focussing on the flux rope model of Lin and Forbes we expect a long CS joining the top of the low-lying loops of the re-forming arcade and the bottom part of the expanding CME bubble, as shown in the left panel of Fig. 15.1. As the CME bubble travels outwards, the CS length increases: A very crude estimate of the CS thickness may be done on the basis of mass conservation into and out of a steady state sheet with incompressible plasma. In this admittedly oversimplified case the ratio between the thickness d and length l of the CS is given by the Alfvén Mach number, i.e. by the ratio of the speed at which plasma is carried into the sheet to the speed at which plasma is ejected out of the ends of the sheet (the latter being on the order of the Alfvén speed). It turns out that d/l is $\approx 10^{-3}$, which leads to a thickness of 1,000 km, for a CS 1.5 solar radii long: a figure order of magnitudes larger than the Larmor radius. Then, if we have some hope to be able to detect CSs in the aftermath of CMEs, what/where should we look at?

The predicted position of the CS suggests us to look for signature of its occurrence in the middle to high corona. There are two experiments suitable for this analysis onboard the Solar Heliospheric Observatory (SOHO): the Ultraviolet Coronagraph Spectrometer

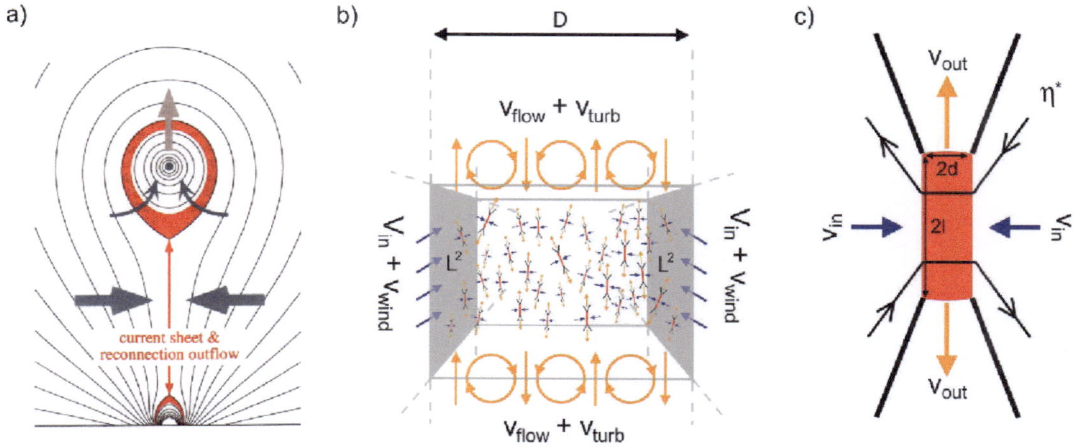

Fig. 15.1 *Left panel* **a**: Cartoon showing the current sheet linking the low lying reconnecting loop system to the outward traveling CME bubble, as envisaged in the model by Lin and Forbes. The *arrows* normal to the CS indicate plasma inflows toward the sheet. *Middle panel* **b**: The internal structure of the CS, in the scenario suggested by Bemporad (2008), showing the mini CSs within the macroscopic sheet of thickness D. Inflows towards the CS occur at a speed $V_{in} + V_{wind}$, outflows occur at a speed $V_{flow} + V_{turb}$. *Right panel* **c**: Cartoon illustrating one of the myriads of tiny sheets within the macro CS. (panels **b** and **c** have been adapted from Fig. 7 of Bemporad 2009)

(UVCS, Kohl et al. 1995) and the Large Angle Spectrometric Coronagraph Experiment (LASCO, Brueckner et al. 1995). The spectroscopic techniques that can be applied to the analysis of UVCS data hold promise for yielding results on the CS physical parameters, while the imaging capabilities of LASCO can provide us, possibly, with CS images at high coronal levels. Let us see whether UVCS data bear evidence of the CS occurrence.

15.3.1 Current Sheets: Temperature and Densities

Ciaravella et al. (2002) first revealed, in UVCS spectra acquired at ≈ 1.5–1.7 solar radii at the time of a CME development, the presence of enhanced emission in the 974 Å line from the [Fe XVIII] ion, at the location – below the CME bubble and above the reforming loop arcade in the low corona – where the CS was expected to show up. Because this ion has a very high formation temperature (≈ 5.10^6 K) and hardly appears even in active regions it should originate in a hot plasma, likely to be identified with the CS overheated region. After Ciaravella et al. paper, several other works analyzed UVCS CME associated spectra and detected emission in the 974 Å line (e.g. Ko et al. 2003; Bemporad et al. 2006; Lin et al. 2007; Schettino et al. 2009) at the CS location: this led to the use of this line at the time of CMEs as a proxy for the CS occurrence.

UVCS spectra show, at the CS position, emission also from other cooler ions, usually detected in coronal spectra. Lines from these ions originate from the ambient corona as well as from the CS plasma: the line ratio technique allows the CS electron temperature to be inferred, whenever the CS contribution to the total line intensity can be isolated and the intensity of the [Fe XVIII] 974 Å line, which forms solely in the CS, is also measured.

The intensity I of a collisionally excited line from the *nth* ion of the element with abundance A depends on the electron temperature T_e and density N as shown by the relationship

$$I = A \int_{LOS} f(T_e) N^2 dx \qquad (15.2)$$

where $f(T_e)$ is a function of the electron temperature that includes the ionization equilibrium of the ion and the integration extends along the line of sight (LOS). Equation 15.2 shows that the ratio of the intensities of two lines of different ions of the same element depends only on temperature, while the ratio of lines from ions of different elements depends also on the ratio of the element abundances. Figure 15.2 summarizes results obtained by different authors (Ciaravella et al. 2002; Ko et al. 2003; Bemporad et al. 2006; Ciaravella and Raymond 2008; Schettino et al. 2009) for the CSs temperatures in different CMEs and at different times after the CME ejection. T_e in CSs turns out to be higher than in the ambient corona by a factor that, early in the CS formation, may possibly be as high as 10, and slowly decreases over a time scale of a few days. This behavior appears to be shared by all events, from about 0.5 days after the CS formation onwards. The initial

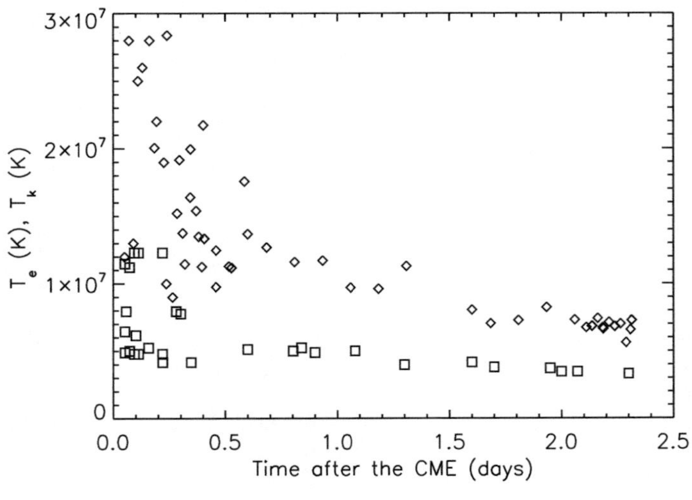

Fig. 15.2 Plot of electron (*squares*) and kinetic (*diamonds*) temperatures vs. time after the CME ejection, inferred from different authors from CME UVCS data. The plot includes all available values

temperature of CSs, though, is not adequately known, as only two CSs have been observed in their earliest stage of formation and the inferred temperature values disagree by $\approx 50\%$. Even if the results of Fig. 15.2 suggest a behavior common to all CME associated CSs, there may be a selection effect favoring long-lasting CSs, whose detection is easier, and, possibly, a temperature/wavelength selection effect, should the CS emit over a temperature/wavelength range not covered by UVCS. Also, the results of Fig. 15.2 refer to a single level in the corona (more precisely: values refer to data acquired at 1.5–1.7 solar radii) and little is known about the profile of T_e vs. the heliocentric distance within the CS. Bemporad et al. (2007) give electron temperatures increasing by $\approx 10\%$, throughout the CME, over the interval 1.6–1.9 solar radii. This result implies heating during the CME expansion and should be confirmed by further observations.

UVCS provided a further valuable information piece about the CSs temperature: the measured width of the [Fe XVIII] 974 Å line profile can be used to calculate the kinetic temperature T_k from the relationship

$$\Delta\lambda = 2\sqrt{ln2}\frac{\lambda_0}{c}\sqrt{\frac{2kT_k}{m}} \tag{15.3}$$

where λ_0 is the wavelength of the line centroid, m is the ion mass and T_k is given by

$$T_k = T_e + \frac{m}{2k}v_{nth}^2 \tag{15.4}$$

where v_{nth} is the non thermal turbulent speed of a plasma whose ion temperature has been assumed equal to the electron temperature. Kinetic temperatures are given in Fig. 15.2 and are sensibly higher than electron temperatures, revealing the CS plasma to be in a turbulent state. The turbulent speed decreases from ≈ 60 to ≈ 30 kms^{-1} over a couple of days, with a similar behavior for all the events. The spectroscopic evidence of turbulence in CSs adds to the evidence provided by WL images (see Section 15.3.2) at higher heliocentric distances and gives a crucial information on the plasma within the CS (although it is fair to say that the large width of the line could be ascribed to, e.g., the presence of waves in the CS). We will come back in Section 15.5 to discuss the implication of this result.

Densities in CSs can be inferred from Eq. (15.2), once T_e is known, from Emission Measure ($EM = \int N^2dx$) values, provided the size of the CS along the LOS is known. This, however, is practically unknown and density values are affected by the uncertainty in this quantity. Inferred values range between 1 and 10, in units of 10^7 cm^{-3}, at heliocentric distances of 1.5–1.7 solar radii, with CS LOS depth on the order of 0.06–0.5 solar radii. Should the thickness values turn out to be realistic, CSs densities will be higher than densities in streamer, at comparable heliocentric distances. Parenti et al. (2000) give densities between 3.3 10^6 and 9.42 10^6 cm^{-3} for equatorial and mid-latitude streamers at 1.6 solar radii. It is interesting to notice that, in the only case where the CS extension along the LOS has been estimated from a combination of UVCS and Mauna Loa Solar Observatory Mark IV K-coronameter data (Ciaravella and Raymond 2008), the CS depth and densities turned out to be within the range of the above cited values.

If we now look for the signatures of CSs in the higher corona, we need to resort to LASCO images. White Light (WL) thin, bright features, with a lifetime on the order of a few days, show up, in the aftermath of CMEs, along the direction joining the reconnecting loop tops to the CME bubble, i.e., along the same direction where the bright UVCS Fe XVIII emission lies. It becomes obvious to interpret these "rays" as structures that form as a consequence of the CME ejection, to be possibly identified with the model predicted CS.

A WL brightness enhancement implies a higher electron density: Vršnak et al. (2009) analyzed several CME associated WL rays and inferred their density vs. heliocentric distance profile from LASCO mass images (e.g. Vourlidas et al. 2000) under the working hypothesis that they appear as a consequence of steady-state Petschek reconnection, wherein two pairs of standing slow-mode shocks (SMSs) originate from the top section of the diffusion region. The excess density of the ray is then caused both by the density enhancement across the SMS and by the upward transport of denser plasma from lower regions along the ray. Over the height interval between 2 and 6 solar radii, rays present an excess density, with respect to the ambient corona, up to one order magnitude. Essentially, rays would be the reconnection outflows jets, and their excess density is consistent with results from models that assume the diffusion region to be located at heliocentric distances between 1.1–1.5 solar radii. Ray densities are expected to decrease in time: data analyzed so far are not enough, though, to draw definite conclusions about this process.

15.3.2 *Current Sheets: Inflows and Outflows*

The description of the physical state of CSs would not be complete, without mentioning the plasma motions in and around them. As we said in Section 15.2.2, evidence for CSs in the solar wind/magnetospheric plasma came from the observations of plasma flows. In the chromosphere, the observations of bidirectional jets (Innes et al. 1997) via the analysis of Doppler shifts in spectra acquired by the Solar Ultraviolet Measurements of Emitted Radiation (SUMER, Wilhelm et al. 1995) on SOHO, confirmed the association between explosive events and reconnection, at the time suggested on the basis of other, more marginal, evidence. However, the small scale of the phenomena did not allow a detailed study of the flow properties.

On the contrary, in most of the WL rays described above (Vršnak et al. 2009) it has been possible to identify the outward motion of blob-like structures, at speeds of the order of 10^2–10^3 kms^{-1}. Outflowing blobs are not unusual and have been observed in other cases as well (e.g. Ko et al. 2003). These features turn out to provide interesting information on the CS plasma and are further discussed in Section 15.4. Sunward flows have been observed in "supra-arcade" (i.e. structures above post-flare X-ray loops) associated with CMEs (Innes et al. 2003a). Innes et al. (2003b) detected high plasma flows (up to $\approx$1,000 kms^{-1}) and high Doppler widths in the Fe XXI line observed by SUMER above the flare arcade that developed on 21 April 2002, when a large CME was observed as well. While these data show the occurrence of plasma with high flows and at higher temperatures than observed by UVCS (that, however, takes data at higher heliocentric distances), their interpretation in terms of reconnecting scenarios is, at this time, not clear and further work is needed to incorporate Innes et al. results in reconnection models.

The detection of flows towards the CS is more difficult. Yokoyama et al. (2001) inferred values of $\approx$5 kms^{-1} and a similar value has been given by e.g. Bemporad et al. (2008). Higher values have been derived by Vršnak et al. (2009) and by Lin et al. (2007), who give speeds of $\approx$ 25–60 kms^{-1}, depending on the observation altitude and the time after CME ejection. Narukage and Shibata (Narukage and Shibata 2006), from EIT data, give values ranging from 2.6 to 38 kms^{-1} and suggested a correlation between the CME speed and the inflow velocity. According to Lin et al. reconnection inflows slow down sensibly, decreasing by a factor $\approx$10 over about 500 s from the CME ejection. Authors adopted diverse indirect technique to infer the inflow speed and it is difficult to compare their results.

15.4 The Size and Internal Structure of Current Sheets

In Section 15.3 we left out any consideration about the dimension of the region with enhanced Fe XVIII emission – that we considered as a CS proxy – or about the transverse dimension of WL rays. Let us now go back to this issue: the section of the UVCS slit occupied by Fe XVIII emission is on the order of a few times 10^4 up to 10^5 km (e.g., Lin et al. 2007 and references herein). If this unexpectedly large size corresponds to the real dimension of the CS, its effective electrical resistivity η_e should be very high. We can calculate its value from measured values of v_i and d, because we have, from the standard reconnection theory

$$v_i = \frac{\eta}{d} \tag{15.5}$$

where v_i is the inflow speed. It turns out that the electrical resistivity $\eta_e = \eta/\mu$ is on the order of 5. 10^5 ohm m. This "observed" value can be compared with the classical resistivity η_c and, for instance, the ion-acoustic anomalous resistivity η_a for temperature on the order of 1 million degree and densities on the order of 10^9 cm^{-3}. It follows that (Priest 1982)

$$\eta_c \approx 4\pi\, 10^2 T^{-3/2} = 10^{-6} \text{ ohm m} \tag{15.6}$$

$$\eta_a \approx 6.4\pi\, 10^6 N^{-1/2} = 60 \text{ ohm m} \tag{15.7}$$

and we conclude that there is a discrepancy of orders of magnitude between the "observed" resistivity and the classical or anomalous resistivity. Are there alternative processes that can raise the resistivity to values that high?

Lin et al. (2009) have recently discussed this issue and pointed out that such large values possibly result from an improper use of Eq. (15.5), valid in conditions where classical/anomalous resistivity are responsible for diffusion processes. However, the authors suggest that Eq. (15.5) might adequately describe average properties of diffusion sites such as those inferred from the UVCS data analyses, most likely representative of average properties of a region embedding smaller CS structures. Lazarian and Vishniac (1999), for instance, propose a a scenario where a stochastic component of the magnetic field is responsible for multiple reconnection within sheets. This turbulent sheet is much broader than predicted by classical values, perhaps only one order of magnitude smaller than the previously inferred observed values.

Turbulence can indeed broaden CSs (Drake et al. 2006) and we found strong indications favoring turbulent CSs, besides the high kinetic temperatures shown in Fig. 15.2. In Section 15.3.2 we described CS outflows inferred from the outward motion of WL bloblike features: these are interpreted in terms of the tearing instability that sets in when the CS length becomes of the order of a few times its width (Furth et al. 1963). Numerical time-dependent MHD models of reconnection show the formation of magnetic islands, likely promoted by tearing instability and to be identified with blobs, confirming reconnection to be unsteady, or "bursty" as it is often referred to (e.g. Riley et al. 2007; Bárta et al. 2007). Time-dependent Petschek reconnection (e.g. Priest and Forbes 2000 and references therein) results in the formation of individual outflow regions, surrounded by slow mode shocks, that also point to turbulent CSs. Recently, Schettino et al. (2009) have shown that UVCS Fe XVIII data provide evidence of transient CSs, beside the main CS, that support the suggestion that a CS may become occasionally observable at the time of formation of tearing instability structures. In conclusion, the large values of the CS thickness d and electrical resisitivity η_e may be indicative of turbulent CSs and be not that absurd.

That CSs can be turbulent, or, more precisely, fractal, has been suggested by Shibata and Tanuma (2001), who pointed out that the tearing instability leads to a thinning of the CS, because of the strong inflow triggered by the plasmoid ejection. The process occurs repeatedly, until the microscopic scale is reached: eventually a fractal CS forms with magnetic islands of different sizes, the largest ones originated via coalescence of smaller structures. An observational check of the fractal nature of CS is possible, looking for the distribution of tracers of the energy release process vs. their size or duration, which should follow a power law (Nishizuka et al. 2009). A power law shape has indeed been observed in spectra of radio data acquired at the time of flares (Aschwanden 2002; Karlický et al. 2005).

Filling the gap between the macroscopic CSs apparently observed and the microscopic scale over which reconnection is expected to occur, has always been a major concern in reconnection studies. The ideas described above of turbulent or fractal CS may help building up a realistic scenario. Recently, Bemporad (Bemporad 2008) worked out a model where multiple micro CSs fill up the region where a CS is observed and found how many of these are needed to comply with the high temperature, thick CME CSs observed by UVCS. Fig. 15.1, panels b and c, show Bemporad model: in case of a turbulent CS, with, e.g., anomalous resistivity induced by ion-acoustic instabilities, 10^{10} micro CSs with sizes of the order of 10 – 80 m (respectively, $2d$ and $2l$, in panel 3 of Fig. 15.1), distributed within a sheet volume of $(10^4)^3$ km^3 can account for the observed CS properties.

15.5 Conclusions and Open Questions

From the observations described in Section 15.2.1 and the discussion in the previous section, it appears safe to conclude that CSs, or at least those associated with CMEs, are turbulent. One of the most turbulent medium we know of, is the magnetosheath, the region downstream of the bow shock that forms from the interaction of the solar wind and the terrestrial magnetosphere (Retinó et al. 2007). We may ask whether we can learn anything from the observations of the magnetosheath.

The region has been explored by the Cluster (Escoubet et al. 1997) mission. In 2002 the four spacecraft of the Cluster mission traversed a number of thin CSs, whose width was on the order of $\approx$100 km, i.e. on the order of the ion inertial length or the ion gyroradius. The signatures detected by Cluster were consistent with those expected from Hall reconnection (e.g. Sonnerup 1979; Ren et al. 2005) that occurs on spatial scale small enough that electron and ion motions decouple. Hall reconnection is fast, as verified by Cluster data that give a dimensionless reconnection rate v_{in}/v_A on the order of 0.1.

This observation is quite relevant in demonstrating that a turbulent medium allows for fast reconnection and raises the issue of whether Hall reconnection may occur in coronal plasma as well. Open questions we have, for which Hall reconnection may provide answers, include what causes fast reconnection in solar events and how to explain the sudden initiation of reconnection within a slow energy build-up process. Longcope et al. (2005) for instance, analyzing Transition Region and Coronal Explorer (TRACE, Handy et al. 1999) observations, pointed to a slow accumulation of magnetic energy over about a day, followed by a sudden onset of a brief reconnection episode which lasted for ≈ 3 h. Cassak et al. (2008) presented a model whereby Sweet–Parker reconnection spontaneously evolves towards Hall reconnection whenever the dissipation region becomes so thin to match the ion skin depth. According to these authors, fast reconnection occurs rapidly when this condition is met.

The role of Hall reconnection in coronal plasma is a very active research area nowadays, both theoretically (e.g. Litvinenko 2009 and references therein), and for the observational viewpoint. Cassak et al. (2008) analyzed flares from a sample of Sun-like stars to search for observations that might support their model. Because the pre-flare reconnection is undoubtedly collisional, reconnection is initially occurring on the slow Sweet-Parker time scale within a sheet whose width is much larger than the ion gyroradius δ_i. Eventually, however, if the reconnecting fields are strong enough, the sheet width δ_{SP} (inversely proportional to the square root of the field) becomes smaller than δ_i. At this stage, Hall reconnection may suddenly initiate, releasing energy stored during the Sweet-Parker regime. If the magnetic field are not that strong, Hall reconnection does not set in, accounting for regions not undergoing eruptions. The authors claim that, calculating δ_{SP} and δ_i from stellar parameters, it turns out that the Sweet-Parker thickness matches, at the time of flares, the ion gyroradius, thus providing an observational test for their model. Although the works of Cassak and co-authors are not exempt from criticisms (e.g. Craig et al. 2008) the inclusion of collisionless effects in resistive models of reconnection holds promise to advance our present knowledge of the reconnection process.

In an alternative way of interpreting observations, the large widths of the Fe XVIII UVCS emission and of the WL rays seen in LASCO images may assumed to be evidence of the region between the slow mode shocks of the Petschek model, rather than being the actual CS, as suggested by Vršnak et al. (2009). In this case the predicted widths are consistent with those that are observed, but it becomes difficult to explain the origin of the Fe XVIII line widths, as Petschek mechanism is not expected to produce turbulence (Ciaravella and Raymond 2008). Also, so far we lack evidence for the presence of the shocks: another open question that waits to be answered.

As mentioned earlier, we may be looking at a plasma sheet, rather than at a current sheet. This scenario avoids searching for exotic ways to enhance the electrical resistivity and, as any simple solution to a problem, is very appealing. Recently, Lui et al. (2009) pointed out that the heliospheric CS is narrower by at least a factor ten than the heliospheric plasma sheet within which is embedded. According to the authors, this suggests an analogy between the heliospheric and the coronal configuration, where the UVCS/WL features – which have about the same width as the heliospheric plasma sheet – are plasma sheet that analogously embed the undetectable CS. Liu et al. work focusses on STEREO data and potential field simulations: further work is needed to see how the high temperatures and broad line widths inferred from spectroscopic data would fit into the proposed scenario.

In the limited space of this review, we did not even mention issues like particle acceleration in CSs, or 3-D reconnection, which are nonetheless no less relevant. Solution to the problems discussed above may be relevant for particle acceleration models as well. Future observations will help also answer other questions: densities inferred from UVCS data, largely affected by the unknown depth of the CS, will be refined if STEREO will be able to measure the 3-D structure of the current/plasma sheet. Knowledge of the profile of the electron temperature T_e vs. the heliocentric distance will tell us whether the CS is roughly isothermal, justifying the tacit assumption of ionization equilibrium adopted by spectroscopic techniques.

Recognizing how wide is the range of astrophysical phenomena where reconnection occurs, makes the analysis of reconnection-related data especially interesting: this an area where astrophysical observations may really help developing more realistic model of the physics of the process.

Acknowledgements The author thanks the meeting organizers for the opportunity to participate in the session "Universal Heliophysical Processes" and give a talk there. Support from ASI/INAF/ I/023/09/0 is gratefully acknowledged.

References

Antiochos SK, DeVore CR, Klimchuk JA (1999) A model for solar coronal mass ejections. ApJ 510:485–493

Aschwanden MJ (2002) Particle acceleration and kinematics in solar flares: a synthesis of recent observations and theoretical concepts. Space Sci Rev 101:1–227

Bárta M, Vršnack B, Karlický MJ (2008) Dynamics of plasmoids formed by the current sheet tearing. A&A 477:649–655

Bemporad A (2008) Spectroscopic detection of turbulence in post-CME current sheets. ApJ 689:572–584

Bemporad A, Poletto G, Suess ST, Ko Y-K, Schwadron NA, Elliott HA, Raymond JC (2006) Current sheet evolution in the aftermath of a CME event. ApJ 638:1110–1128

Bemporad A, Raymond J, Poletto G, Romoli M (2007) A comprehensive study of the initiation and early evolution of a coronal mass ejection from ultraviolet and white-light data. ApJ 655:576–590

Biernat HK, Heyn MF, Semenov VS (1987) Unsteady Petschek reconnection. JGR 92:3392–3396

Brueckner G, Howard RA, Koomen MJ et al (1995) The large angle spectroscopic coronagraph (LASCO). Solar Phys 162:357–402

Cassak PA, Drake JF, Shay MA (2006) A model for spontaneous onset of fast magnetic reconnection. ApJ 644:L145–L148

Cassak PA, Mullan DJ, Shay MA (2008) From solar and stellar flares to coronal heating: theory and observations of how magnetic reconnection regulates coronal conditions. ApJ 676:L69–L72

Ciaravella A, Raymond JC, Li J, Reiser P, Gardner LD, Ko Y-K, Fineschi S (2002) Elemental abundances and post-coronal mass ejection current sheet in a very hot active region. ApJ 575:1116–1130

Ciaravella A, Raymond JC (2008) The current sheet associated with the 2003 November 4 coronal mass ejection: density, temperature, thickness, and line width. ApJ 686:1372–1382

Craig IJD, Litvinenko YE (2008) Influence of the Hall effect on the reconnection rate at line-tied magnetic X-points. A&A 484:847–850

Drake JF, Swisdak M, Schoffler KM, Rogers BN, Kobayashi S (2006) Formation of secondary islands during magnetic reconnection. Geophys Res Lett 33:L13105

Escoubet CP, Schmidt R, Goldstein ML (1997) Cluster – Science and mission overview. Space Sci Rev 79:11–32

Furth HP, Killeen J, Rosenbluth MN (1963) Finite-resistivityinstabilities of a sheet pinch. Phys Fluids 6:459–484

Gosling JT, Eriksson S, Skoug RM, McComas DJ, Forsyth RJ (2006) Petschektype reconnection exhausts in the solar wind well beyond 1 AU: Ulysses. ApJ 644:613–621

Handy BM, Acton LW, Kankelborg CC et al (1999) The transition region and coronal explorer. Solar Phys 187:229–260

Innes D, Inhester B, Axford WI, Wilhelm K (1997) Bi-directional plasma jets produced by magnetic reconnection on the Sun. Nature 386:811–813

Innes D, McKenzie DE, Wang T (2003a) SUMER spectral observations of post-flare supra-arcade inflows. Solar Phys 217:247–265

Innes D, McKenzie DE, Wang T (2003b) Observations of observations of post-flare supra-arcade inflows. Solar Phys 217:247–265

Karlický M, Bárta M, Mészárosová H, Zlobec P (2005) Time scales of the slowly drifting pulsating structure observed during the April 12, 2001 flare. A&A 432:705–712

Ko Y-K, Raymond JC, Lin J, Lawrence G, Li J, Fludra A (2003) Dynamical and physical properties of a post-coronal mass ejection current sheet. ApJ 594:1068–1084

Kohl JL, Esser R, Gardner LD et al (1995) The ultraviolet coronagraph spectrometer for the solar and heliospheric observatory. Solar Phys 162:313–356

Lazarian A, Vishniac ET (1999) Reconnection in a weakly stochastic field. ApJ 517:700–718

Lin J, Forbes TG (2000) Effects of reconnection on the coronal mass ejection process. JGR 105:2375–2392

Lin J, Li J, Forbes TG, Ko Y-K, Raymond JC, Vourlidas A (2007) Features and properties of coronal mass ejection/flare current sheets. ApJ 658:L123–L126

Lin J, Li J, Ko Y-K, Raymond J (2009) Investigation of thickness and electrical resistivity of the current sheet in solar eruptions. ApJ 693:1666–1677

Linton MG, Longcope D (2006) A model for patchy reconnection in three dimensions. ApJ 642:1177–1192

Litvinenko YE, Chae J, Park S-Y (2007) Flux pile-up magnetic reconnection in the solar photosphere ApJ 662:1302–1308

Litvinenko YE (2009) Analytical description of steady magnetic reconnection in Hall magnetohydrodynamics. ApJ 694:1464–1470

Longcope DW, McKenzie DE, Cirtain J, Scoot J (2005) Observations of separator reconnection to an emerging active region. ApJ 630:596–614

Lui Y, Luhmann JG, Lin RP, Bale SD, Vourlidas A, Petrie GJD (2009) Coronal mass ejections and global coronal magnetic field reconfiguration. ApJ 698:L51–L55

Narukage N, Shibata K (2006) Statistical analysis of reconnection inflows in solar flares observed with SOHO EIT. ApJ 637:1122–1134

Nishizuka N, Asai A, Takasaki H, Kurokawa H, Shibata K (2009) The power-law distribution of flare kernels and fractal current sheets in a solar flare. ApJ 694:L74–L78

Ono Y, Inomoto M, Okazaki T, Ueda Y (1997) Experimental investigation of threecomponent magnetic reconnection by use of merging spheromaks and tokamaks. Phys Plasma 5:1953–1963

Parenti S, Bromage BJI, Poletto G, Noci G, Raymond JC, Bromage GE (2000) Characteristics of solar coronal streamers. A&A 363:800–814

Parker EN (1963) The solar-flare phenomenon and the theory of reconnection and annihiliation of magnetic fields. ApJS 8:177–210

Parker EN (1987) Magnetic reorientation and spontaneous formation of tangential discontinuities in deformed magnetic fields. ApJ 318:876–887

Petschek HE (1964) Magnetic field annihilation. In: Hess WN (ed) The physics of solar flares, NASA SP Symposium,Washington, DC, p. 425

Phan TD, Gosling JT, Davis MS, Skoug RM, Øieroset M, Lin RP, Lepping RP, McComas DJ, Smith CW, Reme H, Balogh A (2006) A magnetic reconnection X-line extending more than 390 Earth radii in the solar wind. Nature 439:175–178

Priest ER (1982) Solar magneto-hydrodynamics. Reidel, Dordrecht

Priest ER, Forbes T (2000) Magnetic reconnection – Magnetohydrodynamics theory and application. Cambridge University Press, New York, NY

Ren Y, Yamada M, Gerhardt S, Ji H, Kulsrud R, Kuritsyn A (2005) Experimental verification of the Hall effect during magnetic reconnection in a laboratory plasma. Phys Rev Lett 95:055003

Retinó A, Sunkvist D, Vaivads A, Mozer F, André M, Owen CJ (2007) In situ evidence of magnetic reconnection in turbulent plasma. Nat. Phys 3:235–238

Riley P, Lionello R, Mikić Z, Linker J, Clark E, Lin J, Ko Y-K (2007) Bursty reconnection following solar eruptions: MHD simulations and comparison with observations. ApJ 655:591–597

Schettino G, Poletto G, Romoli M (2009) UV transient brightenings associated with a coronal mass ejection. ApJ 697:L72–L76

Shibata K, Tanuma S (2001) Plasmoid-induced-reconnection and fractal reconnection. Earth, Planets Space 53:473–482

Sonnerup BUO (1979) Solar system plasma physics, vol 3. North-Holland, Amsterdam

Sweet PA (1958) The neutral point theory of solar flares. In: Lehnert B (ed) IAU Symp. Electromagnetic phenomena in cosmical physics. Cambridge University Press, Cambridge, pp 123–133

Veronig A, Temmer M, Hanslmeier A, Otruba W, Messerotti M (2002) Temporal aspects and frequency distributions of solar soft X-ray flares. A&A 382:1070–1080

Vourlidas S, Subramanian P, Dere KP, Howard RA (2000) Large-Angle spectrometric coronagraph measurements of the energetics of coronal mass ejections. ApJ 534:456–467

Vršnak B, Poletto G, Vujić E, Vourlidas A, Ko Y-K, Raymond JC, Ciaravella A, Žic T, Webb DF, Bemporad A et al (2009) Morphology and density structure of post-CME current sheets. A&A 499:905–916

Wilhelm K, Curdt W, Marsch E et al (1995) SUMER – Solar ultraviolet measurements of emitted radiation. Solar Phys 162:189–231

Wood P, Newkirch T (2005) Electron acceleration in reconnecting crurrent sheets. Solar Phys 226:73–95

Yokoyama T, Akita K, Morimoto T, Inoue K, Newmark J (2001) Clear evidence of reconnection inflow of a solar flare. ApJ 546:L69–L72

Chapter 16

Solar Energetic Particles

Eino Valtonen

Abstract Some observational aspects of solar energetic particles (SEPs) are briefly discussed. The shift from the flare paradigm of SEP origin through the two-class paradigm of impulsive and gradual SEP events to hybrid events is described. Comprehensive investigations of coronal mass ejections (CMEs) and solar energetic particles during the SOHO era have shown that there is a good correlation between the occurrence of large SEP events and fast and wide CMEs. Shocks driven by these CMEs in the solar corona and interplanetary space accelerate the solar wind suprathermal particles. Observations of ionic charge states and elemental composition of high-energy solar particles indicate that also flare material is present in the seed populations of large gradual SEP events. Whether this material is remnant from previous flares and accelerated by CME-driven shocks or is directly accelerated in the flare processes, is still under vigorous research. Significant progress is expected from future solar missions reaching close to the sun and leading to better understanding of the processes in the inner heliosphere.

16.1 Introduction

A continuous flow of solar wind emanates from the sun. In addition to these low-energy charged particles, enhanced fluxes of high-energy particles can be occasionally observed. These solar energetic particles (SEPs) are manifestations of isolated solar energy release processes. According to the present understanding these high-energy particles originate either from solar flare processes or are accelerated by shock waves driven by coronal mass ejections (CMEs) in the solar corona and interplanetary space. In spite of five decades of in situ measurements in space, the exact nature of processes accelerating particles up to relativistic energies is not understood.

The discipline of solar energetic particles is wide, covering topics from particle acceleration to interplanetary transport and including experimental investigations as well as development of theories and modeling. In this chapter, I will review only some selected observational aspects of solar energetic particles. Section 16.2 includes a brief history of SEP research. In Sections 16.3–16.5 some recent results are reviewed. Section 16.3 concentrates on the relations between solar energetic particles and coronal mass ejections. Observations of ionic charge states, elemental composition, and energy spectra of SEPs are briefly discussed in Sections 16.4, and 16.5 presents examples of the significance of the SEP seed populations. In Section 16.6, some observations that do not fit in the current paradigm of SEP events are summarized. Future prospects to enlighten the open issues of solar energetic particle physics are pointed out in Section 16.7, and conclusions are presented in Section 16.8.

16.2 From the Solar Flare Paradigm to the Two-Class Paradigm

First solar energetic particle observations were reported by Forbush (1946) based on ionization chamber

E. Valtonen (✉)
Department of Physics and Astronomy, Space Research Laboratory, University of Turku, Turku, NSW FI-20014, Finland
e-mail: eino.valtonen@utu.fi

M.P. Miralles, J. Sánchez Almeida (eds.), *The Sun, the Solar Wind, and the Heliosphere*, IAGA Special Sopron Book Series 4, DOI 10.1007/978-90-481-9787-3_16, © Springer Science+Business Media B.V. 2011

measurements during the ground level events (GLEs) on February 28 and March 7, 1942 and on July 25, 1946. Forbush (1946) noted that the increases of cosmic-ray intensities occurred nearly simultaneously with solar flares. Further significant observations were carried out by Meyer et al. (1956) by using a neutron monitor network covering a wide range of geomagnetic latitudes. Meyer et al. (1956) studied the February 23, 1956 event, the largest event observed by then. Like the events observed by Forbush (1946), also this event was associated with a well-identified solar flare.

Significant progress in solar cosmic-ray studies was achieved in early 1960s by using rockets and, in particular, satellite instruments. Measurements of solar protons were carried out, e.g., on Explorer XII and Explorer XIV (Bryant et al. 1965). First observations of heavy ions from the sun were made with nuclear emulsions on a rocket flight (Fichtel and Guss 1961). The origin of high-energy particles in all these investigations was associated with solar flares, readily observable at optical wavelengths. The view of solar flare origin of SEPs was long-standing, and to explain how particles from solar flares at solar longitudes magnetically not well connected to Earth were able to rapidly reach Earth, as observed, required development of models for fast coronal transport of particles (e.g., Schatten and Mullan 1977; Kunow et al. 1991, and references therein). For a broader picture of early SEP observations, the reader is referred to, e.g., Fichtel and McDonald (1967) and Shea and Smart (1995).

Early-on views were also presented suggesting a role for solar mass ejections in solar particle production (e.g., Obayashi 1962). Based on comprehensive analysis of radio observations Wild et al. (1963) proposed a two-phase acceleration process, where protons observed in association with large flares had a second phase of acceleration in a shock by the Fermi mechanism. Later Kahler et al. (1978) concluded by using X-ray and white light coronagraph images and particle measurements that during the May 1973 to January 1974 period near solar minimum the occurrence of coronal mass ejections was a necessary requirement for observation of prompt solar proton events. They also suggested that energetic protons were accelerated in the shock front ahead of coronal mass ejections. This established the connection between solar energetic particles and coronal mass ejections. More comprehensive statistics on the association between SEP events and CMEs were provided by Kahler et al. (1984). They also found that peak proton fluxes correlated with both the speeds and the angular sizes of the associated CMEs.

Already in 1970 Hsieh and Simpson (1970) reported on a special class of SEP events with enhanced ^{3}He abundance by factors of ~100 compared to solar wind values. Later, events with ^{3}He enhancement by a factor of ~1,000 or more have been measured. It was also found that at low energies (~1 MeV/n) these small ^{3}He-rich events also had enhanced heavy ion abundances, in particular iron, compared to large solar particle events (e.g., Mason et al. 1986).

In a study of morphological features of limb X-ray flares Pallavicini et al. (1977) presented a classification of the flares into impulsive and gradual events based on their duration. The gradual flares were associated with coronal mass ejections (white light coronal transients). The connection between these two types of flares and solar energetic particles was established by Cane et al. (1986). They found that SEP events associated with impulsive flares were proton-poor, short-duration, and rarely associated with interplanetary shocks. SEP events associated with gradual flares were long-duration, rich in protons and reaching high energies, and in many cases strong interplanetary shocks were involved. Later Cane et al. (1988) found that the existence of an interplanetary shock was the major controlling factor of the time profiles of gradual events. Reames and Stone (1986) associated ^{3}He enhancements with impulsive flares, and subsequently Reames (1988) concluded that observed abundances of heavy elements showed evidence of the presence of two distinct populations of solar energetic particles.

These findings, together with measurements of ionization states of energetic ions indicating different source temperatures of particles in small ^{3}He-rich and in large events, led from the flare paradigm of SEP origin to the two-class paradigm of impulsive and gradual SEP events. Impulsive SEP events were defined to be small, short-duration events with particles accelerated in a flare-related resonant stochastic process. Gradual events were long-duration with high peak intensities. Particles in gradual events were accelerated in interplanetary shocks driven by coronal mass ejections. Impulsive events were confined to the western solar longitudes, where the magnetic connection to the observer at 1 AU was good. Gradual events, due to the extent of the interplanetary shock, had much wider distribution in solar longitudes reaching to the eastern limb and beyond the western limb. For a comprehensive review of impulsive and gradual solar particle events, their characteristics and differences, the reader is referred to Reames (1999) and references therein.

16.3 Solar Energetic Particles and Coronal Mass Ejections

Extensive statistical studies on the relations between coronal mass ejections and solar energetic particles have been carried out during recent years by applying the comprehensive CME database provided by the SOHO LASCO coronagraph (Brueckner et al. 1995). It has been shown that large SEP events are associated with fast and wide CMEs. The average speed of CMEs associated with large SEP events has been reported to be 1524 km/s with frequent association with halo CMEs (Gopalswamy et al. 2005). Kahler (2001) has shown that there is a correlation between the peak intensities of SEPs and the speed of the associated CMEs. This can be understood resulting from the strength of the CME-driven shock accelerating the particles. There are, however, variations of several orders of magnitudes in the SEP peak intensities at a given speed of the associated CMEs. Kahler (2001) concluded that high-intensity SEP events could result from enhanced ambient SEP population serving as seed particles for fast CME shocks. Gopalswamy et al. (2003, 2004) have further investigated the possible causes of these variations. Gopalswamy et al. (2003) found that high-intensity SEP events are likely to be associated with fast and wide CMEs preceded by other wide CMEs from the same solar source region. A possible explanation was that because the magnetic field lines of the preceding CME were still connected to the sun, the accelerated particles were able to return back to the succeeding shock for repeated acceleration.

A more comprehensive study was performed by Gopalswamy et al. (2004) investigating 57 large SEP events and dividing them in two main categories, one preceded and the other not preceded by wide CMEs from the same source region within 24 h. They found that although the properties of CMEs associated with SEP events in these two groups were similar, the SEP intensities were different. When a CME was preceded by another wide CME from the same source region, the corresponding SEP event usually had higher proton peak intensity as shown in Fig. 16.1. The extent of scatter in the CME speed versus SEP intensity plots was reduced when the two subgroups were analysed separately, and for frontside events the proton peak intensity with preceding CMEs showed a better correlation with the CME speed. In both of these two groups the SEP intensity showed poor correlation with

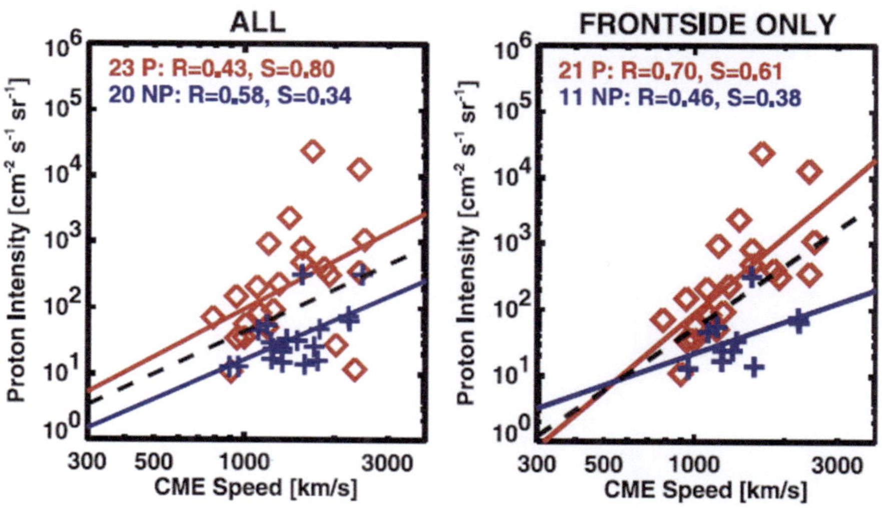

Fig. 16.1 Proton peak intensity vs. CME speed for events with a preceding CME (P, *diamond* symbols) and for events not preceded by another CME (NP, *plus* symbols). Note that the P and NP events form two distinct populations. For frontside P events (*right panel*) the correlation between CME speed and SEP peak intensity is better than for NP events. R is the correlation coefficient for each type of events and S the standard deviation of proton intensity. Reproduced from Gopalswamy et al. (2004). Copyright 2004, American Geophysical Union

the flare size. Gopalswamy et al. (2004) concluded that the inherent properties of the CMEs were not the reason for the differing intensities of SEP events in the two groups, but that the presence of preceding CMEs seemed to be the discriminating characteristic of the high- and low-intensity SEP events. Possible mechanisms of increased particle intensities were attributed to the effects related to the modification of the primary shock strength propagating in the environment of enhanced density produced by the preceding CME, effects due to changed magnetic configuration of the interplanetary magnetic field allowing particles to experience repeated acceleration, or to the existence of favorable seed particles as a consequence of the preceding CME.

Kahler and Vourlidas (2005) examined interactions of fast (>900 km/s) and wide (>60°) CMEs with the coronal environment with the purpose of determining possible factors contributing in resulting SEP peak intensities. They studied SEP events associated with CMEs from longitudes >W30°. From these they selected two extreme groups of SEP events: 15 SEP-rich events with the highest ~20 MeV proton intensities and 16 SEP-poor events with the lowest intensities associated with fast and wide western hemisphere CMEs within their observation period. They found the major differences to be that the SEP-rich CMEs were brighter and more likely to be streamer blowouts and to follow colocated CMEs within 12 or 24 h. Kahler and Vourlidas (2005) confirmed the result of Gopalswamy et al. (2004) that SEP event peak intensities are higher when the associated driver CMEs are preceded within a day by wide CMEs at the same locations. They also concluded that CME brightness, and hence mass, is the most obvious difference between SEP-rich and SEP-poor CMEs. The difference was clearly seen at low coronal heights, and Kahler and Vourlidas (2005) suggested that further SEP-related CME analyses should be focused on heights $\lesssim 6\,R_s$, where shock acceleration of SEPs is most efficient. Their finding that the widths and angular brightness of the SEP-rich CMEs significantly exceeded those of the SEP-poor CMEs indicated that SEP-rich CMEs must be broad in latitude and longitude. Based on the observation that a large fraction of SEP-rich CMEs occurred as streamer blowouts, an alternative requirement could be that SEP-rich CMEs occur in intrinsically dense coronal regions, such as streamers.

Metric type II solar radio bursts are considered as manifestations of coronal shock waves moving outward through the solar atmosphere. Because particles in large SEP events are assumed to be accelerated by coronal and interplanetary shocks, it might then be expected that there is a good correlation between the occurrences of metric type II bursts and large SEP events. Cliver et al. (2004) investigated metric type II bursts with solar western hemisphere sources, and found out that less than half of the metric type II bursts were associated with SEP events observed at Earth. When the metric type II bursts were accompanied by decametric-hectometric (DH) type II emission, the association with SEP events was much higher (90%). Small SEP events were more likely to be associated with metric type IIs than with DH type IIs and the association of SEP events with DH type II bursts increased rapidly with SEP event size (Fig. 16.2). For ~20 MeV proton events with peak fluxes >0.1 protons $cm^{-2}s^{-1}sr^{-1}MeV^{-1}$ the

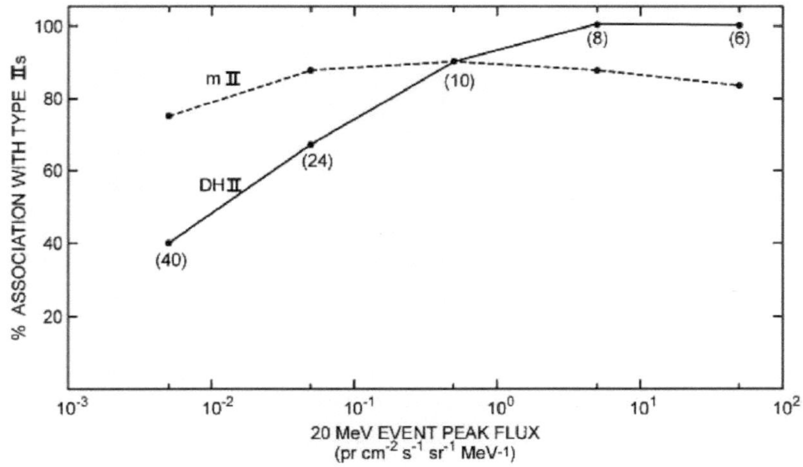

Fig. 16.2 Percentage association of ~20 MeV SEP events with metric type II (*dashed line*) and DH type II (*solid line*) bursts as a function of SEP event peak intensity. Reproduced from Cliver et al. (2004) by permission of the AAS

association with DH type II bursts was 96%. The interpretation of Cliver et al. (2004) for their observations was that large ~20 MeV SEP events result from strong shocks that are capable of persisting well beyond ~3 R_S. The shock acceleration is most efficient above this height and the shocks that survive beyond ~3 R_S are more likely to have broad longitudinal extents, enabling less well connected shocks to intercept open field lines connecting to Earth and enabling particle observations.

Gopalswamy et al. (2005) have shown that the highest-energy CMEs produce type II bursts that have counterparts in all the spectral domains from metric through decameter-hectometric to kilometric wavelengths. These m-to-km type II bursts imply that the shock, presumably driven by a CME, is present and strong enough to produce radio emission over a large range of distances from the sun. They also found a high degree of association between m-to-km type II bursts and SEP events, and the speed and width distributions of CMEs associated with SEP events and m-to-km type II bursts were nearly identical. The small number of m-to-km bursts that were not associated with SEP events had their solar sources at locations, which were magnetically poorly connected to the observer near Earth.

Gopalswamy et al. (2008a, b) studied association of SEP events with radio-quiet and radio-loud fast and wide CMEs. Radio-quiet CMEs were defined as those with no detectable type II radio emission in the metric or DH wavelengths. They found that some radio-quiet CMEs were associated with small SEP events suggesting that these CMEs may drive weak shocks. Weak shocks might result from these CMEs propagating in coronal and interplanetary environments, where the Alfven speed was exceptionally high. On the other hand, some radio-loud CMEs were not associated with SEP events, which was explained mainly by poor magnetic connectivity of the type II burst location to Earth. When the source location on the sun was taken into account, essentially all type II bursts in the DH wavelength range were associated with SEP events. The SEP association of type II bursts increased with increasing wavelength range from metric to decameter-hectometric and longer wavelengths indicating the importance of the longer wavelength type II bursts as indicators of SEP events (Gopalswamy et al. 2008b). This also supports the findings of Cliver et al. (2004) presented in Fig. 16.2.

16.4 Ionic Charge States, Composition, and Energy Spectra

The observed large variations of the ionization states of heavy ions, in particular of Fe, at energies below ~1 MeV/n have been one of the strong arguments in dividing SEP events into the two classes of gradual and impulsive events resulting from source regions with distinctly different temperatures. At low energies (~200–250 keV/n) the ionization states of iron in gradual and impulsive SEP events are completely different. The mean ionic charge distributions peak at ~9 in gradual events and between 15 and 17 in impulsive events (Klecker et al. 2007). Although the mean ionic charge of Fe in gradual SEP events below energies of a few 100 keV/n is usually similar to that of solar wind, at higher energies large differences have been found. At energies $\gtrsim$ 10 MeV/n the mean ionic charge of Fe in gradual SEP events has been observed to be significantly higher (~15–20) than at low energies (e.g., Oetliker et al. 1997; Labrador et al. 2005). Such results indicate that heavy ion charge states would not be in a simple way related only to the plasma temperature of the source region. In impulsive events there is a strong and monotonic increase of the mean ionic charge of Fe in the narrow energy range of ~0.1–0.55 MeV/n. This rapid increase can be explained by additional ionization of ions with increasing energy in a dense environment (Kocharov et al. 2000). Thus, the observed energy dependence of the ionic charge indicates that the acceleration region of impulsive SEPs is low in the corona at altitudes below ~0.2 R_S (Klecker et al. 2007).

Measurements with sensitive instruments have shown that the shape of the energy spectra of ions in ^{3}He-rich SEP events can be very different on event-by-event basis. In some events the spectra of ions from helium to iron can be power laws or broken power laws for all species. In other cases ^{3}He and Fe spectra can be curved at low energies and significantly different from ^{4}He spectra (Mason et al. 2002). These different spectral shapes also lead to very different ^{3}He/^{4}He ratios for different events and as function of energy in the same event. Therefore, no single ^{3}He/^{4}He ratio can be used to characterize the event, but the energy range under investigation must be defined (Mason 2007). The differences in spectral shapes may indicate energization of various ions in different processes. It has also been found that in many gradual events ^{3}He/^{4}He ratios are well above the solar wind ratio previously assumed for gradual SEP events (e.g., Torsti et al. 2003). At energies >10 MeV/n many gradual events are also enriched in heavy elements (Cohen et al. 1999), which was earlier expected to occur only in impulsive events. As in impulsive events, the abundance ratios in gradual events depend on the spectral shapes, which can lead to large event-to-event variation in, e.g., Fe/O ratio.

Mewaldt et al. (2005) studied the energy spectra of ions from hydrogen to iron in the large SEP events of October–November 2003. They found that all species had a relatively sharp break in their spectra at energies ranging from a few MeV/n to ~30 MeV/n. Lighter species had higher break energies than heavier ones like iron. They also found that the spectra could be fitted either by a double power-law or by a single power-law with exponential roll-over at high energies. The latter form follows from the theory of Ellison and Ramaty (1985). The break energies found by Mewaldt et al. (2005) are presented in Fig. 16.3 as a function of Q/M for the two spectral shapes. The break energies obey $(Q/M)^{1.56}$ or $(Q/M)^{1.75}$ dependence, similar to the theory of Li et al. (2005), which predicts the break energy to be proportional to the square of the charge-to-mass ratio. Recently, however, Li et al. (2009) studied the effect of shock geometry on the (Q/M) dependence of the spectral breaks and found that while the dependence for ions accelerated in quasi-parallel shocks was in accordance with the previous theory (Li et al. 2005), it was much weaker in quasi-perpendicular shocks. Spectral breaks indicate less efficient acceleration of particles in a shock above the break energies, perhaps due to less efficient scattering of the ions from the proton-amplified Alfven waves (Mewaldt et al. 2005).

16.5 Significance of Seed Populations

The observed large variability in the composition and energy spectra of large (gradual) SEP events and in ^{3}He-rich (impulsive) events raises the question of the origin of such variations. In particular, ^{3}He and heavy ion enhancements and increasing Fe/O ratio and ionic charge of Fe with energy in large SEP events challenge the view that large SEP events are caused by the acceleration of ambient coronal or solar wind ions at CME-driven shocks. It has been suggested (Mason et al. 1999) that CME-driven shocks might occasionally encounter and reaccelerate suprathermal flare-associated material enriched in ^{3}He and heavy ions. Acceleration of particles from such a source region could cause the observed compositional characteristics of some large SEP events.

Desai et al. (2006) surveyed the heavy-ion abundances in 64 large SEP events in the energy range ~0.1–10 MeV/n. In search of the source populations they compared the heavy-ion abundances with those measured in slow and fast solar wind and in particle populations accelerated near the sun and in interplanetary space. Desai et al. (2006) found event-to-event M/Q-dependent enhancements in heavy-ion abundances relative to ambient coronal values. These enhancements were similar to those seen in ^{3}He-rich SEP events. On the other hand, they also found that

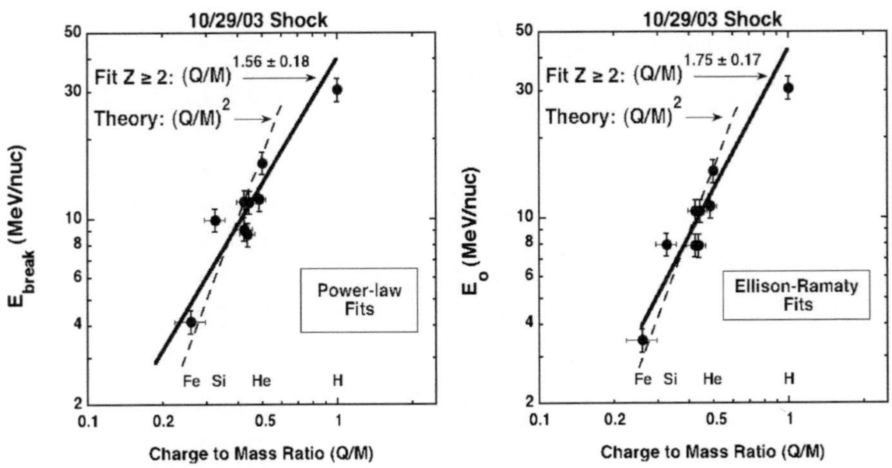

Fig. 16.3 Break energies as a function of Q/M for double power-law fit (*left*) and single power-law fit with exponential roll-over (*right*) to the observed energy spectra of October–November 2003 SEP events. In both cases the Q/M dependence is somewhat weaker than given by the theory of Li et al. (2005) (*dashed line*). Reprinted with permission from Mewaldt et al. (2005). Copyright 2005, American Institute of Physics

the Fe/O ratio in most large SEP events decreased with energy (Fig. 16.4). This would indicate acceleration mechanism in which ions with higher M/Q ratios were accelerated less efficiently. Desai et al. (2006) concluded that the M/Q-dependent enhancements of heavy-ion abundances in large SEP events are an inherent property of a suprathermal seed population that may be produced by processes similar to those occurring in impulsive SEP events. This heavy-ion-enriched material could then be accelerated at CME-driven shocks by rigidity-dependent processes causing species-dependent spectral breaks and consequently the Fe/O ratios to decrease with increasing energy.

Tylka et al. (2005) have stressed, in addition to the nature of the seed population, also the significance of the shock geometry in producing the observed large variations in the heavy ion abundances in large SEP events. As examples of extreme variations in elemental composition in large SEP events Tylka et al. (2005) considered two events: one (21 April 2002) with strongly decreasing Fe/C ratio with energy and another one (24 August 2002) with Fe/C ratio increasing above $\sim$10 MeV/n. The former behavior can be easily explained by the acceleration of material from the suprathermal tail of solar wind with Q/M-dependent spectral breaks as discussed above in Section 16.4. For the latter behavior a different explanation is required. Tylka et al. (2005) proposed that this behavior could be explained by the acceleration of suprathermals from flare activity. The acceleration would be by a quasi-perpendicular shock requiring for efficient acceleration a seed population with higher speeds, like those of the flare suprathermals exhibiting also an elemental composition with enhanced ^{3}He and heavy ions. The complete picture would then be acceleration by a quasi-parallel shock from the seed population of solar wind suprathermals corresponding to low ($\lesssim$10 MeV/n) observed SEP energies and by a quasi-perpendicular shock from the seed population of flare suprathermals at higher energies in events where the flare suprathermals happen to be present in significant amounts.

16.6 Hybrid Events

Large increase of ionic charge states at energies above 10 MeV/n and increase of the heavy-ion abundances at high energies suggest contributions of particles from various sources. The main seed populations seem to be the ambient solar wind and coronal material on one hand and flare material from the low corona on the other hand. Based on the two-class paradigm of the SEP origins it is usually assumed that also the flare material is further accelerated by a CME-driven coronal shock. There is, however, also a significant amount of evidence that the flare component contributing to SEP events would be directly injected from the flare.

Klein and Trottet (2001) carried out a detailed study of individual high energy particle events and found strong evidence that time-extended particle acceleration occurs in the corona after the impulsive flare and contributes to particle fluxes in space. They argued that although the diffusive shock acceleration works well up to some MeV, it is unlikely that the shock

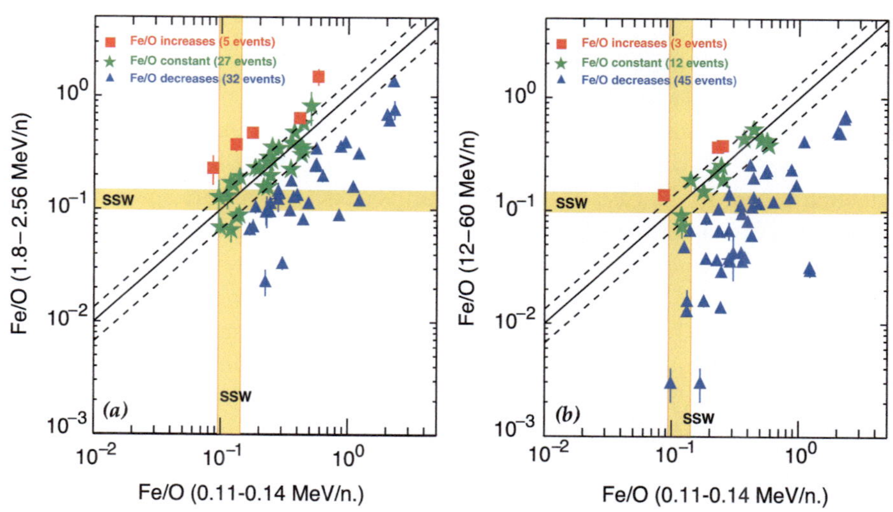

Fig. 16.4 Energy dependence of Fe/O ratio in large SEP events. The ratio is shown at 1.8–2.56 MeV/n (*left*) and at 12–60 MeV/n (*right*) vs. the ratio at 0.11–0.14 MeV/n. In most events the ratio is decreasing with increasing energy (*blue triangles*). *Dashed lines* indicate constant Fe/O ratio within ±30%. The *yellow bands* show the error limits of the average Fe/O ratio measured in the slow solar wind. Reproduced from Desai et al. (2006) by permission of the AAS

mechanism applies to high energy particles. Based on common features identified in the temporal evolution of particles interacting in the solar atmosphere and particles observed in the interplanetary space Klein and Trottet (2001) concluded that coronal acceleration processes downstream of the CME contribute significantly to proton populations from tens of MeV to some GeV in space.

Cane et al. (2002) have shown that essentially all SEP events producing protons above 20 MeV are preceded by long-duration type III radio bursts and all are preceded by CMEs. They concluded that the type III bursts were caused by streams of electrons traveling from close to the solar surface out to 1 AU. These electrons were unlikely shock-accelerated, but instead probably originated in the reconnection regions below fast CMEs. Cane et al. (2002) further concluded that the existence of these type III bursts proved that open field lines extend from within $\sim 0.5\,R_s$ into the interplanetary medium and that particles originating from flares could be expected to contribute in all solar particle events in addition to those accelerated by CME-driven shocks.

In studying 29 large SEP events in the energy range 25–80 MeV/n Cane et al. (2003) identified 19 events with intensities rapidly rising at the time of an associated flare, which had Fe/O ratios above coronal values and constant with time and energy. The remaining 10 events had Fe/O ratios that varied with time and energy with event-averaged values at or below coronal. Of these 10 events, four had two peaks in the intensity-time profiles: the first near the time of the associated flare with high Fe/O ratio and the other at the time of the shock passage with a lower Fe/O. Cane et al. (2003) concluded that SEP events have two components and that at high rigidities the first component, which was assumed to be flare generated, usually dominates and interplanetary shock-accelerated particles forming the second component make only a minor contribution except in the case of unusually fast shocks. Their observations above 25 MeV/n indicated that a population of flare particles was present in most major solar particle events.

Torsti et al. (2003) reported of significant intensities of ^{3}He in all SEP events with ^{4}He intensity exceeding a certain threshold. The abundance ratio ^{3}He/^{4}He was found to vary in the range $\sim$0.003–2, well above the solar wind value. They argued that $^3\mathrm{He}/^4\mathrm{He} \approx 0.015$ should be regarded as a normal composition in high-energy SEP events. Torsti et al. (2003) considered the reported common overabundance of ^{3}He as a signature that impulsive flare-type processes always participate in SEP production.

Based on SEP event onset timing Klein and Posner (2005) found a number of flare-related SEP events in which proton release started simultaneously with the release of electron beams observed through their type III radio emission at decametric-to-kilometric wavelengths. These events were studied to identify the coronal processes that occurred when the particle release started at the sun. Klein and Posner (2005) concluded that the first escaping protons were accelerated roughly between $0.1\,R_s$ and $0.5\,R_s$ above the photosphere, a height range well behind the front of the associated CMEs at the time of acceleration. The magnetically stressed corona in the aftermath of the CME was identified as a plausible site of acceleration. Similar conclusions of acceleration of electrons and protons very low in the corona behind a rising CME were reached by Maia et al. (2007).

Observations of large SEP events with characteristics that have been interpreted as direct contributions from flare processes to particle intensities at 1 AU have led to extensions of the two-class paradigm of SEP events. For example, Kocharov and Torsti (2002) have presented a classification system emphasizing the importance of CME liftoff/aftermath processes in the solar corona as well as the possible role of seed particle re-acceleration, which may explain the observed hybrid solar energetic particle events.

16.7 Future Prospects

One of the principal reasons that the acceleration processes and sources of solar energetic particles have remained puzzles in spite of observations with state-of-the-art space instruments for many years is that the processes and sources mainly occur very close to the sun, while the observations are usually carried out at or near 1 AU. When traveling the distance from the sun to Earth through the interplanetary magnetic medium much of the essential information initially carried by the particles is washed out and distorted. Taking into account all the propagation effects and deducing the characteristics of the original acceleration processes and physical conditions of the source regions is extremely difficult. Therefore, as far as particle measurements are concerned, the only solution is to go close to the sun. Two such missions, the ESA Solar Orbiter and NASA Solar Probe Plus, are currently in preparation.

Both Solar Orbiter and Solar Probe Plus will carry a comprehensive energetic particle instrument suite. Solar Orbiter will have an elliptical orbit around the sun with closest approach to the sun of 0.28 AU reaching this close distance several times during the nominal mission lifetime. Solar Probe Plus will go even closer to the sun, to the minimum distance of 9.5 R_S. It is expected that both missions will contribute in revolutionizing our knowledge and understanding of the processes in the inner heliosphere. They will provide the unique opportunity to disentangle SEP propagation and transport effects from the acceleration and injection processes active at the sun and in the inner heliosphere. These missions will allow for the first time to unambiguously determine the sources of energetic particles, and to use SEPs to map out the large-scale connectivity of the unexplored region of the innermost heliosphere.

16.8 Conclusions

Statistical studies have shown that large solar energetic particle events are associated with fast and wide coronal mass ejections. In particular, western hemisphere fast and wide CMEs associated with decametric-hectometric type II radio bursts produce large SEP events with high probability. Preconditioning of SEP source region by a preceding CME may be important in SEP production leading to high correlation between CME speeds and SEP event peak intensities.

Ion charge states in gradual SEP events are compatible with solar wind values at low energies, but have large variations at energies above 10 MeV with Fe charge states up to the range of 15–20. ^{3}He-rich events can have extreme variations in ^{3}He/^{4}He ratio with energy and large enhancements in heavy ion abundances. Large variations in heavy ion abundances also occur in gradual SEP events. In large SEP events, spectral breaks usually occur at Q/M-dependent energies ranging from a few MeV/n to ~30 MeV/n. Seed populations and shock geometry can have significant effects on the elemental composition in gradual SEP events.

The two-class paradigm of SEP events seems not to be sufficient to explain all recent observations. Extended classification systems are required for the interpretation of observed characteristics of SEP events. To answer many open questions on the acceleration mechanisms and source regions of SEPs, measurements close to the sun with advanced instruments are needed in order to minimize the propagation effects obscuring the present observations at or near 1 AU.

References

Brueckner GE, Howard RA, Koomen MJ et al (1995) The large angle spectroscopic coronagraph (LASCO). Solar Phys 162:357–402

Bryant DA, Cline TL, Desai UD, McDonald FB (1965) Studies of solar protons with Explorers XII and XIV. Astrophys J 141:478–499

Cane HV, McGuire RE, von Rosenvinge TT (1986) Two classes of solar particle events associated with impulsive and long-duration soft X-ray flares. Astrophys J 301:448–459

Cane HV, Reames DV, von Rosenvinge TT (1988) The role of interplanetary shocks in the longitude distribution of solar energetic particles. J Geophys Res 93:9555–9567

Cane HV, Erickson WC, Prestage NP (2002) Solar flares, type III radio bursts, coronal mass ejections, and energetic particles. J Geophys Res 107(A10):1315

Cane HV, von Rosenvinge TT, Cohen CMS, Mewaldt RA (2003) Two components in major solar particle events. Geophys Res Lett 30(12):8017

Cliver EW, Kahler SW, Reames DV (2004) Coronal shocks and solar energetic proton events. Astrophys J 605:902–910

Cohen CMS, Mewaldt RA, Leske RA, Cummings AC, Stone EC,Wiedenbeck ME, Christian ER, von Rosenvinge TT (1999) New observations of heavy-ion-rich solar particle events from ACE. Geophys Res Lett 26:2697–2700

Desai MI, Mason GM, Gold RE, Krimigis SM, Cohen CMS, Mewaldt RA, Mazur JE, Dwyer JR (2006) Heavy ion elemental abundances in large solar particle events and their implications for the seed population. Astrophys J 649:470–489

Ellison DC, Ramaty R (1985) Shock acceleration of electrons and ions in solar flares. Astrophys J 298:400–408

Fichtel CE, Guss DE (1961) Heavy nuclei in solar cosmic rays. Phys Rev Lett 6:495–497

Fichtel CE, McDonald FB (1967) Energetic particles from the sun. Annu Rev Astron Astrophys 5:351–398

Forbush SE (1946) Three unusual cosmic-ray increases possibly due to charged particles from the sun. Phys Rev 70:771–772

Gopalswamy N, Yashiro S, Lara A, Kaiser ML, Thompson BJ, Gallagher BT, Howard RA (2003) Large solar energetic particle events of cycle 23: a global view. Geophys Res Lett 30:8015

Gopalswamy N, Yashiro S, Krucker S, Stenborg G, Howard RA (2004) Intensity variation of large solar energetic particle events associated with coronal mass ejections. J Geophys Res 109:A12105

Gopalswamy N, Aguilar-Rodriguez E, Yashiro S, Nunes S, Kaiser ML, Howard RA (2005) Type II radio bursts and energetic solar eruptions. J Geophys Res 110:A12S07

Gopalswamy N, Yashiro S, Xie H, Akiyama S, Aguilar-Rodriguez E, Kaiser ML, Howard RA, Bougeret J-L (2008a) Radio-quiet fast and wide coronal mass ejections. Astrophys J 674:560–569

Gopalswamy N, Yashiro S, Akiyama S, Mäkelä P, Xie H, Kaiser ML, Howard RA, Bougeret JL (2008b) Coronal mass ejections, type II radio bursts, and solar energetic particle events in the SOHO era. Ann Geophys 26:3033–3047

Hsieh KC, Simpson JA (1970) The relative abundances and energy spectra of ^{3}He and ^{4}He from solar flares. Astrophys J 162:L191–L196

Kahler SW (2001) The correlation between solar energetic particle peak intensities and speeds of coronal mass ejections: Effects of ambient particle intensities and energy spectra. J Geophys Res 106:20947–20955

Kahler SW, Hildner E, van Hollebeke MAI (1978) Prompt solar proton events and coronal mass ejections. Solar Phys 57:429–443

Kahler SW, Sheeley NR Jr, Howard RA, Koomen MJ, Michels DJ, McGuire RE, von Rosenvinge TT, Reames DV (1984) Associations between coronal mass ejections and solar energetic proton events. J Geophys Res 89:9683–9693

Kahler SW, Vourlidas A (2005) Fast coronal mass ejection environments and the production of solar energetic particle events. J Geophys Res 110:A12S01

Klecker B, Moebius E, Popecki MA (2007) Ionic charge states of solar energetic particles: A clue to the source. Space Sci Rev 130:273–282

Klein K-L, Posner A (2005) The onset of solar energetic particle events: prompt release of deka-MeV protons and associated coronal activity. Astron Astrophys 438:1029–1042

Klein K-L, Trottet G (2001) The origin of solar energetic particle events: coronal acceleration versus shock wave acceleration. Space Sci Rev 95:215–225

Kocharov L, Kovaltsov GA, Torsti J, Ostryakov VM (2000) Evaluation of solar energetic Fe charge states: effect of proton-impact ionization. Astron Astrophys 357:716–724

Kocharov L, Torsti J (2002) Hybrid solar energetic particle events observed on board SOHO. Solar Phys 207:149–157

Kunow H, Wibberenz G, Green G, Müller-Mellin R, Kallenrode M-B (1991) Energetic particles in the inner solar system. In: Schwenn R, Marsch E (eds) Physics and chemistry in space, Space and solar physics, vol 21, Physics of the inner heliosphere II, Particles, waves and turbulence, pp 243–342 Springer-Verlag, Berlin Heidelberg

Labrador AW, Leske RA, Mewaldt RA, Stone EC, von Rosenvinge TT (2005) High energy ionic charge state composition in the October/November 2003 and January 20, 2005 SEP events. Proceedings of 29th international cosmic ray conference, Pune, India, 1, pp 99–102

Li G, Zank GP, Rice WKM (2005) Acceleration and transport of heavy ions at coronal mass ejection-driven shocks. J Geophys Res 110:A06104

Li G, Zank GP, Verkhoglyadova O, Mewaldt RA, Cohen CMS, Mason GM, Desai MI (2009) Shock geometry and spectral breaks in large SEP events. Astrophys J 702:998–1004

Maia DJF, Gama R, Mercier C, Pick M, Kerdraon A, Karlicky M (2007) The radio-coronal mass ejection event on 2001 April 15. Astrophys J 660:874–881

Mason GM (2007) ^{3}He-rich solar energetic particle events. Space Sci Rev 130:231–242

Mason GM, Reames DV, Klecker B, Hovestadt D, von Rosenvinge TT (1986) The heavy-ion compositional signature in ^{3}He-rich solar particle events. Astrophys J 303:849–860

Mason GM, Mazur JE, Dwyer JR (1999) ^{3}He enhancements in large solar energetic particle events Astrophys J 525:L133–L136

Mason GM, Wiedenbeck ME, Miller JA et al (2002) Spectral properties of He and heavy ions in ^{3}He-rich solar flares. Astrophys J 574:1039–1058

Mewaldt RA, Cohen CMS, Mason GM et al (2005) Solar energetic particle spectral breaks. In: Li G, Zank GP, Russell CT (eds) The physics of collisionless shocks, AIP conference proceedings 781, pp 227–232 American Institute of Physics, USA

Meyer P, Parker EN, Simpson JA (1956) Solar cosmic rays of February, 1956 and their propagation through interplanetary space. Phys Rev 104:768–783

Obayashi T (1962) Propagation of solar corpuscles and interplanetary magnetic fields. J Geophys Res 67:1717–1729

Oetliker M, Klecker B, Hovestadt D, Mason GM, Mazur JE, Leske RA, Mewaldt RA, Blake JB, Looper MD (1997) The ionic charge of solar energetic particles with energies of 0.3–70 MeV per nucleon. Astrophys J 477:495–501

Pallavicini R, Serio S, Vaiana GS (1977) A survey of soft X-ray limb flare images: the relation between their structure in the corona and other physical parameters. Astrophys J 216:108–122

Reames DV (1988) Bimodal abundances in the energetic particles of solar and interplanetary origin. Astrophys J 330:L71–L75

Reames DV (1999) Particle acceleration at the sun and in the heliosphere. Space Sci Rev 90:413–491

Reames DV, Stone RG (1986) The identification of solar ^{3}He-rich events and the study of particle acceleration at the sun. Astrophys J 308:902–911

Schatten KH, Mullan DJ (1977) Fast azimuthal transport of solar cosmic rays via a coronal magnetic bottle. J Geophys Res 82:5609–5620

Shea MA, Smart DF (1995) History of solar proton event observations. Nucl Phys B (Proc Suppl) 39A:16–25

Torsti J, Laivola J, Kocharov L (2003) Common overabundance of ^{3}He in highenergy solar particles. Astron Astrophys 408:L1–L4

Tylka AJ, Cohen CMS, Dietrich WF, Lee MA, Maclennan CG, Mewaldt RA, Ng CK, Reames DV (2005) Shock geometry, seed populations, and the origin of variable elemental composition at high energies in large gradual solar particle events. Astrophys J 625:474–495

Wild JP, Smerd SF, Weiss AA (1963) Solar bursts. Annu Rev Astron Astrophys 1:291–366

Chapter 17

From Micro- to Macro-scales in the Heliosphere and Magnetosphere

Dastgeer Shaikh, Igor S. Veselovsky, Quanming M. Lu, and Gary P. Zank

Abstract From a broader perspective, the heliosphere and planetary magnetospheres provide a test bed to explore the plasma physics of the Universe. In particular, the underlying nonlinear coupling of different spatial and temporal scales plays a key role in determining the structure and dynamics of space plasmas and electromagnetic fields. Plasmas and fields exhibit both laminar and turbulent properties, corresponding to either well organized or disordered states, and the development of quantitative theoretical and analytical descriptions from physics based first principles is a profound challenge. Limited observations and complications introduced by geometry and physical parameters conspire to complicate the problem. Dimensionless scaling analysis and statistical methods are universally applied common approaches that allow for the application of related ideas to multiple physical problems. We discuss several examples of the interplay between the scales in a variety of space plasma environments, as exemplified in the presentations of the IAGA session *From Micro- to Macro-scales in the Heliosphere and Magnetospheres.*

17.1 Turbulent Spectra in the Solar Wind and Interstellar Medium

The solar wind and interstellar medium is predominantly in a turbulent state (Marsch and Tu 1995; Goldstein et al. 1995; Bruno and Carbone 2005) in which low frequency fluctuations are described typically by a magnetohydrodynamics (MHD) description of plasma. Nonlinear interactions amongst these fluctuations lead to a migration of energy in the inertial range that is characterized typically by a Kolmogorov-like 5/3 spectrum (Frisch 1995). The 5/3 power spectrum is observed frequently, both in the interstellar medium (ISM) and solar wind (SW). The ubiquity of the turbulence spectrum on a variety of length scales, leading to a Kolmogorov-like 5/3 law, is one of the long standing puzzles of classical statistical theories of turbulence, the origin and nature of which remains a topic of considerable debate. Owing to its complexity, magnetized plasma turbulence in general is not only lacking substantially in theoretical developments because of its analytically intractable nature, but it also poses computationally a challenging task of resolving multiple scale flows and fluctuations that are best described statistically. The fields of plasma and hydrodynamic turbulence have grown tremendously with the advent of high speed supercomputing and efficient numerical algorithms. It is not possible to cover all aspects of the field in this chapter, and so we concentrate mainly on the physical processes that lead to the 5/3 spectra in both ISM and SW plasmas. Understanding energy cascade processes is important particularly from the point of view of nonlinear interactions across disparate scales, turbulence transport, wave propagation, heating processes in the solar wind, structure formation, cosmic ray scattering, and particle acceleration throughout the heliosphere.

D. Shaikh (✉)
Center for Space Plasma and Aeronomic Research and Physics Department, University of Alabama, Huntsville, AL 35899, USA
e-mail: dastgeer.shaikh@uah.edu

M.P. Miralles, J. Sánchez Almeida (eds.), *The Sun, the Solar Wind, and the Heliosphere*, IAGA Special Sopron Book Series 4, DOI 10.1007/978-90-481-9787-3_17, © Springer Science+Business Media B.V. 2011

17.1.1 Turbulence Spectra in the Interstellar Medium

It is a curious observation (Fig. 17.1) that electron density fluctuations in the interstellar medium (ISM) exhibit an omnidirectional Kolmogorov-like (Kolmogorov 1941) power spectrum $k^{-5/3}$ (or 11/3 spectra index in three dimensions) over a 4 to 6 decade range (Armstrong et al. 1981; Higdon 1984, 1986; Armstrong et al. 1990). The observed turbulence spectrum extends over an extraordinary range of scales i.e. from an outer scale of a few parsecs to scales of few AUs or less. Interstellar scintillation, describing fluctuations in the amplitude and phase of radio waves caused by scattering in the interstellar medium, exhibit the power spectrum of the interstellar electron density that follows a 5/3 index (Armstrong et al. 1995). The origin and nature of this big power law is described in an extensive review by Elmegreen and Scalo (2004). Chepurnov and Lazarian (2010) used the data of the Wisconsin Hα Mapper (WHAM) and determined that the amplitudes and spectra of density fluctuations can be matched to the data obtained for interstellar scintillations and scattering that follow a Kolmogorov-like spectrum spanning from 106 to 1,017 m scales. Angular broadening measurements also reveal, more precisely, a Kolmogorov-like power spectrum for the density fluctuations in the interstellar medium with a spectral exponent slightly steeper than $-5/3$ (Mutel et al. 1998; Spangler 1999). Regardless of the exact spectral index, the density irregularities exhibit a power-law spectrum that is essentially characteristic of a fully developed isotropic and statistically homogeneous incompressible fluid turbulence, described by Kolmogorov (1941) for hydrodynamic and Kraichnan (1965) for magnetohydrodynamic fluids. Turbulence, manifested by interstellar plasma fluid motions, therefore plays a major role in the evolution of the ISM plasma density, velocity, magnetic fields, and the pressure. Radio wave scintillation data indicates that the rms fluctuations in the ISM and interplanetary medium density, of possibly turbulent origin and exhibiting Kolmogorov-like behavior, are only about 102001). This suggests that ISP density fluctuations are only weakly compressible. Despite the weak compression in the ISP density fluctuations, they nevertheless admit a Kolmogorov-like power law, an ambiguity that is

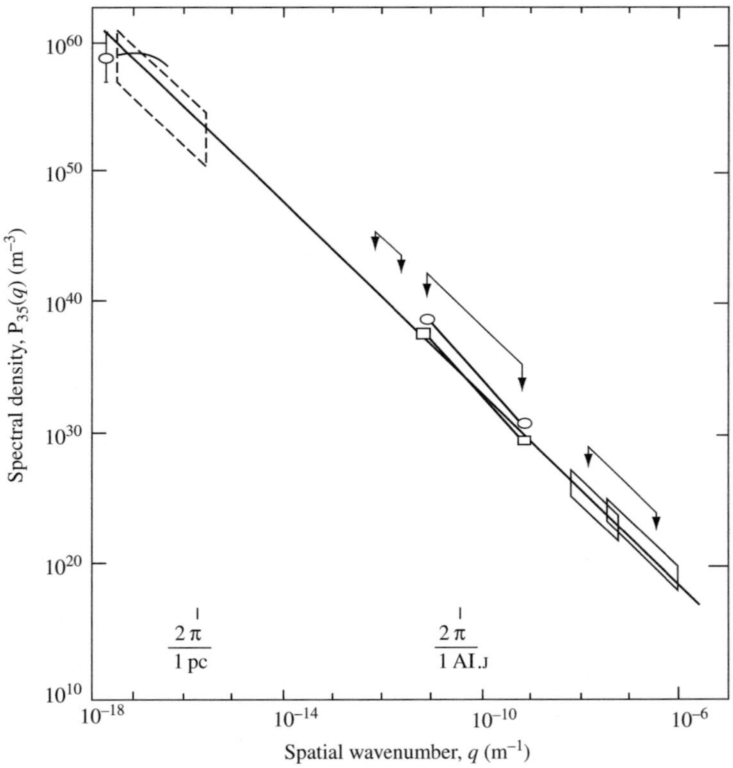

Fig. 17.1 ISM turbulence spectrum exhibiting a 5/3 power law (Armstrong et al. 1981)

not yet completely resolved by any fluid/kinetic theory or computer simulations. That the Kolmogorov-like turbulent spectrum stems from purely incompressible fluid theories (Kolmogorov 1941; Kraichnan 1965) of hydrodynamics and magnetohydrodynamics offers the simplest possible turbulence description in an isotropic and statistically homogeneous fluid. However, since the observed electron density fluctuations in the ISM possess a weak degree of compression, the direct application of such simplistic turbulence models to understanding the ISM density spectrum is not entirely obvious. Moreover, the ISM is not a purely incompressible medium and can possess many instabilities because of gradients in the fluid velocity, density, magnetic field etc. where incompressibility, inhomogeneity and even isotropy are certainly not good assumptions. This calls for a fully self-consistent description of ISM fluid, one that couples incompressible modes with weakly compressible modes and deals with the strong nonlinear interactions amongst the ISM density, temperature, velocity and the magnetic field. Note that the coupling of different modes is an intrinsic property of MHD perturbations of finite amplitude.

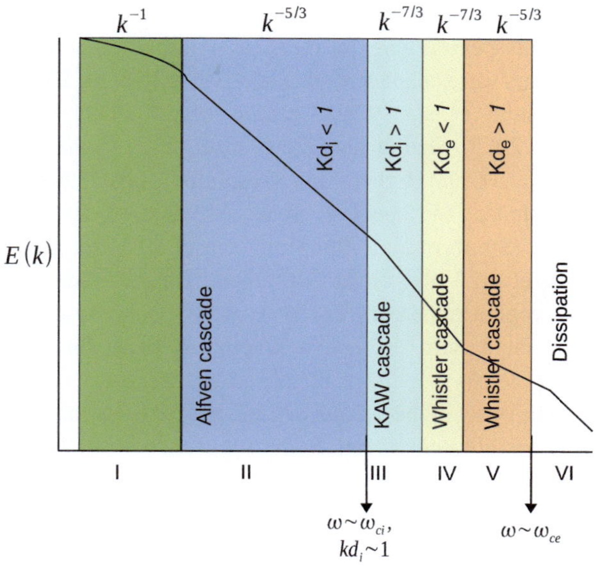

Fig. 17.2 Schematic of the power spectral density (PSD) composite spectrum in the solar wind turbulent plasma as a function of frequency (wavenumber). Several distinct regions are identified with what is thought to be the dominant energy transfer mechanism for that particular region. The nonlinear processes associated with the transition from region II (MHD regime) to region III (kinetic or Hall MHD regime) are not yet fully understood. The power spectra in region III vary from k^{-2} to k^{-4}. The boundary of regions III and IV identifies where electron and ion motions are decoupled. Regions IV and V are identified as whistler cascade regimes. The outerscale of MHD turbulence corresponds to the smaller k mode in region II which can possibly extend over a few parsecs in the context of ISM (Armstrong et al. 1981)

17.1.2 Solar Wind Turbulence Spectra

Solar wind plasma, on the other hand, occurs on much smaller scales, i.e. few thousands of kilometers, compared to the ISM scales. A wealth of data from in-situ observations is available from numerous spacecraft and reveals the nonlinear turbulent character of the magnetized solar wind plasma fluid. It is evident from these observations that the solar wind plasma yields a multitude of spatial and temporal length-scales associated with an admixture of waves, fluctuations, structures and nonlinear turbulent interactions. In-situ measurements (Matthaeus and Brown 1988, Goldstein et al. 1994, 1995; Ghosh et al. 1996) indicate that solar wind fluctuations, extend over several orders of magnitude in frequency and wavenumber. The fluctuations can be described by a power spectral density (PSD) spectrum that can be divided into three distinct regions (Goldstein et al. 1995; Leamon et al. 1999) depending on the frequency and wavenumber. This is shown in the schematic of Fig. 17.2. The first region corresponds to a flatter spectrum, associated with lower frequencies consistent with a k^{-1} (where k is wavenumber) power law. A second identifiable region follows and extends to the ion/proton gyrofrequency, with a spectral slope that has an index ranging from $-3/2$ to $-5/3$. This region is identified with fully developed turbulence, and is generally described on the basis of the incompressible magnetohydrodynamic (MHD) equations. The turbulent interactions in this regime are thought to be governed entirely by Alfvenic cascades. Spacecraft observations (Leamon et al. 1999; Bale et al. 2005; Alexandrova et al. 2007, 2008; Sahraoui et al. 2007, 2009) further reveal that at length scales beyond the MHD regime, i.e. length scales less than ion gyro radius $k\rho_i < 1$ and temporal scales greater than the ion cyclotron frequency $\omega > \omega_{ci} = eB_0/m_e c$ (where k,ρ_{ci},e,B_0,m_e,c are respectively characteristic mode, ion gyroradius, ion cyclotron frequency, electronic charge, mean magnetic field, mass of electron, and speed of light), the spectrum exhibits a spectral break, and the spectral index of the solar wind turbulent fluctuations varies between -2 and -5 (Smith et al. 2006; Goldstein et al. 1994; Leamon et al. 1999; Bale et al. 2005; Shaikh and Zank 2008, 2009; Sahraoui et al. 2009). Higher time resolution observations find that at the spectral break, Alfvenic MHD cascades (Smith et al. 2006; Goldstein et al. 1994; Leamon et al. 1999) close. The characteristic modes in this region appear to evolve typically on timescales associated with dispersive kinetic Alfvenic fluctuations.

The onset of the second or the kinetic Alfvén inertial range is not understood. Some suggestions have however been made. The spectral break may result from energy transfer processes associated with possibly kinetic Alfven waves (KAWs) (Hasegawa 1976), electromagnetic ion- cyclotron-Alfven (EMICA) waves (e.g., Gary et al. 2008) or by fluctuations described by a Hall MHD (HMHD) plasma model (Alexandrova et al. 2007, 2008; Shaikh and Zank 2008, 2008a). Stawicki et al. (2001) argue that Alfven fluctuations are suppressed by proton cyclotron damping at intermediate wavenumbers so the observed power spectra are likely to comprise weakly damped dispersive magnetosonic and/or whistler waves. Beinroth and Neubauer (1981) and Denskat and Neubauer (1982) have reported the presence of whistler waves based on Helios 1 and 2 observations in this high frequency regime. A comprehensive data analysis by Goldstein et al. (1994), based on correlations of sign of magnetic helicity with direction of magnetic field, indicates the possible existence of multiscale waves (Alfvénic, whistlers and cyclotron waves) with a single polarization in the dissipation regime. Counter-intuitively, in the $\omega < \omega_{ci}$ regime, or Alfvenic regime, Howes et al. (2008) noted the possibility that highly obliquely propagating KAWs are present (with $\omega > \omega_{ci}$) making it questionable that damping of ion cyclotron waves is responsible for the spectral breakpoint.

Fluid (Shaikh and Zank 2010) and kinetic (Howes et al. 2008) simulations, in qualitative agreement with spacecraft data described as above, have been able to obtain the spectral break point near the characteristic turbulent length scales that are comparable with the ion inertial length scale (d_i). These simulations find Kolmogorov-like $k^{-5/3}$ spectra for length scales larger than ion inertial length scales, where MHD is typically a valid description. By contrast, smaller (than di) scales were shown to follow a steeper spectrum that is close to $k^{-7/3}$ (Howes et al. 2008; Shaikh and Zank 2009). Spacecraft data and simulations thus reveal that migration of turbulent energy proceeds essentially through different regions in k-space, i.e. k^{-1},$k^{-5/3}$ and $k^{-7/3}$. Of course, the turbulent cascade does not entirely terminate immediately beyond the $k^{-7/3}$ regime. Fluid and kinetic simulations (Biskamp et al. 1996; Galtier 2006; Galtier and Buchlin 2007; Cho and Lazarian 2003; Shaikh and Zank 2005; Shaikh 2009; Gary et al. 2008; Saito et al. 2008; Howes et al. 2008) show that spectral transfer of energy extends even beyond the $k^{-7/3}$ regime and is governed predominantly by small scale, high frequency, whistler turbulence. The latter also exhibits a power law.

17.1.3 Extended Composite Spectra of the Solar Wind Plasma

Theory and simulations indicate that turbulent fluctuations in the high frequency and $k\rho_i > 1$ regime correspond to electron motions that are decoupled from the ion motions (Kingsep et al. 1990; Biskamp et al. 1996; Shaikh et al. 2000a, b; Shaikh and Zank 2003; Cho and Lazarian 2003; Saito et al. 2008; Gary et al. 2008). Correspondingly, ions are essentially unmagnetized and can be treated as an immobile neutralizing background fluid. This regime corresponds to the whistler wave band of the spectrum and comprises characteristic scales that are smaller than those that describe MHD, KAW or Hall MHD processes. An extended composite schematic describing the whistler mode spectra is also shown in Fig. 17.2. Specifically, regions IV and V in Fig. 17.2 identify characteristic modes that are relevant for the description of whistler wave turbulence (Biskamp et al. 1996; Shaikh and Zank 2005; Shaikh 2009). The boundary of regions III and IV represents a wavenumber band in spectral space that corresponds to the decoupling of electron and ion motions. Wavenumbers above this boundary characterize the onset of whistler turbulence. The spectral cascades associated with whistler turbulence are described extensively by Biskamp et al. (1996), Shaikh et al. (2000a), Shaikh et al. (2000b), Shaikh and Zank (2003, 2005), Shaikh (2009). Cho and Lazarian 2004 describe scale dependent anisotropy that is mediated by whistler waves in the context of electron MHD plasma. Gary et al. (2008) and Saito et al. (2008) have reported two-dimensional electromagnetic particle-in-cell simulations of an electron MHD model to demonstrate the forward cascade of whistler turbulence. Their work shows that the magnetic spectra of the cascading fluctuations become more anisotropic with increasing fluctuation energy. Interestingly, whistler turbulence associated with longer wavelengths in region IV exhibits a power spectrum $k^{-7/3}$ that is similar to the short wavelength spectrum of kinetic Alfven waves (KAW), as shown in region III of Fig. 17.2. The underlying physical processes responsible for the spectrum differ significantly for KAW and whistler waves.

The Hall MHD description of magnetized plasma is valid up to region III where characteristic turbulent scales are smaller than ion inertial length scales ($kd_i > 1$). Beyond this location, high frequency motion of plasma is governed predominantly by electron motions, and ions form a stationary neutralizing background. Consequently, the ion motions decouple significantly from electron motion. These aspects of the spectra, depicted by regions IV and V in Fig. 17.2, can be described adequately by whistler wave model. The Hall MHD models are therefore not applicable in regions IV, V and beyond. Neither can they describe kinetic physics associated with the dissipative regime. Since the high frequency regime (i.e. regions IV and V) is dominated by electron motions, there exists an intrinsic length scale corresponding to the electron inertial length scale $d_e = c/\omega_{p_e}$ (where ω_{p_e} is the electron plasma frequency). The characteristic turbulent length scales in regions IV and V are comparable with de and therefore describe scales larger (i.e. $kd_e < 1$ in region IV) and smaller (i.e. $kd_e > 1$ in region V) than the electron inertial scale. While whistler wave models can describe nonlinear processes associated with length scales as small as the electron inertial length scale, they fail to describe finite electron Larmor radius effects for which a fully kinetic description of plasma must be used.

Beside those issues described above, we do not understand what leads to the decoupling of ion and electron motions near the boundary of regions III and IV for example. Although the turbulent spectra are described by similar spectral indices, the nonlinear processes are fundamentally different in regions III and IV.

17.1.4 A Nearly Incompressible Description of the SW and ISM Spectra- the 5/3 MHD Regime

Earlier fluid models, describing the turbulent motion of a compressible ISM fluid, have been based mostly on isothermal and adiabatic assumptions, due largely to their tractability in terms of mathematical and numerical analysis. Unfortunately, such models cannot describe the complex nonlinear dynamical interactions amongst ISM fluctuations self-consistently. For instance, density fluctuations, in the context of related solar wind work, were thought to have originated from nonlinear Alfven modes (Spangler 1987). A simple direct relationship of density variations with Alfvenic fluctuations is not entirely obvious as the latter are not fully self-consistent and are incompressible by nature thereby ignoring effects due to magnetoacoustic perturbations for example. On the other hand, fully compressible nonlinear MHD solutions, for both high- and low-cases, show that Alfven and slow modes exhibit a $k^{-5/3}$ spectrum, while fast modes follow a $k^{-3/2}$ spectrum (Cho and Lazarian 2002, 2003). The formation of density power spectrum in the simulations of isothermal MHD turbulence was studied in Cho and Lazarian (2003), Beresnyak et al. (2005), Kowal et al. (2007). In particular, in Beresnyak et al. (2005), the logarithm of density was shown to follow the Goldreich-Sridhar scaling in terms of both density and anisotropy. This is an important finding that sheds the light onto the nature of the density fluctuations.

One of the most debated issues in the context of solar wind turbulence is the non-equipartition between the kinetic and magnetic part of the energies that leads to a discrepancy between the two spectra. The kinetic as well as magnetic energy spectra for fast or slow modes nevertheless do not relate to a Kolmogorov-like density spectrum. The latter modes have been suggested as candidates for generating density fluctuations (**? ?**) in the interstellar medium. Alternate explanations are that density structures (anisotropic) in the ISM emerge from pressure-balance stationary modes of MHD (also called Pressure Balance Structures, PBS) (Higdon 1986), or from inhomogenities in the large-scale magnetic field via the four-field model of Bhattacharjee et al. (1998). These descriptions are inadequate for a general class of ISM problems. The PBSs form a special class of MHD solutions and are valid only under certain situations when the magnetic and the pressure fluctuations exert equal forces in the stationary state. These structures, limited in their scope to the general ISM conditions, nevertheless do not offer an entirely self-consistent explanation to the observed density spectrum. Similarly, an inertial range turbulent cascade associated with the low turbulent Mach number four field MHD model is not yet known. Moreover this isothermal inhomogeneous fluid model is valid only for a class of MHD solutions and yields a linear Mach number (M) scaling, $O(M)$, amongst the various fluctuations (Bhattacharjee et al. 1998).

One of the earlier attempts to understand the ISM density fluctuations, and relate it to an incompressible fluid turbulence model dates back Montgomery et al. (1987) who used an assumed equation of state to relate ISM density fluctuations to incompressible MHD. This approach, called a pseudosound approximation, assumes that density fluctuations are proportional to the pressure fluctuations through the square of sound speed. The density perturbations in their model are therefore "slaved" to the incompressible magnetic field and the velocity fluctuations. This hypothesis was further contrasted by Bayly et al. (1992) on the basis of their 2D compressible hydrodynamic simulations by demonstrating that a spectrum for density fluctuations can arise purely as a result of abandoning a barotropic equation of state, even in the absence of a magnetic field. The pseudosound fluid description of compressibility, justifying the Montgomery et al. approach to the density-pressure relationship, was further extended by Matthaeus and Brown (1988) in the context of a compressible magnetofluid (MHD) plasma with a polytropic equation of state in the limit of a low plasma acoustic Mach number (Matthaeus and Brown 1988). The theory, originally describing the generation of acoustic density fluctuations by incompressible hydrodynamics (Lighthill 1952), is based on a generalization of Klainerman and Majda's work (Klainerman and Majda 1981, 1982; Majda 1984) and accounts for fluctuations associated with a low turbulent Mach number fluid, unlike purely incompressible MHD. Such a nontrivial departure from the incompressible state is termed "nearly incompressible." The primary motivation behind NI fluid theory was to develop an understanding and explanation of the interstellar scintillation observations of weakly compressible ISM density fluctuations that exhibit a Kolmogorov-like power law. The NI theory is, essentially, an expansion of the compressible fluid or MHD equations in terms of weak fluctuations about a background of strong incompressible fluctuations. The expansion parameter is the turbulent Mach number. The leading order expansion satisfies the background incompressible hydrodynamic or magnetohydrodynamic equations (and therefore fully nonlinear) derived on the basis of Kreiss principle (Kreiss, 1982), while the higher order yields a high frequency weakly compressible set of nonlinear fluid equations that describe low turbulent Mach number compressive HD as well as MHD effects. Zank and Mathaeus derived the unified self-consistent theory of nearly incompressible fluid dynamics for non-magnetized hydrodynamics as well as magnetofluids, with the inclusion of the thermal conduction and energy effects, thereby identifying different and distinct routes to incompressibility (Zank and Matthaeus 1991, 1993). In the NI theory, the weakly perturbed compressive fluctuations (denoted by subscript 1) are expanded about the incompressible modes (denoted by superscript 1) for velocity and pressure variables as $U = U^{\infty} + \varepsilon U_1$, $p = p_0 + \varepsilon^2(p^{\infty} + p_1)$ respectively. Here ε is a small parameter associated with the turbulent fluid Mach number M_s through the relation $C_s^2 = \gamma p/\rho$, $M_s = U_0/C_s$ and γ is the ratio of the specific heats, U_0 is the characteristic speed of the turbulent fluid, and Cs is the acoustic speed associated with sound waves. Due to a lack of uniqueness in the representation of the fluid density and temperature fields, either of the choices $T = T_0 + \varepsilon T_1$ or $T = T_0 + \varepsilon^2 T_1$ is consistent. The first choice corresponds to a state where temperature fluctuations dominate both the incompressible and compressible pressure and is referred to as the heat fluctuation dominated (HFD) regime. On the other hand the second choice in which all the variables are of similar order is described as the heat fluctuation modified (HFM) regime. Since the thermal fluctuations in HFD regime appear at an order $O(\varepsilon)$ as compared with the pressure $O(\varepsilon^2)$, they dominate the NI ordering. By contrast, the thermal fluctuations have the 2 same ordering with respect to the other fluctuations (density, pressure etc.) in a HFM regime. The NI theory introduces a further fundamentally different explanation for the observed Kolmogorov-type density spectrum in that the ISM density fluctuations can be a consequence of passive scalar convection due to background incompressible fluctuations as well as a generalized pseudo-sound theory. The theory further predicts various correlations between the density, temperature and the acoustic as well as convective pressure fluctuations (Zank and Matthaeus 1991, 1993; Shaikh and Zank 2005, 2006, 2007).

The validity and nonlinear aspects of the NI model, within the context of the interstellar medium, has recently been explored by Shaikh and Zank (2003, 2005, 2006, 2007, 2010). The theory of nearly incompressible (NI) fluids, developed by Matthaeus, Zank and Brown, based on a perturbative expansion technique is a rigorous theoretical attempt to understand the origin of weakly compressible density fluctuations in the interstellar medium, and one that provides formally a complete fluid description of ISM turbulence with the inclusion of thermal fluctuations and the full energy equation self-consistently, unlike the previous models described above (Zank and Matthaeus 1991, 1991, 1993; Matthaeus and Brown 1988). Owing to its broad perspective and wide range of applicability for interstellar medium problems, we use here a nearly incompressible description of fluids to investigate interstellar turbulence with a view to explaining the observed Kolmogorov-like ISM density spectrum. A central tenant of the homogeneous NI theory is that the density fluctuations are of higher order, of higher frequency and possess smaller length-scales than their incompressible counterparts to which they are coupled through passive convection and the low frequency generation of sound. Most recently, Hunana et al. (2006) and Hunana and Zank (2006, 2010) have extended the NI hydrodynamic and MHD theory to inhomogeneous flows, finding that the density fluctuations can also be of order Mach number, in agreement with a slightly different approach advocated by Bhattacharjee et al. (1998). The NI fluid models, unlike fully incompressible or compressible fluid descriptions, allow us to address weakly compressible effects directly in a quasi-neutral ISM fluid. Furthermore, NI theory has enjoyed notable success in describing fluctuations and turbulence in the supersonic solar wind. The NI model has recently been solved numerically and compared to observations in an effort to understand the Kolmogorov-like density spectrum in the ISM (Shaikh and Zank 2004, 2005, 2006, 2007, 2008, 2009, 2010). One of our results, shown in Fig. 17.3, describes the evolution of density fluctuations from a fully compressible initial state. We find from our three-dimensional (decaying turbulence) simulations that a $k^{-5/3}$ density fluctuation spectrum emerges in fully developed compressible MHD turbulence from non-linear mode coupling interactions that lead to the migration of spectral energy in the transverse (i.e. $k \perp U$) Alfvenic fluctuations, while the longitudinal "compressional modes" corresponding to $k \parallel U$ fluctuations make an insignificant contribution to the spectral transfer of inertial range turbulent energy. The explanation, in part, stems from the evolutionary characteristics of the MHD plasma that governs the evolution of the non-solenoidal velocity field in the momentum field. It is the non-solenoidal component of plasma motions that describes the high-frequency contribution corresponding to the acoustic time-scales in the modified pseudo-sound relationship (Montgomery et al. 1987; Matthaeus and Brown 1998; Zank and Matthaeus 1990, 1993). What is notable in the work of Shaikh and Zank is that they find a self-consistent evolution of a Kolmogorov-like density fluctuation spectrum in MHD turbulence that results primarily from turbulent damping of non-solenoidal modes that constitute fast and slow propagating magnetoacoustic compressional perturbations. These are essentially a higher frequency (compared with the Alfvenic waves) component that evolve on acoustic timescales and can lead to a "pseudo-sound relationship" as identified in the nearly incompressible theory (Zank and Matthaeus 1990, 1993; Bayly et al. 1992; Matthaeus et al. 1998; Shaikh and Zank 2004a, b, c, 2006, 2007). The most significant point to emerge from the simulation is the diminishing of the high-frequency component that is related to the damping of compressible plasma motion. This further leads to the dissipation of the small-scale and high-frequency compressive turbulent modes. Consequently, the MHD plasma relaxes toward a nearly incompressible state where the density is convected passively by the velocity field and eventually develops a $k^{-5/3}$ spectrum. This physical picture suggests that a nearly incompressible state develops naturally from a compressive MHD magnetoplasma in the solar wind.

Among other work, describing a Kolmogorov-like 5/3 spectrum in the context of MHD turbulence, are Cho and Vishniac (2000), Maron and Goldreich (2001), Cho et al. (2002, 2003), Muller and Biskamp (2002), Cho and Lazarian (2002, 2003), Kritsuk et al. (2009). Our results, describing a Kolmogorov-like 5/3 spectrum in the solar wind plasma, are thus consistent with these work. It is noted, however, that Maron and Goldreich (2001) report a

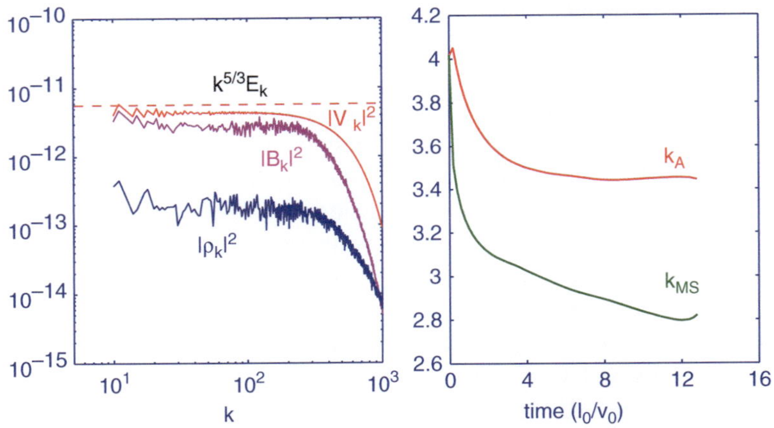

Fig. 17.3 (*Left*) Velocity fluctuations are dominated by shear Alvenic motion and thus exhibit a Kolmogorov-like $k^{-5/3}$ spectrum. The *middle curve* shows the magnetic field spectrum. Density fluctuations are passively convected by the nearly incompressible shear Alfvenic motion and follow a similar spectrum in the inertial range. The numerical resolution in 3D is 512^3. (*Right*) The evolution of Alfvenic (kA) and fast/slow magnetosonic (kMS) modes demonstrates that the spectral cascades are dominated by Alfvenic modes

Kraichnan-like 3/2 spectrum in the inertial range for velocity fluctuations. The controversy of 5/3 or 3/2 is nonetheless beyond the scope of our review article.

17.1.5 Hall MHD Model of SW Turbulence-Extended Spectra

To describe the extended solar wind spectra in Fig. 17.2, time dependent, fully compressible three dimensional simulations of Hall MHD plasma in a triply periodic domain have been developed. This represents a local or regional volume of the solar wind plasma or ISP. The turbulent interactions in region II were described above by a 3D MHD model which is a subset of Hall MHD model since it does not contain the $\mathbf{J} \times \mathbf{B}$ term in the magnetic field induction equation. Note that the dynamics of length-scales associated with region III, i.e. corresponding to the KAW modes, cannot be described by the usual MHD models as they do not describe turbulent motions corresponding to the characteristic frequencies larger than an ion gyro frequency. At 1 AU, ion inertial length scales are smaller than ion gyro radii in the solar wind (Goldstein et al. 1995). Plasma effects due to finite Larmor radii can readily be incorporated in MHD models by introducing Hall terms to accommodate ion gyro scales up to scales as small as ion inertial length scales.

The Hall model in the limit of a zero ion-inertia converges to the usual MHD model, and assumes that the electrons are inertial-less, while the ions are inertial (Krishan and Mahajan 2004). Hence, the electrons and ions have a differential drift, unlike the one fluid MHD model for which the electron and ion flow velocities are identical. The Hall MHD description of magnetized plasma has previously been employed to investigate wave and turbulence processes in the context of solar wind plasma. Sahraoui et al. (2007) extended the ordinary MHD system to include spatial scales down to the ion skin depth or frequencies comparable to the ion gyrofrequency in an incompressible limit. They further analyzed the differences in the incompressible Hall MHD and MHD models within the frame work of linear modes, their dispersion and polarizations. Galtier (2006) developed a wave turbulence theory in the context of an incompressible Hall MHD system to examine the steepening of the magnetic fluctuation power law spectra in the solar wind plasma. Furthermore, Galtier and Buchlin (2007) have developed 3D dispersive Hall magnetohydrodynamics simulations within the paradigm of a highly turbulent shell model and demonstrated that the large-scale magnetic fluctuations are characterized by a $k^{-5/3}$-type spectrum that steepens at scales smaller than the ion inertial length d_i to $k^{-7/3}$. The observed spectral break point in the solar wind plasma, shown by the regime III in Fig. 17.2, has been investigated using 3D simulations of a two fluid nonlinear Hall MHD plasma model (Shaikh and Zank, 2009).

In the inertial-less electron limit the electron fluid does not influence the momentum of solar wind plasma directly except through the current. Since the electron fluid contributes to the electric field, plasma currents and the magnetic field are affected by electron oscillations. The combination of electron dynamics and ion motions distinguishes the Hall MHD model from its single fluid MHD counterpart. Thanks to the inclusion of electron dynamics, Hall MHD can describe solar wind plasma fluctuations that are associated with a finite ion Larmor radius and thus a characteristic plasma frequency is $\omega > \omega_{ci}$. Because Hall MHD contains both ion and electron effects, there is a regime at which the one set of plasma fluctuations no longer dominates but instead is dominated by the other. This introduces an intrinsic scale length/timescale (frequency) that separates ion dominated behavior in the plasma from electron dominated. It is the Hall term corresponding to the $\mathbf{J} \times \mathbf{B}$ term in Faraday's equation that is primarily responsible for decoupling electron and ion motion on ion inertial length and ion cyclotron time scales (and introducing an intrinsic length scale). It is this feature that makes Hall MHD useful in describing dissipative solar wind processes when single fluid MHD is not applicable (the MHD model breaks down at $\omega > \omega_{ci}$). Hall MHD allows us to study inertial range cascades beyond $\omega > \omega_{ci}$, and can be extended to study dissipative heating processes where ion cyclotron waves are damped. However to study MHD processes, once can put $d_i = 0$ in region II. The extreme limit of fluid modeling applied to solar wind processes (even beyond the limit of the Hall MHD regime) is to use of an electron MHD model in which high frequency electron dynamics is treated by assuming stationary ions that act to neutralize the plasma background.

Turbulence involves nonlinear interactions of modes in all three spatial directions. Three dimensional computations are numerically expensive, but, with the advent of high speed vector and parallel distributed memory clusters, and efficient numerical algorithms such as those designed for Message Passing Interface (MPI) libraries, it is now possible to perform magnetofluid turbulence studies at substantially higher resolutions. Based on MPI libraries, three dimensional, time dependent, compressible, non-adiabatic, driven and fully parallelized Hall magnetohydrodynamic (MHD) nonlinear codes have been developed that run efficiently on both distributed memory clusters like distributed-memory supercomputers or shared memory parallel computers. This allows for very high resolution in Fourier spectral space. Shaikh and Zank (2010) have developed a 3D periodic code that is scalable and transportable to different cluster machines, and extends earlier MHD codes of theirs (Shaikh and Zank 2006, 2007, 2008, 2009, 2010a, b). Their code treats the solar wind plasma fluctuations as statistically isotropic, locally anisotropic, homogeneous and random, consistent with ACE spacecraft measurements (Smith et al. 2006). The numerical algorithm accurately preserve the ideal rugged invariants of fluid flows, unlike finite difference or finite volume methods. The conservation of ideal invariants (energy, enstrophy, magnetic potential, helicity) in inertial range turbulence is an extremely important feature because these quantities describe the cascade of energy in the inertial regime, where turbulence is, in principle, free from large-scale forcing as well as small scale dissipation. Damping of plasma fluctuations may nonetheless occur as a result of intrinsic non-ideal effects such those introduced by the finite Larmor radius. An example of the plasma velocity, magnetic field and current is shown in Fig. 17.4.

In the simulations of Shaikh and Zank, the nonlinear spectral cascade in the modified KAW regime leads to a secondary inertial range in the vicinity of $kd_i \simeq 1$, where the turbulent magnetic and velocity fluctuations form spectra close to $k^{-7/3}$. This is displayed in Fig. 17.5, which also shows that for length scales larger than the ion thermal gyroradius, an MHD inertial range spectrum close to $k^{-5/3}$ is formed. The characteristic turbulent spectrum in the KAW regime is steeper than that of the MHD inertial range. Identifying the onset of the secondary inertial range has been the subject of debate because of the presence of multiple processes in the KAW regime that can mediate the spectral transfer of energy. These processes include, for instance, the dispersion and damping of EMICA waves, turbulent dissipation, etc.

Regimes IV and V, shown in the schematic of Fig. 17.2, requires that we invoke a whistler model for the plasma. Whistler modes are excited in the solar wind plasma when the characteristic plasma fluctuations propagate along a mean or background magnetic field with frequency $\omega > \omega_{ci}$ and the length

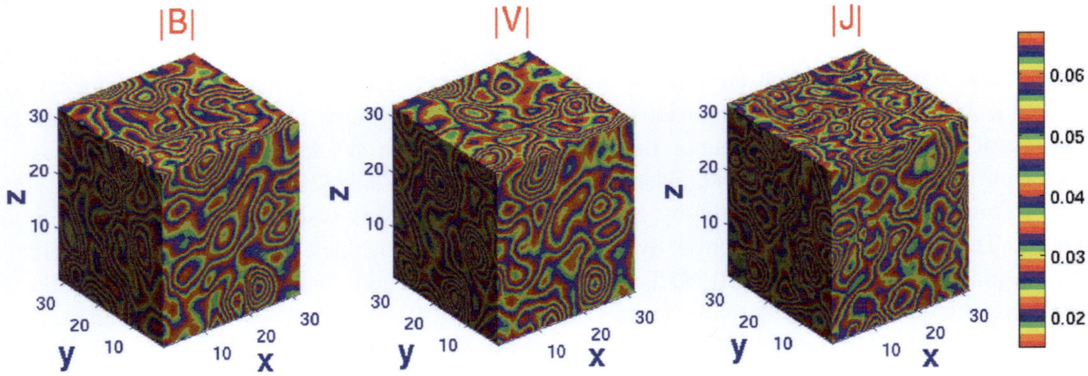

Fig. 17.4 Three dimensional structures of magnetic, velocity and current fields in Hall MHD turbulence. Turbulent equipartition between velocity and magnetic field leads to almost similar large scale structures in the two fields, while current is more intermittent

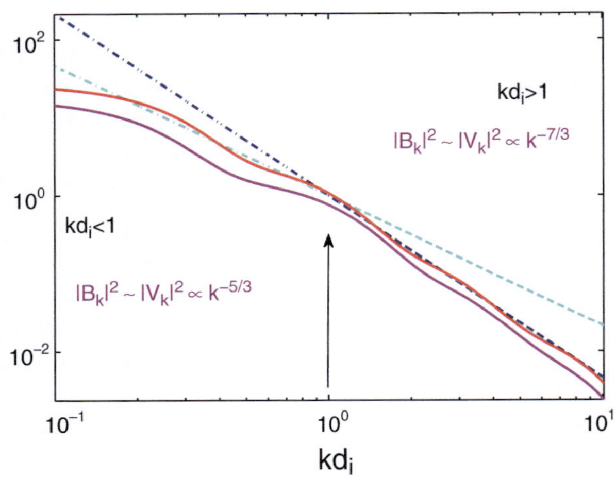

Fig. 17.5 Inertial range turbulent spectra for magnetic and velocity field fluctuations. The fluctuations closely follow respectively $k^{-5/3}$ and $k^{-7/3}$ scaling in the $kd_i < 1$ and $kd_i > 1$ KAW regimes. $kd_i = 0.05$ and 1.0 respectively in the $kd_i < 1$ and $kd_i > 1$ regimes. The *dash-dot straight lines* correspond to a $k^{-5/3}$ and $k^{-7/3}$ power law

scales are $c/\omega_{pi} < \ell < c/\omega_{pe}$, where ω_{pi},ω_{pe} are the plasma ion and pi pe electron frequencies respectively. The electron dynamics plays a critical role in determining the nonlinear interactions while the ions provide a stationary neutralizing background against fast moving electrons and behave as scattering centers. Whistler wave turbulence can be described by an electron magnetohydrodynamics (EMHD) model for the plasma (Kingsep et al. 1990), utilizing a single fluid description of quasi neutral plasma. The EMHD model has been discussed in considerable detail in earlier work (Kingsep et al. 1990; Biskamp et al. 1996; Shaikh et al. 2000a, b; Shaikh and Zank 2003, 2005). In whistler modes, the currents carried by the electron fluid are important (Shaikh 2000, 2009, 2010). Turbulent interactions mediated by the coupling of whistler waves and inertial range fluctuations have been studied in three dimensions based on a nonlinear 3D whistler wave turbulence code (Shaikh and Zank, 2010).

Electron whistler fluid fluctuations, in the presence of a constant background magnetic field, evolve by virtue of nonlinear interactions in which larger eddies transfer their energy to smaller eddies through a forward cascade. The Kolmogorov model postulates that the cascade of spectral energy occurs exclusively between neighboring Fourier modes (i.e. local interaction) until the energy in the smallest turbulent eddies is finally dissipated. This leads to a damping of small scale motions. By contrast, the large-scales and the inertial range turbulent fluctuations remain unaffected by direct dissipation of the smaller scales. In the absence of a mechanism to drive turbulence at the larger scales in the (Shaikh and Zank 2009) simulations, the large-scale energy simply migrates towards the smaller scales by virtue of nonlinear cascades in the inertial range and is dissipated at the smallest turbulent length-scales. The spectral transfer of turbulent energy in the neighboring Fourier modes in whistler wave turbulence follows a Kolmogorov phenomenology (Kolmogorov 1941; Iroshnikov 1963a, b; Kraichnan 1965) that leads to Kolmogorov-like energy spectra. Thus, the 3D simulations of whistler wave turbulence in the $kd_e < 1$ and $kd_e > 1$ regimes exhibits respectively $k^{-7/3}$ and $k^{-5/3}$ (see Fig. 17.6) spectra. The inertial range

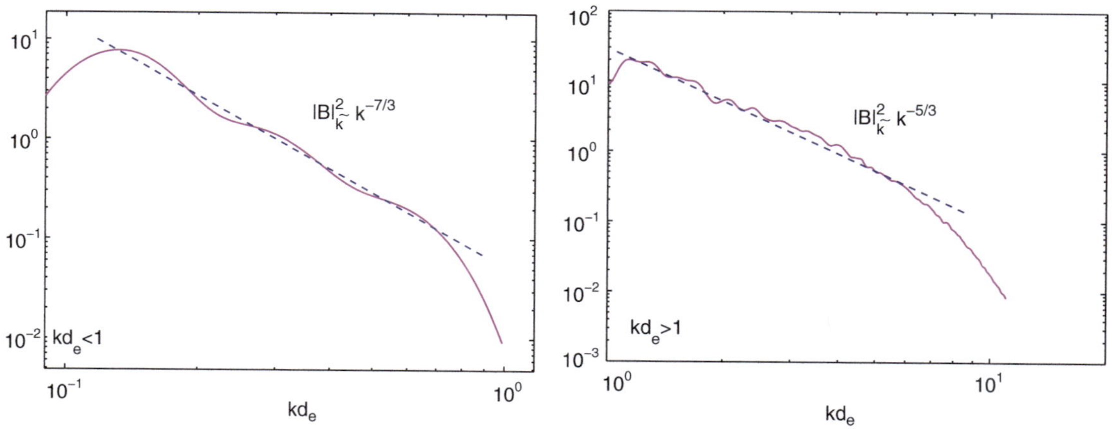

Fig. 17.6 (*Left*) 3D simulation of whistler wave turbulence in the $kd_e < 1$ regime exhibits a Kolmogorov-like inertial range power spectrum close to $k^{-7/3}$. (*Right*) The small scales magnetic field fluctuations in the $kd_e > 1$ regime depicts a Kolmogorov-like $k^{-5/3}$ spectrum which is a characteristic of hydrodynamic fluid

turbulent spectra obtained from 3D simulations are consistent with 2D models (Shaikh and Zank 2005). The whistler wave dispersion relation shows that wave effects dominate at the large scale, i.e. the $kd_e < 1$ regime, and the inertial range turbulent spectrum exhibits a Kolmogorov-like $k^{-7/3}$ spectrum. On the other hand, turbulent fluctuations on smaller scales (i.e., in the $kd_e > 1$ regime) behave like non-magnetic eddies in a hydrodynamic fluid and yield a $k^{-5/3}$ spectrum. The wave effect is weak, or negligibly small, in the latter. Hence the nonlinear cascades are determined essentially by the hydrodynamic-like interactions. Thus, the observed whistler wave turbulence spectra in the $kd_e < 1$ and $kd_e > 1$ regimes (Fig. 17.6) can be understood on the basis of Kolmogorov-like arguments that describe the inertial range spectral cascades. In the electron whistler wave regime, the fluid simulations describing a 7/3 spectrum are also reported by Ng et al. (2003), Cho and Lazarian (2004). Their results are consistent with our simulations described in Fig 17.6a.

We note that 7/3 regime of whistler turbulence is different from the usual 5/3 regime in the MHD turbulence. The 5/3 regime does not terminate sharply beyond the inertial range MHD fluctuations, but there is another cascade regime, not describable by the MHD equations, that deviates significantly from the 5/3 regime and is describable by whistler mode turbulence.

17.2 Perpendicular Shocks

Diffusive shock acceleration (DSA) is considered to be the mechanism responsible for the acceleration of energetic particles and the consequent generation of power-law spectra observed at quasi-parallel shocks (Axford et al. 1977; Bell 1978; Blandford and Ostriker, 1978). At a quasi-parallel shock, energetic ions can be scattered by the self-excited and pre-existing waves and turbulence upstream and downstream of the shock, so leading to their multiple crossing of the shock. Because these ions can stream far upstream along the magnetic field and excite low-frequency plasma waves, the turbulence responsible for particle scattering ahead of the shock is present. In this way, energetic particles can be accelerated by DSA to high energies and form a power-law spectrum (Lee 1983; Zank et al. 2000).

At a quasi-perpendicular shock, no self-consistent plasma wave excitation occurs upstream, which therefore limits particle scattering. Because of this, DSA cannot be used to explain the observed power-law spectra of energetic particles at quasi-perpendicular shock waves in the usual way. Lu et al. (2009) investigated the interaction of Alfven waves with a perpendicular shock using a two-dimensional hybrid simulation. Alfven waves are injected from the left boundary, and they have no obvious effects on the propagation speed of the shock. After the upstream Alfven waves are transmitted into the downstream, their amplitude is enhanced by about 10–30 times. Consistent with the fluid theory (McKenzie and Westphal 1969), the transmitted waves can be separated into two parts: one that propagates along the direction parallel to the background magnetic field, and the other along the direction anti-parallel to the background magnetic field. In addition, we also find obvious ripples in the shock front due to the interaction of the Alfven waves and the perpendicular shock.

In a realistic shock, of course, the structure of a quasi-perpendicular shock is considerably more complicated than described above, and the meandering magnetic field lines can cross the shock front more than once. Kóta (2009) has discussed the efficiency of ion acceleration at a perpendicular shock using an analytical approximation and numerical simulations. Energetic ions are generated at places where the field lines cross into the upstream region and soon re-cross the shock back to the downstream region. These ions may be accelerated to very high energies through multiple mirroring at the stronger downstream field.

Umeda et al. (2009) also discussed the effect of the rippling of perpendicular shock fonts on electron acceleration in the shock-rest-frame using a full particle simulation. The cross-scale coupling between ion-scale mesoscopic shock ripples and an electron-scale microscopic instability was found to play an important role in energizing electrons at quasi-perpendicular shocks. At the shock front, the ions reflected by the shock experience considerable acceleration upstream at a localized region where the shock-normal electric field of the rippled structure is polarized upstream. The current-driven instability is unstable and large-amplitude electrostatic waves grow upstream. As a result, electrostatic waves can trap electrons upstream, and then energetic electrons are generated via a form of surfing acceleration at the leading edge of the shock transition region.

Shinohara and Fujimoto (2009) discussed non-stationary behavior of the shock front, since it has been thought to play an important role for dissipation mechanism in collision-less shocks. Using JAXA's new super-computer facility allowed them to perform a three-dimensional simulation of a quasi-perpendicular shock. The simulation parameters were selected to simulate specific Cluster observational results. The full ion to electron mass ratio, $M/m = 1{,}840$, was used, and almost (one ion inertia length)^2 square plane perpendicular to the upstream flow direction was allocated for this simulation. The 3D results of Shinohara and Fujimoto (2009) showed that both self-reformation and whistler emission are present. By comparing their 3D results with 2D simulations based on the same simulation parameters (Fig. 17.7), they confirmed that the 3D result is not simply a superposition of 2D behavior but instead identified new wave activity in the front of the shock foot region. Because of the enhanced wave activity, electrons are much more efficiently heated in the 3D simulations than in 2D simulations. That the shock structure is changed significantly in adding a further degree of freedom with the third spatial dimension emphasizes the importance of fully multi-dimensional studies. The simulation of Shinohara and Fujimoto (2009) also identifies the importance of using the full mass ratio in simulations.

17.3 Global Magnetospheric Modeling and Observations

A systematic evaluation of ground and geostationary magnetic field predictions generated by a set of global MHD models shows that a metrics analysis of two different geospace parameters, the geostationary and ground magnetic field, yields surprising similarities. However, the parameters reflect rather different properties of geospace (Pulkkinen et al. 2010). More specifically, by increasing the spatial resolution and including more realistic inner magnetospheric physics made the model predictions by the BATS-R-US model more accurate. By contrast, the OpenGGCM model had a tendency to generate larger differences to observations than BATS-R-US in terms of the prediction efficiency, but the model provided more accurate representation of the observed spectral characteristics of the ground and geostationary magnetic field fluctuations. This suggests that both models capture some of the intrinsic physical elements necessary to realistic modeling, but the complexity of identifying realistic boundary conditions and the capturing of the physics between different plasma regimes in the Earth-magnetospheric

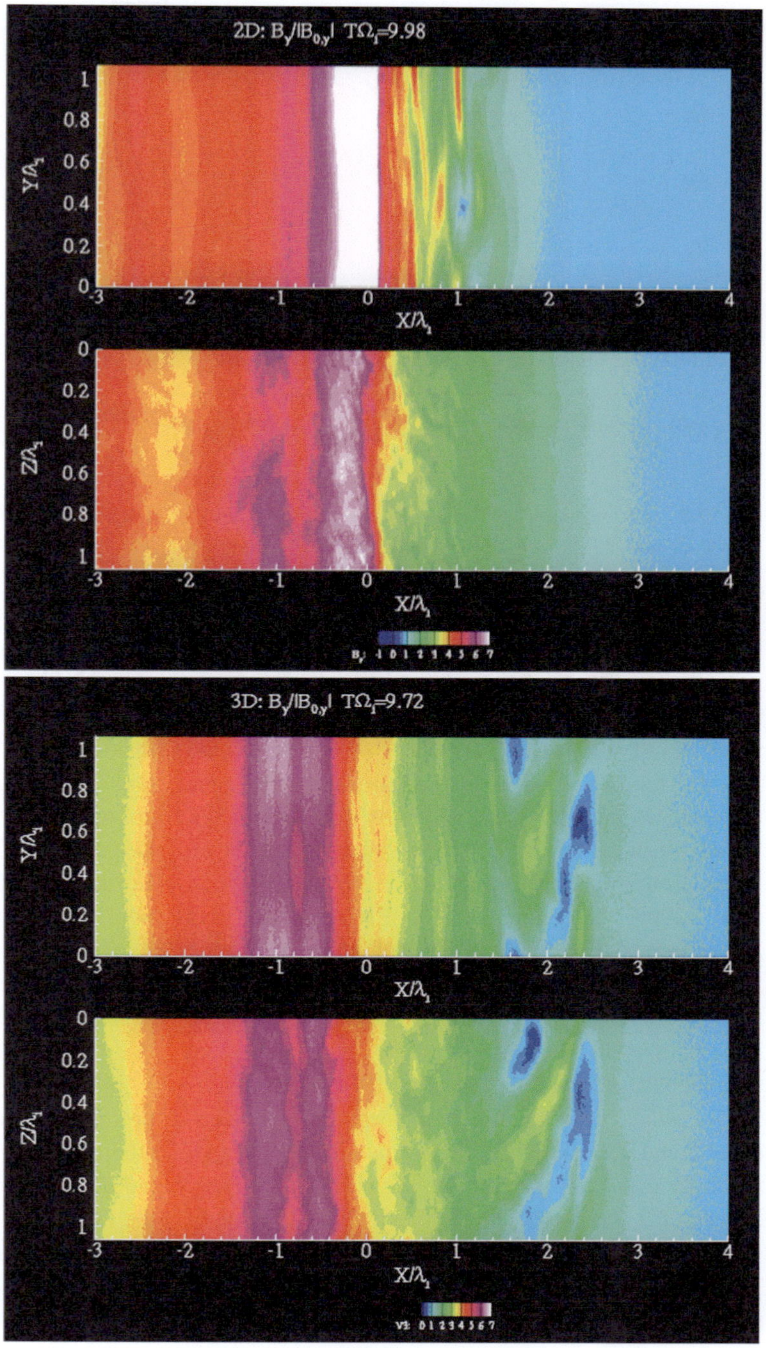

Fig. 17.7 Color contours of the By component (the major component of the magnetic field), showing (**a**) two 2D results (in the XY and XZ planes), and (**b**) a 3D result. The 3D results show that large amplitude wave active exists persistently (independent of the reformation phase) in the furthest front of the shock. [Shinohara and Fujimoto (2009).]

interaction means that this will remain an outstanding problem for years to come.

It is well known that the southward component of the interplanetary magnetic field (IMF) Bz is the primary heliospheric parameter responsible for geomagnetic storms. Yakovchouk et al. (2009) performed a statistical analysis of the peak values of the IMF B_z component with different combinations of plasma parameters and the hourly Dst (omniweb.gsfc.nasa.gov) and Dxt/Dcx (Karinen and Mursula 2005, 2006) geomagnetic indices for all identified perturbations in 1963–2009 (Fig. 17.8a). Storms without available interplanetary data were not included in the database. Yakovchouk et al. (2009) concluded that

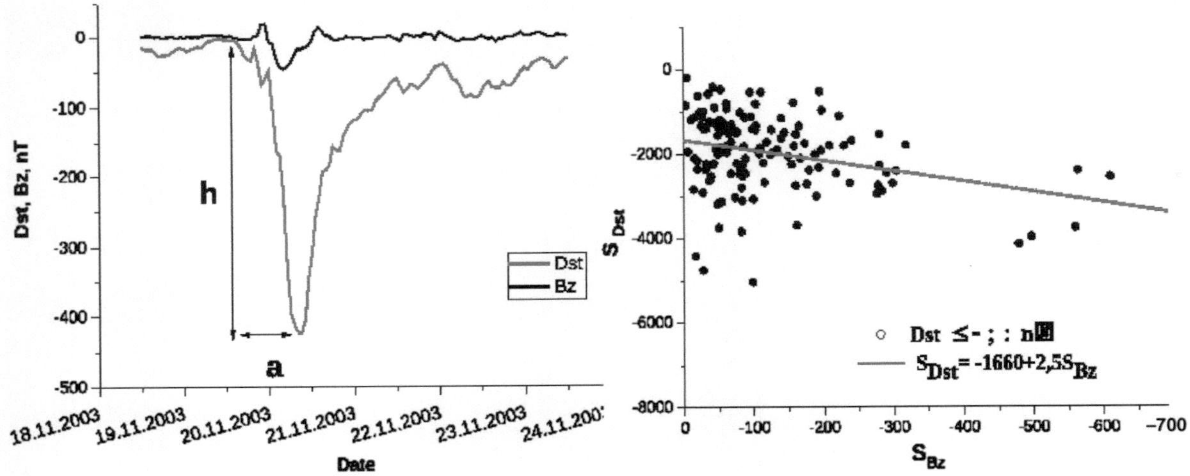

Fig. 17.8 **a** An example of a large geomagnetic storm showing the Bz and Dst profiles as a function of time. **b** The dependence of area (S = h × a) for the development phase duration of a storm for Dst and Bz peaks in 1963–2009 and its approximation by a linear fit (Yakovchouk et al. 2009)

the storms occur more often (twice as often) during the development phase of the solar cycle than during the rising phase. The average waiting time between consecutive Dst peaks is 11 days for Dst < −50 nT and 50 days for $Dst < -100nT$. The average delay time between Dst and Bz peak values is 4–6 h. A semi-annual variation of the Dst peak values exists for all levels. Empirical formulae are derived by Yakovchouk et al. (2009) that relate $Dst/Dcx/Dxt < -50nT$ and B_z/E_y values ($E_y = U_x B_z$ - the peak value of electric field, whereUx is the radial velocity component of the solar wind) based on their analysis of the observations. The relations that they present are in a good agreement with the Akasofu relation (Akasofu 1981), and are useful for quick estimates and reconstruction of heliospheric and geomagnetic parameters with accuracy of the order of a few tens percent (Fig. 17.8a). A dependence of the area for the development phase duration of storms with Dst and Bz peaks also exists (Fig. 17.8b). The accuracy of reconstruction is less when only fragmentary geomagnetic data are available.

17.4 Distribution Functions of Protons and Interstellar Hydrogen in the Inner and Outer Heliosheath

The Interstellar Boundary EXplorer (IBEX) (McComas et al. 2006, 2009a, b), launched on 19 October, 2008, is measuring the energetic neutral atom (ENA) flux from the boundary regions of our heliosphere. Contemporaneously, Voyager 1 and 2 (V1, V2) are making in situ measurements of plasma, energetic particles, and magnetic fields along two trajectories in the heliosheath (Stone et al. 2008; Richardson et al. 2008; Decker et al. 2008; Burlaga et al. 2008; Gurnett and Kurth 2008). The interpretation of the IBEX observations will depend critically on global simulations of the solar wind-local interstellar medium (LISM) interaction (e.g., Heerikhuisen et al., 2008), informed by in situ data returned by the Voyager spacecraft. Underlying the determination of the ENA flux observed at 1 AU is the form of the proton distribution function in the inner and outer heliosheath. ENAs are created by charge exchange of interstellar neutral H and heliosheath (inner and outer) protons or ions. Because the inner heliosheath is hot, a population of energetic neutral atoms is created. The flux of ENAs will therefore depend quite sensitively on the number of particles in the wings of the hot proton population downstream of the heliospheric termination shock (TS), something recognized by both Prested et al. (2008) and Heerikhuisen et al. (2008) in their introduction of a κ-distribution to model the inner heliosheath proton distribution. In particular, in an important extension of their earlier work, Heerikhuisen et al. (2008) developed a fully self-consistent 3D MHD-kinetic neutral hydrogen (H) model describing the solar wind-LISM interaction (Pogorelov et al. 2006, 2007, 2008) using a κ-distribution to describe the underlying proton distribution in the inner heliosheath. Previous models assumed a Maxwellian description for the protons with self-consistent coupling to interstellar neutral H – the self-consistent coupling being crucial in determining the global heliospheric structure (Baranov and Malama, 1993; Pauls et al. 1995; Zank et al. 1996a; Pogorelov et al. 2006) (see Zank 1999; Zank et al. 2009; Pogorelov et al. 2009a, b, c for extensive reviews). Prested et al. (2008) by contrast used a test particle approach to model the neutral H production based on an ideal MHD model.

The treatment of the heliospheric proton distribution function as a κ-distribution yields important differences in both the global structure of the heliosphere (decreasing the overall extent of the inner heliosheath between the TS and the heliopause) and the predicted ENA flux at 1 AU (Heerikhuisen et al. 2008). By assuming a κ-distribution with index $\kappa = 1.63$ (this motivated by the observed spectral index associated with energetic particles downstream of the heliospheric termination shock, Decker et al. 2005), Heerikhuisen et al. (2008) find that the ENA flux at 1 AU is substantially higher than for a corresponding Maxwellian proton distribution with the same temperature. This is not especially surprising of course because the κ-distribution contains many more particles in the wings of the distribution than the corresponding Maxwellian, thereby giving higher fluxes of ENAs at higher energies. Why the heliosheath proton distribution function should be like a kappa distribution with a spectral index close to 1.63 is however quite unclear. The answer may well reside in the processing of the upstream pickup ion distribution by the TS and the subsequent statistical relaxation of the processed distribution in the heliosheath (Livadiotis and McComas 2009). IBEX will provide definitive observations of the ENA flux at 1 AU that will allow us to estimate the proton distribution in the inner heliosheath.

Related to the question of the heliosheath proton distribution are the plasma and magnetic field observations made by Voyager 2 on the second crossing of the TS. V2 has a working plasma instrument and the coverage was sufficient to identify three distinct crossings of the TS and make in situ measurements of the microstructure. The identified TS-3 crossing revealed an almost classical perpendicular shock structure (Burlaga et al. 2008; Richardson et al. 2008). However, plasma measurements revealed that the solar wind proton temperature changed from 20,000 K upstream to 180,000 K downstream (Richardson et al. 2008; Richardson 2009). Although hot solar wind plasma is sometimes observed, the average downstream proton plasma temperature is an order of magnitude smaller than predicted by the MHD Rankine-Hugoniot conditions, and the global self-consistent models all yield downstream proton temperatures of $\sim 2 \times 10^6$ K (Zank et al. 2009). The downstream shock heated solar wind ion temperature observed by V2 is in fact so low that the downstream flow appears to remain supersonic (Richardson et al. 2008). Furthermore, the transmitted solar wind proton distribution appears to be essentially a broadened/heated Maxwellian (with a somewhat flattened peak), and there is no evidence of reflected solar wind ions being transmitted downstream (Richardson 2009). Richardson et al. (2008) and Richardson (2009) concluded that pickup ions (PUIs) experienced preferential heating at the TS and thus provided both the primary shock dissipation mechanism and the bulk of the hot plasma downstream of the TS. Unfortunately, the Voyager spacecraft were not instrumented to measure PUIs directly. That PUIs provide the TS dissipation and heated downstream plasma had in fact been predicted by Zank et al. (1996b) in their investigation of the interaction of PUIs and solar wind ions with the TS. They concluded that "P[U]Is may therefore provide the primary dissipation mechanism for a perpendicular TS with solar wind ions playing very much a secondary role." Thus the basic model of Zank et al. (1996b) for the microstructure of the TS appears to be supported by the V2 observations. However, both the observed solar wind proton distribution and a shock dissipation mechanism based on PUIs means that the downstream proton distribution function is a (possibly complicated) function of the physics of the TS. Zank et al. (2010) have extended their basic model of the quasi-perpendicular TS, mediated by PUIs, to derive the complete downstream proton distribution function in these regions, identifying the partitioning of energy between solar wind protons and PUIs, and infered potential implications of these results for the ENA flux observed at 1 AU in terms of spectra and skymaps. They did not attempt to synthesize a complete description of the inner heliosheath proton distribution at this point, preferring instead to elucidate the physics of the quasi-perpendicular termination shock, and relate that physics to the production of ENAs.

Other regions of the TS, notably the high polar regime and possibly the heliotail region of the TS, may require the introduction of distinctly different physical processes for shock dissipation, and a complete model of the heliosheath proton distribution will therefore need to account for multiple shock regimes. Katushkina and Izmodenov (2009) have begun to explore different aspects of this.

The model developed by Zank et al. (2010) describes the basic plasma kinetic processes and microphysics of the quasi-perpendicular TS in the presence of an energetic PUI population. They find that the solar wind protons do not experience reflection at the cross-shock potential of the TS, and are transmitted directly into the heliosheath. PUIs, by contrast, can be either transmitted or reflected at the TS, and provide the primary dissipation mechanism at the shock, and dominate the downstream temperature distribution. An inner heliosheath proton distribution function was derived that is (1) consistent with V2 solar wind plasma observations, and (2) is similar to a κ-distribution with index 1.63. The composite inner heliosheath proton distribution function is a superposition of cold transmitted solar wind protons, a hot transmitted PUI population, and a very hot PUI population that was reflected by the cross-shock electrostatic potential at least once before being transmitted downstream. The composite spectrum possesses more structure than the κ-distribution but both distributions have approximately the same number of protons in the wings of the distribution (and therefore many more than a corresponding Maxwellian distribution). Finally, ENA spectra from various directions at 1 AU generated by either the composite (TS) heliosheath proton distribution or the κ-distribution are very similar in intensity, although some structure is present in the composite case. The spectral shape is a consequence of the contribution to the ENA flux by primarily heliosheath transmitted and reflected PUIs. The ENA spectrum is dominated by transmitted PUI created ENAs in the energy range below 2 keV and reflected PUI created ENAs in the range above 2 keV. This may give us an opportunity to use IBEX data to directly probe the microphysics of the TS. The skymaps are dominated by ENAs created by either transmitted PUIs or reflected PUIs, depending on the energy range.

IBEX, in completing its first full scan of the sky, created maps of energetic neutral atom (ENA) flux for energies between 100 and 6 keV (McComas et al. 2009a; Schwadron et al. 2009; Funsten et al. 2009; Fuselier et al. 2009). The overall flux intensities appear to be generally within about a factor of two or three of those predicted by global models of the interaction between the solar wind (SW) and local interstellar medium (LISM). A most unexpected feature was the presence in the IBEX ENA maps of a "ribbon" that encircles the sky, passing closer to the heliospheric nose direction in the south and west than in the north and east. The ribbon represents a nearly threefold enhancement in ENA flux compared to adjacent parts of the sky, but the shape and magnitude of the energy spectrum is primarily ordered by ecliptic latitude rather than its location inside or outside of the ribbon (Funsten et al. 2009). This suggests that ENAs inside the ribbon come from the same population of parent ions. 3D global heliospheric models make it possible to simulate the flux of ENAs at 1 AU (Fahr and Lay 2000; Gruntman et al. 2001; Heerikhuisen et al. 2007; Sternal et al. 2008; Prested et al. 2008; Heerikhuisen et al. 2008; Izmodenov et al. 2009). The assumptions made by global models have been refined as new observational data emerged. For example, the termination shock (TS) crossing by the Voyager 1 and 2 spacecraft, in 2004 and 2007 respectively (Stone et al. 2005, 2008), suggested a north-south asymmetry of the heliosphere. The inferred asymmetry led to new global models with larger than previously thought interstellar magnetic field (ISMF) strengths (Pogorelov et al. 2007, 2009; Izmodenov et al. 2009). Measurements of Lyman-alpha back-scattered photons in the nearby SW (Lallement et al. 2005), suggest asymmetries in the outer heliosheath (OHS) that can be linked to the plane of the LISM magnetic and velocity vectors the so-called hydrogen deflection plane. Models confirmed (Izmodenov et al. 2005; Pogorelov et al. 2008, 2009) that the deflection of interstellar hydrogen from helium due to the shape of the OHS does indeed take place primarily in the hydrogen deflection plane. The IBEX observations enable the first global validation of these models and their components and, thus, yield insight into the physical processes that drive the structure and dynamics of the outer heliosphere. The fact that the ribbon was not predicted by any models suggests that it is generated by physical processes that have so far been omitted from models.

The relationship between the ribbon and the region just outside the heliopause where the ISMF is perpendicular to radial vectors from the sun, was discovered by the IBEX team (McComas et al. 2009a; Funsten et al. 2009; Schwadron et al. 2009) using model results from Pogorelov et al. (2009). Several possible explanations for this correlation were given in those papers, some of which rely on stresses created by the ISMF near the heliopause to generate regions of enhanced density which, combined with a local population of non-isotropic PUIs, may lead to enhanced ENA emissions. However, plots of the total pressure (magnetic plus thermal) on the surface of the simulated heliopause display no banded structures related to magnetic forces or density enhancements at all. Self-consistently coupled MHD-neutral solution indicate that enhancements in magnetic pressure and thermal pressure are somewhat anti-correlated, resulting in a relatively smooth total pressure profile. The underlying physics for generating the ribbon discovered in the IBEX data must explain a number of observed features. Firstly, the ribbon appears to be closely related to the orientation of the magnetic field just outside the heliopause, in a way that links enhanced ENA flux to regions where the outer heliosheath magnetic field BOHS is perpendicular to the heliocentric radial vector r (see Fig. 17.4 in McComas et al. (2009a); Fig. 17.3 in Funsten et al. (2009); and Fig. 17.2 in Schwadron et al. (2009)) such that $\mathbf{B}_{\mathrm{OHS}} \cdot \mathbf{r} \sim 0$. Secondly, it needs to explain why the spectrum of ENAs is very nearly the same inside and outside the ribbon (Fig. 17.2 in McComas et al. (2009a)). Thirdly, it must be based on physical processes that are excluded from all previous heliospheric models, thereby explaining why no ENA ribbon feature has been seen in any models of the SW-LISM interaction.

Heerikhuisen et al. (2010) considered the possibility that solar wind-created neutrals could create pick-up ions (PUIs) in the outer heliosheath to explain the ribbon of enhanced ENA flux observed by IBEX. Their approach relies on the fact that the average velocity of ions in the SW and inner heliosheath (IHS) is anti-sunward, so that the majority of ENAs propagate away from the sun into the outer heliosheath. In the region of enhanced interstellar plasma density surrounding the heliopause, some of these ENAs charge-exchange and create PUIs in the slow warm subsonic plasma of the outer heliosheath. These PUIs will initially form a ring-beam distribution, with a velocity component along the magnetic field. Over time this distribution-will isotropize by wave-particle interactions (Williams and Zank 1994). However, the ring distributed PUIs may charge-exchange with the fairly dense interstellar hydrogen (> 0.2 cm^{-3}), resulting in a new "secondary" ENA. These secondary ENAs have been included in models before (Izmodenov et al. 2009), but only in an isotropic way for an axially symmetric heliosphere without an interstellar magnetic field (ISMF). If "re-neutralization" occurs quickly, the PUI will not have had time to scatter to some random direction over a complete shell, but rather the secondary ENA will be directed to some random vector on a partial shell. Furthermore, in locations where $\mathbf{B}_{\mathrm{OHS}} \cdot \mathbf{r} \sim 0$, the plane of the ring about which the shell distribution is forming intersects the Sun, and leads to an increased ENA flux from these locations (see Fig. 17.9). This mechanism could thus explain the link between the ribbon and the orientation of the ISMF.

Heerikhuisen et al. (2010) used a 3D steady-state MHD-plasma/kinetic neutral model of the heliosphere (Heerikhuisen et al. 2008, 2009; Pogorelov et al. 2008), with uniform SW conditions and a 3 G ISMF in the hydrogen deflection plane pointed towards ecliptic coordinate (224,41). The LISM boundary conditions are consistent with the analysis of Slavin and Frisch (2008). They used a Lorentzian (or "kappa") distribution for IHS protons. Plotted in Fig. 17.10 are all-sky maps of ENA flux for both the simulated and observed data. The simulated ribbon does not line up exactly, but the offset is almost certainly due to a slightly "incorrect" choice of the ISMF orientation in the simulation. The observed ribbon, particularly the southern-most portion, moves slightly at high energies. The Heerikhuisen et al. simulation reproduces this effect, which can be attributed to the larger mean free path of high energy primary ENAs resulting in a ribbon from PUIs at a larger distance into the outer heliosheath, where the magnetic field orientation is slightly different.

A second important observation is the absence of a unique spectral signature associated with the ribbon. The all-sky spectrum predicted by the Heerikhuisen et al. simulations shows that the ribbon appears to have a locally steeper spectrum, while the observed spectrum shows almost no change across the ribbon. One reason for a steeper ribbon spectrum in the simulation is that a "bump" is formed at the SW energy, which is uniform and constant in their simulation while in reality the bump should be spread over a time-averaged SW energy profile. This deficiency may be addressed

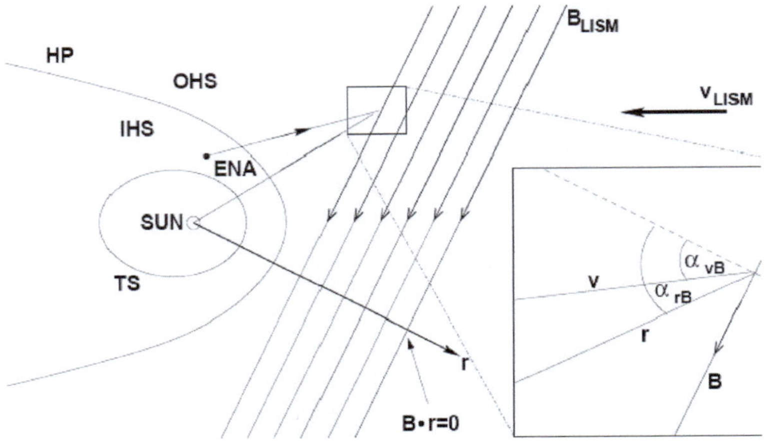

Fig. 17.9 Schematic of the heliosphere in the plane containing the ISMF and velocity vectors (BLISM and VLISM). A primary energetic neutral atom (ENA) created in the inner heliosheath (IHS) region between the termination shock (TS) and the heliopause (HP) is shown as it moves into the outer heliosheath (OHS) whereupon it ionizes and becomes an outer heliosheath pickup ion (PUI) that can re-neutralize to form a secondary ENA. Note that the OHS magnetic field becomes highly warped close to the heliopause (see Pogorelov et al. 2009). (Heerikhuisen et al. 2010)

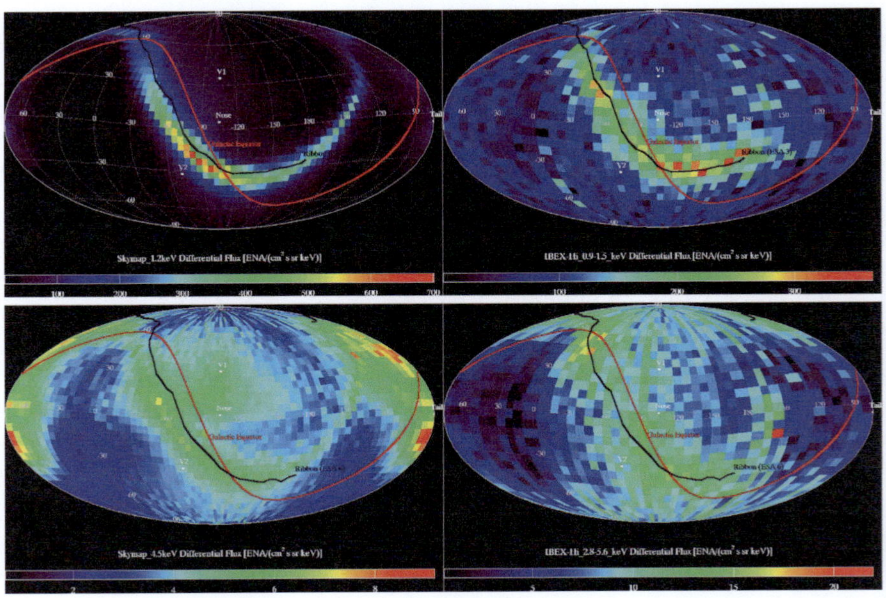

Fig. 17.10 All-sky maps of simulated (*left*) and observed (*right*) ENA flux at 1.1 keV (*top*) and 4.5 keV (*bottom*). The simulation uses a = 1.63 spectral index for IHS protons and has assumed that all PUIs retain partial shell distributions long enough to re-neutralize before they isotropize. The *red curve* is the galactic plane, and a best fit to the observed ribbon is shown as a *black line*. Note that the ribbon shifts down slightly at high energies. Units of ENA flux are (cm^2 s sr keV)$^{-1}$. (Heerikhuisen et al. 2010)

in future models by using a spectrum that depends on physical processes of energization at the termination shock as experienced by core SW ions and PUIs (Zank et al. 2010). Using such a composite spectrum would also allow for spectral indices of less than 1.5 over the IBEX energy range, something that is not possible with a κ-distribution (Livadiotis and McComas 2009).

Finally, A careful comparison with the observed ribbon suggests that if the Heerikhuisen mechanism is correct, then the ISMF is directed close to the ecliptic coordinates (224,41) used in their model, and close to the value (221,39) corresponding to the center of the ribbon observed in Funsten et al. (2009).

17.5 Conclusions

We have considered several illustrative examples of the complicated interplay and coupling between large and small space-time scales, and slow and fast processes in space plasmas of magnetospheric, heliospheric and interstellar origin. Laminar and turbulent processes coexist, and energy transfer manifests itself from small to large and large to small scales – essentially direct and inverse cascades ensuring that a full understanding of complex space plasma systems requires the proper coupling of disparate scales.

Acknowledgements This study was fulfilled as a part of the Programs of the Russian Academy of Sciences: "Origin and Evolution of Stars and Galaxies" (P-04), "Solar Activity and Space Weather" (P-16, Part 3) and Plasma Processes in the Solar System (OFN-15), MSU Interdisciplinary Scientific Project and supported by the RFBR grants 07-02-00147, NSh-1255.2008.2. DS and GPZ acknowledge the partial support of NASA grants NNX09AB40G, NNX07AH18G, NNG05EC85C, NNX09AG63G, NNX08AJ21G, NNX09AB24G, NNX09AG29G, and NNX09AG62G. The authors are grateful to participants of the IAGA meeting in Sopron who provided their published materials used for the preparation of this chapter.

References

Alexandrova O, Carbone V, Veltri P, Sorriso-Valvo L (2007) Planet Space Sci 55:2224
Alexandrova O, Carbone V, Veltri P, Sorriso-Valvo L (2008) Astrophys J 674:1153
Armstrong JW, Cordes JM, Rickett BJ (1981) Nature 291:561
Armstrong JW, Coles WA, Kojima M, Rickett BJ, (1990) Astrophys J 358:685
Armstrong JW, Rickett BJ, Spangler SR (1995) ApJ 443:209
Axford WI, Leer E, Skadron G (1977) In: Christou CYa (ed) Proceedings of the 15th international conference on cosmic rays, vol 11. Bulgarian Academy of Science, Sofia, p 13
Bale S et al (2005) Phys Rev Lett 94:215002
Baranov VB, Malama YuG (1993) J Geophys Res 98:15157–15163
Bayly BJ, Levermore CD, Passot T (1992) Phys Fluids A 4:945954
Beinroth HJ, Neubauer FM (1981) J Geophys Res 86:7755
Bell AR (1978) MNRAS 182:147–156
Beresnyak A, Lazarian A, Cho J (2005) ApJL 624:L93
Biskamp D (2003) Magnetohydrodynamic turbulence. Cambridge University Press,Cambridge.
Biskamp D, Schwarz E, Drake JF (1996) Phys Rev Lett 76:1264
Bhattacharjee A, Ng CS, Spangler SR (1998) Astrophys J 494:409
Blandford RD, Ostriker JP (1978) Astrophys J 221:L29–L32
Brandt PC, Roelof EC, Wurz P, Barabash S, Bazell D, DeMajistre R, Sotirelis T, Decker R (2009) Geophys Res Abstr 11:EGU2009-9588
Bruno R, Carbone V (2005). Living Rev Solar Phys 2:4. URL (cited on ¡25.07.2009¿): http://www.livingreviews.org/lrsp-2005-4
Burlaga LF et al (2008) Nature 454:75
Chepurnov A, Lazarian A (2010) ApJ 710:853
Cho J, Lazarian A (2002) Phys Rev Lett 88:245001
Cho J, Lazarian A (2003) MNRAS 345:325
Cho J, Lazarian A (2004) ApJ 615:L41–L44
Cho J, Lazarian A (2009) ApJ 701:236
Cho J, Lazarian A, Vishniac ET (2002) ApJ 564:291
Cho J, Lazarian A, Vishniac ET (2003) ApJ 595:812
Cho J, Vishniac ET (2000) ApJ 539:273
Decker RB et al (2005) Science 309:2020
Decker RB et al (2008) Nature 454:67
Denskat KU, Neubauer FM (1982) J Geophys Res 87:2215
Elmegreen BG, Scalo J (2004) ARAA 42:211
Frisch U (1995) Turbulence: the legacy of A. N. Kolmogorov. Cambridge University Press, Cambridge
Fuselier SA et al (2009) Science 326(5955):962–964. doi:10.1126/science.1180981
Funsten H et al (2009) Science 1(10):1126
Galtier S (2008) J Geophys Res 113:A01102. doi:10.1029/2007JA012821
Galtier S (2006) J Plasma Phys 72:721
Galtier S, Buchlin E (2007) Astrophys J 656:560
Gary SP et al (2008) Geophys Res Lett 35:L02104
Ghosh S, Goldstein ML (1997) J Plasma Phys 57:129
Ghosh S, Siregar E, Roberts DA, Goldstein ML (1996) J Geophys Res 101:2493
Goldstein ML, Roberts DA, Matthaeus WH (1995) Annu Rev Astron Astrophys 33:283
Goldstein ML, Roberts DA, Fitch CA (1994) J Geophys Res 99:11519
Gurnett DA, Kurth WS (2008) Nature 454:67
Hasegawa A, Chen L (1976) Phys Rev Lett 36:1362
Heerikhuisen J, Florinski V, Zank GP (2006) J Geophys Res 111:A06110
Heerikhuisen J, Pogorelov NV, Zank GP, Florinski V (2007) ApJ 655(1):L53–L56
Heerikhuisen J, Pogorelov NV, Florinski V, Zank GP, le Roux J (2008) Astrophys J 682:679
Heerikhuisen J, Pogorelov NV, Florinski V, Zank GP, Kharchenko V (2009) Astronomical Society of the Pacific 406:189
Heerikhuisen, J., N.V. Pogorelov GP Zank GB Crew PC Frisch HO Funsten PH Janzen DJ 2010, Astrophys. J., 708L, 126.
Higdon JC (1984) Astrophys J 285:109
Higdon JC (1986) Astrophys J 309:342
Howes GG, Dorland W, Cowley SC, Hammett GW, Quataert E, Schekochihin AA, Tatsuno T (2008) Phys Rev Lett 100:065004
Hunana P, Zank GP, Shaikh D, (2006) Phys Rev E 74(2):026302
Hunana P, Zank GP (2010) Inhomogeneous nearly incompressible description of magnetohydrodynamic turbulence. ApJ, Issue 1, pp. 148–167
Iroshnikov PS (1963a) Astron Zh 40:742
Iroshnikov PS (1963b) Sov Astron 7 (4):566–571
Izmodenov VV, Malama YG, Ruderman MS, Chalov SV, Alexashov DB, Katushkina OA, Provornikova EA (2009) Space Sci Rev 146:329–351
Izmodenov V, Alexashov D, Myasnikov A (2005) A & A 437:L35
Karinen A, Mursula K, (2006) J Geophys Res 111(A8) CiteID A08207

Katushkina O, Izmodenov V (2009) Paper 402-TUE-01145-0862 presented at the IAGA 11th scientific assembly, August 23–30, 2009, Sopron, Hungary, Programme, p 51
Kolmogorov AN (1941) Dokl Acad Sci URSS 30:301
Klainerman S, Majda A (1981) Commun Pure Appl Math 34:481
Klainerman S, Majda A (1982) Commun Pure Appl Math 35:629
Kraichnan RH (1965) Phys Fluids 8:1385
Kingsep AS, Chukbar KV, Yankov VV (1990) Reviews of plasma physics, vol 16. Consultant Bureau, New York, NY
Krishan V, Mahajan SM (2004) J Geophys Res 109:A11105
Kreiss H-O (1982) Commun Pure Appl Math 33:399439
Kóta J (2009) Paper 402-TUE-01345-0626 presented at the IAGA 11th scientific assembly, August 23–30, 2009, Sopron, Hungary, Programme, p 47
Kowal G, Lazarian A, Beresnyak A (2007) ApJ 658:423
Krimigis SM, Mitchell DG, Roelof EC, Brandt PC (2009) Presentation at "Voyagers in the Heliosheath Observations, models, and plasmas physics", January 914, Kauai, Hawaii, USA
Kritsuk AG, Ustyugov SD, Norman ML, Padoan P (2009) J Phys Conf Ser 180:012020
Lallement R, Bertaux JL, Dalaudier F (1985) Astron Astrophys 150:21–32
Lallement R et al (2005) Science 307:1447
Leamon RJ, Ness NF, Smith CW, Wong HK (1999) AIPC 471:469
Lee MA (1983) J Geophys Res 88:6109
Livadiotis G, McComas DJ (2009) J Geophys Res 114(A11):CiteID A11105
Lighthill MJ (1952) Proc R Soc Lond Ser A 211:564587
Lu QM, Hu Q, Zank G (2009) Astrophys J 706(1):687–692
Majda A (1984) Compressible fluid flow and systems of conservation laws in several space variables. Springer, New York, NY
Malama YG, Izmodenov VV, Chalov SV (2006) Astron Astrophys 445(2):693701
Maron J, Goldreich P (2001) ApJ 554:1175
Marsch E, Tu C-Y (1995) MHD structures, waves and turbulence in the solar wind: observations and theories. Kluwer, Netherlands
Matthaeus WH, Brown M (1988) Phys Fluids 31:3634
Matthaeus WH et al (2005) Phys Rev Lett 95:231101
McComas DJ, Reisenfeld DB, Schwadron NA (2010) Astrophys J 708(2) L126–L130
McComas DJ, Allegrini F, Bochsler P, Bzowski M, Collier M, Fahr H, Fichtner H, Frisch P, Funsten H, Fuselier S et al (2004) In: Florinski V, Pogorelov NV, Zank GP (eds) The interstellar boundary explorer (IBEX), physics of the outer heliosphere, Third annual IGPP conference, AIP CP719, pp 162–181
McComas D et al (2006) In: Heerikhuisen J, Florinski V, Zank GP (eds) The physics of the inner heliosheath, American Institute of Physics Conference Proceedings, vol 858, p 241
McComas D et al (2009a) Space Sci Rev 146:11–33. doi:10.1007/s11214-009-9499-4
McComas D et al (2009b) Science 2(10):1126
McKenzie JF, Westphal KO (1969) Planet Space Sci 17:1029–1037
Montgomery DC, Brown MR, Matthaeus WH (1987) J Geophys Res 92:282284
Mutel RL,Molnar LA, Spangler SR (1998) American astronomical society, 192nd AAS Meeting, #46.13. Bull Am Astron Soc 30:879
Mursula K, Karinen A (2005) Geophys Res Lett 32(14):L14107
Ng CS, Bhattacharjee A, Germaschewski K, Galtier S (2003) Phys Plasmas 10:1954
Parnell C et al (2009) Astrophys J 698(1):75–82
Pauls HL, Zank GP, Williams LL (1995) J Geophys Res 100(21):595
Pauls HL, Zank GP (1997) J Geophys Res 102(19):779
Pogorelov NV, Zank GP, Ogino T (2006) Astrophys J 644:1299
Pogorelov NV, Stone EC, Florinski V, Zank GP (2007) Astrophys J 668:611
Pogorelov NV Heerikhuisen J, Zank GP (2008) Astrophys J 675:L41
Pogorelov NV, Heerikhuisen J, Zank GP, Borovikov SN (2009a) Space Sci Rev 143:31. doi:10.1007/s11214-008-9429-x
Pogorelov NV, Borovikov SN, Zank GP, Ogino T (2009b) Astrophys J 696:1478
Pogorelov NV, Heerikhuisen J, Mitchel J, Cairns IH, Zank GP (2009c) Astrophys J 695:L31
Prested C, Schwadron N, Passuite J, Randol B, Stuart B, Crew G, Heerikhuisen J, Pogorelov N, Zank G, Opher M, Allegrini F, McComas DJ, Reno M, Roelof E, Fuselier S, Funsten H (2008) J Geophys Res 113(A6) CiteID A06102
Podesta JJ, Chandran BDG, Bhattacharjee A, Roberts DA, Goldstein ML (2009) J Geophys Res 114:A01107. doi:10.1029/2008JA013504
Pulkkinen A, Rasttter L, Kuznetsova M, Hesse M, Ridley A, Raeder J, Singer HJ, Chulaki A (2010) J Geophys Res 115(A3) CiteID A03206
Richardson JD et al (2008) Nature 454:63
Richardson JD (2009) Geophys Res Lett 35 CiteID L23104
Saito S, Gary SP, Li H, Narita Y (2008) Phys Plasmas 15:102305
Sahraoui F et al (2007) J Plasma Physics 73:723
Sahraoui F et al (2009) Phys Rev Lett 102:231102
Schwadron N et al (2009) Science 326 (5955):966. doi:10.1126/science.1180986
Smith CW, Isenberg PA, Matthaeus WH, Richardson JD (2006) Astrophys J 638:508–517
Shaikh D (2004) Phys Scripta 69(3):216
Shaikh D (2009a) J Plasma Phys 75:117
Shaikh D (2009b) MNRAS 395:2292
Shaikh D (2010a) MNRAS 405, 2521–2528. doi:10.1111/j.1365-2966.2010.16625.x
Shaikh D (2010b) Phys Lett A 374(25):2551–2554
Shaikh D, Shukla PK (2008) FRONTIERS IN MODERN PLASMA PHYSICS: 2008 ICTP international workshop on the frontiers of modern plasma physics. AIP conference proceedings, vol 1061, pp 66–75
Shaikh D, Shukla PK (2008) AIPC 1061:6675
Shaikh D, Shukla PK (2009) Phys Rev Lett 102:045004
Shaikh D, Zank GP (2003) Astrophys J 599:715
Shaikh D, Zank GP (2004) 602:L29–L32
Shaikh D, Zank GP, (2005) Phys Plasmas 12:122310
Shaikh D, Zank GP (2006) Astrophys J 640:195
Shaikh D, Zank GP (2007) Astrophys J 656:17
Shaikh D, Zank GP (2008) 688(1):683–694
Shaikh D, Zank GP (2009) Monthly notices of the royal astronomical society 400(11):1881–1891
Shaikh D, Zank GP (2010a) AIPC 1216:180

Shaikh D, Zank GP, (2010b) AIPC 1216:168
Shaikh D Das A, Kaw P, Diamond PH (2000a) Phys Plasmas 7:571
Shaikh D, Das A, Kaw P (2000b) Phys Plasmas 7:1366 5
Shinohara I, Fujimoto M (2009) Paper 402-TUE-01415-1055 presented at the IAGA 11th scientific assembly, Aug 23–30, 2009, Sopron, Hungary, Programme, p 54
Spangler S (1987) Phys Fluids 30:1104–1109
Spangler S (1999) Astrophys J 522:879–896
Spangler SR (2001) Space Sci Rev 99:261–270
Stawicki O, Gary SP, Li H (2001) JGR 106(A5):8273
Stone EC, Cummings AC, McDonald FB, Heikkila BC, Lal N, Webber WR (2005) Science 309(5743):2017–2020
Stone EC et al (2008) Nature 454 71
Umeda T, Yamao M, Yamazaki R (2009) Astrophys J 695(1): 574–579
Veselovsky IS (2007) Adv Space Res 40(7):1087–1092
Veselovsky IS, Prokhorov AV (2008) Solar Syst Res 42(2):177–182
Washimi H, Zank GP, Hu Q, Tanaka T, Munakata K (2007) Astrophys J Lett 670:L139.
Williams LL, Zank GP (1994) J Geophys Res 99:19229
Yakovchouk OS, Veselovsky IS, Mursula K (2009) Adv Space Res 43(4):634–640
Zank GP, Matthaeus WH (1991) Phys Fluids A3:69–82
Zank GP, Matthaeus WH (1993) Phys Fluids A5:257–273
Zank GP, Muller H-R (2003) JGR 108(A6):1240. doi:10.1029/2002JA009689
Zank G P,Rice WKM, Wu CC (2000) J Geophys Res 105(A11):25079–25095
Zank GP, Heerikhuisen J, Pogorelov NV, Burrows R, McComas D (2010) Astrophys J 708(2):1092–1106. doi:10.1088/0004-637X/708/2/1092
Zank GP, Pauls HL, Williams LL, Hall DT (1996a) J Geophys Res 101(21):639
Zank GP, Pauls HL, Cairns IH, Webb GM (1996b) J Geophys Res 101:457
Zank GP et al (2009) Space Sci Rev 146(1):295.
Zank GP (1999) Space Sci Rev 89:413

Chapter 18

Selected Solar Influences on the Magnetosphere: Information from Cosmic Rays

Karel Kudela and Leonid L. Lazutin

Abstract The state of the magnetosphere is influenced by the effects driven from the solar surface. The models of geomagnetic field are parametrized by the magnetopshere activity indices which are related to IMF and solar wind characteristics. Cosmic rays could serve as a tool for "remote sensing" of the redistribution of IMF structure in interplanetary space and for checking validity of geomagnetic field models with external current systems. The anisotropy of cosmic rays observed on the ground is influenced by superposition of (a) interplanetary anisotropy due to transitional effects and by (b) variable transmissivity of magnetosphere itself. The possibilities to deconvolute the two dependences is discussed. Anisotropy observed at neutron monitors and muon telescopes just before the onset of some geomagnetic storms is reviewed. The changes of geomagnetic cut-off, structure of the transmissivity function and asymptotic directions for various geomagnetic field models during strong geomagnetic storms are discussed. Low altitude polar orbiting satellites with large geometric factors for high energy particles (e.g. CORONAS-F) are suitable for (a) estimates of energy spectra of solar or interplanetary accelerated particles by checking the outer zone boundary of trapping and for (b) checking how the different geomagnetic field models are fitting the observed trapped particle profiles in different local time sectors. Independently on the state of magnetosphere, the measurements of energetic "neutral emissions" (gammas and neutrons) near the Earth or on the ground, serve as indicator of acceleration processes on solar surface.

K. Kudela (✉)
IEP SAS, Watsonova 47, 040 01 Kosice, Slovakia
e-mail: kkudela@upjs.sk

18.1 Introduction – Cosmic Rays and Space Weather

Cosmic rays including the particles accelerated to high energy in solar flares or at the CMEs are affecting the magnetosphere especially at high latitudes and the atmosphere at high altitudes. The onset of high energy particle flux at the Earth is the first indication of possible radiation hazard storm near the Earth. The cosmic ray (CR) anisotropy observed by the ground based stations can serve as one of the elements of the alert before the geoeffective events. The book by Dorman (2009) provides a comprehensive review of the cosmic ray interaction with the magnetosphere. The relations of the cosmic ray studies to the space weather effects are reviewed e.g. in Kudela et al. (2000); Kudela (2009) among the other papers.

18.1.1 *Short Time Forecast of Radiation Storms*

The particles with the energy of several tens to hundreds of MeV are the most important for the radiation hazard effects during solar radiation storms with the electronic element failures on satellites, communication and with biological consequences. Before their massive arrival, the detectors of the CRs observing secondaries above the atmospheric threshold and at locations with various cutoff rigidity can provide useful alerts several minutes to tens minutes in advance, if the good temporal resolution and network by many stations is in real time operation. The neutron monitor (NM) at single site (high latitude,

M.P. Miralles, J. Sánchez Almeida (eds.), *The Sun, the Solar Wind, and the Heliosphere*, IAGA Special Sopron Book Series 4, DOI 10.1007/978-90-481-9787-3_18,

good statistics) allows to obtain a real time energy spectrum. For January 20, 2005 event it was shown at the South Pole by combination of NM64 and by that lacking usual lead shielding (Bieber et al. 2006). The ground level enhancement (GLE) real-time alarm based on the 8 high latitude NMs including those at high mountain is described in the paper Kuwabara et al. (2006). A three level alarm system (by number of NMs exceeding threshold value above that of the baseline) is suggested. Out of 10 GLEs in 2001–2005 archived data the system produced 9 correct alarms. The GLE system gives earlier warning than the satellite (SEC/NOAA) alert. Recently the paper by Su et al. (2009) checked the potential of the South Pole NM data and the data from monitor lacking the shielding for prediction of radiation storm intensity measured by the GOES. The data from the two groups of the GLE used (12 compared with P4–P8, 7 additionally compared with high energy channels of the GOES) have shown that the South Pole GLE observations can be used to predict radiation intensity of the higher energy proton channels from the GOES.

Recently, also the progress in using the NM at low and middle latitudes is reported before the radiation storms. Several steps of the GLE alert algorithm using the NM network have been described by Mavromichalaki et al. (2009) in the NMDB project (http://www.nmdb.eu).

The paper by Anashin et al. (2009) describes another type of alert signal for GLEs which can be found in real time at http://cr0.izmiran.ru/GLE-AlertAndProfiles and forecast of Solar Neutron Alert at: http://cr0.izmiran.ru/SolarNeutronMonitoring.

In addition to the early GLE alerts by the ground based NMs the forecasts from satellite data are reported. The paper by Posner (2007) demonstrates the important possibility of short-term forecasting of the appearance and intensity of solar ion events by means of relativistic electrons measured on satellites.

When high energy particles strike the atmosphere, they produce the secondary population (and the tertiary one in the NMs) and change the ionisation and contribute to the dose at airplane altitudes and above. The longest data set of ionizing component of secondaries at different altitudes has been collected in FIAN Moscow (Stozhkov et al. 2007). While the ionization measured by Geiger counters has strong solar activity cycle variation at high altitudes, it is not the case for low altitudes (Bazilevskaya et al. 2008). During solar proton events (SPE) the ionisation is increasing especially at high latitudes. The SPE occuring during Forbush decreases of galactic CR, however, have rather complex effect on that (Usoskin et al. 2009).

18.1.2 The CR Precursors of Geoeffective Events

The CMEs have various size, geometry, speed and direction of motion with respect of the Earth (see http://cdaw.gsfc.nasa.gov/CME_list/) and they differ in geo-effectiveness (Gopalswamy et al. 2009).

Analysis of the CR measurements at the NM energies showed long time ago the existence of precursors before the arrival of an interplanetary shock to the Earth and before the onset of the Forbush decrease (Dorman 1963). Due to the high CR velocity, parallel mean free path λ_{par} and gyroradius, the information about precursory anisotropies related to the IMF inhomogenity, is transmitted fast to remote locations: intensity deficit of the CR can be observed up to the distance of 0.1 $\lambda_{par} \cos(\boldsymbol{\Phi})$, $\boldsymbol{\Phi}$ - cone angle of IMF (Ruffolo 1999). Precursors to the Forbush decrease (FD) are proposed in the frame of the pitch-angle transport near oblique, plane-parallel shock. Assuming different values of the power-law index of magnetic turbulence, mean free path and decay length for typical primary energies to which the NM and muon detectors (MD) are sensitive. The loss cone precursors should be observed by NM $\sim$4 h prior to shock arrival, and by MD $\sim$15 h prior to shock arrival (Leerungnavarat et al. 2003).

Recently there have been reported several case and statistical type of studies on the precursors before the geomagnetic storms based on the CR anisotropy or specific features of the counting rate variability. The muon detectors (MD) are used for multidirectional measurements. The MD at Sao Martinho, Brazil, showed that subtracting contribution from the diurnal anisotropy determined by the Global Muon Detector Network (GMDN), the clear signatures of the precursor before the storm on December 14, 2006, were found (Fushishita et al. 2009). The loss cone precursor (deficit of the CR flux at small pitch angles) appeared only ~6.6 h after the CME eruption on the Sun, when the interplanetary shock was expected to be located 0.2 AU from the Sun.

On the September 2005 the Forbush decrease was investigated and a clear modulation in about 8-h periodicity was emerging from the pre-Forbush subsets. The analyzed case study suggests that the CR datasets, containing seven days of data with 5-min time resolution, can give a signal for interplanetary storms approaching the Earth up to 9 hours before the onset of the FD-main phase at two NMs with different cutoff rigidities (Diego and Storini 2009).

New muon measurements were reported. Data from a large muon track detector – the hodoscope URAGAN (surface 34 m^2) around the heliospheric disturbances in 2007–2008 were analyzed. Each track was reconstructed with accuracy $< 1°$. Among 63 events, when URAGAN data existed, in 53 events (84%) disturbances of anisotropy vector had been observed. Although the distribution of time differences of perturbation between the ACE and the URAGAN were rather wide, the mean value of the onset time of perturbation by the two measurements was -13.6 ± 2.6 h (Timashkov et al. 2009).

The statistical study of CR precursors in 2001–2007 before different storms using the Global Muon Detector Network (GMDN) was done in the paper Da Silva et al. (2009). The storms were divided into three groups, namely the super storms (Dst< -250 nT); the intense storms (-250 nT <Dst< -100 nT); and the moderate storms (-100 nT <Dst< -50 nT). The percentage of the events accompanied by the precursors prior to the SSC increases with increasing peak Dst is: 15% of MSs, 30% of ISs and 86% of SSs were accompanied by CR precursors observed on average 7.2 h in advance of the SSC.

18.2 Using Magnetospheric Filter for the Energy Spectra of Accelerated Particles

Measurements within the magnetosphere can provide information about the solar and/or interplanetary acceleration of particles by using the geomagnetic field filter on charged particles and no effect on high energy photons and neutrons.

18.2.1 Charged Particles, Low Orbits

Having large geometrical factor for energetic particles measured at low, nearly polar orbiting satellites, the arrival of solar flare particles can be checked according to its boundary position and the flux at four segments of trajectory per one orbit. The CORONAS-F was a low altitude satellite (detailed description of the complex of measurements can be found in the paper Kuznetsov 2008) and one of the devices, namely SONG (described e.g. in Kuznetsov et al. 2004), had such possibility.

Checking value of proton flux at different L shells (4 times per orbit at selected Ls from 1.75 to 3) and assuming the simple shape of energy spectra of the type $J(>E) = J_0 E^{-\gamma}$, the spectra in Fig.18.1 was obtained (Kuznetsov et al. 2007) and compared with the NMs (Vashenyuk et al. 2005; Miroshnichenko et al. 2005).

Recently the PAMELA experiment provided important information on the energy spectra of the SEP during the GLE on December 13, 2006. Combining the low energy measurements by the GOES (3 channels covering 30–500 MeV), three energy channels by the PAMELA (from 0.1 to 1 GeV) and NM data, the authors obtained the time evolution of the fit of spectra over long time period (De Simone et al. 2009a).

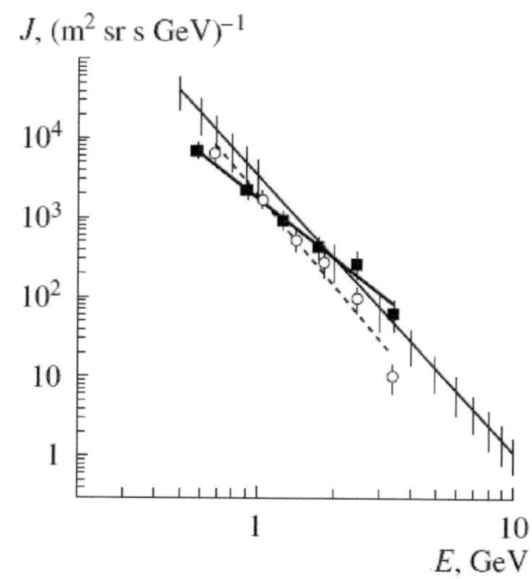

Fig. 18.1 Energy spectra of SEP on October 28, 2003 at 1,142–1,146 UT evening sector (*black squares*) and at 1,204–1,209 UT morning sector (*circles*) (Kuznetsov et al. 2007). Comparison with NM data (line >400 MeV) according to (Vashenyuk et al. 2005; Miroshnichenko et al. 2005)

For practical purposes the position of penetration boundary of the SEP was fitted from large amount of observations during different geomagnetic activity levels (Smart et al. 2006; Smart and Shea et al. 2009).

However, the position of the boundary of the SEP penetration to low orbits was not known exactly for the given geomagnetic activity level. The large spread of magnetic latitude at fixed Kp and Dst is illustrating that situation in the Fig. 18.2.

The boundary position during the penetration of the SEP on low orbits had rather complicated character especially during strong geomagnetic events. One of

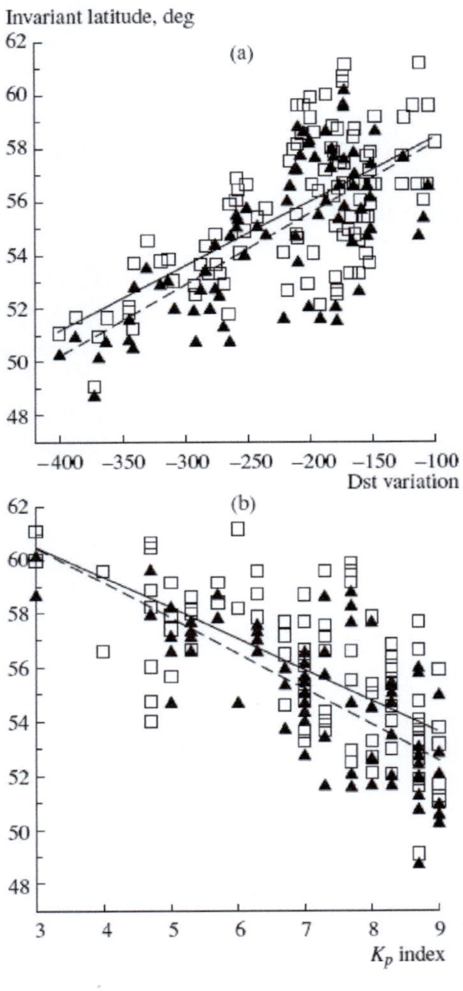

Fig. 18.2 Dependences of the SEP penetration boundaries on (**a**) Dst and (**b**) Kp in the evening and night MLT sectors: invariant latitudes of the penetration boundaries for protons (*squares*) 1–5 and (*triangles*) 50–90 MeV and linear regression for (1–5)-MeV and (50–90)-MeV protons (*solid* and *dashed lines*, respectively) CORONAS-F. Interval 2001–2005. (Myagkova et al. 2009)

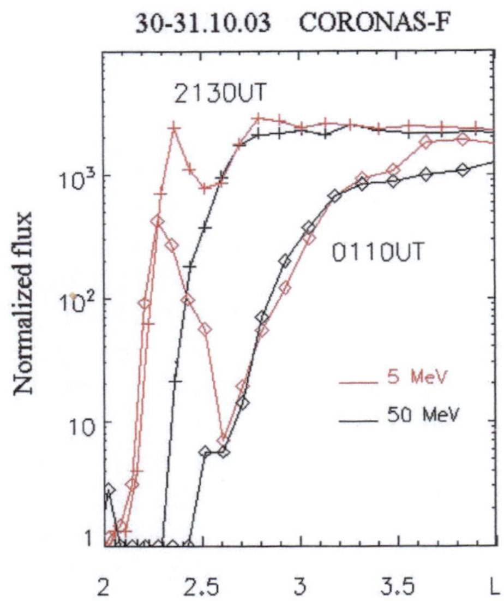

Fig. 18.3 Solar protons with energy 1–5 MeV might be trapped, creating temporary solar CR belts on $L = 2 - 3$ or providing additional flux to the previously existed population. This trapping action is observed as a double boundary effect in the Coronas-F measurements. Double boundary effect of 1–5 MeV protons is seen during the magnetic storm recovery phase, 30–31 October, 2003. The *dotted lines* indicate the penetration boundary of the 50–90 MeV protons. The *solid lines* indicate the penetration boundary of 1–5 MeV protons. (Lazutin et al. 2009)

specific features which had not been understood quite well was the double structure of the boundary position, This is shown in Fig. 18.3. Value of L at given position during the storms depends on the geomagnetic field model used.

Two more questions obtained from the observations remain not understood well, namely (a) 1–100 MeV SEP penetrate into the magnetosphere to lower latitudes as deep as it is not allowed by any magnetic field models, and (b) during some strong magnetic storms penetration boundaries coincide for wide energy range in comparison with normal (expected) penetration structure. This is seen from the Fig. 18.4.

18.2.2 Neutral High Energy Emissions

The measurements of high energy photons not affected by geomagnetic field provide important information about the timing of proton acceleration in solar flares. Clear increase in the energy spectra of photons at energy 50–100 MeV, associated with π^o decay was

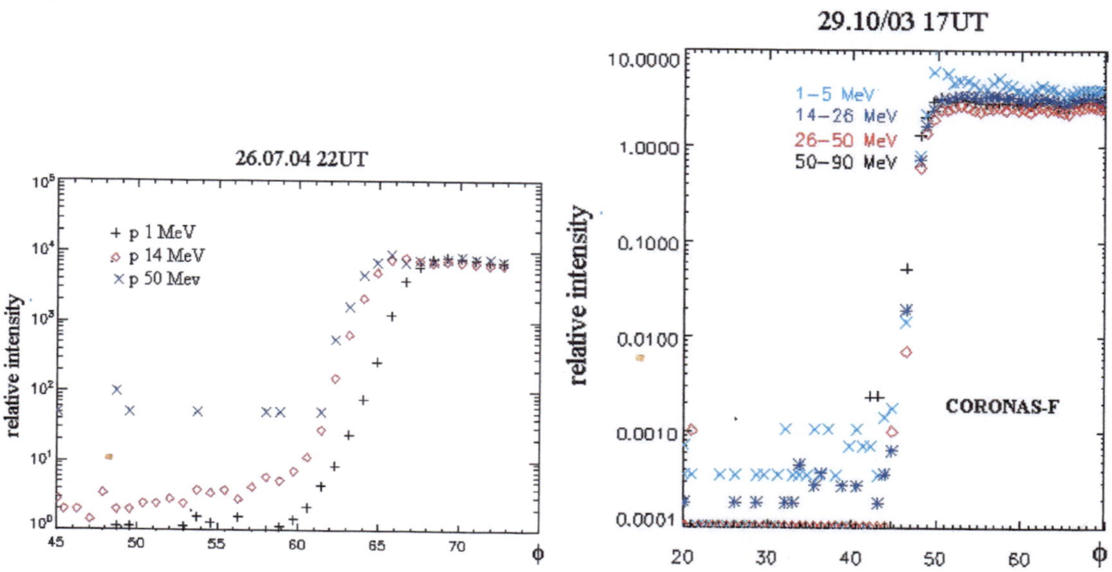

Fig. 18.4 While the boundary position during the event on July 26, 2004 was organized acording to the kinetic energy of protons (*left*), it was practically coinciding for the wide energy range in the strong geomagnetic disturbance on October 29, 2003. It is not clear: was it created by some special "transparency" of the magnetospheric boundary during strong magnetic storms or was it a consequence of some changes of the magnetic field structure of the inner magnetosphere?

Table 18.1 From (Kurt et al. 2009). The time onset of pion-decay gamma rays and the flux observed by the SONG on the CORONAS-F during strong solar flares for the period July 2001–January 2005

Date	Location/importance	Onset of π-decay γ emission, UT	γ-ray flux at 100 MeV, $[\mathrm{MeV^{-1}cm^{-2}s^{-1}}]$	Particles
25 August 2001	S17E34, 3B/X5.3	16:30:16±2 s	7.3 10^{-4}	n
28 October 2003	S16E08, 4B/X17.2	11:03:51±2 s	6.8 10^{-3}	GLE65, n
04 November 2003	S19W83, X28.9	19:42:38±4 s	1.0 10^{-3}	n
20 January 2005	N14W61, 3B/X7.1	06:45:34±4 s	3.6 10^{-3}	GLE69

reported during some flares (Kurt et al. 2009). It indicated the exact time of energetic proton appearance in the solar atmosphere. This allows to compare the proton acceleration time with the start time of the GLE recorded by the ground NMs, and to calculate the time interval when the GLE particle escaped from the corona. It is shown for the four large flares observed by the SONG instrument on the CORONAS-F in Table 18.1.

18.3 Transmissivity Function and Albedo Cosmic Rays

The only possibility to obtain predictions of cosmic ray transmissivity through the magnetosphere is numerical tracing of particle motion in the given geomagnetic field model. The equation describing the particle motion in a static magnetic field leads to the system of 6 linear differential equations with unknown values (position, velocity vector) which is usually solved numerically (e.g. McCracken et al. 1965; Bobik 2001, among the others). The review of the progress of the 50 years trajectory calculation can be found in the paper (Smart et al. 2009). For the trajectory computations with the step dR summarized over larger rigidity interval DR, the useful approach is the transmissivity function TF(R,DR) – the probability that a particle of the rigidity (R,R+DR) can access the given point in the model field (Kudela and Usoskin 2004). Similar concept was introduced earlier – the cutoff probability (Heinrich and Spill 1979).

The application of the TF was used e.g. to estimate the contribution of the secondary CR population – the reentrant albedo particles at low earth orbits. The product of energy spectra of the galactic CR (CREME 96) and the TF for different bands of geomagnetic latitudes separated at the energies below the vertical cut-off the albedo particles from the measurements for the AMS experiment (Bobik et al. 2006) and recently also for PAMELA experiment (De Simone et al. 2009b).

18.4 Transmissivity of the CR During Geomagnetic Disturbances

For geomagnetically disturbed periods it was necessary to use the geomagnetic field models with external current systems. Out of them the three were used here for comparison of predictions of the TF for a strong geomagnetic storm when improvement of the transmissivity during large Dst depression has been observed as a combined effect – starting the Forbush decrease seen at low cut-off stations and the increase apparent at the middle and low latitude NMs. Depression of the *Dst* on November 20, 2003 to −475 nT was accompanied by strong increase of count rate on several NMs, especialy Rome (6.3 GV cutoff) and Athens (8.3 GV).

The three different geomagnetic field models, namely (i) Tsyganenko' 89 (Tsyganenko 1989); (ii) the Boberg model (Boberg et al. 1995) and (iii) Tsyganenko 2004 (Tsyganenko and Sitnov 2005) provided different TF functions for that period by trajectory computations for a middle latitude station (Fig. 18.5).

The asymptotic directions for acceptance of the CR were similar for the three models before the storm. However, for the period of the minimum Dst, the structure of asymptotics was significantly different for different models. The third difference was in comparison of time of minimum cut-off rigidity (time of peak of the CR during the storm). For this particular case the better correspondence with measurements had been provided by the Tsyganenko 2004 model than by the other two. However, for another storm, namely November 7–8, 2004, it was not the case and the Boberg model (Boberg et al. 1995) including the Dst provided better alignement with maximum CR intensity than the Tsyganenko 2004 model. More details on comparison of different models with the CR is e.g. in the papers

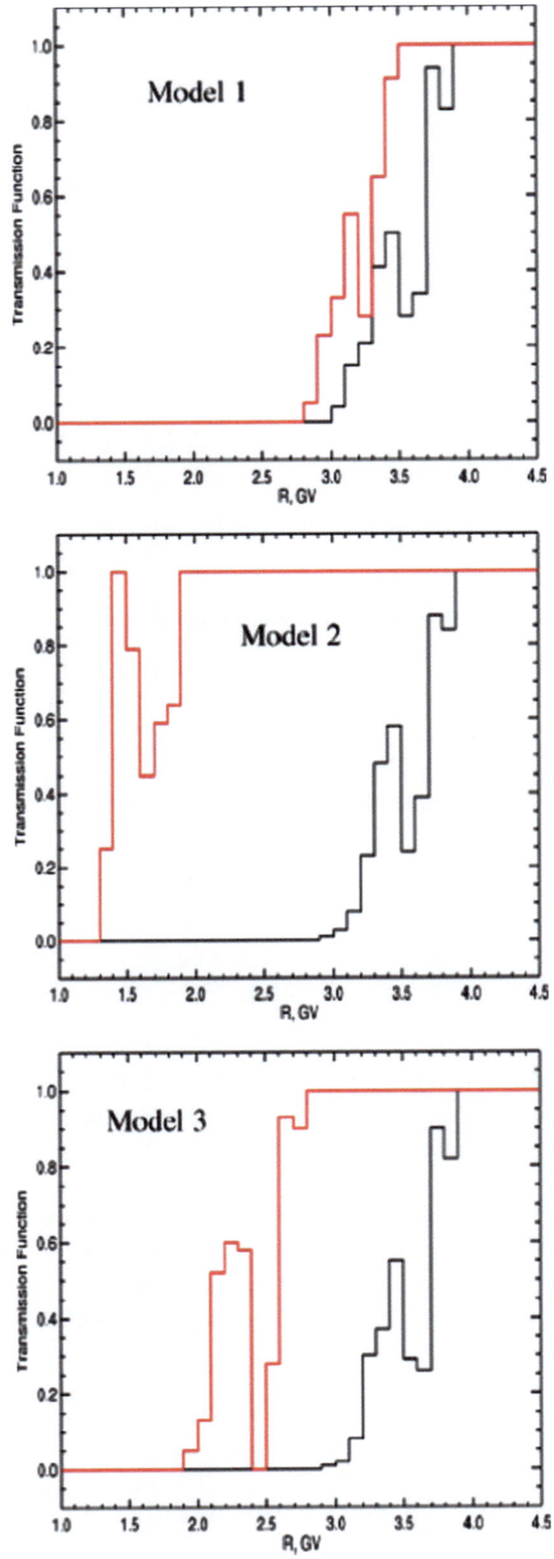

Fig. 18.5 TF (for vertical direction) for Lomnický Štít before the onset of the storm (November 20, 2003, 02 UT, *black*) and during the *Dst* minimum (19 UT, *red*) for three models. Adopted from (Kudela et al. 2008)

Kudela et al. (2008); Tyasto et al. (2008) among the others.

How to check correctly the validity of geomagnetic field models by the CR during the geomagnetic storms? It has to be assumed there are two effects superimposed during the geomagnetic storm, namely (i) the pure anisotropy of the CR in interplanetary medium during the CME propagation and "seen" by the CR coming to the magnetosphere, and (ii) the reconstruction of asymptotic directions; of the transmissivity and timing of the cut-off depression due to the changes of external current systems in the magnetosphere during disturbances. Thus, it is necessary to have an independent estimate of the interplanetary anisotropy if middle and low latitude NMs are used for checking the validity of the field models existing. There may be useful an approach from both sides of the primary spectra. The Spaceship Earth (Bieber and Evenson 1995), the ring of the NMs at high latitudes providing the anisotropy at low energies (not strongly affected by magnetospheric disturbance with asymptotic directions in narrow interval of longitudes, close to the ecliptic), and not strongly changing the system of asymptotics due to the magnetosphere's reconstruction (below the atmospheric cut-off) may serve as a potential estimate for the interplanetary CR anisotropy at low energies. On the other hand, the network of muon directional telescopes (Munakata et al. 2000) studies anisotropy at energies above NM ($\sim$50 GeV), not strongly affected by the changes in the magnetosphere. This may serve as an estimate of the anisotropy at high energies. If consistent picture of anisotropies in interplanetary space at low and high energies is obtained for a geomagnetic storm, and the estimate of anisotropy at "middle energies" is estimated, the validity of various geomagnetic field models during geomagnetic storms can be done with using the world wide network of the NMs as well as with new instruments as e.g. Chilingarian et al. (2009). In addition, energization of electrons to relativistic energies during the substorms puts also constraints on magnetospheric topology and on the geomagnetic field models (Antonova et al. 2009; Antonova 2009).

18.5 Summary

For radiation hazard events the alerts from both single point measurements of the CR at different energies (South Pole) and the NM network progressed in recent years – alerts constructed earlier than those from the satellites (for GLEs), are providing estimation of the fluence and peak intensity. High potential of electron measurements is stressed. New space instruments for that are in preparation (e.g. Grimani and Fabi 2009).

Progress in geoeffective events precursor studies is reported, namely new case and statistical studies, especially at high energies (GMDN), and the new instruments (e.g. URAGAN hodoscope) are showing potential possibilities of the alert. The CMEs have various size, geometry, speed and direction of motion with respect of the Earth and they differ in geoeffectiveness. The problem with using only CR signatures remains in the large variability of precursory timing from the anisotropy onset to the onset of geomagnetic storm. Anisotropy depends on geometry, velocity and direction of CME motion, and on magnetic field structure. Thus information from the CR can be used as an additional parameter for the forecast of geoeffective events.

Energy spectra of the SEP at low nearly polar orbits were obtained (e.g. CORONAS-F, PAMELA) based on the knowledge of cut-offs. Detailed empirical models of cut-offs (with parameter Kp) exist. There is, however, observed large variability of the boundary position for the geomagnetic activity given by the Kp and Dst only. Double boundary with the trapping of protons was observed during the strong storms. Boundary of penetration and reconfiguration of magnetospheric fluxes during the storms may serve as a source for verification of geomagnetic field models (also for the trapped populations when fluxes are described by the adiabatic invariants). Some puzzles remain: e.g. during strong storms the penetration boundaries are almost identical in wide energy range of protons. High energy gamma rays and neutrons measured at low orbits provide unique additional information about the time of acceleration of protons.

During strong geomagnetic storms at the NM energies the different field models provide different TF, asymptotic directions and timing of the CR variations. An "overlap" of interplanetary anisotropy and the changes in conditions of particle access during the geomagnetic disturbances remains a problem for testing the validity of geomagnetic field models during these events by the worldwide network of the NMs. Comparison of interplanetary anisotropy estimates at high energies (by GMDN) and at the low ones (e.g. Spaceship Earth) is needed, if middle and low latitude NMs are used for the geomagnetic field model verifications.

Acknowledgements Copyright permission of re-use of figure 18.1 by Springer as well as for re-use of figures 18.3 and 18.5 by Elsevier is acknowledged. KK wishes to acknowledge VEGA grant agency, Project 2/0081/10 for support.

References

Anashin V, Belov A, Eroshenko E et al (2009) The ALERT signal of ground level enhancements of solar cosmic rays: physics basis, the ways of realization and development. Proceedings of the 31st ICRC, Lodz, icrc1104, 2009

Antonova EE, Kirpichev IP, Stepanova MV et al (2009) Topology of the high latitude magnetosphere during large magnetic storms and the main mechanisms of relativistic electron acceleration, Adv Space Res 43(4, 16):628–633

Antonova EE (2009) Regular and turbulent mechanisms of relativistic electron acceleration in the magnetosphere of the Earth: theoretical treatment and results of experimental observations. Proceedings of the 21st ECRS, Kosice, p 17–26

Bieber JW, Evenson P (1995) Spaceship Earth - an optimized network of neutron monitors. Proceedings of the 24th ICRC, Rome, vol 4, pp 1316–1319

Bieber JW et al (2006) AOGS 3rd Annual Meeting, Singapore, 2006

Bazilevskaya GA, Usoskin IG, Flckiger EO et al (2008) Cosmic ray induced ion production in the atmosphere. Space Sci Rev 137:149–173

Boberg PR, Tylka AJ, Adams JH Jr et al (1995) Geomagnetic transmission of solar energetic protons during the geomagnetic disturbance of October 1989. Geophys Res Lett 22(9):1133–1136

Bobik P (2001) PhD Thesis, P.J. Safarik University, Kosice

Bobik P, Boella G, Boschini MJ et al (2006) Magnetospheric transmission function approach to disentangle primary from secondary cosmic ray fluxes in the penumbra region. J Geophys Res 111(A5):A05205 10.1029/2005JA011235

Chilingarian A, Hovsepyan G, Arakelyan K et al (2009) Space environmental viewing and analysis network (SEVAN). Earth, Moon Planets 104(1–4):195–210

Da Silva MR, Dal Lago A, Gonzalez WD et al (2009) Global muon detector network observing geomagnetic storm's precursor since March 2001. In: Proceedings of the 31st ICRC, Lodz, icrc0739, 2009

De Simone N, Adriani O, Barbarino GC et al (2009a) Study of protons of solar origin in the events of 13 and 14 December 2006 with Pamela detector. In: Proceedings of the 31st ICRC, Lodz, icrc0794, 2009

De Simone N, Adriani, O, Barbarino GC et al (2009b) Comparison of models and measurements of protons of trapped and secondary origin with PAMELA experiment. In: Proceedings of the 31st ICRC, Lodz, icrc0255, 2009

Diego P, Storini M (2009) Modulation signatures on cosmic-ray periodicities before a forbush decrease. In: Proc. 31st ICRC, Lodz, icrc0044, 2009

Dorman LI (1963) Geophysical and astrophysical aspects of cosmic rays. North-Holland, New York, NY

Dorman L (2009) Cosmic Rays in magnetospheres of the Earth and other planets. Lightning Source UK Ltd, Milton Keynes Springer

Flückiger EO (1982) Rep. No. AFGL-TR-82-0177

Fushishita A, Munakata K, Miyasaka E et al (2009) Precursors of the Forbush decrease on December 14, 2006 observed with the Global Muon Detector Network (GMDN). In: Proceedings of the 31st ICRC, Lodz, icrc0502, 2009

Gopalswamy N, Maekelae P, Xie H et al (2009) CME interactions with coronal holes and their interplanetary consequences. J Geophys Res 114:A00A22

Grimani C, Fabi M (2009) Short-term forecasting of solar energetic ions on board LISA. Proceedings of the 31st ICRC, Lodz, icrc0255

Heinrich W, Spill A (1979) Geomagnetic shielding of cosmic rays for different satellite orbits. J Geophys Res 84(A8):4401–4404

Kassovicova J, Kudela K (1998) On the computations of cosmic ray trajectories in the geomagnetic field, Preprint IEP SAS, Kosice, 1998, pp 1–12

Kudela K (2009) Cosmic rays and space weather: direct and indirect relations. In: D'Olivo JC, Medina-Tanco G, Valdés-Galicia JF (eds) Proceedings of the 30th international cosmic ray conference, Rogelio Caballero. Universidad Nacional Autonóma de México, Mexico City, Mexico, vol 6, pp 195–208

Kudela K, Bucik R, Bobik P (2008) On transmissivity of low energy cosmic rays in disturbed magnetosphere. Adv Space Res 42:1300–1306

Kudela K, Storini M, Hofer MY, Belov A (2000) Cosmic rays in relation to space weather. Space Sci Rev 93(1–2):153–174

Kudela K, Usoskin IG (2004) On magnetospheric transmissivity of cosmic rays. Czech. J. Phys. 54, 239–254

Kurt VG, Yushkov BY, Kudela K et al (2009) High-energy gamma-ray emission of solar flares as an indicator of acceleration of high-energy protons. In: Proceedings of the 31st ICRC, Lodz, icrc0589, 2009

Kuwabara T, Bieber JW, Clem J et al (2006) Development of a ground level enhancement alarm system based upon neutron monitors. Space Weather 4, 10, S10001, Oct 11, 2006

Kuznetsov VD (2008) Observations of the Sun-Earth system within the CORONAS-F mission (July 31, 2001 to December 6, 2005). J Atmos Solar-Terr Phys 70(2–4):234–240

Kuznetsov SN, Kudela K, Myagkova IN et al (2004) First experience with SONGM measurements on board CORONAS-F satellite. Indian J Radio Space Phys 33(6): 353–357

Kuznetsov SN, Yushkov BY, Kudela K (2007) Measurement of the spectrum of relativistic protons from solar flares on October 28 and November 2, 2003 onboard The CORONAS-F satellite. Cosmic Res 45(4) 373–375

Lazutin LL, Kuznetsov SN, Panasyuk YM (2009) Solar cosmic rays as a source of the temporary inner radiation belts. Adv Space Res 44:371–375

Leerungnavarat K, Ruffolo D, Bieber JW (2003) Loss cone precursors to Forbush decreases and advance warning of space weather effects. Astrophys J 593(1 Pt 1):587–596

Mavromichalaki E, Souvatzoglou G, Sarlanis C et al (2009) Using the real-time NeutronMonitorDatabase to establish an Alert signal. In: Proceedings of the 31st ICRC, Lodz, icrc1381, 2009

McCracken KG, Rao UR, Fowler BC, Shea MA, Smart DF (1965) IQSY Instruction manual No 10, Cosmic Ray Tables (Asymptotic directions, variational coefficients and cut-off rigidities), Issued by IQSY Committee, 6 Cornwall Terrace, London NW1, 183 pp, May 1965

Miroshnichenko LI, Klein K-L, Trottet G et al (2005) Relativistic Nucleon and Electron Production in the 2003 October 28 Solar Event, J Geophys Res vol 110:A09S08. doi:10.1029/2004JA010936

Munakata K, Bieber JW, Yasue S-I et al (2000) Precursors of geomagnetic storms observed by the muon detector network. J Geophys Res 105(27):457

Myagkova IN, Bogomolov AV, Yushkov BY et al (2009) Study of the extreme location of the penetration boundary of solar energetic particles (protons) into the Earth's magnetosphere during the magnetic storms in 2001–2005. Bull Russian Acad Sci: Phys 73(3):322–324

Posner A (2007) Up to 1-hour forecasting of radiation hazards from solar energetic ion events with relativistic electrons. Space Weather 5(5):S05001

Ruffolo D (1999) Transport and acceleration of energetic charged particles near an oblique shock. Astrophys J 515(2):787–800

Smart DF, Shea MA (2009) Fifty years of progress in geomagnetic cutoff rigidity determinations. Advances in Space Research, accepted, online ScienceDirect

Smart DF, Shea MA, Tylka AJ (2006) A geomagnetic cutoff rigidity interpolation tool: accuracy verification and application to space weather. Adv Space Res 37:1206–1217

Stozhkov YI, Svirzhevsky NS, Bazilevskaya GA et al (2007) Fluxes of cosmic rays in the maximum of absorption curve in the atmosphere and at the atmosphere boundary (1957–2007). Preprint FIAN, pp 77

Timashkov DA, Barbashina NS, Chernov DV et al (2009) Analysis of heliospheric disturbances during solar minimum using data of muon hodoscope URAGAN. In: Proceedings of the 31st ICRC, Lodz, icrc0891, 2009

Tsyganenko NA (1989) A magnetospheric magnetic field model with a warped tail current sheet. Planet Space Sci 37(1):5–20

Tsyganenko NA, Sitnov MI (2005) Modelling the dynamics of the inner magnetosphere during strong geomagnetic storms. J Geophys Res 110:A03208. doi:10.129/2004JA010798

Tyasto MI, Danilova OA, Dorman LI et al (2008) On the possibility to check the magnetosphere's model by CR: the strong geomagnetic storm in November 2003. Adv Space Res 42:1556–1563

Su Yeon Oh, Bieber JW, Clem J et al (2009) Neutron monitor forecasting of radiation storm intensity. In: Proceedings of the 31st ICRC, Lodz, icrc0602, 2009

Usoskin IG, Tylka AJ, Kovaltsov GA et al (2009) Ionization effect of strong solar particle events: low-middle atmosphere. In: Proceedings of the 31st ICRC, Lodz, icrc0162, 2009

Vashenyuk EV, Miroshnichenko LI, Balabin YV et al (2005) Dynamics of relativistic SCR in the events of October–November 2003. Izv Akad Nauk Ser Fiz 69(6):808–811 http://cdaw.gsfc.nasa.gov/CMElist/

Chapter 19

Radio Emission Processes as Tracers of Heliospheric Weather: An Ontological Approach

Mauro Messerotti

Abstract The heliosphere is a complex physical system composed of a set of coupled plasma sub-systems typically in a state of marginal stability. Hence a variety of perturbations can be triggered by instabilities occurring from large to small spatial and temporal scales. This characterizes the heliospheric weather, i.e. the physical state of the heliosphere on short- to mid-timescale, which is a key aspect for the study of space weather and space climate. Interacting plasmas in the heliosphere originate a variety of radio emissions according to processes that, in turn, are signatures of kinetic and magnetohydrodynamic plasma processes occurring at different scales. Space- and ground-based observations of such radio emissions represent a fundamental tool for deriving the associated emission processes and, therefore, the underpinning plasma processes that are tracers of the plasma state, i.e. of the heliospheric weather. In this work, by means of an ontological approach, we illustrate the present observational and interpretative scenario of the heliospheric radio emission processes and we highlight the expected improvements by the forthcoming next generation instruments.

M. Messerotti (✉)
INAF-Astronomical Observatory of Trieste, Loc. Basovizza n. 302, 34012 Trieste, Italy; Department of Physics, University of Trieste, Via A. Valerio n. 2, 34127 Trieste, Italy;
INFN-Trieste Division, Via A. Valerio n. 2, 34127 Trieste, Italy
e-mail: messerotti@oats.inaf.it

19.1 Introduction

Heliophysics is a new discipline which studies the heliosphere and the physical processes occurring in this space region. Radio emission processes represent a specific class of such processes, which are intrinsic to the physical state of the plasma both at the source and along the propagation path and have a universal character. Therefore, the observed radio features are suitable to be used as plasma probes and tracers respectively, provided that adequate emission and propagation models are available. Heliospheric weather refers to the physical state of the heliosphere and can be fruitfully traced by observing and analyzing heliospheric radio emissions.

The aim of this chapter is to review at the highest level of abstraction the most well known heliospheric radio emission processes as universal physical processes in order to identify their probing and tracing capabilities for space weather. Considering the large variety of processes and the need to focus on key aspects, we use an ontological approach to provide conceptual schemes from a selection of related literature.

The work is organized as follows. In Section 19.2 we introduce the use of concept maps in defining ontologies that are a key tool in advanced data handling. In Section 19.3 we describe a domain ontology for the heliosphere. Radio emission processes in astrophysical plasmas are considered in Section 19.4. Heliospheric radio emission sources and features are schematized in Section 19.5. Radio diagnostics of heliospheric weather are commented in Section 19.6, and heliospheric data mining in Section 19.6.3. Conclusions are drawn in Section 19.7.

M.P. Miralles, J. Sánchez Almeida (eds.), *The Sun, the Solar Wind, and the Heliosphere*, IAGA Special Sopron Book Series 4, DOI 10.1007/978-90-481-9787-3_19, © Springer Science+Business Media B.V. 2011

19.2 Domain Ontologies and Concept Mapping

Advanced data handling in heliophysics involves search, retrieval, processing and analysis of data from different instruments and physical domains. The architecture of an infrastructure capable of successfully operating in this context is typically based on semantic ontologies, as this allows the advanced definition of data and metadata models for the registries on which the search engine operates. Semantic ontologies provide the knowledge framework on one or more domains that makes possible to exploit the data and data relationships at both human and machine level (e.g. Messerotti 2006a).

An ontology describes knowledge on a domain as a formulation of a conceptual scheme constructed by: (a) defining the precise meaning of domain entities (*semantics*); (b) identifying the relationships among entities (*associativity*); (c) stating the rules between entities and sets of entities (*operativity*). The foundation of the ontology is aimed at overcoming a set of issues typical of any discipline such as e.g.: (a) ambiguous definition of specific physical domains; (b) non-univocal terminologies for the domain entities; (c) fragmentary (and/or limited to sub-domains) definitions of relationships among entities. These issues have to be resolved to the maximum achievable extent in order to facilitate complex data searching through complementary domains.

With reference to a specific domain, *patterns of regularities* identified in objects are expressed by *concepts* that are *descriptive knowledge elements*. *Logical action links* among concepts are expressed by *relationships* that are *inference knowledge elements*. Concepts linked by relationships constitute a set of *propositions* that code the domain knowledge according to a *semantic model* (see Fig. 19.1).

Concept maps are useful means to graphically represent knowledge and can be used in different ways to manage knowledge at both human and machine level as required by data handling infrastructures (e.g. Novak and Cañas 2008). Concept maps are diagrams where blocks and connectors represent concepts and relationships characterized by the related linking actions respectively. In Fig. 19.2 an example is reported for the generic definition of a discipline that studies a set of entities obeying physical laws. Concept and linking relations form propositions whose set represents semantic knowledge on that field.

Unlike standard graphical diagrams, concept maps embed a structure implementable as eXtensible HyperText/eXtensible Markup Language (XHTML/XML) documents that can manage different entities associated with concepts like scripts, hyperlinks, etc. Furthermore, a specific Connection Mapping Language (CXL) has been implemented to specifically cope with concept mapping compliant with machine level handling.

Concept maps are not to be considered as static entities that code absolute knowledge, but have to be continuously updated according to the state-of-the-art domain knowledge derived from the most recent achievements of the scientific community. Their intrinsic structure makes them very suitable for this continuous process, leading to an update of the on-top semantic ontology.

19.3 A Domain Ontology for the Heliosphere and for Heliophysics

To minimize ambiguities, a key requirement for any new discipline is the definition of the operational domain and scope by appropriate terminology. This is of particular relevance when trying to identify *universal processes*. The most effective method to accomplish such a goal is the foundation of a domain ontology which codes the related knowledge by describing concepts and relationships for that specific domain (see Section 19.2). An effective methodology is the use of concept maps to graphically represent domain knowledge (e.g. Novak and Cañas 2008). In particular, we used a platform-independent software tool developed by the Institute for Human and Machine Cognition (Pensacola, Florida, USA) for knowledge modelling and sharing (Cañas et al. 2004). This approach proved to be successful in setting the basis of a domain ontology for space weather and space climate (e.g. Messerotti 2006b) used in reviewing solar activity models for space weather (Messerotti et al. 2009).

A concept map describing the heliosphere, its physical components and relevant processes is presented in Fig. 19.3. The heliosphere is defined as the region of space that embeds the Sun, the planets, the minor bodies and dust. It is permeated by the solar wind,

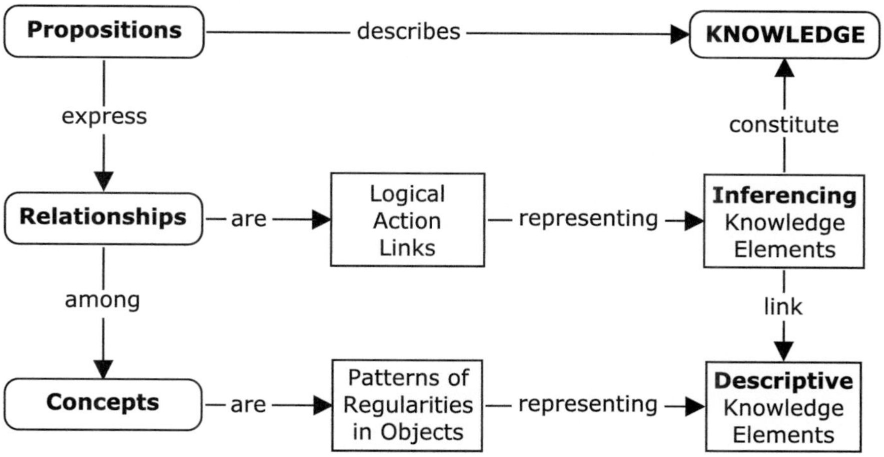

Fig. 19.1 The semantic model of knowledge described by a concept map

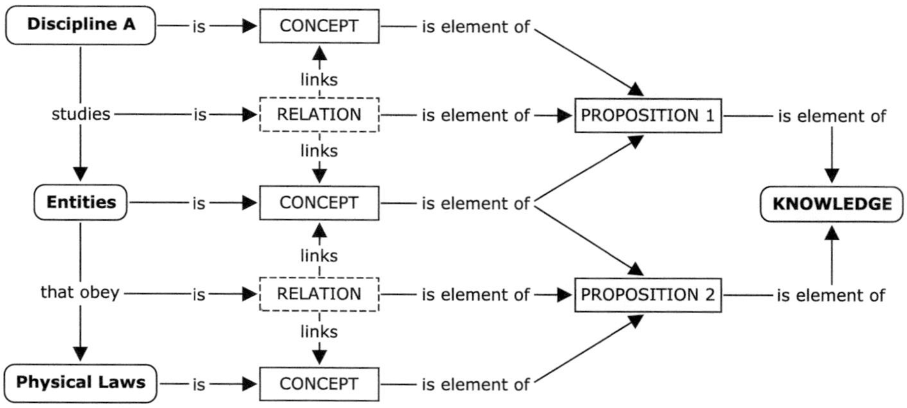

Fig. 19.2 The elements of a concept map and their relation to semantic knowledge

the inner moving magnetized plasma, and is confined by the local neighbourhood interstellar wind, the outer moving magnetized plasma. The interaction interface is the heliopause, which separates the bow shock and the termination shock, where the interstellar wind and the solar wind change the regime from supersonic to subsonic, respectively. Magnetic fields transported by solar wind, plasma streams, plasmoids and particle beams typically originate at the Sun and can interact via both kinetic and magnetohydrodynamic processes of various kind.

Similarly, it is fundamental to define heliophysics and heliospheric meteorology, a neologism inferred from space meteorology (Fig. 19.4). In fact, a clear definition avoids inappropriate use as when heliophysics is referred to as the discipline which studies heliospheric perturbations, as this is only a subset of its objectives. In particular, heliophysics models the physical processes occurring in the heliosphere which determine its physical state at small and large spatial scales. Processes occurring on short and long time scales characterize respectively heliospheric weather and heliospheric climate, both studied by heliospheric meteorology.

19.4 Radio Emission Processes in Astrophysical Plasmas

Radio emission processes in astrophysical plasmas have been extensively reviewed. We refer to, e.g., Dulk (1985) for the radio emission from the Sun and stars, Bastian et al. (1998) for the radio emission from solar flares, and Treumann (1997) for the heliospheric radio emission theory.

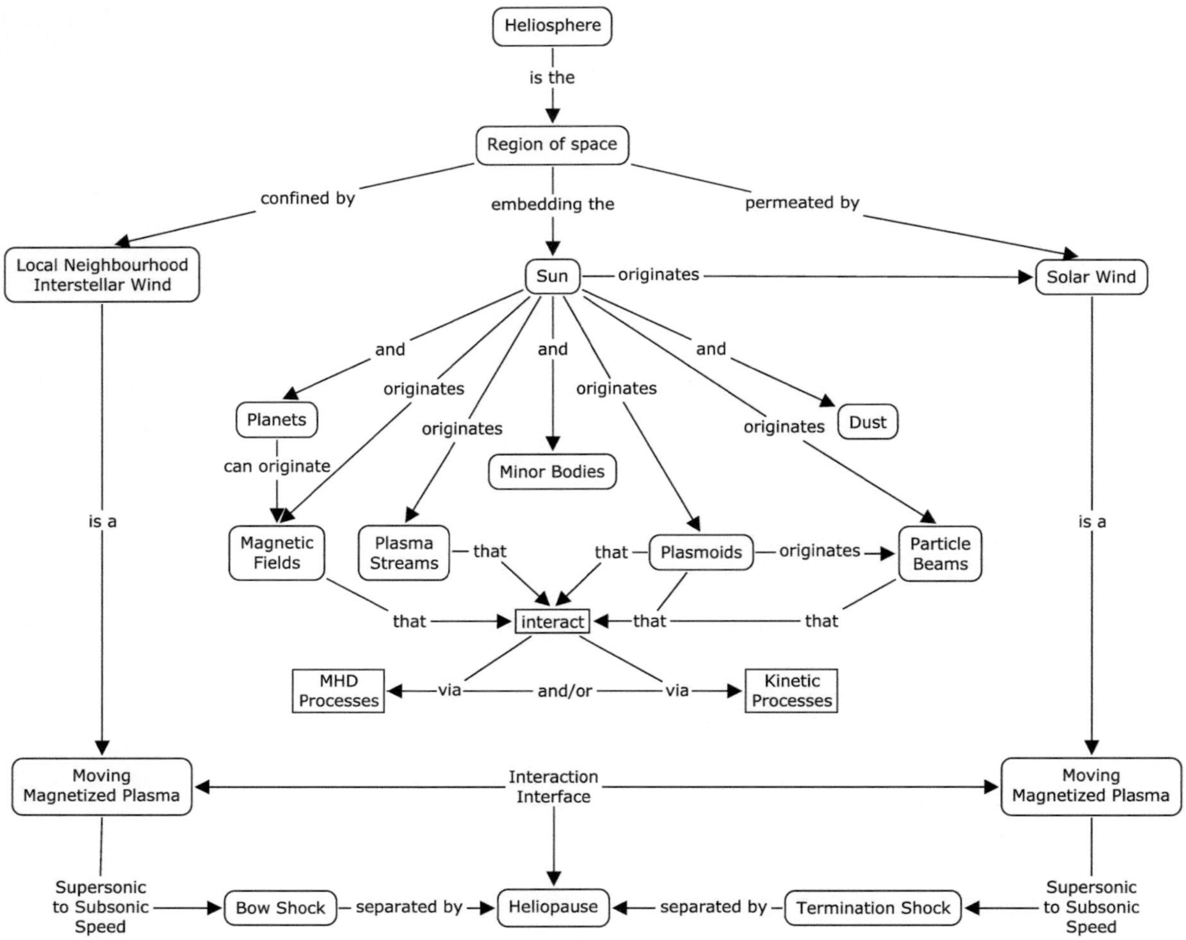

Fig. 19.3 Concept map that describes the heliosphere, its physical components and relevant processes at the highest level of abstraction

A scheme of the radio emission processes in the solar atmosphere is presented in Fig. 19.5. The direct generation of radio emission involves electrons deflected by ions or gyrating along magnetic field lines, and an electron distribution with population inversion in a strong magnetic field or low density plasma, which originates respectively thermal free-free emission, incoherent gyroresonance/gyrosynchrotron emission, and coherent electron-cyclotron maser emission, characterized by different circular polarization states. The indirect generation of radio emission involves wave growth and coupling in an unstable plasma configuration that originates coherent plasma radiation emission according to the scheme reported in Fig. 19.6: a plasma in thermodynamic equilibrium becomes unstable upon the action of a perturbation that originates an instability; an increased turbulence level sets in and electromagnetic radiation is generated via a nonlinear process involving wave coalescence or scattering. The beam-driven radio emission is a typical example of coherent radio emission produced by a multi-step process, as outlined in Section 19.4.1. Typically, coherent radio emission exhibits high brightness temperatures ($T_b > 10^9 - 10^{10}$ K), which is a characterizing parameter.

19.4.1 Shock- and Beam-Driven Radio Emission

Typical radio emission processes in the Sun (Bastian et al. 1998) and in the heliosphere (Treumann 1997)

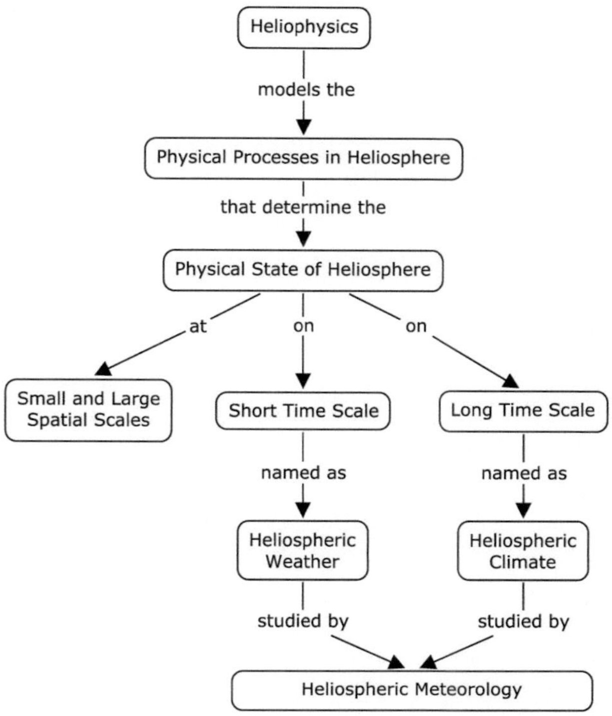

Fig. 19.4 Concept map that describes the operational framework of heliophysics and heliospheric meteorology

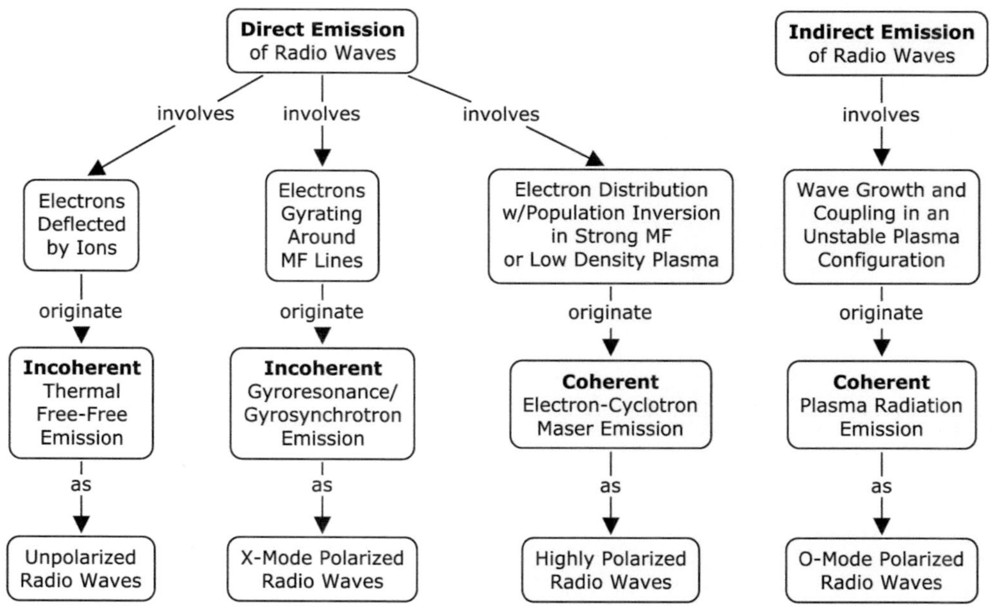

Fig. 19.5 Concept map of the typical radio emission processes occurring in the solar atmosphere. (MF - Magnetic Field; X-Mode - eXtraordinary wave Mode; O-Mode - Ordinary wave Mode)

are originated by moving agents that trigger plasma perturbations when traversing background plasma layers, and generate frequency drifting emissions by plasma radiation mechanism (Fig. 19.7). In fact, solar and interplanetary Type III radio bursts exhibit a high frequency drift and are associated with fast

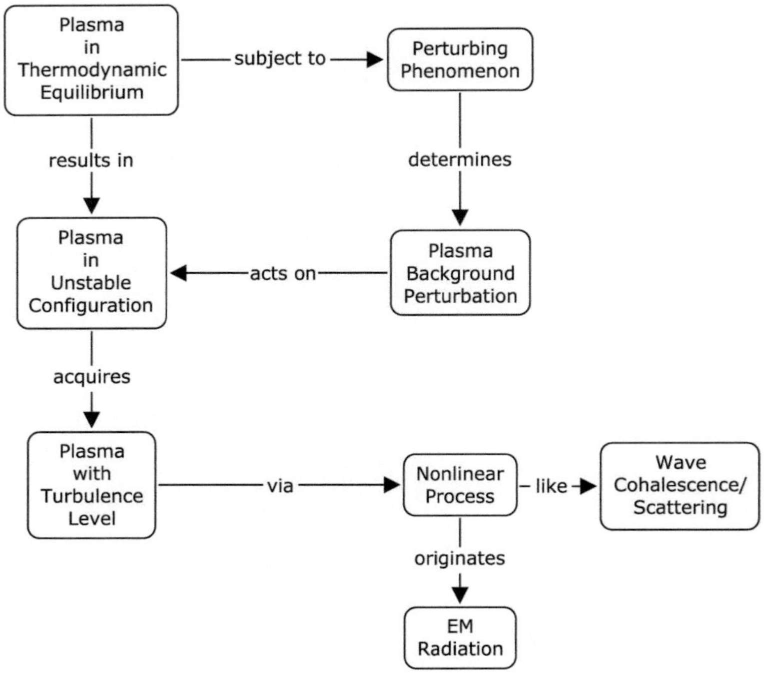

Fig. 19.6 Concept map schematizing the indirect generation of radio waves by plasma radiation mechanism via a multi-step process that involves the production of longitudinal plasma waves by plasma instability and their conversion into transverse radio waves via non-linear coupling or scattering processes. (EM - ElectroMagnetic)

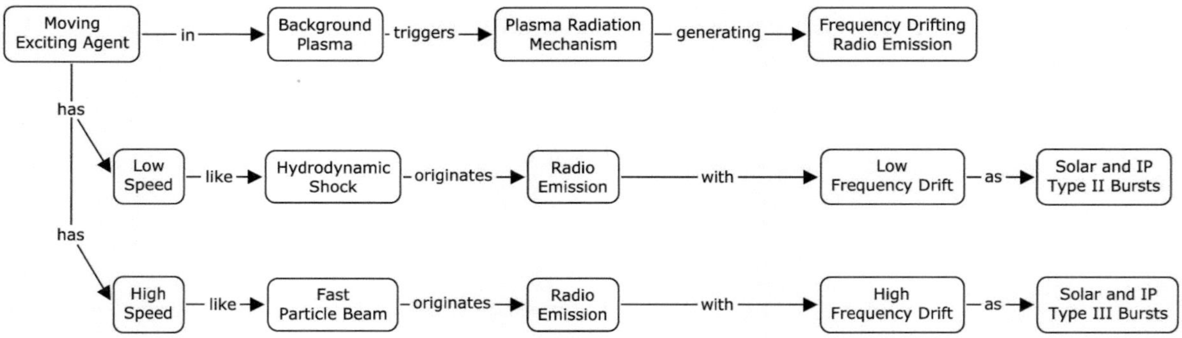

Fig. 19.7 Concept map that summarizes the typical radio emission processes involving shocks and particle beams

propagating particle beams, whereas solar and interplanetary Type II radio bursts are characterized by low frequency drift and are associated with slow propagating hydrodynamic shocks.

As detailed in Section 19.6, both radio features directly and indirectly carry information on various parameters of the source and propagation medium as well. This is sketched in Fig. 19.8 for particle beams. Beams can form by direct or shock acceleration related to the localized physical state of the plasma, and propagate through the background plasma, whose structure and dynamics influence beam propagation, stabilization, radiation and disruption.

19.5 Heliospheric Radio Emission Sources and Features

A possible synopsis of heliospheric radio emission sources and features (Treumann 1997; Cairns et al. 2000; Bastian 2001) is reported in Fig. 19.9 and is classified according to the behavior in the time-frequency domain. Solar radio emissions originate in the chromosphere and in the corona and are classified as slowly-varying (S) component, Type I, II, III, IV, and V radio bursts, each of them with a variety of fine structures that define a large set of

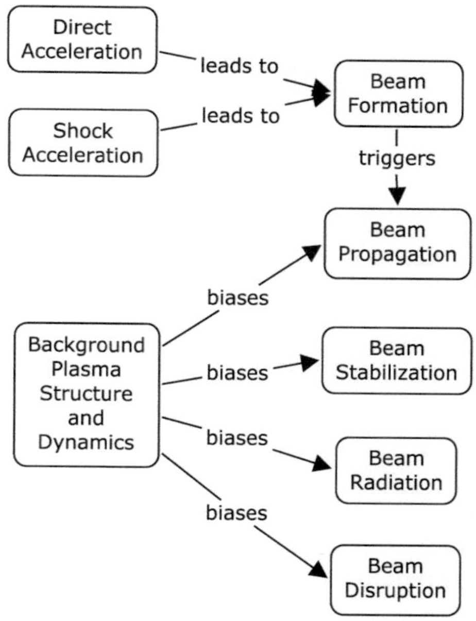

Fig. 19.8 Concept map of the role of particle beams as radio exciters

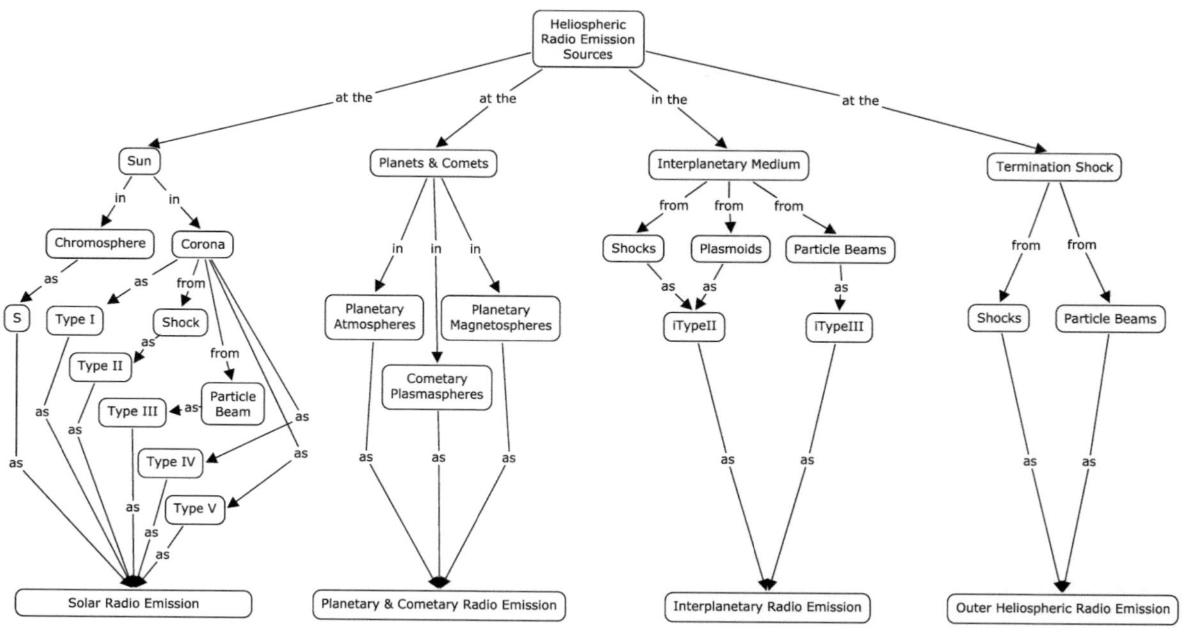

Fig. 19.9 Concept map that reports a synopsis of heliospheric radio sources and related radio emission features

sub-classes. Planetary and cometary radio emissions occur in planetary atmospheres and magnetospheres, and in cometary plasmaspheres. Interplanetary radio emissions are originated by shocks and plasmoids as interplanetary Type II radio bursts (iTypeII) and by particle beams as interplanetary Type III radio bursts (iTypeIII). Outer heliospheric radio emissions occur at the termination shock by shocks and particle beams.

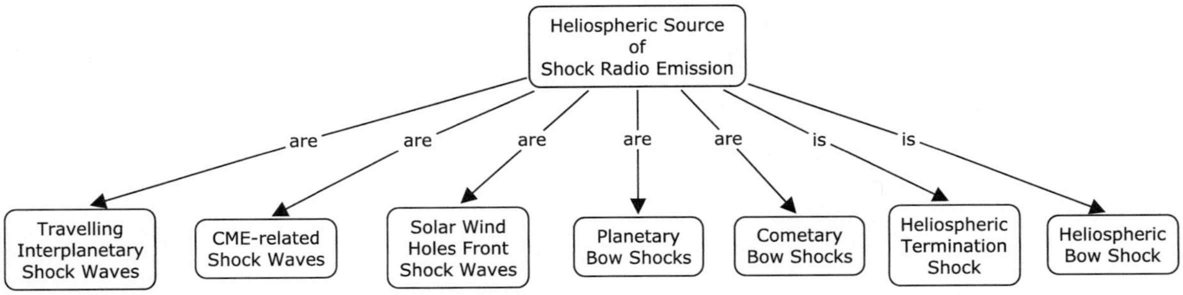

Fig. 19.10 Concept map that lists the sources of heliospheric radio emissions by shock

19.5.1 Heliospheric Shock Radio Emission Sources

At large spatial scales that characterize the heliosphere, a set of radio emissions sources by shock are identifiable (Treumann 1997; Cairns et al. 2000). They are reported in Fig. 19.10 as, respectively, travelling interplanetary shock waves, CME-related shock waves, solar wind holes front shock waves, planetary bow shocks, cometary bow shocks, the heliospheric termination shock, and the heliospheric bow shock. In fact, radio emission in the heliosphere is mainly due to shocks of various kinds that are often weak and do not radiate efficiently apart from planetary bow shocks and strong CME-piston driven shocks (Treumann 1997).

19.6 Radio Diagnostics of Heliospheric Weather

By exploiting the information carried by heliospheric radio events, various inferences can be derived on the perturbed state of the heliospheric plasma that characterizes heliospheric weather (e.g. Bastian 2001). In the following sections we summarize 2.5D and 3D radio diagnostics by stressing the role of new generation ground-based radio instruments such as LOFAR (LOw Frequency ARray).

19.6.1 2.5D Radio Diagnostics

The new generation radio spectrometers like the CALLISTO class (Benz et al. 2005) for ground-based observations or the space-based ones aboard spacecrafts like e.g. the STEREO/WAVES experiment (Bougeret et al. 2008) can provide a wealth of physical information by reverse modelling. Such instruments provide information on radio features (radio flux density and circular polarization) in the time-frequency domain but typically with no spatial resolution. Spatial information is inferred upon specific plasma and emission models. Hence radio diagnostics can be considered as 2.5D ones, as spatial information is not intrinsic to the measure. Despite of such limitation, a refined set of diagnostic inferences has been set up, as summarized in the following paragraphs.

Inferences from Radio Event Parameters in the Time-Frequency Domain: The total duration of the event is related to global geometry, energetics, and evolution of the underpinning process(es). The instantaneous duration depends on the local geometry, energetics, and process(es) in the perturbed layer. The instantaneous duration evolution gather information on the density along trajectory, local process(es) along trajectory, and propagation effects in the medium relevant to the exciting agent dynamics.

Inferences from Radio Event Parameters in the Time Domain at Fixed Frequency: The flux density time profile depends on the response of the layer, evolution of the instability, and the underpinning process(es) in the perturbed layer. The polarization time profile is related to the local geometry, magnetic field, and propagation effects in the perturbed layer. The polarization mode delay evolution carries information about propagation effects along trajectory, local process(es) along trajectory, propagation effects in the medium, all of which are dependent on the density along trajectory.

Inferences from Radio Event Parameters in the Frequency-Time Domain: The maximum bandwidth is determined by the global geometry, density model, and process(es) in the perturbed region. The instantaneous bandwidth is related to the local geometry, density and process(es) in the perturbed layer. The frequency drift depends on the density along trajectory, local process(es) along trajectory, and propagation effects in the medium that influence the exciting agent dynamics. Similarly for the frequency drift evolution that is related to the instability evolution.

Inferences from Radio Event Parameters in the Energy Domain: The total power density depends on the process(es) energetics, source characteristics, and instability characteristics, i.e., on the global energetics of the perturbed region. The spectral power density at fixed time is related to the instantaneous energetics of plasma layers determined by the global response of layers at fixed time and local process(es). The spectral power density at fixed frequency provides information about the energetics of a single layer via its response at fixed frequencies and operating process(es).

19.6.2 3D Radio Diagnostics

Imaging radio instruments provide spatial information that allows the reconstruction of 3D models of the radio source when combined with radio spectrometer observations.

This approach has been successfully used e.g. by Aschwanden et al. (1992) who investigated the 3D reconstruction of Type III exciters trajectories, combining high spatial resolution VLA radio maps and radio spectral data to infer the 3D trajectories of electron beams, trace the magnetic field lines, and derive a 3D model of coronal structures.

Direct radio imaging of features like Coronal Mass Ejection (CME) emissions (e.g. Bastian et al. 2001) can be used to infer CME onset, formation and propagation, shock physics, beam physics, magnetic field, and flare association.

The indicative location of solar radio burst sources with respect to flare geometry can be estimated on such methodological bases. Moreover, new ground-based instruments like LOFAR (e.g. Best et al. 2008) will offer key features to refine and extend such measurements. In fact, LOFAR will operate in the low frequency range associated with electron densities 10^8–10^7 cm^{-3} corresponding to 1.15–2.5 solar radii from the Sun. The high sensitivity of the instrument will extend the diagnostic capabilities to faint radio signatures, the arcsec spatial resolution will resolve the radio source spatial structure, and the subsecond time evolution will enable to track the source evolution. Combined radio imaging capabilities and radio imaging spectroscopy will provide the 3D time evolution of the radio source.

Furthermore, LOFAR will be able to perform low-frequency radar diagnostics of heliospheric plasma turbulence by detecting low-frequency echoes of radar pulses directed to the solar and heliospheric plasma from a suitable trasmitting system. The turbulence spectrum of plasma waves in the solar and heliospheric plasma can be determined and this is a key factor for understanding radio wave generation, scattering and coupling in plasmas.

Joint analysis of ground- and space-based radio observations was carried out to characterize the interplanetary and heliospheric plasma. For example, Hoang et al. (1997) used Ulysses and Artemis observations inside and outside the ecliptic to study the directivity of interplanetary Type III bursts. Using ISEE 1 data, Cairns (1994) studied the fine structure in plasma waves and radiation by electron beams reflected from the front near the plasma frequency in Earth's foreshock. Further progress in the observation and interpretation of Type II and III bursts in the solar corona, in the solar wind and in the Earth's foreshock has been considered in Cairns et al. (2000).

Radio emissions at the heliopause were studied by Gurnett and Kurth (1996) and radio emissions from the heliopause crossed by global merged interaction regions by Cairns et al. (2000). Both were ascribed to similar mechanisms involving shocks and particle beam reflection at the front (Treumann 1997; Cairns et al. 2000).

Finally we mention an effective heliospheric radio diagnostics based on the use of interplanetary scintillation data to reconstruct the solar wind 3D structure (Hick and Jackson 2001).

19.6.3 Heliospheric Data Mining

Heliophysics requires the joint exploitation of solar, heliospheric, magnetospheric and ionospheric observations in every available spectral band. Anyway, the huge volume of inhomogeneous data continuously ingested from ground- and space-based instruments that are organized in geographically distributed repositories makes this achievement a difficult goal. To approach this issue, the European Commission funded the FP7 Project HELIO (Heliophysics Integrated Observatory; http://www.helio-vo.eu/). HELIO, started in 2009 and with a duration of 36 months, will benefit from the expertise acquired in developing EGSO (European Grid of Solar Observations) (e.g. Aboudarham et al. 2006). It will provide the most comprehensive integrated information system of this domain by building a software infrastructure that will guide the final user in transparently placing complex searches on multiple heliospheric data sets. A key requirement for the optimum data search and retrieval by such an infrastructure is setting up data and propagation models relevant to the available data sets together with an extended domain ontology capable to support the heliophysics semantics in a Virtual Observatory (VO) framework. In this context, concept maps are a useful tool for coding knowledge at human level and generate machine readable ontologies as outlined in Section 19.2. The optimization of HELIO features has been occurring by the active involvement of the scientific communities dealing with different heliospheric subfields that collaborate by providing specific science use cases (e.g. Bentley et al. 2009).

19.7 Conclusions

In this work we presented a series of basic concept maps that schematize various aspects of heliospheric radio emission processes as universal processes. This set of concept maps is propaedeutic to the foundation of a domain ontology for heliophysics capable of defining and describing the involved physical systems and processes and their interrelationships at different levels of complexity (Novak and Cañas 2008).

As reported in the relevant concept maps, radio emission in the heliosphere (Treumann 1997) is mainly due to shocks of various kinds that are often weak and do not radiate efficiently apart from planetary bow shocks and strong CME-piston driven shocks.

Interplanetary Type II and Type III radio bursts have been observed up to large distances and represent effective plasma diagnostics. In situ measurements, like the STEREO/WAVES ones, will provide new insights into such phenomena as heliospheric plasma probes (see Bougeret et al. 2008).

The termination shock radiation is detectable in situ and is driven by heliospheric weather (Gurnett and Kurth 1996; Cairns et al. 2000).

Refined radio diagnostics provide significant information on the plasma features, but they must be complemented with 3D observations from space and in situ measurements to exploit all their potentialities. LOFAR will play a key role due to its observational peculiarities like low observing frequency, high spatial and time resolution, imaging and spectroscopy capabilities which will allow to track CMEs and to probe the plasma turbulence via radar techniques (Best et al. 2008). Interplanetary scintillation techniques allow the derivation of the solar wind 3D structure (Hick and Jackson 2001).

A VO approach for effective heliospheric data search and retrieval, like the one that will be provided by the EC FP7 Project HELIO, is a promising tool for advancing the modelling of heliospheric processes via successful data mining.

Acknowledgements This work has been carried out in the framework of the Projects "Exploration of the Solar System" and "Attività Scientifica per l'Analisi Dati Sole e Plasma" funded by the Italian Space Agency (ASI), the EC FP7 Project "HELIO - Heliophysics Integrated Observatory", and the activities of Working Group 1 of the EC COST Action ES0803 "Developing Space Weather Products and Services in Europe". The author wishes to thank the Conveners of Session IV0.3 for their kind invitation to present this work, the Editor for the untiring support, an anonymous referee whose comments significantly contributed to improve the quality of the paper, as well as Mrs. G. Schiulaz (INAF-OATS) for careful proofreading.

References

Aboudarham J, Scholl I, Fuller N, Csillaghy A, Bentley RD, Antonucci E, Ciminiera L, Finkelstein A, Ipson S, Messerotti M, Pike D, Vial JC, Zharkova V (2006) A new way to look at observations with EGSO. In: Bothmer V, Hady AA (eds) Solar activity and its magnetic origin, Proceedings of 233rd IAU Symposium. Cambridge University Press, Cambridge, pp 229–235

Aschwanden MJ, Bastian TS,White SM (1992) 3D reconstruction methods of coronal structures by radio observations. ESA, Proceedings of the First SOHO workshop: coronal streamers, coronal loops, and coronal and solar wind composition (SEE N93-31343 12-92), pp 217–220

Bastian TS, Benz AO, Gary DE (1998) Radio emission from solar flares. Annu Rev Astron Astrophys 36:131–188

Bastian TS (2001) Radio wave propagation in the corona and the interplanetary medium. Astrophys Space Sci 277(1–2): 107–116

Bastian TS, Pick M, Kerdraon A, Maia D, Vourlidas A (2001) The coronal mass ejection of 1998 April 20: direct imaging at radio wavelengths. Astrophys J 558:(1):L65–L69

Bentley RD, Aboudarham J, Csillaghy A, Jacquey C, Hapgood MA, Messerotti M, Gallagher P, Bocchialini K, Hurlburt NE, Roberts D, Sanchez Duarte L (2009) Addressing science use cases with HELIO. In: American Geophysical Union, Fall Meeting 2009, abstract #SH54A-06

Benz AO, Monstein C, Meyer H (2005) Callisto a new concept for solar radio spectrometers. Solar Phys 226(1):143–151

Best PN (2008) LOFAR-UK Consortium, the: LOFAR-UK White Paper: A Science case for UK involvement in LOFAR. eprint arXiv:0802.1186

Bougeret JL et al (2008) S/WAVES: The radio and plasma wave investigation on the STEREO mission. Space Sci Rev 136(1–4):487–528

Cairns IH (1994) Fine structure in plasma waves and radiation near the plasma frequency in Earth's foreshock. JGR 99(A12):23505–23513

Cairns IH, Robinson PA, Zank GP (2000) Progress on coronal, interplanetary, foreshock, and outer heliospheric radio emissions. PASA 17(1):22–34

Cañas AJ, Hill G, Carff R, Suri N, Lott J, Eskridge T et al (2004) CmapTools: A knowledge modeling and sharing environment. In: Cañas AJ, Novak JD, González FM (eds) Concept maps: Theory, methodology, technology. Proceedings of the first international conference on concept mapping, vol I. Universidad Pública de Navarra, Pamplona, Spain, pp 125–133

Dulk GA (1985) Radio emission from the sun and stars. Annu Rev Astron Astrophys 23:169–224

Gurnett DA, Kurth WS (1996) Radio Emissions from the outer heliosphere. Space Sci Rev 78(1/2) 53–66

Hick PP, Jackson BV (2001) Three-dimensional solarwind modeling using remote-sensing data. Space Sci Rev 97(1/4) 35–38

Hoang S, Poquerusse M, Bougeret J-L (1997) The directivity of solar kilometric Type III bursts: Ulysses-artemis observations in and out of the ecliptic plane. Solar Phys 172(1/2):307–316

Messerotti M (2006) Virtual observatories and virtual grids: the interplay in fully exploiting solar-terrestrial data. 20th international CODATA conference scientific data and knowledge within the information society, 22–25 Oct 2006, Beijing. Online presentation

Messerotti M (2006) Advances in space meteorology modeling and predicting – the key factor of life evolution. In: Burdyuzha V (ed) The future of life and the future of our civilization, ISBN 978-1-4020-4967-5. Springer, Berlin, pp 133–143

Messerotti M, Zuccarello F, Guglielmino SL, Bothmer V, Lilensten J, Noci G, Storini M, Lundstedt H (2009) Solar weather event modelling and prediction. Space Sci Rev 147:121–185

Novak JD, Cañas AJ (2008) The theory underlying concept maps and how to construct and use them. Technical Report IHMC CmapTools 2006-01 Rev 01-2008. Institute for Human and Machine Cognition, Florida

Treumann RA (1997) Heliospheric radio emission theory. In: Rucker HO, Bauer SJ, Lecacheux A (eds) Planetary radio emissions IV (ISBN 978-3-7001-3933-1 Online Edition). Austrian Academy of Sciences, Vienna, 383–394

Part VI
Radio Emissions

Chapter 20

Solar and Interplanetary Radio Emissions

Bo Li, Dalmiro J. F. Maia, and Milan Maksimovic

Abstract Recent progress on radio emissions from the Sun and in the interplanetary medium is reviewed, with particular emphases on type II and III solar radio bursts, and emissions from the Earth's foreshock. Observations from single or multi spacecraft and ground-based instruments have provided new insights into interactions of beam-wave, wave-wave, and wave-plasma in the radiation source regions, and into the relations between type II and III bursts, flares, and coronal mass ejections. Theoretical and numerical models have been developed to realistically predict electron acceleration by flares or shocks of various origins, source emissions, radiation propagation, and remote radio emissions. The better understanding of such radio phenomena will help us answer important questions related to electron acceleration during flares and coronal mass ejections, and the plasma conditions of the solar corona, solar wind, and the interplanetary medium.

20.1 Introduction

During the most violent solar activity, i.e. solar flares and coronal mass ejections (CMEs), radio emission is radiated from the Sun and the interplanetary (IP) medium. In fact, during such activity electromagnetic radiation may be seen across the whole spectrum from radio emission to X-ray, and geomagnetic storms and modifications of the space weather may occur. Major solar radio emissions include type II and type III solar radio bursts, which covers several orders of magnitude in wavelength from meter to kilometer, and even decimeter for type III bursts. They are produced by electron beams energized by CME/flare-driven shocks, and flares, respectively, and reach Earth prior to CMEs and beams. The ability to understand type II and III bursts, which serve as proxies for CMEs and flares, thus has great impacts on predicting space weather. Other relevant classes of solar radio bursts include type I, IV and V bursts, which are less well-understood than type II and III bursts.

At shorter metric to centimeter wavelengths solar radio emissions often show diverse fine structures besides continuum emission. The fine structures are collectively called pulsation (Nindos and Aurass 2007), which includes, e.g., spikes and zebra patterns. Pulsations usually relate to flares and occur in regions very close to the site of flare energy release.

Radio emission has also been detected in the foreshock region of Earth's bow shock. In fact, energetic electron and plasma wave activities have been observed in the foreshock regions of all the planets visited so far in the solar system (e.g., (Anderson et al. 1981; Scarf et al. 1979)), suggesting radio emission may be a common phenomenon to all the planetary foreshocks. Due to its close proximity to us the Earth's bow shock, in particular, provides important information on shock physics in other astrophysical contexts (Burgess 2007).

Furthermore, radio emissions at frequencies near 2–3 kHz have also been observed to occur approximately once every solar cycle (Gurnett 1995). The power emitted by the source is at least 10^{13} W, leading it to be the strongest radio source in the solar system. It

B. Li (✉)
University of Sydney, Sydney, NSW 2006, Australia
e-mail: boli@physics.usyd.edu.au

M.P. Miralles, J. Sánchez Almeida (eds.), *The Sun, the Solar Wind, and the Heliosphere*, IAGA Special Sopron Book Series 4, DOI 10.1007/978-90-481-9787-3_20, © Springer Science+Business Media B.V. 2011

is believed that the radiation is driven by a global shock reaching the vicinity of the heliopause, the boundary between the heliosphere and the interstellar medium (Gurnett 1995; Mitchell et al. 2005).

The contributions of radio observations since its beginning more than 60 years ago to our understanding of solar and solar terrestrial physics has been recently reviewed by Pick and Vilmer (Pick and Vilmer 2008). Besides, as topical reviews in this volume cover areas of input of radio imaging to our understanding of flares, CMEs, and electron acceleration (Vilmer Chapter 22, this volume), observations of high frequency fine/superfine structures (Yan 2009), and observations and theory of 2–3 kHz radiation (Cairns Chapter 23, this volume), this review will focus on type II and III bursts, and emissions from Earth's foreshock. The Chapter is organized as follows: In Section 20.2 a summary of the basic mechanisms for solar and IP radio emissions is made, followed by an outline of the effects of radiation propagation in Section 20.3. Section 20.4 highlights recent progress on coronal and IP type II bursts, from the perspectives of observations, theory, and numerical modelling; and Section 20.5 discusses from the same aspects of progress on coronal and IP type III bursts. Section 20.6 reviews some recent results on emissions from the Earth's foreshock.

20.2 Emission Mechanisms

Mechanisms that are important to solar and IP radio emissions can be classified as coherent and incoherent. Among various mechanisms we review those that are generally believed to be important for radio emissions from the Sun and in the IP medium: coherent plasma emission, linear mode conversion and cyclotron maser emission, and incoherent gyrosynchrotron emission (Bastian et al. 1998; Cairns et al. 2000; Melrose 1980).

Type II and III bursts, Earth's foreshock emission, and 2–3 kHz emissions are generally believed to be due to the plasma emission mechanism (Bastian et al. 1998; Cairns et al. 2000; Gurnett 1995). According to plasma emission, a beam of nonthermal electrons drive Langmuir waves near the fundamental of the local plasma frequency f_p, which are then converted via nonlinear wave-wave interactions to transverse electromagnetic waves with frequencies near f_p, or $2f_p$, or both. Here $f_p = \omega_p/(2\pi)$, $\omega_p = (n_e e^2/m_e \varepsilon_0)^{1/2}$, and n_e is the electron number density. Plasma emission is also responsible for type V bursts, and may be for type I bursts.

Linear mode conversion may be relevant to type II and III bursts, and Earth's foreshock emission (Kim et al. 2007; Yin et al. 1998). In this scenario, wave modes in inhomogeneous plasma are often coupled to each other, and one mode can be converted to another for some ranges of frequencies and angles of propagation. Langmuir waves encountering density gradients can be directly converted into radio waves near f_p. Nonlinear interaction between primary and reflected Langmuir waves lead to emission of $2f_p$ radiation. Mode conversion is also important for radio emissions in the magnetosphere and ionospheric auroral regions.

Energetic electrons moving in a magnetic field can also produce radio emissions near the electron cyclotron frequency f_{ce} and its harmonics, by the coherent cyclotron maser emission due to loss-cone distribution function. The cyclotron frequency f_{ce} depends on the magnetic field strength B through $f_{ce} = \Omega_{ce}/(2\pi)$, and $\Omega_{ce} = eB/m_e$. Cyclotron maser emission is the accepted mechanism for solar microwave spikes, and planetary radio emission (e.g., Earth's auroral kilometric radiation (Melrose 1980)).

In addition, incoherent gyrosynchrotron emission can be produced due to spiralling motion of energetic electrons in a magnetic field, with radiation frequencies at the harmonics of f_{ce}/γ, where γ is the Lorentz factor. This mechanism is responsible for centimetric and millimetric solar radio emissions (Bastian et al. 1998).

20.3 Propagation Effects

Radio emissions seen by a remote observer is modified from the source emission due to propagation effects. On one hand, large-scale density gradients refract and reflect radio emissions and lead to angular focusing. On the other hand, small-scale density fluctuations scatter radio waves, produce directional isotropization, angular and spectral broadening, and wave damping (Bastian 2001; Poquerusse and McIntosh 1995). For radio emission that originates in the corona, radiation loss by free-free absorption due to electron-ion collisions is particularly severe for f_p emission (Benz 1993).

Propagation effects complicate the analysis of radio data, like the radiation pattern. However, propagation effects can be exploited to extract the properties of the plasma medium through which radiation propagates, such as spatial spectrum of the density fluctuations (Bastian 2001; Celnikier et al. 1983).

20.4 Type II Solar Radio Emissions

Type II bursts are produced by electrons accelerated by propagating shocks in the corona and the IP medium. The shocks are usually associated with CMEs and flares. The emissions are characterized by slow drift from high to low frequencies with time (e.g., drift rate $< 1\,\mathrm{MHz\,s^{-1}}$ in metric range), harmonic bands, and band splitting (Nelson and Melrose 1985). Type II bursts in the IP medium are generated by plasma emission, evidenced by in situ observation of the source region of an IP type II burst (Bale et al. 1999). Bale et al. (1999) observed energetic electrons and Langmuir waves upstream of the associated CME-driven shock, analogous to the situations in the Earth's foreshock.

Coronal type II bursts often occur in the frequency range ~20−400 MHz. Occasionally, coronal type II bursts with low starting frequencies are observed down to frequencies as low as 100 kHz. IP type II bursts are observed at frequencies ~20 kHz−20 MHz. Metric type II bursts are observed more frequently than IP type II bursts. Although it is generally accepted that IP type IIs are produced by CME-driven shocks, the relations between coronal type IIs and CMEs and flares, and between coronal and IP type IIs remain controversial (Cane and Erickson 2005; Cliver et al. 1999; Maia et al. 2000).

20.4.1 Coronal Type II Bursts

Using unprecedented high-cadence observations from STEREO/SECCHI Liu et al. (Liu et al. 2009) examined the relationship between a metric type II event and a shock driven by the 2007 December 31 CME. Liu et al. found a causal relationship between the metric and decametric-hectometric type II bursts, and that the shock height-time curve determined from the type II spectrum is consistent with the shock propagation obtained from deflection of a coronal streamer (Fig. 20.1). These results demonstrated unambiguously that the metric type II burst was caused by the CME-driven shock, which was formed in the low corona and propagated into the IP medium.

Based on a MHD model for CME dynamics and a radiation model for type II emission from the CME-driven shocks, Schmidt and Gopalswamy (Schmidt and Gopalswamy 2008) developed a synthetic MHD and kinetic model, which includes major MHD large-scales and kinetic radiative small-scales. Schmidt and Gopalswamy presented synthetic radio maps of CMEs in the corona below about $4R_S$, including CME evolution, CME shock, and type II dynamic spectra, where R_S is the solar radius. The model, with further improvement, may help with interpretation of future observations with radio imaging telescope LOFAR.

Traditionally the coronal density profile is obtained by using eclipse and coronagraph observations of path-integrated white light and UV radiation, and the

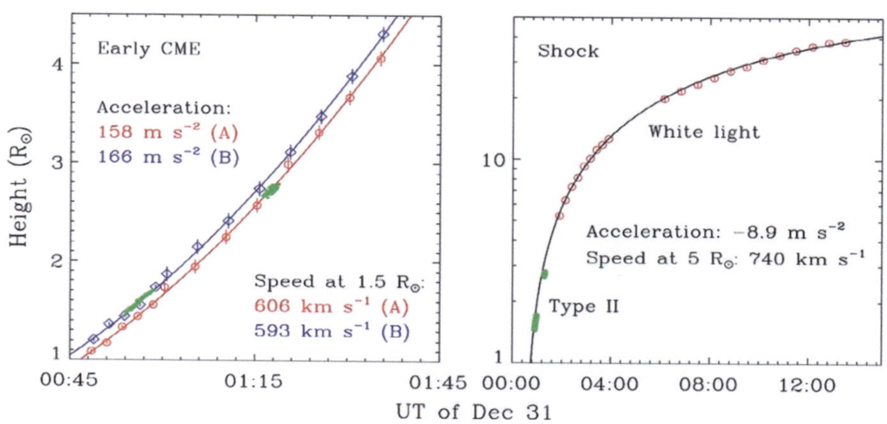

Fig. 20.1 Height of the 2007 December 31 CME front (*left*) and shock (*right*) relative to the center of the Sun measured by STEREO A (*red*) and B (*blue*). *Green dots* indicate the height determined from the associated type II burst (from Liu et al. 2009)

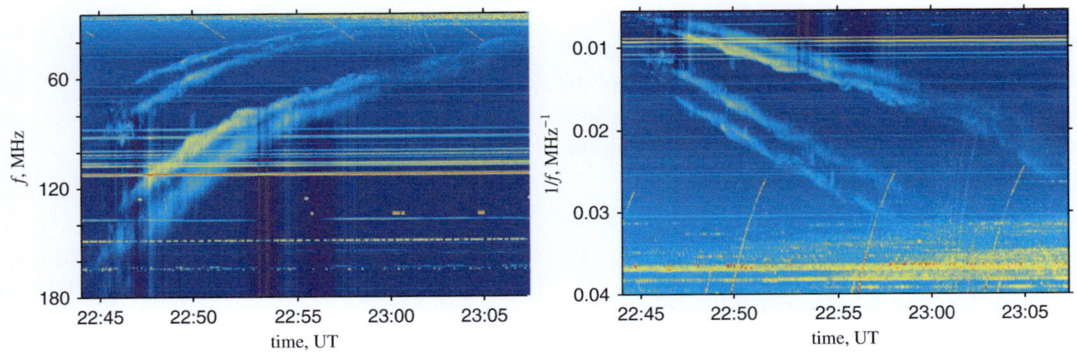

Fig. 20.2 Dynamic spectrum of the 2002 October 4 type II burst: (*left*) frequency f versus time t and (*right*) $1/f$ versus t (from Lobzin et al. 2008)

assumption of power-law sums for the density $n_e = \sum_i a_i r^{-\alpha_i}$ to allow analytical solution, where a_i and α_i are constants. The different indices α_i are larger than 2 (e.g., the Baumbach-Allen model; Baumbach 1937; Allen 1947), and the high index term dominates at sufficiently small r. For 8 well-defined coronal type II bursts Lobzin et al. (2008) found that the radio bursts in the form of $1/f$ versus t dynamic spectra closely follows straight lines (Fig. 20.2), where f and t are frequency and time, respectively. The straight line results of the coronal type IIs are very similar to some IP type IIs (Reiner et al. 1998), where IP shocks moved with constant speed through the solar wind plasma with $n_e \propto r^{-2}$. Assuming constant shock speeds in the corona, the new results of Lobzin et al. suggest that the density profile in the type II source regions closely resembles the solar wind, or, the solar wind starts within $1R_S$ above the photosphere. These results thus offer new insight to the origins of the corona and solar wind. Further, Cairns et al. (2009) have also reported similar coronal density profiles for coronal type III bursts from observations and simulations.

20.4.2 Interplanetary Type II Bursts

Since the first and sole report of an IP type II source region nearly 10 years ago (Bale et al. 1999), recently Pulupa and Bale (2008) provided the first detailed study of IP type II source regions. Using Wind/WAVES data during the period of 1999–2000 about 400 IP shock events were analysed. Pulupa and Bale found a total of 8 events that show in situ type II emission that was correlated with interplanetary magnetic field (IMF)-parallel electron beams. They confirmed the existence of a foreshock source region on the IP shock fronts found by Bale et al. (1999), and observed loss-cone features in the electron distribution functions, which indicate fast Fermi acceleration was at work. They further made the first estimate of the lateral scale size of the IP foreshock regions, which was $\sim$1-3$\times$ 10^8 m. The new results will help to understand better the shock structure and fine structures of type II spectra, and to improve type II modelling.

Realistic prediction of an IP type II burst observed during 24–26 August 1988 has been made by Florens et al. (2007). They developed a data-driven solar wind model based on spacecraft data at 1 AU, and combined it with the type II theory of Knock et al. (2003). The predicted timing, frequencies, and radio fluxes agreed reasonably well with the observations (Fig. 20.3). This predictive work will be potentially suitable for interpreting type II data from STEREO; however, both the solar wind model and the type II theory require further extension before this is the case. For instance, data-driven, realistic modelling of the IMF is needed, rather than assuming the nominal Parker prediction.

20.5 Type III Solar Radio Emissions

Type III bursts are observed more frequently than type II bursts, they are characterized by fast frequency drift (e.g., $\sim$100 MHz s^{-1} in metric range) due to the high speed ($0.1-0.3c$) of electron beams (Suzuki and

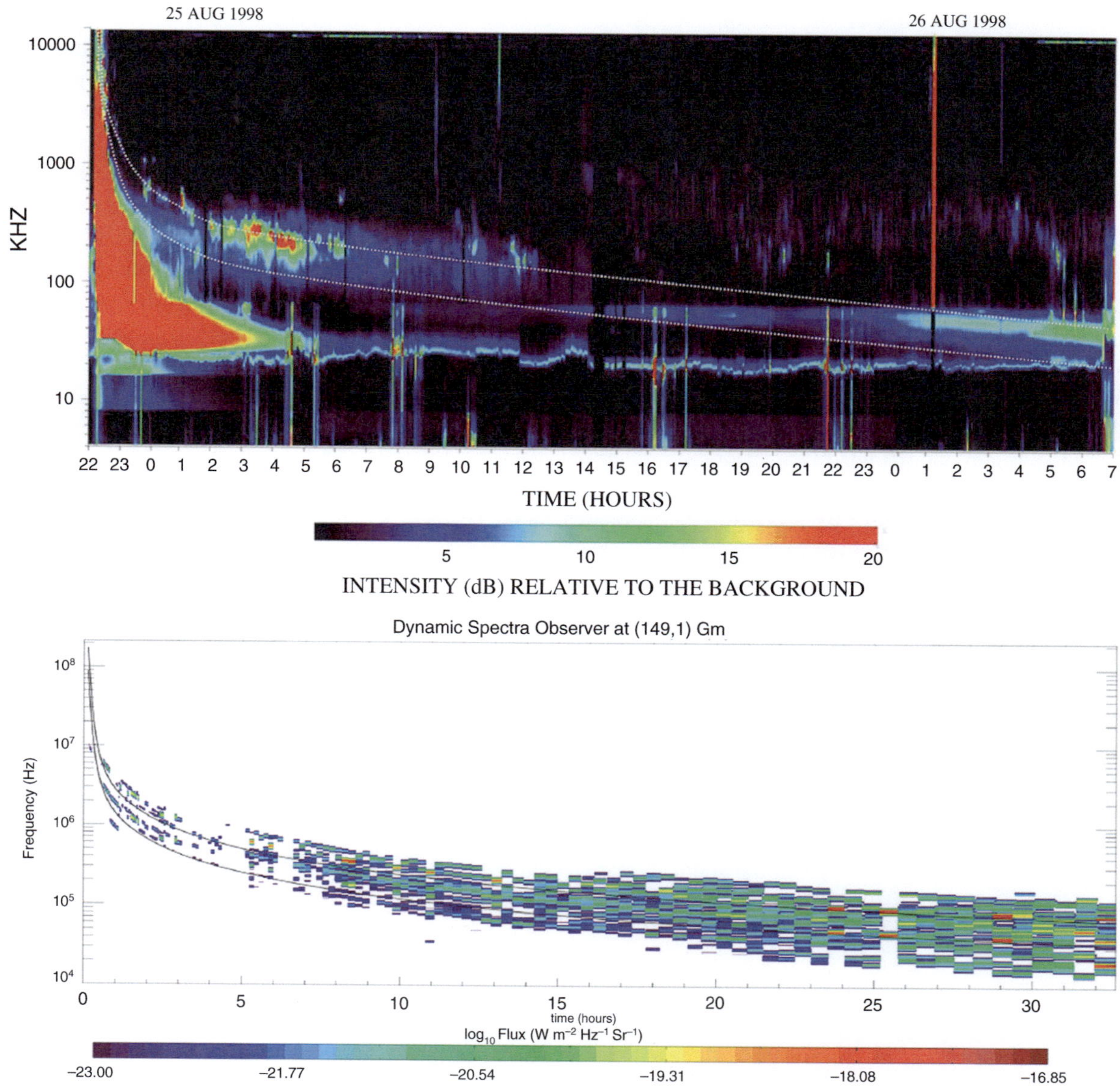

Fig. 20.3 Dynamic spectrum of the 1998 August 24–26 type II burst: (*top*) observation and (*bottom*) prediction (from Florens et al. 2007)

Dulk 1985). They are one of the most studied form of solar system radio emissions, and better understood than type IIs and other classes of solar radio bursts.

Observations of type III bursts over decades of frequency (1 GHz–10 kHz) furnish the most effective remote probes of coronal and IP medium properties over distances from solar surface to beyond Earth. Type III bursts are used to diagnose electron acceleration during flares, and to extract the magnetic field configuration along which the beams propagate and the plasma density along their trajectory (Bastian et al. 1998; Fainberg and Stone et al. 1974).

As to the relation between coronal and IP type III bursts, the most detailed study of some 200 large events recorded by ARTEMIS and Ulysses/URAP was presented by Poquerusse et al. (1996). They showed that not every coronal type III burst, even if strong, produces an IP type III burst. However, almost every IP type III burst is found to be rooted in a group of coronal type III bursts. These results are important for estimating the number of electrons accelerated during flares using electron data in the solar wind when tracing back electron beam path towards the Sun (Lin 1985).

20.5.1 Coronal Type III Bursts

Type III bursts originated in the corona show drift to low or high frequencies, which are called the normal burst and reverse-slope (RS) burst, respectively. They are produced by electron beams propagating upward along open magnetic field lines, and downward toward the Sun, respectively.

Bidirectional type III bursts have been observed in the decimetric and metric frequency ranges, consisting of a pair of normal and RS bursts with a common frequency-time origin. Bidirectional type IIIs provide strong evidence of magnetic reconnection and may provide the most stringent geometric constraints on the acceleration region in flares (Aschwanden 2002; Bastian et al. 1998). For instance, the starting frequencies of bidirectional type III bursts indicate that the vertical extents of an elementary acceleration region and the entire acceleration region are about 100–1,000 km and 5,000–50,000 km, respectively (Aschwanden et al. 1995; Xie et al. 2000) (Fig. 20.4).

The first realistic simulations of the source dynamics of coronal and IP type III bursts were recently developed by Li et al. (2006). The simulations include microscale quasilinear and nonlinear processes, intermediate-scale driven ambient density fluctuations, and large scale evolution of beams, Langmuir and ion sound waves, and fundamental and harmonic radiation. Further, Li et al. have developed a realistic simulation model for prediction of coronal type IIIs observed remotely (Li 2007; Li et al. 2008a). The simulation model includes 3D source structure, beam and wave source dynamics, radiation propagation, and radiation dynamic spectrum measured at Earth.

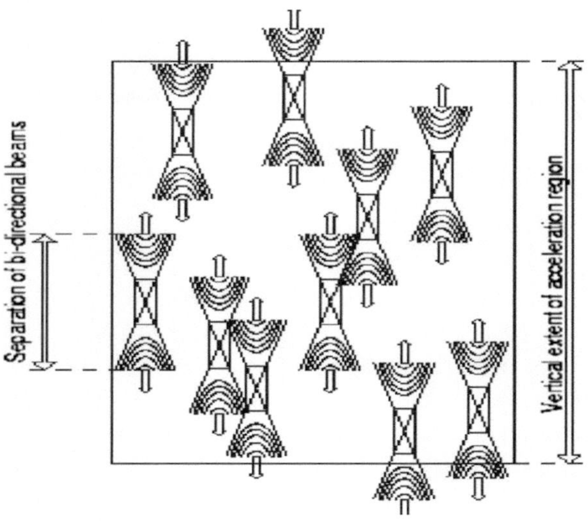

Fig. 20.4 Schematic diagram showing starting frequencies of bidirectional type III bursts with respect to the vertical geometry of the acceleration region (from Aschwanden 2002)

The simulations of normal, bidirectional, and a group of coronal bursts by Li et al. (Li 2007; Li et al. 2008b, 2008c, 2009) showed that $2f_p$ emission dominates f_p emission. The predicted radio characteristics, including flux, brightness temperature, drift rate, and half-power duration, agree quantitatively with typical observations. Some of the important results are: (1) The simulations confirm quantitatively the standard interpretation of the drift rate of $2f_p$ emission in terms of the plasma density profile and a characteristic beam speed. This interpretation is routinely used to infer beam speed from observation of drift rate (e.g., Wild 1950). (2) The apparent intersection frequency of a pair of bidirectional bursts agrees quantitatively with the true central frequency at the acceleration site. This result justifies the approximation of using the apparent intersection frequencies as the central frequencies of the acceleration made in observations (Aschwanden et al. 1995; Xie et al. 2000). (3) The simulations also show that a *single* coronal $2f_p$ emission is unlikely to continue into the solar wind, although some emissions are very strong and will extend into the upper corona. However, the simulations also imply that a *group* of coronal type III bursts may extend to the solar wind as an IP type III burst, consistent qualitatively with observations (Poquerusse et al. 1996). (4) More importantly, the simulations provide constraints on conditions of the corona and electron acceleration during flares (Fig. 20.5).

Motivated by the need to predict space weather based on radio observations, an automatic identification method for coronal type III bursts has been developed (Lobzin et al. 2009). The method is successfully implemented in a new Automated Radio Burst Identification System, which works in real time, compared with the previous non-automatic identification of real-time data (usually by eye). Preliminary results show that the method has a performance nearly 85%.

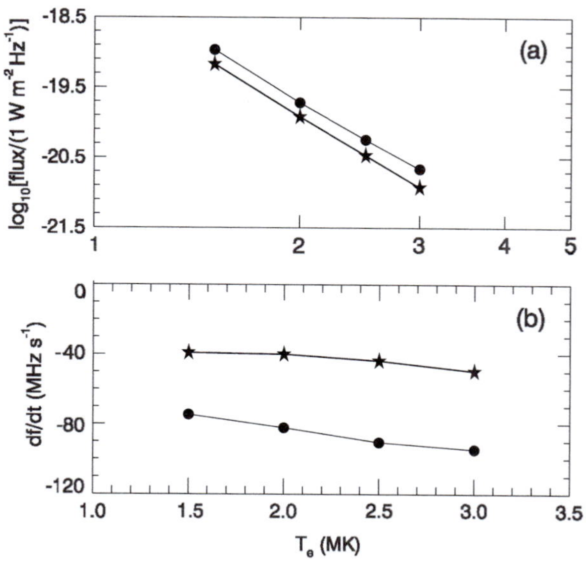

Fig. 20.5 Effects of varying the electron temperature on spectral characteristics of coronal type III bursts: (**a**) peak flux and (**b**) drift rate at 200 MHz (*stars*) and 300 MHz (*circles*). (Modified from Li et al. 2009)

20.5.2 Interplanetary Type III Bursts

In situ observations of energetic electron beams, Langmuir waves, and ion sound waves have confirmed that plasma emission is responsible for most IP type III bursts (Cairns and Robinson 1995; Lin et al. 1986). IP type III bursts show as isolated, complex, and storm type IIIs.

Isolated IP type III bursts have been used to derive the topology of IMF by assuming a global density model, and have confirmed the Parker spiral structure of the field (Fainberg and Stone 1974). An IP type III burst sometimes undergo sudden intensity enhancement and diminution when the electron beam passing through plasma turbulence in the vicinity of an IP shock (MacDowall 1989). Such type IIIs thus provide an important diagnostics to track weak IP CME shocks that are type II quiet.

Complex type IIIs at kilometric wavelength show unusually high intensity, and complex and long-lasing intensity-time profiles (Cane et al. 1981). The origin of the associated electrons has been a subject of hot debate. It remains unclear whether they are accelerated by flares, or CME-driven shocks, or others (Cane et al. 1981; Klassen et al. 2002; Reiner et al. 2000). A recent case study (Reiner et al. 2008) showed that a complex type III burst was likely resulted from an electron beam that was accelerated from a coronal magnetic field reconfiguration caused by an erupting CME, and subsequently propagated through the turbulent region of the overlying CME.

The new capabilities of waveform analyses provided by STEREO/WAVES instruments offer better understanding of IP type III source and Langmuir waves in the solar wind. Intense and isolated Langmuir wave packets are sometimes observed in the solar wind (Nulsen et al. 2007; Thejappa et al. 1999). Using Langmuir spectral and waveform data from STEREO, Ergun et al. (2008) interpreted these wave packets as eigenmodes trapped in density cavities (Fig. 20.6). This work provided the first observational confirmation of Langmuir eigenmodes in space plasmas. These linear eigenmodes may be the starting point of nonlinear evolution, which is critical for producing type II and III bursts.

STEREO/WAVES data also offer, for the first time, a complete set of direct evidence of coherent Langmuir wave-ion sound wave coupling, which occurred during one IP type III burst (Henri et al. 2009). By using complementary methods of Fourier, wavelet, and bicoherence analyses Henri et al. (2009) showed that the electrostatic decay of a Langmuir wave packet follows the parametric version, rather the kinetic version as in most observations (Cairns and Robinson 1995). The coupling region was estimated to have a spatial

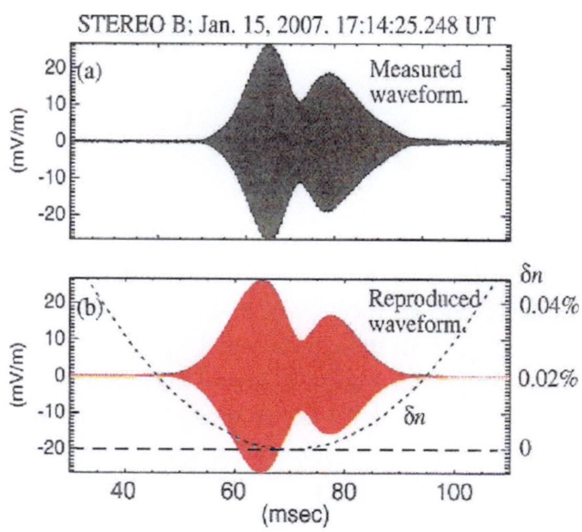

Fig. 20.6 Langmuir waveform of the 2007 January 14 event: (**a**) STEREO observation and (**b**) prediction (from Ergun et al. 2008)

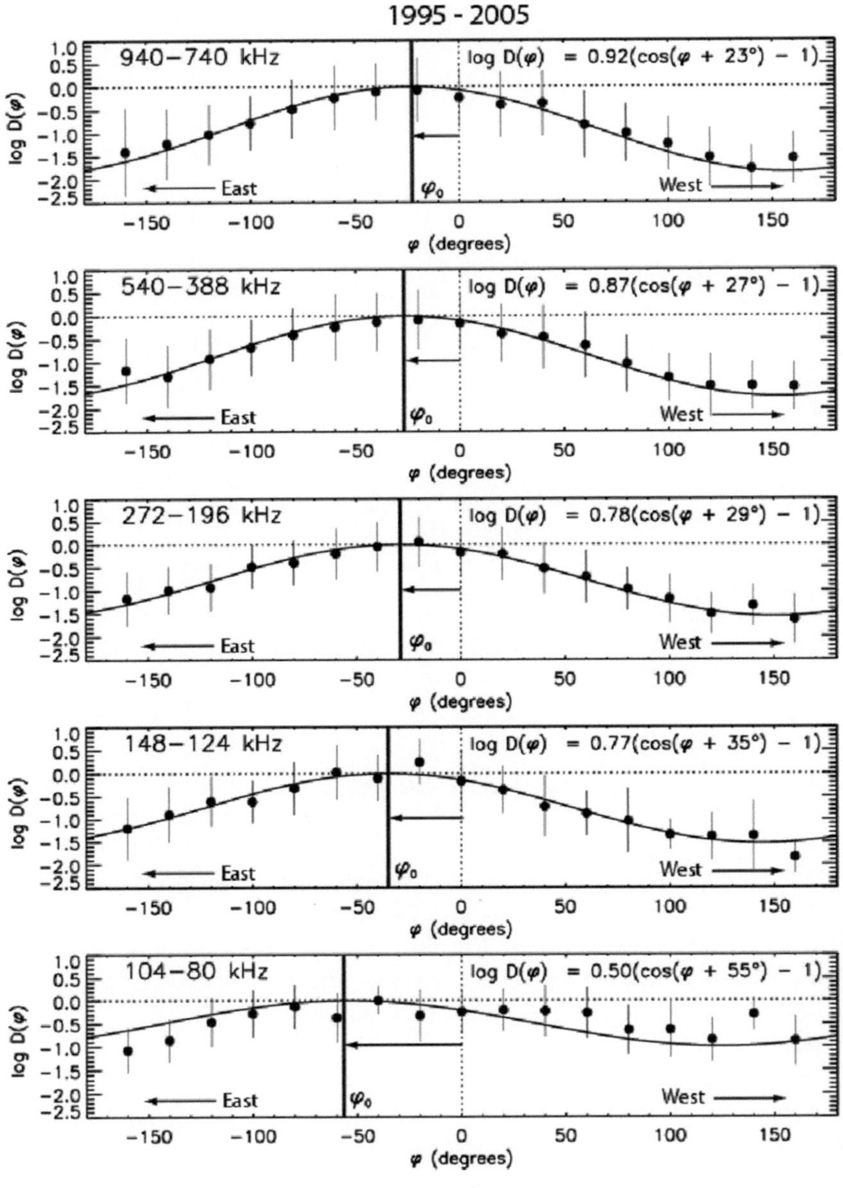

Fig. 20.7 Radiation diagram D as function of longitude φ relative to the source in different frequency ranges (from Bonnin et al. 2008)

length of about 20 km. Further knowledge of the evolution of the Langmuir wave packet undergoing the parametric decay process will be important to predict source emission of the type III burst.

Knowledge of the intrinsic emission pattern of radio source is important for constraining radio emission mechanisms and propagation effects, especially for IP type III bursts. Hoang et al. (1997) reported the first 2D radiation pattern of kilometric type III bursts, using radio data recorded by ARTEMIS and Ulysses/URAP and a number of approximations. Recently, using radio data simultaneously recorded by Wind and Ulysses and assuming that the radio source were located on spiral magnetic field lines connecting to the associated flare sites, Bonnin et al. (2008) obtained the average directivity pattern of hectometric and kilometric type III bursts in 2D, latitude and longitude relative to the emission source (Fig. 20.7).

Monte Carlo simulations of radio propagation effects have long been carried out (e.g., Riddle 1974; Steinberg et al. 1971). However, all these studies assumed that the spatial power spectrum of density fluctuations are Gaussian, whereas it has been established that the spectra in the corona and solar wind are power laws (e.g., Spangler 2002). Using the Kolmogorov power spectrum Thejappa et al. (2007) simulated the directivity, size, and position of the apparent sources of f_p and $2f_p$ emissions of IP type II and III bursts. Thejappa et al. demonstrated that the simulation results account well for the widespread visibility of radio bursts observed by Wind and Ulysses.

20.6 Radiation from Earth's Foreshock

The Earth's foreshock is the region upstream of the Earth's bow shock but downstream of the tangential of IMF lines. The foreshock plasma includes convected solar wind plasma and electrons reflected and accelerated by, or leaking through, the bow shock. Electron beams are naturally formed which drive Langmuir waves (Cairns 1987; Filbert and Kellogg 1979). Radiation near both f_p and $2f_p$ has been observed (Cairns 1986; Hoang et al. 1981). The conversion of Langmuir waves to radio emission appears most likely via nonlinear processes, although linear mode conversion may also operate under certain conditions (Kellogg et al. 1999; Yin et al. 1998).

Since the extensive modelling for electron beams in Earth's foreshock by Fitzenreiter et al. (1990), a quantitative theoretical model has been developed by Kuncic et al. (2002, 2004), which includes electron beams, Langmuir waves, and f_p and $2f_p$ emissions. The model self-consistently incorporates all key aspects of foreshock physics: reflection and acceleration of solar wind electrons at the bow shock, generation of unstable electron beams in the foreshock, growth of beam-driven Langmuir waves and their conversion to electromagnetic waves via nonlinear processes. The predicted radio fluxes agree closely with spacecraft observations, and the f_p emission is typically 2–3 orders lower than the $2f_p$ emission (Fig. 20.8).

Kuncic and Cairns (2005) further extended their model and applied it to the other planetary foreshocks, by taking into account the variation of solar wind conditions with heliocentric distance and differences in bow shock geometry. They predicted that the Jovian foreshock emissions are intrinsically the strongest of all planetary foreshock emissions, and suggested that there was evidence that Jovian foreshock emissions were in fact detected by Ulysses. They also suggested that Cassini may be capable of detecting foreshock emissions from Venus and Saturn under favorable solar wind conditions.

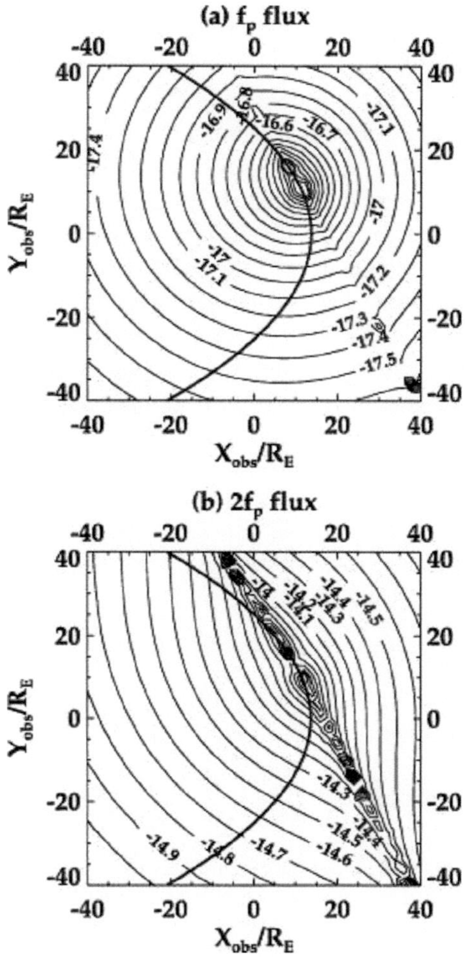

Fig. 20.8 Flux levels (in logarithmic scale) of (**a**) f_p and (**b**) $2f_p$ radiation within $40R_E$ from Earth (from Kuncic et al. 2002)

The unique orbits of STEREO spacecraft offers observation of Earth's foreshock Langmuir waves at much larger distances from Earth, up to $\sim$1,000R_E, than the previous maximum distances of $\sim$100R_E with ISEE-3, where R_E is the Earth radius (Malaspina et al. 2009). Using these new Langmuir data Malaspina et al. (2009) found that the maximum Langmuir field strength falls via a power-law with the distance between the spacecraft and the tangent point. This trend of variation was found to quantitatively agree with predictions from the foreshock model of Kuncic et al. (2004) and quasilinear simulations of Li et al. (2006), provided Langmuir wave scattering off density fluctuations is included. These results indicate that ambient density fluctuations played an important role in the Langmuir wave dynamics in the foreshock.

20.7 Summary

Solar and IP radio emissions afford a unique diagnostics of electron acceleration by flares, coronal, IP and planetary shocks. They also provide important information on plasma conditions in the solar corona, solar wind and the IP medium. Our understanding of source emissions, radiation propagation, and radio emissions seen by a remote observer have improved due to the joint efforts of observations and theoretical and numerical modelling. Future radio imaging spectroscopy coupled with observations in situ and at other wavelengths, and with more advanced modelling, will enable detailed comparisons between data and predictions. This will thus improve our knowledge of the electron acceleration and the conditions in the corona and solar wind.

Acknowledgements Li was support by The Australian Research Council.

References

Allen CW (1947) Mon Not R Astron Soc 107:426
Anderson RR, Eastman TE, Gurnett DA, Frank LA, Parks GK (1981) J Geophys Res 86:4493
Aschwanden MJ (2002) Space Sci Rev 101:1
Aschwanden MJ, Benz AO, Dennis BR, Schwartz RA (1995) Astrophys J 455:347
Bale SD, Reiner MJ, Bougeret J-L, Kaiser ML, Krucker S, Larson DE, Lin RP (1999) Geophys Res Lett 26:1573
Bastian TS (2001) Astrophys Space Sci 277:107
Bastian TS, Benz AO, Gary DE (1998) Annu Rev Astron Astrophys 36:131
Baumbach S (1937) Astron Nachr 263:131
Benz AO (1993) Plasma astrophysics: kinetic processes in solar and stellar coronae. Kluwer, Dordrecht
Bonnin X, Hoang S, Maksimovic M (2008) Astron Astrophys 489:419
Burgess D (2007) In: Klein K-L, MacKinnon A (eds) The high energy solar corona: waves, eruptions, particles. Lecture Notes in Phys vol 725. Springer, Berlin, p 161
Cairns IH (1986) J Geophys Res 92:2315
Cairns IH (1987) J Geophys Res 91:2975
Cairns IH, Robinson PA (1995) Astrophys J 453:959
Cairns IH, Robinson PA, Zank GP (2000) Pub Astron Society Aust 17:22
Cairns IH, Lobzin VV, Warmuth A, Li B, Robinson PA, Mann G (2009) Astrophys J 706:L265
Cane HV, Erickson WC (2005) Astrophys J 623:1180
Cane HV, Stone RG, Fainberg J, Steinberg JL, Hoang S, Stewart RT (1981) Geophys Res Lett 8:1285
Celnikier LM, Harvey CC, Jegou R, Moricet P, Kemp M (1983) Astron Astrophys 126:293
Cliver EW, Webb DF, Howard RA (1999) Solar Phys 187:89
Ergun RE, Malaspina D, Cairns IH et al (2008) Phys Rev Lett 101:051101
Fainberg J, Stone RG (1974) Space Sci Rev 16:145
Filbert PC, Kellogg PJ (1979) J Geophys Res 84:1369
Fitzenreiter RJ, Scudder JD, Klimas AJ (1990) J Geophys Res 95:4155
Florens MSL, Cairns IH, Knock SA, Robinson PA (2007) Geophys Res Lett 34:04104
Gurnett DA (1995) Space Sci Rev 72:243
Henri P, Briand C, Mangeney A, Bale SD, Califano F, Goetz K, Kaiser M (2009) J Geophys Res 114:A03103
Hoang S, Steinberg JL, Stone RG, Zwickl RH, Fainberg J (1981) J Geophys Res 86:4531
Hoang S, Poquerusse M, Bougeret J-L (1997) Sol Phys 172:307
Kellogg PJ, Goetz K, Monson SJ, Bale SD (1999) J Geophys Res 104:17069
Kim E-H, Cairns IH, Robinson PA (2007) Phys Rev Lett 99:015003
Klassen A, Bothmer V, Mann G, Reiner MJ, Krucker S, Vourlidas A, Kunow H (2002) Astron Astrophys 385:1078
Knock SA, Cairns IH, Robinson PA (2003) J Geophys Res 108:1361
Kuncic Z, Cairns IH, Knock SA, Robinson PA (2002) Geophys Res Lett 29:1161
Kuncic Z, Cairns IH, Knock SA (2004) J Geophys Res 109:A02108
Kuncic Z, Cairns IH (2005) J Geophys Res 110:A07107
Li B (2007) Astrophys J 666:L129
Li B, Robinson PA, Cairns IH (2006) Phys Rev Lett 96:145005
Li B, Cairns IH, Robinson PA (2008a) J Geophys Res 113:A06104
Li B, Cairns IH, Robinson PA (2008b) J Geophys Res 113:A06105
Li B, Robinson PA, Cairns IH (2008c) J Geophys Res 113:A10101
Li B, Cairns IH, Robinson PA (2009) J Geophys Res 114:A02104
Lin RP (1985) Solar Phys 100:537
Lin RP, Levedahl WK, Lotko W, Gurnett DA, Scarf FL (1986) Astrophys J 308:954
Liu Y, Luhmann JG, Bale SD, Lin RP (2009) Astrophys J 691:L151
Lobzin VV, Cairns IH, Robinson PA (2008) Astrophys J 677:L129
Lobzin VV, Cairns IH, Robinson PA, Steward G, Patterson G (2009) Space Weather 7:S04002
MacDowall RJ (1989) Geophys Res Lett 16:923
Maia D, Pick M, Vourlidas A, Howard R (2000) Astrophys J 528:L49
Malaspina D, Li B, Cairns IH, Robinson PA, Kuncic Z, Ergun RE (2009) J Geophys Res 114:A12101
Melrose DB (1980) Plasma astrohysics. Nonthermal processes in diffuse mgnetized plasmas - vol 1: The emission, absorption and transfer of waves in plasmas; vol 2: Astrophysical applications. Gordon and Breach, New York, NY
Mitchell JJ, Cairns IH, Muller HR, Zank GP (2005) Geophys Res Lett 32:07101

Nelson GJ, Melrose DB (1985) In: McLean DJ, Labrum NR (eds) Solar radiophysics: studies of emission from the sun at metre wavelengths. Cambridge University Press, Cambridge, p 333

Nindos A, Aurass H (2007) In: Klein K-L, MacKinnon A (eds) The high energy solar corona: waves, eruptions, particles. Lecture Notes in Phys. vol 725. Springer, Berlin, p 251

Nulsen AL, Cairns IH, Robinson PA (2007) J Geophys Res 112:A05107

Pick M, Vilmer N (2008) Astron Astrophys Rev 16:1

Poquerusse M, McIntosh PS (1995) Solar Phys 159:301

Poquerusse M, Hoang S, Bougeret J-L, Moncuquet M (1996) In: Winterhalter D, Gosling JT, Habbal SR, Kurth WS, Neugebauer M (eds) Solar wind eight, AIP CP-382. American Institute of Physics, New York, NY, p 62

Pulupa M, Bale SD (2008) Astrophys J 676:1330

Reiner MJ, Kaiser ML, Fainberg J, Stone RG (1998) J Geophys Res 103:29651

Reiner MJ, Karlicky M, Jiricka K, Aurass H, Mann G, Kaiser ML (2000) Astrophys J 530:1049

Reiner MJ, Klein K-L, Karlicky M, Jiricka K, Klassen A A, Kaiser ML, Bougeret J-L (2008) Solar Phys 249:337

Riddle AC (1974) Solar Phys 35:153

Scarf FL, Gurnett DA, Kurth WS (1979) Science 204:991

Schmidt JM, Gopalswamy N (2008) J Geophys Res 113:A08104

Spangler SR (2002) Astrophys J 576:997

Steinberg JL, Aubier-Giraud M, Leblanc Y, Boischot A (1971) Astron Astrophys 10:362

Suzuki S, Dulk GA (1985) In: McLean DJ, Labrum NR (eds) Solar radiophysics: studies of emission from the sun at metre wavelengths. Cambridge University Press, Cambridge, p 289

Thejappa G, Goldstein ML, MacDowall RJ, Papadopoulos K, Stone RG (1999) J Geophys Res 104:28279

Thejappa G, MacDowall RJ, Kaiser ML (2007) Astrophys J 671:894

Wild JP (1950) Aust J Sci Res A3:541

Xie RE, Fu QJ, Wang M, Liu YY (2000) Solar Phys 197:375

Yan Y (2009) 11th IAGA scientific assembly, Sopron, abstract 406-WED-01130-0345

Yin L, Ashour-Abdalla M, El-Alaoui M, Bosqued JM, Bougeret J-L (1998) Geophys Res Lett 25:2609

Chapter 21

Radiation in the Solar System Through Converted Electrostatic Waves

Paul J. Kellogg and David M. Malaspina

Abstract Most electromagnetic radiation which is of interest to radio astronomy comes either from electrons rotating in magnetic fields, from near mode crossing of modes, or from conversion of the much slower electrostatic waves generated by various plasma processes. Conversion of electostatic waves is least well understood. Here is reported recent progress on understanding how Langmuir waves generated in the foreshocks of planets and in Type II and III solar bursts are converted to electromagnetic waves.

21.1 Introduction

Natural electromagnetic radiation is often generated by accelerated electrons, but conversion of other modes to electromagnetic radiation is also often important. Much of the radiation of interest in astrophysics is generated by electrons rotating in magnetic fields, and astronomers frequently do not consider any other process. In the solar system, however, we are able to observe, in-situ, the regions where radiation is generated. It is found that much of the electromagnetic radiation is generated by conversion of electrostatic waves. Since electrostatic waves are slow, of the same order of speed as particle motions, they can be generated by instabilities resonating with particle motions. A large body of work deals, more or less successfully, with generation by such instabilities. In some cases the conversion process is much less well understood. Other processes, in particular linear and nonlinear conversion of Langmuir waves to electromagnetic radiation has been a subject of uncertainty for nearly 50 years. There has been recent progress in understanding this process, which will be the subject of this report.

There is another mode conversion process, conversion of modes at or near intersection of an electromagnetic dispersion relation and an electrostatic dispersion relation, mode crossing. Mode conversion of Bernstein modes ($n + 1/2$ modes) near the upper hybrid frequency in the Earth magnetosphere is one process which was worked out some time ago and seems well understood (Jones 1980, 1987). This is not the main topic here.

21.2 History

Conversion of Langmuir waves is considered to be the essential process in radiation associated with the Earth bow shock and in Type II and III solar radio bursts. In both of these, radiation is observed at two frequencies, one near the local plasma frequency (the fundamental) and the other at twice the plasma frequency (the harmonic). These two often have amplitudes which are roughly comparable, though one may also be absent. Other harmonics are much weaker or not observed.

Figure 21.1 shows a typical dynamic spectrum for radiation from the Earth foreshock, data from Wind (Bougeret et al. 1995). Of the two prominent traces in the lower panel, the upper one represents the harmonic electromagnetic radiation. The lower one is the Langmuir waves, whose lower boundary is at the local plasma frequency. This figure, like Figs. 21.2 and 21.7,

P.J. Kellogg (✉)
University of Minnesota, Minneapolis, MN, USA
e-mail: pauljkellogg@gmail.com

M.P. Miralles, J. Sánchez Almeida (eds.), *The Sun, the Solar Wind, and the Heliosphere*, IAGA Special Sopron Book Series 4, DOI 10.1007/978-90-481-9787-3_21, © Springer Science+Business Media B.V. 2011

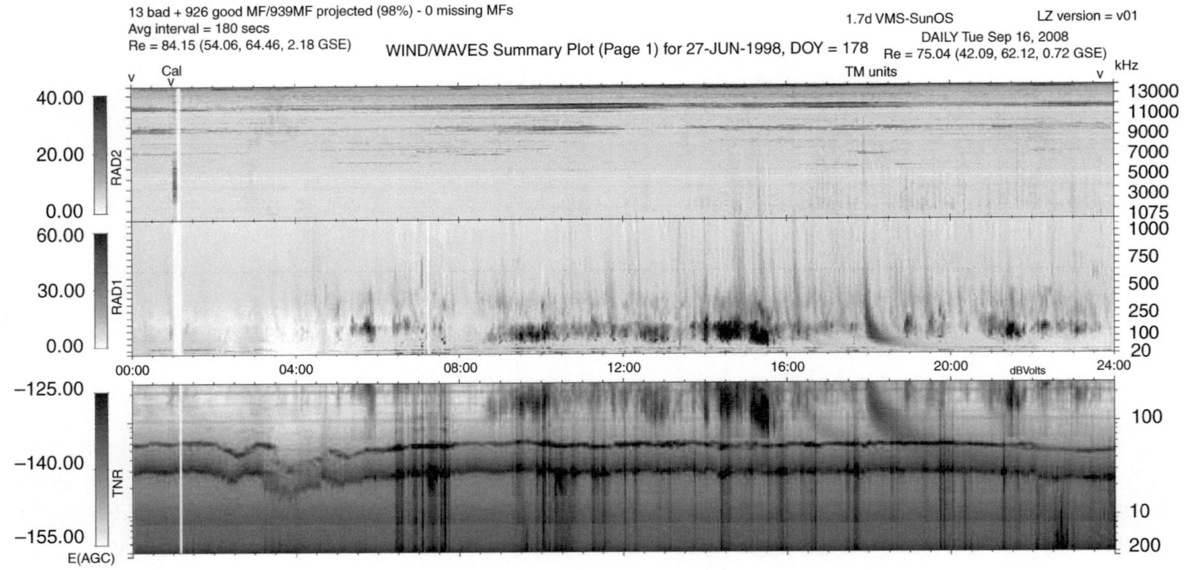

Fig. 21.1 Strong fundamental and harmonic emission

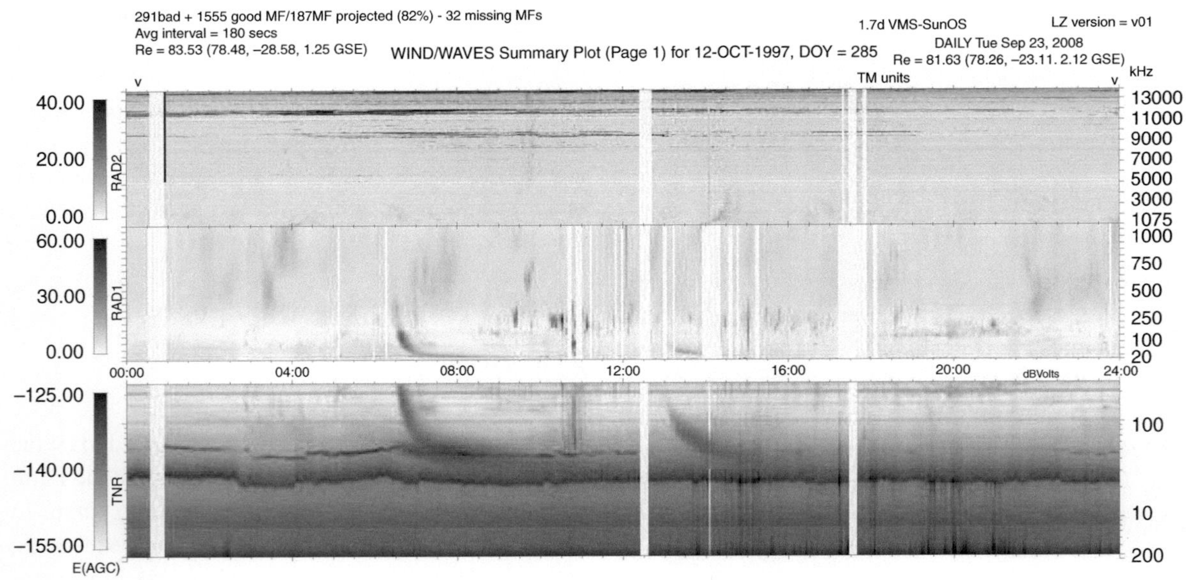

Fig. 21.2 Delayed harmonic

are taken from the Wind web site, and much of the caption is too small to read. The data are for a whole day, and the lower panel is a dynamic spectrum from 4 to 250 kHz. Most of the lower trace is simply the weak electrostatic wave resulting from the thermal energization of the states of the Langmuir mode at kT per degree of freedom (Kellogg 1981). It is called the quasi-thermal line, quasi because the solar wind is not in thermal equilibrium. The darker vertical lines mark Langmuir waves excited by a beam instability, the beam coming from points on the Earth's bow shock where the solar wind magnetic field is tangent to the shock (Filbert and Kellogg 1979). It is these intense waves that generate the harmonic emission. These instability waves saturate the instrument, and the resulting distortion makes the signal appear at frequen-

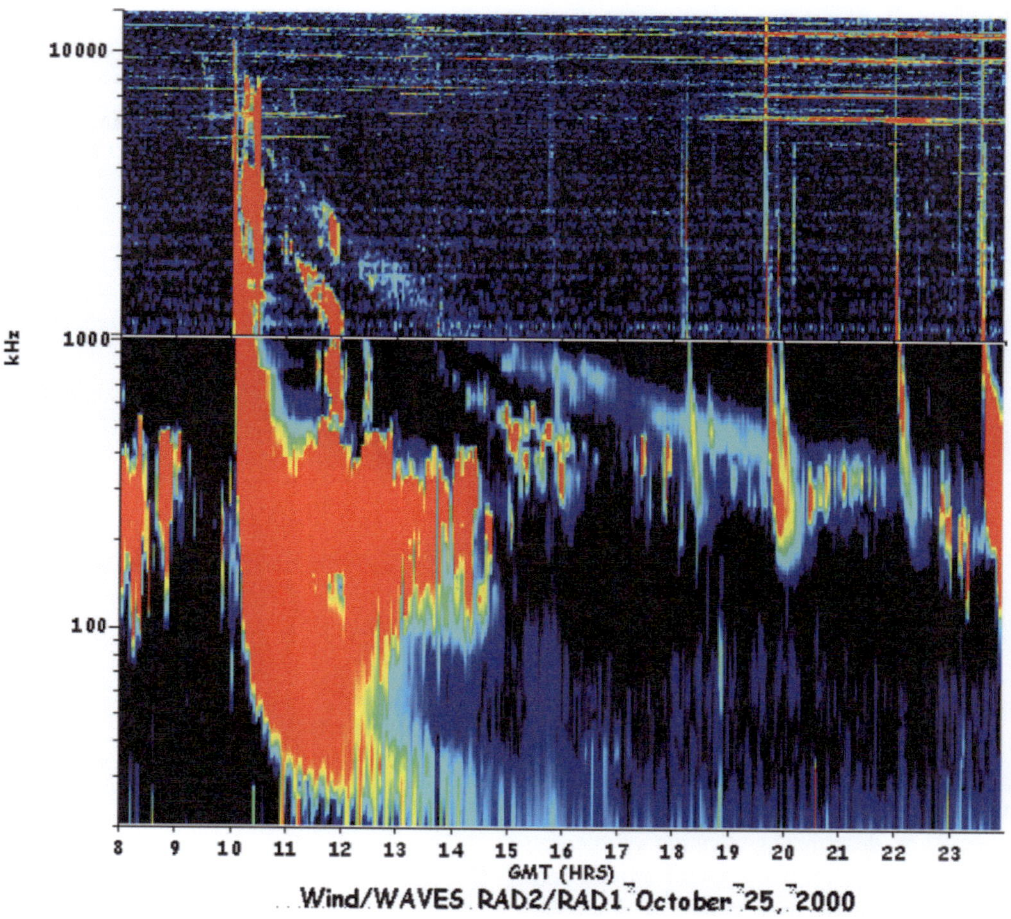

Fig. 21.3 An example of a solar Type II burst

cies where there is no actual power. Because of the saturation, these signals appear only modestly above the quasi-thermal signal, but in reality they are orders of magnitude stronger.

Figure 21.2 serves to confirm the interpretation above. During the observations for Figure 21.2, the Wind satellite was about 80 RE upstream from the Earth and its bow shock. It will be seen that the upper trace, the electromagnetic radiation, is delayed with respect to the lower trace by about half an hour. As has been said, the lower trace represents the local Langmuir waves, and at about 0245 and again at about 0400, there are jumps in the frequency and hence in the plasma density at Wind. These take about half an hour to be convected to the region, close to the bow shock, where they generate the electromagnetic radiation.

It will be seen also that the harmonic radiation disappears about noon. Generally we have found that this occurs when the tangent point, where the interplanetary magnetic field is tangent to the bow shock, is far back on the shock where the shock is weak.

In Fig. 21.3 is shown a Type II burst from the sun. These are thought to be generated by the same process as in the Earth's bow shock, though the shock originates in the sun (Cane 1984). This is nearly a textbook example, with the harmonic visible for most of the event. The fundamental is intermittent, probably due to shielding of the source by regions of higher density.

A more common Type II is shown in Fig. 21.4. Here it appears that most of the emission is probably in the fundamental. When the harmonic is absent there is probably no part of the shock front which is tangent to the interplanetary magnetic field.

A very early attempt at a theory for these two processes, fundamental and harmonic, was made by

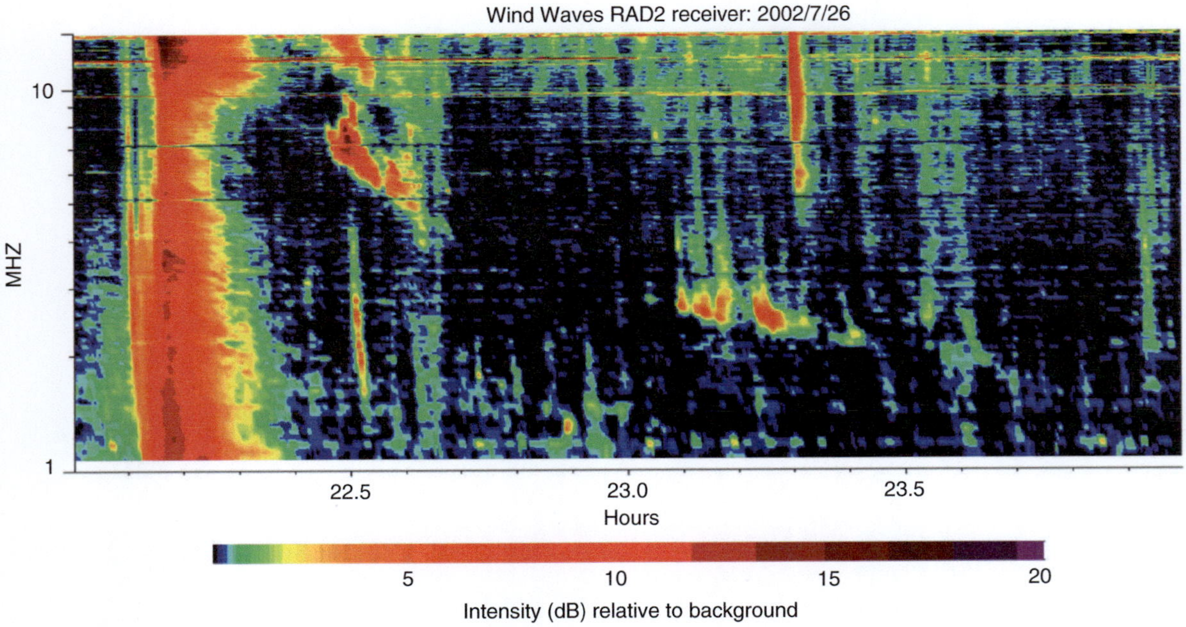

Fig. 21.4 A more common fragmentary Type II burst

Ginzberg and Zheleznyakov (1958) and this chapter must always be quoted in discussions like the present one. They described generation of the fundamental as scattering from the cloud around ions, and that of the harmonic as a nonlinear interaction involving two oppositely directed Langmuir waves. Their theory for the fundamental has since been superceded, perhaps twice, but the generation of the harmonic is still believed to be as they described, except that the origin of the opposing Langmuir wave is now believed to be different, a change which we will discuss. The necessity to modify the theory of Ginzberg and Zheleznyakov became evident from the following observations (Kellogg 1986 and unpublished).

Figure 21.5 shows typical density or plasma frequency fluctuations, together with estimates of the frequencies of Langmuir waves. The left side shows an old attempt (Kellogg 1986) to reconstruct realtime fluctuations from published spectra, and the right side shows actual measurements from STEREO.

The Langmuir waves were assumed to be resonant with beam electrons, that is, they satisfied the condition:

$$\omega/k = v_{beam} \tag{21.1}$$

giving ω slightly above the plasma frequency:

$$\omega^2 = \omega_p^2 + k^2 v_{th}^2 \tag{21.2}$$

In the left hand figure, horizontal straight lines above the shaded areas mark the resonant frequency just calculated for two beam speeds, energies 2 and 10 keV, that are known to be typical of Type III solar bursts. It will be seen that these frequencies are only slightly above the plasma frequency. Because the fluctuations of plasma frequency in the solar wind are larger than this frequency difference, Langmuir waves will suffer reflection and trapping when they encounter an increase of plasma frequency which matches their own frequency, and enhanced absorption when they encounter a region of reduced plasma frequency. This changes the considerations of Ginzberg and Zheleznyakov. For the generation of the harmonic, the oppositely directed Langmuir waves now come from reflection or trapping, not from a nonlinear interaction. Otherwise, the process is the same as they envisioned.

On the other hand, a different process becomes possible for the generation of the fundamental. Oblique reflection generally involves mixing modes, for example, the change of polarization of light when reflected at oblique incidence. Here it can involve mixing to an electromagnetic mode.

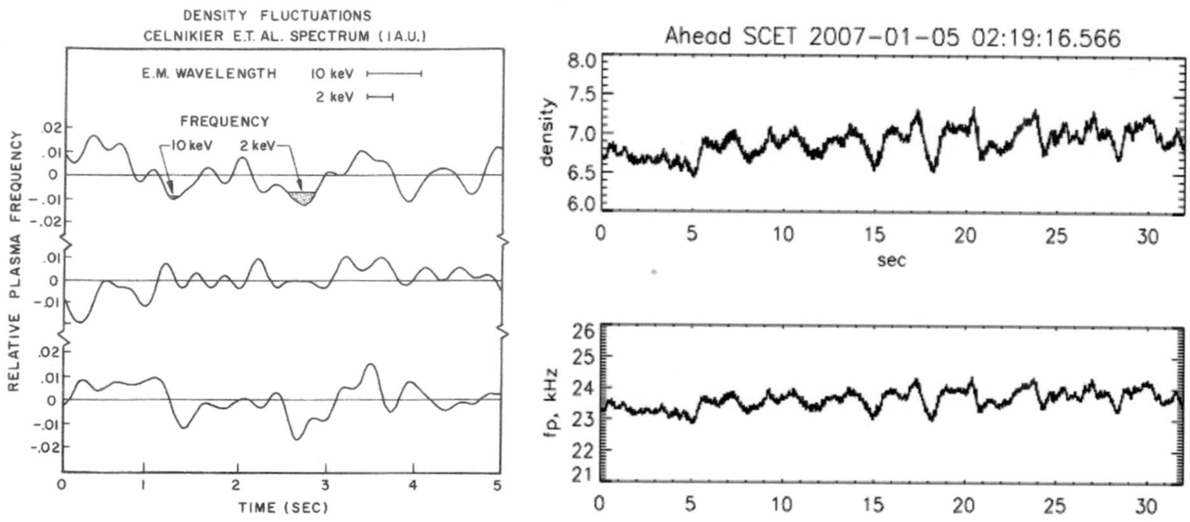

(a) from spectra (b) STEREO observations

Fig. 21.5 Fluctuations of density and plasma frequency in the solar wind

21.3 Recent Advances

There has been recent progress in the theory of such mode mixing at reflection, though the theory is perhaps not yet quite complete. An early attempt was that of Field (1956) who, however, calculated the change for propagation into a vacuum. Hinkel-Lipsker et al. (1992) calculated the conversion analytically for a plasma without magnetic field. Yin et al. (1998, 1999) included a magnetic field, but considered a very large total change in density on a ramp, finding conversion efficiencies as high as 50%. Willes and Cairns (2001), pointed out that the dispersion relations could be matched for only a very small range of angles of incidence when a more usual change of density was considered. This greatly reduced the maximum available efficiency.

Recently, Kim et al. (2008) have done numerical calculations using a rather complete description of the interaction, limited only by their approximations and by requiring that the density gradient be parallel to the magnetic field. They emphasized an important difference between energy and power conversion efficiencies, with energy conversion efficiency being appreciably lower, because of the different group velocities and therefore power fluxes of the modes. The conversion efficiencies depend on angle of incidence and plasma parameters. The maximum value that they find is about 8% for energy, but the corresponding power conversion efficiency is 50–70%, in line with values found by other authors.

Figure 21.6 shows the principal results of their work. The results are expressed in terms of dimensionless variables, of which the vertical axis contains the angle of Langmuir wave incidence and the horizontal axis contains the ratio of electron cyclotron frequency to Langmuir wave frequency.

The SWaves experiment on STEREO (Bougeret et al. 2008) was intended to measure conversion efficiency of the Langmuir waves of Type III solar bursts. Unfortunately the sun has been so quiet that no very good results are yet available. However, the following shows what eventually will be possible. The object of these calculations is to consider the question: is there enough energy in the Langmuir waves to power the electromagnetic emission, given the low conversion efficiency found by Kim et al. (2008). We have chosen an example from Wind data. The problem with the Wind data, which we had hoped to remedy on STEREO, is that only a limited number of the most intense Langmuir waves are recorded, and we are not able to estimate that part of the total energy due to periods with waves of smaller amplitude. Hence the observations certainly underestimate the energy in Langmuir waves. Figure 21.7 shows a Type III burst (starting at about 0610) together with the Langmuir waves

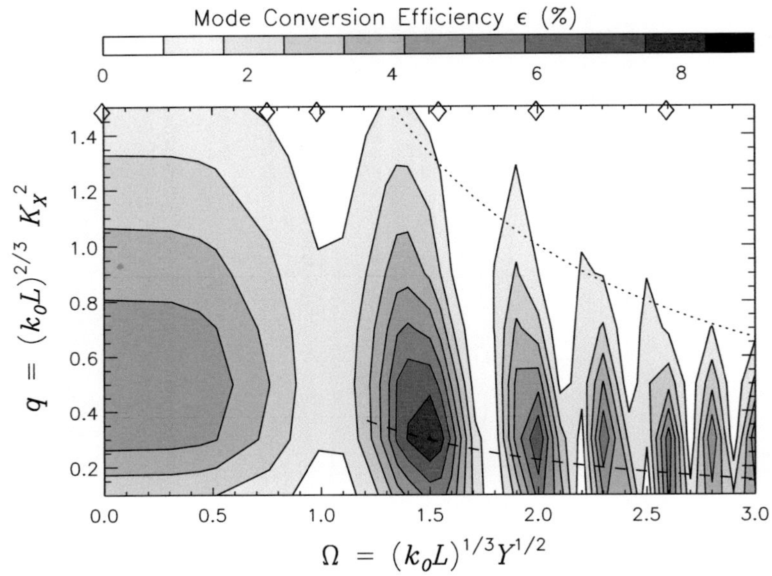

Fig. 21.6 Dependence of conversion efficiency on parameters of the reflection

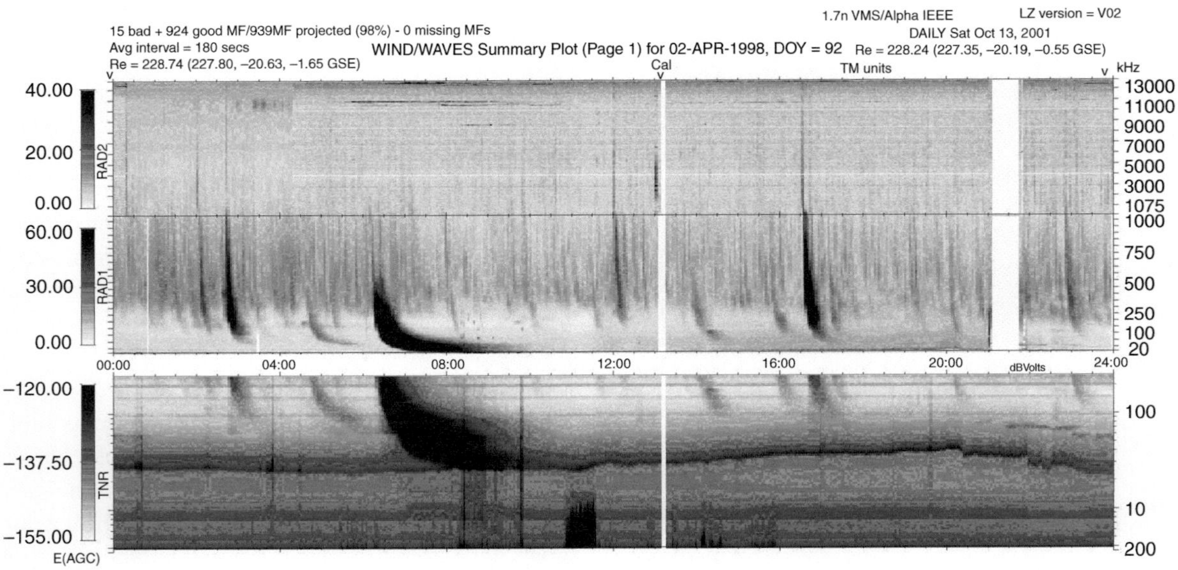

Fig. 21.7 A Type III burst with local Langmuir waves

generated as the electron beam passes over Wind, at about 0630. The Time Delay Sampler (TDS) of Wind-Waves captured twelve 17 ms bursts of the Langmuir waves, and one is shown as Fig. 21.8. Note that this event is usually interpreted as nonlinear decay because the modulation of the envelope indicates interference between a Langmuir wave and its daughter.

In order to estimate the energy in Langmuir waves we suppose that each is a blob with a transverse area A and a thickness given by the observed passage time and the assumption that the blobs are convected with the solar wind. The 12 events occurred in a time interval of T = 3,200 s, so the collision rate with Langmuir wave blobs having a density n blobs per m^3 is:

$$\mathrm{n\,A\,V}_{sw} = 12/3{,}200 \text{ collisions/s}$$

The total energy contained in a radial column of area 1 m^2 and length Vsw T is then:

$$\mathrm{n\,W\,V}_{sw}\,\mathrm{T}$$

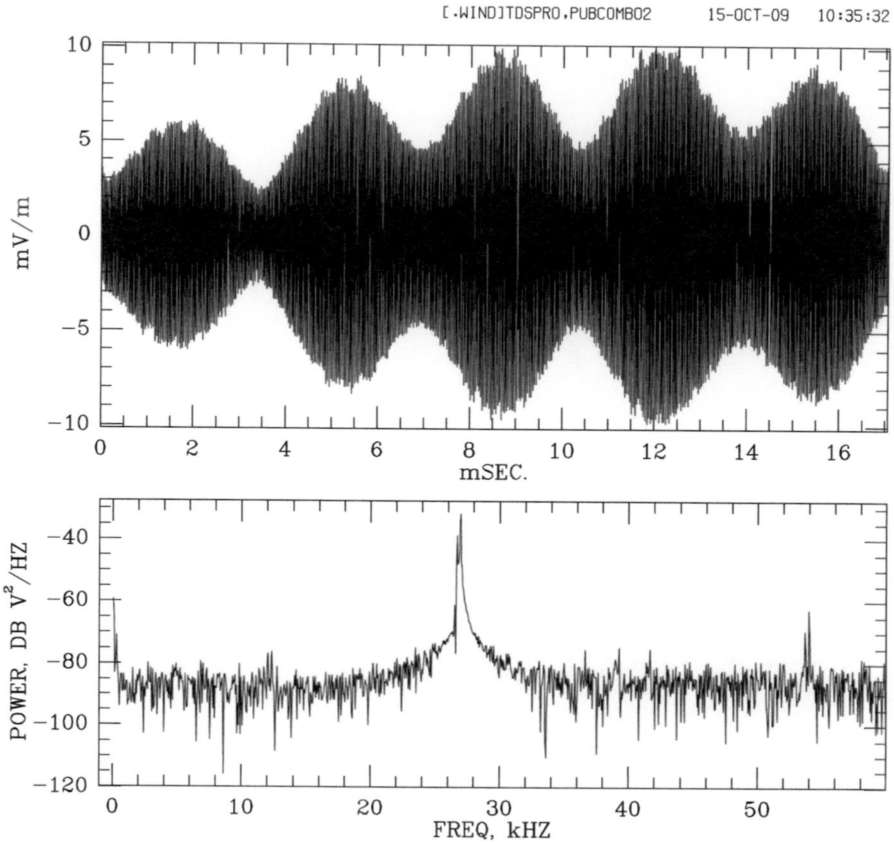

Fig. 21.8 One of the 12 Langmuir wave events of Fig. 21.7

where $W = (A/\epsilon_0) \int E^2 dx$ with $dx = V_{sw}dt$ is the energy in a blob.

The unknown area A cancels out, and so we can calculate the energy in the column from the observations.

Next the radiated energy was calculated. In order to distinguish between Langmuir waves and electromagnetic radiation, we have calculated the radiated energy at 1.5 times the local plasma frequency. The radiated energy does not vary very fast with frequency, so this should give a quite accurate estimate of the radiated energy at the frequency of the Langmuir waves. Figure 21.9 shows the time sequence of the electromagnetic power, with an interpolation for a period of saturation due to Langmuir waves. The radiated power in the bandwidth 800 Hz, corresponding to the bandwidth occupied by the Langmuir waves, is $7\ 10^{-12}$J.

This corresponds to about 15% of the (underestimated) Langmuir wave energy. As said earlier, we hope to do his more accurately with STEREO, but so far the sun has been too reticent. However, the low efficiencies calculated by Kim et al. (2008) make it appear that conversion on reflection may not explain the observed intensities. We shall, however, revisit this, and conclude that this process is not yet ruled out.

Some progress has been made recently in understanding the generation of the harmonic also. This is principally the work of the second author, D. Malaspina. First, Ergun et al. (2008) showed that a significant number of the waveforms observed by the Time Domain Sampler (TDS) on STEREO, (and presumably also on Wind) represent trapped eigenmodes of Langmuir waves in modes corresponding to waves in a potential well. Often more than one mode is observed in a given density cavity. Figure 21.10 shows some examples, showing the original data, the results of the fit, and the fitted spectrum: Since these eigenmodes are standing waves, they are the oppositely directed Langmuir waves which are required for the generation of the harmonic by a nonlinear process. Furthermore, their fields are accurately known from

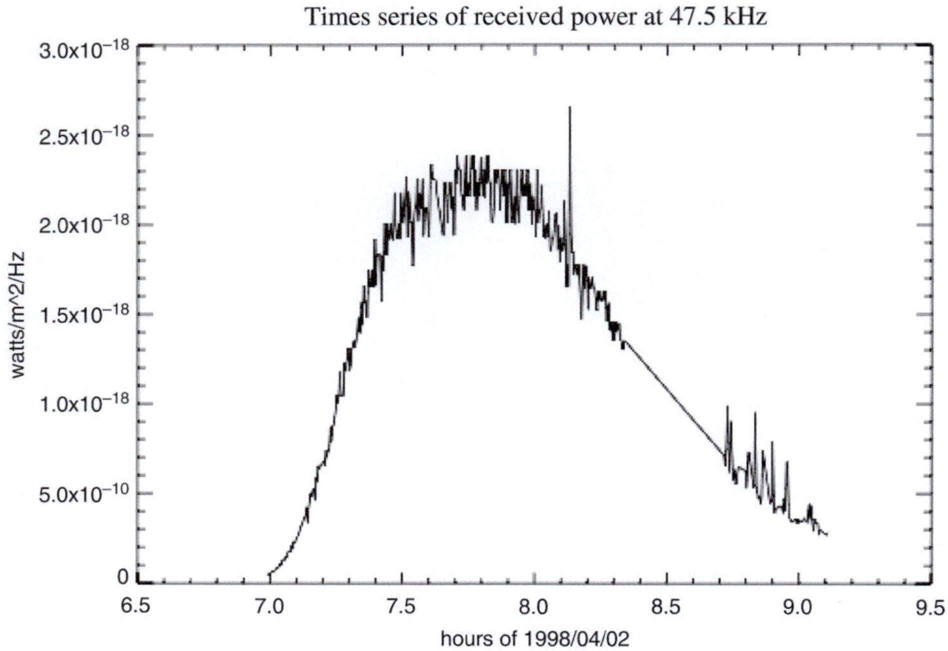

Fig. 21.9 Power radiated at 1.5 fp

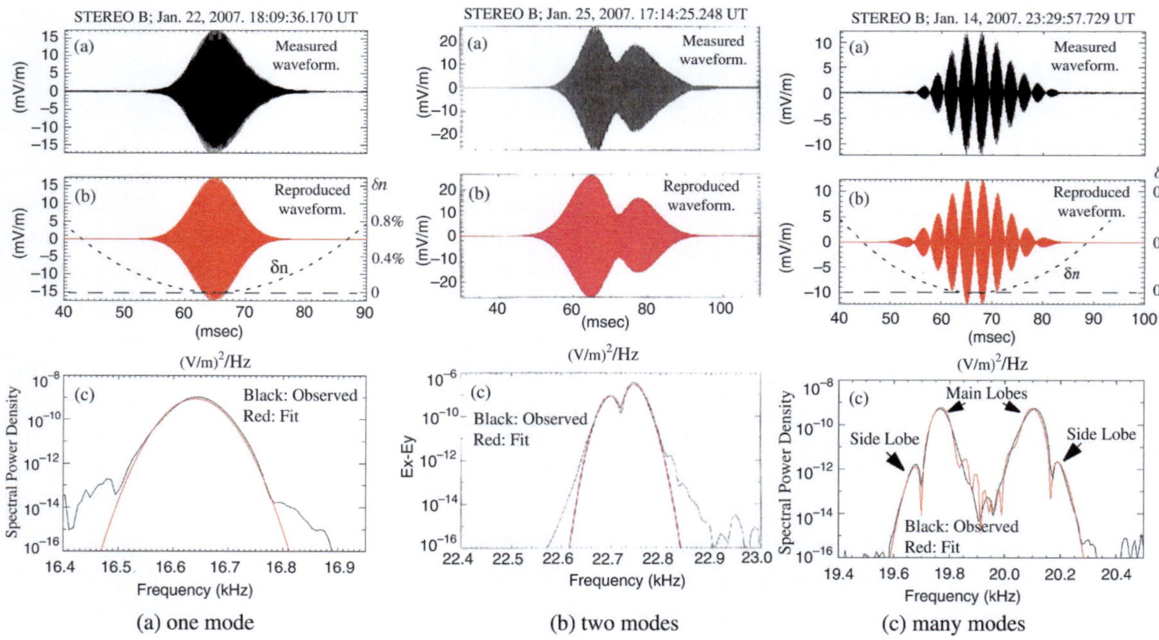

Fig. 21.10 Trapped Langmuir wave eigenmodes

the Gauss-Hermite functions used in the fit, and so the harmonic current and therefore the radiated power can be calculated. Figure 21.11, a spectrum of the fields in a typical Intense Localized Structure (ILS), shows that harmonic fields at 28 kHz are actually present in the interior of a cavity which has trapped

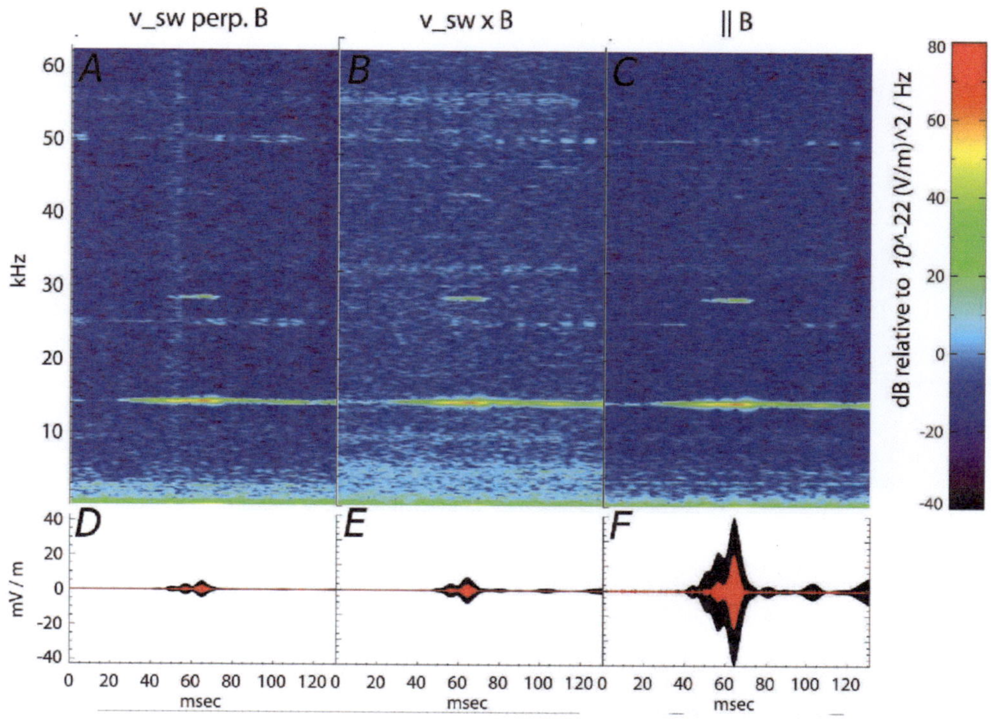

Fig. 21.11 Dynamic spectrum of the fields trapped in a cavity

Langmuir waves. The spectra observed on each of the three nearly orthogonal antennas on STEREO are shown. The electronics for the STEREO Time Domain Sampler are accurately linear up to saturation and the saturation level for the individual antennas is about 150 mV/m, so the shown harmonic is genuinely natural.

It is difficult to measure the radiation from a single trapped wave bunch. The TDS, which is designed to measure intense Langmuir waves, is not sufficiently sensitive for the weak radiated fields. On the other hand, the radio experiment is saturated during most Langmuir wave events, and so any observed harmonics are suspect. Therefore one of us (Malaspina) has estimated the total radiation from a collection of Langmuir waves in the Earth foreshock. Unfortunately, this cannot be done as exactly as the accurate knowledge of the fields of an observed trapped bunch ought to allow. The radiation from the opposing Langmuir waves, and the harmonic current which they imply can be calculated according to standard electromagnetic theory. For this, however, some information is necessary concerning the three dimensional structure of the events, as the spacecraft traverses the structure and gives information in only one dimension.

The three dimensional structure of these events has been studied by Malaspina and Ergun (2008). However, for the present calculation the model of what we call an ILS is taken as that of a plane wave confined to a spherical cavity. The calculation requires a volume containing the emitting ILS's, the density of ILS's within that volume, and an integration over the emission of the ILS, which depends on their size and field strength. The essence of the calculation will be described here. The volume containing the emitting structures is taken from the extensive study of the foreshock by Kasaba et al. (2000) using Geotail data. It is taken to be a long cylinder of radius 10 RE and length 100 RE as Kasaba et al. have shown that the Langmuir waves decrease very slowly in amplitude.

Next, an extensive study of the emission of these structures as functions of amplitude, size, electron and ion temperatures and beam speed was undertaken. Since the structures are often smaller than a wavelength of the electromagnetic radiation, which reduces their efficiency as antennas, the largest structures are

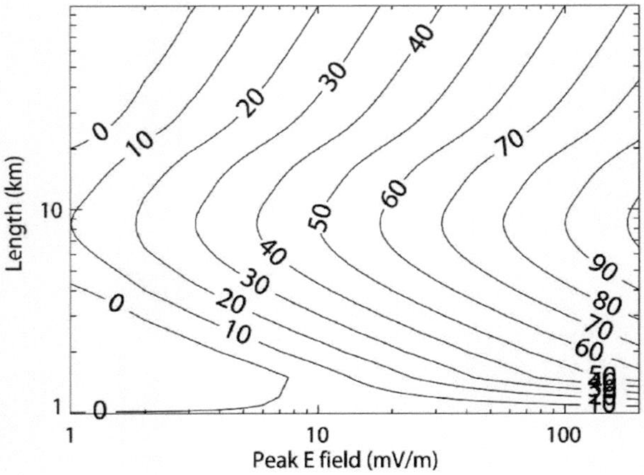

Fig. 21.12 Relative radiation efficiency as a function of E and size

Parameter	Unit	Most radiation	Least radiation	Nominal
T_e	(eV)	7.7	6.5	7.0
T_i	(eV)	7.5	5.2	6.0
f_p	(kHz)	26.15	30.41	27.64
v_b	(v_{th})	10	5	7.5
I_{Total}	(Wm^{-2})	7.3×10^{-16}	2.3×10^{-18}	1.7×10^{-16}

the most important emitters. The radiated power is approximately proportional to E^4, as is well known from theory, but the dependence on size shows the usual resonance effects of size and wavelength of an antenna. Figure 21.12 shows results for size and field strength. Similar studies have been done for other parameters, where the dependence is less strong. As shown in Fig. 21.12, the range of lengths considered in this study is from 1 to 100 km. For a typical solar wind speed of 500 km/s this corresponds to observation times of 2–200 ms, which in turn requires measurements in the frequency range up to 500 Hz. Measurements of density fluctuations have not been published in this range, so Malaspina et al. (to be submitted) have also undertaken an evaluation of the density spectrum using the variation of observed frequency during single events. This also will be the subject of another paper. Here we wish only to point out that the measurment of density measurements at higher frequencies is important for understanding electromagnetic emission.

These calculations, using Langmuir wave data from STEREO, are compared with measured harmonic intensities from the Wind radio experiment (Reiner et al. 1996). in the following table, which gives calculated emitted power as a function of various parameters. The upper four lines give the range of values used in calculating the effects of various parameters other than those of Fig. 21.12. The bottom line gives the range of emitted power for these parameters, together with the observation using Wind data by Reiner et al. It will be seen that the observed emission (nominal) falls within the range of calculated values and that therefore the trapped wave process can power the emission of the harmonic.

In addition to the radiation at the harmonic, there will also be a contribution to the fundamental, As the trapped waves are being reflected continuously at each side of the density cavity, there may also be conversion to the fundamental via the process discussed above in which the report by Kim et al. (2008) may represent the most accurate theory at present. The fundamental has difficulty in escaping of course, as the parent Langmuir wave at nearly the same frequency is totally reflected. However the electromagnetic wave has a much longer wavelength which will facilitate tunneling, and there may be some frequency shift if the boundary is moving. This way of calculating the conversion (by multiple reflections) has not yet been carried out to the authors knowledge. Hence the process of conversion to the fundamental electromagnetic signal at reflection has not yet been treated with adequate accuracy.

21.4 Conclusion

Langmuir waves are converted to electromagnetic waves principally at two frequencies, the frequency of the Langmuir wave and its harmonic. The fundamental may be generated by a linear conversion mechanism involving oblique reflection at a density increase, but the data are not yet sufficient to establish this. The harmonic is quite likely to be generated by opposing Langmuir waves which are eigenfunctions trapped in density depressions. Calculations of this process for the Earth foreshock agree with observations.

Acknowledgements This work was supported by the (PJK) National Aeronautics and Space Administration under grants NNX07AF23G.(STEREO) and NNX07AM97G and NNX08AT81G (Wind) and (DMM) STEREO and NASA Earth and Space Sciences Fellowship 08-Helio08F-0008.

References

Bougeret J-L, Kaiser ML, Kellogg PJ, Manning R, Goetz K, Monson SJ, Monge N, Friel L, Meetre CA, Perche C, Sitruk L, Hoang S (1995) WAVES: The radio and plasma wave investigation on the Wind spacecraft. Space Sci Rev 71:231

Bougeret J-L, et al (2008) S/Waves: The radio and plasma wave investigation on the STEREO mission. Space Sci Rev 136:487

Cane HV (1984) The relationship between coronal transients, Type II bursts and interplanetary shocks. Astron Astrophys 140(1 pt 1):205–209

Ergun RE, Cairns IH, Goldman MV, Newman DL, Robinson PA, Eriksson S, Bougeret J-L, Briand C, Bale SD, Cattell CA, Kellogg PJ, Kaiser ML (2008) Eigenmode structure in solar-wind Langmuir waves. Phys Rev Lett 101(5):051101

Field GB (1956) Radiation by plasma oscillations. ApJ 124:555–570

Filbert PC, Kellogg PJ (1979) Electrostatic noise at the plasma frequency beyond the Earth's bow shock. J Geophys Res 84:1369–1381

Ginzberg VL, Zheleznyakov VV (1958) On the possible mechanisms of sporadic radio emission. Soviet Astron. 2:653 Astron Zh. 35:694

Hinkel-Lipsker, DE, Fried BD, Morales GJ (1992) Analytic expressions for mode conversion in a plasma with a linear density profile. Phys Fluids B 4:559

Jones D (1980) Latitudinal beaming of planetary radio emissions. Nature 288:225. doi:10.1038/288225a0

Jones D, Calvert W, Gurnett DA, Huff RL (1987) Observed beaming of terrestrial myriametric radiation. Nature 328:391–395. doi:10.1038/328391a0

Kasaba Y, Matsumoto H, Omura Y, Anderson, RR, Mukai T, Saito Y, Yamamoto T, Kokubun S (2000) Statistical studies of plasma waves and backstreaming electrons in the terrestrial electron foreshock observed by Geotail. J Geophys Res 105(A1):79–103

Kellogg PJ (1981) Calculation and observation of thermal electrostatic noise in solar wind plasma. Plasma Phys 23(8):735–751

Kellogg PJ (1986) Observations concerning the generation and propagation of Type III Solar Bursts. Astron Astrophys 169:329–335

Kim E-H, Cairns IH, Robinson PA (2008) Mode conversion of Langmuir to electromagnetic waves at magnetic field-aligned density inhomogeneities: simulations, theory, and applications to the solar wind and the corona. Phys Plasmas 15(10) 102110

Malaspina DM, Ergun RE (2008) Observations of three-dimensional Langmuir wave structure J Geophys Res 113(A12):A12108

Reiner MJ, Kaiser ML, Fainberg J, Desch MD, Stone RG (1996) 2fp radio emission from the vicinity of the Earth's foreshock:WIND observations. Geophys Res Lett 23(10) 1247–1250

Willes AJ, Cairns IH (2001) Mode conversion and reflection of Langmuir waves in an inhomogeneous solarwind. Pub Astron Soc Aust 18:355

Yin L, Ashour-Abdalla M, El-Alaoui M, Bosqued JM, Bougeret JL (1998) Generation of electromagnetic fpe and 2fpe waves in the Earth's electron foreshock via linear mode conversion. Geophys Res Lett 25(14):2609–2612

Yin L, Ashour-Abdalla M (1999) Mode conversion in a weakly magnetized plasma with a longitudinal density profile. Phys Plasmas 6:449

Chapter 22

Contributions of Radioheliograph Observations to the Understanding of Solar Flares, Coronal Mass Ejections, Electron Beams in the Corona and in the Interplanetary Medium

Nicole Vilmer

Abstract In this chapter, I shall review observations showing how radio astronomy (and in particular radio images at metric/decimetric wavelengths) contributed to our knowledge on solar flares, coronal mass ejections and electron beams in the corona and in the interplanetary medium.

22.1 Introduction

This chapter will review some aspects of the input of radio observations to our understanding of solar and solar-terrestrial physics. It will be focussed on the spatially resolved radio observations of phenomena linked to solar activity. More material on this topic can be found in the recent review of the input of 65 years of radio observations to solar flare physics by Pick and Vilmer (2008). Radio emission from electron beams propagating in the corona and in the interplanetary medium can provide information about the electron acceleration sites and their evolution in the course of flares as well as information on the location of injection of electrons in the corona and in the interplanetary medium. Radio images obtained with a high temporal cadence provide crucial information on the initiation, development and propagation of Coronal Mass Ejections (CMEs) in the low corona and on the importance of the interplay between different spatial scales for the development of CMEs. Radio observations also allow to understand the relationship and connections between the energetic electrons in the corona and the electrons measured in situ as well as to improve our knowledge on the origin and propagation of energetic particles (SEPs) in the interplanetary medium.

22.2 Some Historical Results on Still Open Questions

The first observation of radio emission from the Sun dates back to February 1942, when a chain of British radar stations recorded a strong radio noise originating from the direction of the Sun (Hey 1942). The connection between the intense radio bursts and large solar flares was soon established after this first observation (e.g. Hey 1946). Radiospectrographs started to be developed in the 1950s, allowing to continuously record the intensity of the solar emission as a function of frequency and time. The first observations in the frequency range from 40 to 70 MHz (Wild et al. 1950) revealed a large variety of narrowband bursts drifting toward lower frequencies. Two subclasses were identified: the type II and the type III bursts respectively characterized by a frequency drift of $\sim$0.25 and $\sim$20 MHz/s. Harmonic features with a 2:1 frequency separation were observed in the spectra of type II and type III bursts (Wild et al. 1954). This was taken as evidence that these emissions originate from a common source producing oscillations at a fundamental frequency and at its second harmonic. The observed narrow bandwidth (2–3% of the central frequency) led to the conclusion that a natural frequency is controlling the oscillations (either electron gyro-frequency or plasma frequency). As it

N. Vilmer (✉)
LESIA-Paris Observatory, 92195 Meudon Cedex, France
e-mail: nicole.vilmer@obspm.fr

M.P. Miralles, J. Sánchez Almeida (eds.), *The Sun, the Solar Wind, and the Heliosphere*, IAGA Special Sopron Book Series 4, DOI 10.1007/978-90-481-9787-3_22,

was already known that emission at the fundamental gyrofrequency does not easily escape from the solar atmosphere (Roberts 1952), it was concluded that the drifting emissions are at the local plasma frequency f_p and its second harmonics. Because the density (thus the plasma frequency) decreases outward from the sun, the emission drifts rapidly from high to lower frequencies. With the development of interferometers allowing to estimate the positions of sources at different frequencies, it was concluded that while type II bursts are excited by a disturbance travelling through the corona at speeds of $\sim 10^3$ km/s, type III bursts are generated by electron beams with typical speeds of 10^5 km/s (Wild et al. 1959a, b). The disturbance associated with the type II burst was later identified as a MHD shock by Uchida (Uchida 1960). In the 1960s, the different radio emissions associated with a solar flare were ideally schematised as shown on Fig. 22.1. The flare consists of an impulsive phase with typical duration of a few minutes, characterized by dm-m activity (type III bursts) and a subsequent gradual phase with typical durations of tens of minutes initiated by the first phase. This second phase which occurs only in large flares may be preceded by a type II burst and is characterized by intense radio continua (type IV bursts) from centimeter to meter wavelengths. This emission can evolve with time towards a smooth continuum observed for periods of several hours in the m-dam wavelengths (i.e. in the high corona). While at centimeter wavelengths these continua are produced by gyrosynchrotron emissions of flare energetic electrons, at lower wavelengths these emissions are usually plasma wave emissions.

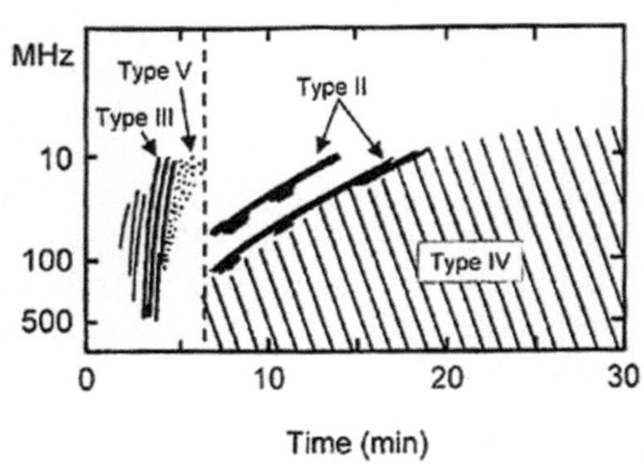

Fig. 22.1 Idealized sketch of a complete radio event associated with a solar flare (from Wild et al. 1963)

22.2.1 Radio Emissions in the Gradual Phase and First Association with Solar Cosmic Rays

The first detection of relativistic protons associated with solar flares was achieved in 1942 using ground level cosmic ray monitors. A striking relationship was observed by (Boischot et al. 1959) between variations of solar cosmic ray intensities of protons of about 170 MeV measured in a balloon flight (Anderson 1958) and a continuum storm intensity at meter/decameter wavelengths (see Fig. 22.2). The association between 10–100 MeV protons detected by Polar Cap Absorption effects and type IV bursts was then systematically studied. A quasi systematic association with energetic protons was found for type IV bursts associated with flares located in the western solar hemisphere in the case of a microwave flux density greater than 10^{-17} J/m^2/Hz and of a second long duration phase at metric wavelengths (Pick 1961). This association led to the idea that solar energetic particles are accelerated by the same process as the fast electrons responsible for the type IV radiation.

22.2.2 Respective Role of the Shock or of a Time Extended Coronal Acceleration in the Production of High Energy Electrons and Protons (SEPs)?

In the model schematised in Fig. 22.1, the shock wave revealed by the type II burst initiated during the impulsive phase creates conditions allowing to accelerate particles to very high energies. These high energy particles are partially trapped in coronal loops where they radiate type IV emissions and some escape in the interplanetary medium. A few years later, a long duration hard X-ray burst (HXR) was observed for the first time on OSO5 (see Fig. 22.3). The HXR emission was attributed to bremsstrahlung of electrons accelerated in the solar atmosphere (Frost et al. 1971). The authors concluded that the time profiles of the X-ray observations and the X-ray spectra support the idea of particle acceleration in two phases as described by (Wild et al. 1963) and proposed that the second phase is due to acceleration in a large scale shock revealed by the type II burst.

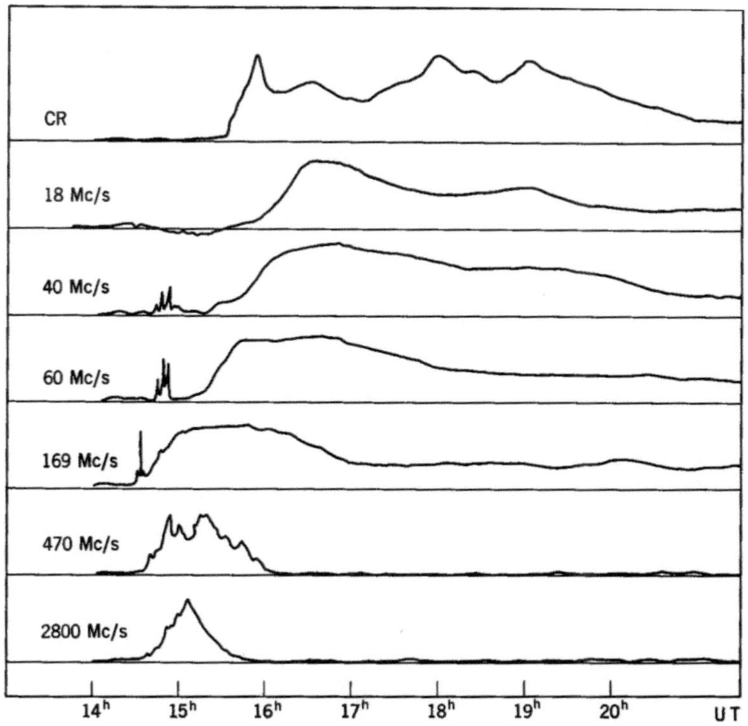

Fig. 22.2 Flux of solar energetic particles (CR) and of radio emission measured at various frequencies (from Boischot et al. 1959)

An alternative interpretation was proposed a bit later and based on the observations that for long duration flares such as the one observed with OSO5, there is a relationship between HXR and metric/decimetric emissions in the gradual phase (e.g. Hudson 1978; Klein et al. 1983) (see Fig. 22.4): hard X-ray and radio emissions start and end within a few minutes and have similar global time evolutions despite the large distance in heights of emitting sources. This shows that electrons are continuously produced and injected in magnetic structures of different scales on time scales of several tens of minutes. Positions of the radio sources in the gradual phase (FC1 and FC2) obtained at one frequency with the Nançay Radioheliograph (NRH) also provided the evidence that there may be many acceleration/interaction sites of energetic electrons. These acceleration/injection sites are not related to the type II shock but to the extrapolation towards the Sun of the coronal loop transient observed by the Coronograph/Polarimeter aboard the SMM satellite (Klein et al. 1983) (see Fig. 22.5). Such observations raised some questions on the respective role of the shock or of a continuous acceleration in large scale coronal structures for the production of the energetic electrons in the gradual phase of long duration flares. This question is still being debated and we will show in the next section how the question can be revisited with new radio observations as well as with more detailed observations of coronal mass ejections.

22.3 Electron Acceleration Sites in Solar Flares as Deduced from Combined Radio and X-Ray Observations

Coherent plasma radiations in the decimetric/metric domains provide sensitive diagnostics of supra-thermal electrons (around a few tens of keV) accelerated in the low and middle corona in connection with solar flares. Indirect evidence of electron acceleration sites in the corona first came from broad band radio spectral observations. Electron beams propagating along magnetic field lines in the corona produce coherent emissions at the local plasma frequency or at its harmonic. The emitted radio bursts (type III, resp. reverse type III bursts) will exhibit characteristic frequency drifts either towards lower, resp. higher frequencies if the beam propagates in the direction opposite to the ambient electron density gradient (upwards), resp. in the direction of the gradient (downwards). Pairs of type

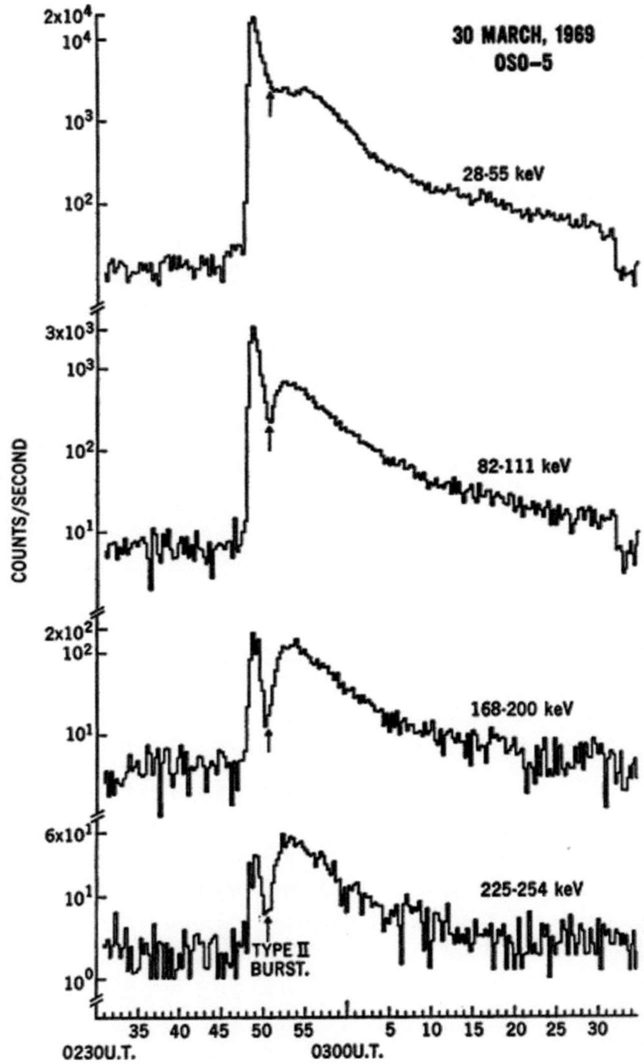

Fig. 22.3 Time evolution of the Hard X-ray flux observed by OSO5 in several energy bands for a long duration flare. The *vertical arrow* indicates the time of appearance of a type II burst in the metric band (from Frost et al. 1971)

III and reverse type III bursts are sometimes observed and their starting frequencies usually between 220 and 910 MHz have been used to deduce a mean density in the electron acceleration site which lies between 6 10^8–10^{10} cm^{-3} for fundamental emission or 1.5 10^8–3 10^9 cm^{-3} for harmonic emission (e.g. Aschwanden et al. 1995; Aschwanden et al. 1997). This implies a much lower density in the acceleration region than the one observed in the bright soft X-ray loops (10^{10}–10^{11} cm^{-3}) and suggests that the acceleration region lies above them, being located e.g. in a cusp reconnection site. Typical heights where electrons of tens to a few hundreds of keV are accelerated can also be estimated from time of flight analysis of hard X-ray emissions leading to estimates of the acceleration heights ranging between 2 10^4 and 5 10^4 km (Aschwanden et al. 1998). Recent observations of a flare for which HXR and radio images at several frequencies in the decimetric/metric domains were obtained using RHESSI and NRH data confirmed that in simple flares the electron associated site may lie in a cusp reconnection site (see e.g. Fig. 22.6) (Vilmer et al. 2002). Figure 22.7 shows the time evolution of the HXR counts (not background subtracted) in 4 energy bands and of the X-ray spectrograms (background subtracted) from 3 to 200 keV measured by RHESSI, together with the radio spectrum observed by PHOENIX-2 in the 150–550 MHz frequency range, the time evolution of the radio flux density at 164, 236, 327, 408, 432 MHz measured with the NRH and the

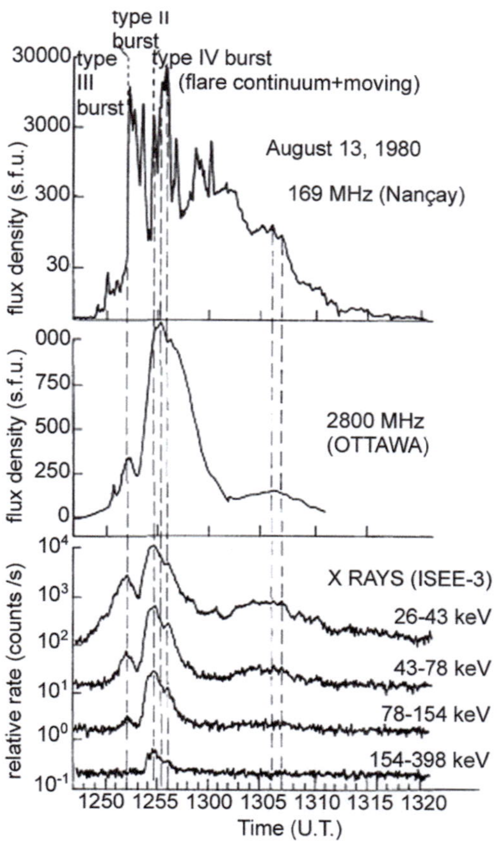

Fig. 22.4 Time evolutions of the radio flux densities at 169 MHz, 2,800 MHz compared with the time evolution of the Hard X-ray flux observed by ISEE3 in several energy bands (from Klein et al. 1983)

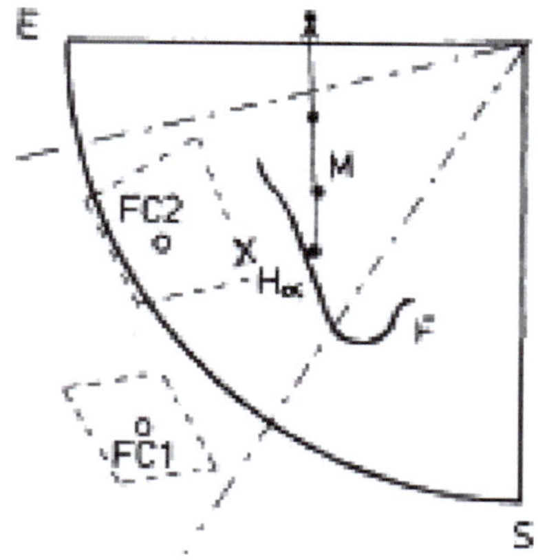

Fig. 22.5 Positions of the sources observed at 169 MHz during August 13, 1980 event. *Circles* indicate the positions of the sources of the radio continuum in the gradual phase, *dashed lines* their extension. F represents a filament and the position of the flare is marked by a *cross*. *Dots* show the trajectory of a moving radio source (M). The two *dashed lines* indicate the latitudinal limits (measured at 1.5 solar radius from the Sun's center) of the coronal loop transient observed by the Coronograph/Polarimeter aboard the SMM satellite (from Klein et al. 1983)

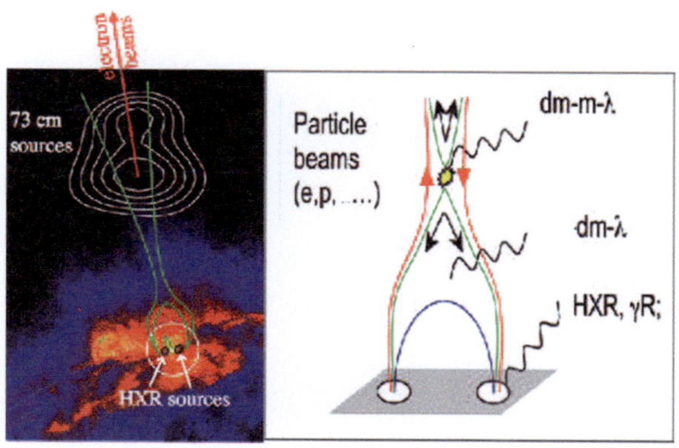

Fig. 22.6 *Left*: Overlay of a hard X-ray source from RHESSI (*black contours* within the *white circle*) and radio sources from the NRH (*white contours*) on an EUV image (SOHO/EIT) of the flaring active region (adapted from Vilmer et al. 2002). The *green lines* are a schematic drawing of a plausible magnetic configuration. *Right*: cartoon scenario for flare-related particle acceleration derived from these observations (adapted from Klein 2007)

radio spectrum from WIND/WAVES in the 1–14 MHz band. The X-ray emission observed with RHESSI extends to energies of 100 keV during a shorter time interval during the flare (40s around 11:06:20 UT). The simultaneous start of intense HXR and metric radio emissions is consistent with the results of previous studies (e.g. Benz et al. 1983; Raoult et al. 1985). The duration of the strongest part of the metric/decimetric radio emission is found to be similar to the one of the HXR peak above 50 keV and the radio emission below 12 MHz (i.e. arising from electron beams injected in the high corona) starts together with the sudden increase of the HXR flux at energies above 50 keV. This shows the link between the injection of electron beams towards the high corona and towards the HXR emitting sites. A very close correspondance is observed between the change on the time scale of a few seconds of the pattern of the HXR source in the 25–40 keV range and the pattern of the radio source at the highest frequency (410 MHz) which can be imaged (Fig. 22.8 from Vilmer et al. 2002). This strongly supports the previous suggestion of common acceleration/injection sites for hard X-ray and decimetric radio emitting electrons and the cartoon schematized in Fig. 22.6.

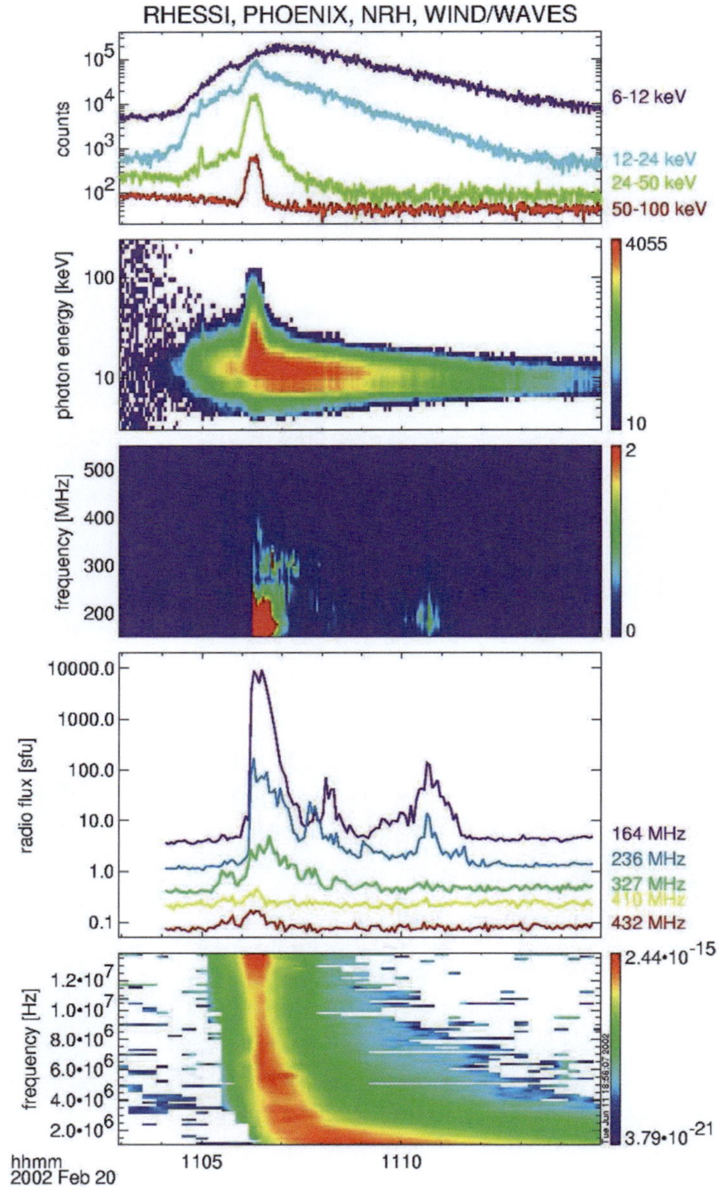

Fig. 22.7 Multi-frequency observations of the radiative signatures of energetic electrons from the chromosphere to the high corona. From top to bottom: -time profiles of hard X-ray count-rates from RHESSI, - dynamic hard X-ray spectrum from RHESSI, - dynamic dm-m wave radio spectrum from PHOENIX-2, - selected dm-m wave time histories (NRH), - dynamic deka-hectometre wave spectrum from Wind/Waves (Vilmer et al. 2002)

In the case of more complex flares, the direct comparisons between the evolution with time of HXR and radio sources at different frequencies, i.e. at different heights allow to probe the electron acceleration sites in the corona and to show that there are several sites of production of electrons for the ones staying in the middle corona and the ones escaping to the interplanetary medium (Vilmer et al. 2003). Figure 22.9 shows indeed that while the first strong hard X-ray peak at energies above 50 keV is not associated with strong radio emission, the second one is associated with intense decimetric/metric (and dekametric emissions). This implies that, during the first peak, energetic electrons are confined in low lying loops (heights of several 10^4 km) with no access to the higher ($>10^5$ km) corona. It is found that the second hard X-ray peak comes from a slightly different position in the active region and is associated with the appearance of new radio components at all frequencies with positions further away from the active region. This second peak thus results from energy release in different magnetic structures

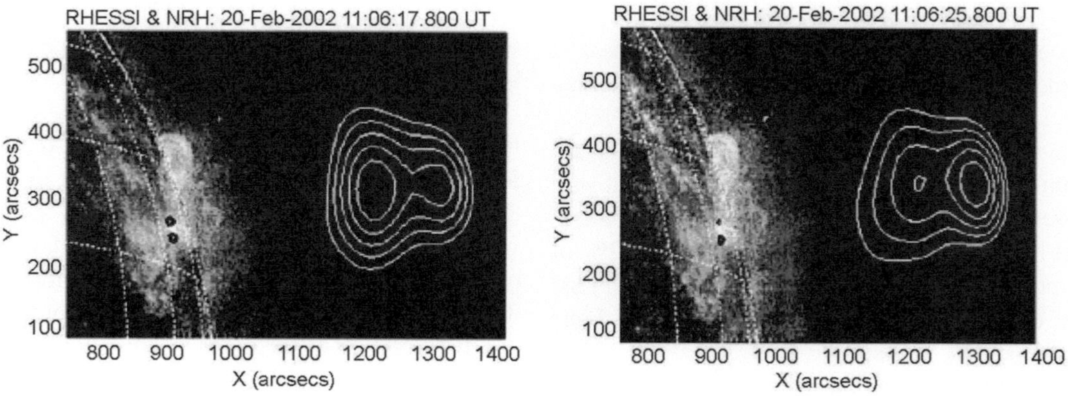

Fig. 22.8 RHESSI isocontours (*black*) (40, 60, 80% of the maximum) at 25–40 keV and NRH contours at 410 MHz (*white*) (50, 60, 70, 80, 90%) observed at a time when similar X-ray fluxes are emitted by the X-ray sources (*left*) and at a time when the southernmost X-ray source is predominant (*right*). The RHESSI and NRH contours are superposed on an EIT image (adapted from Vilmer et al. 2002)

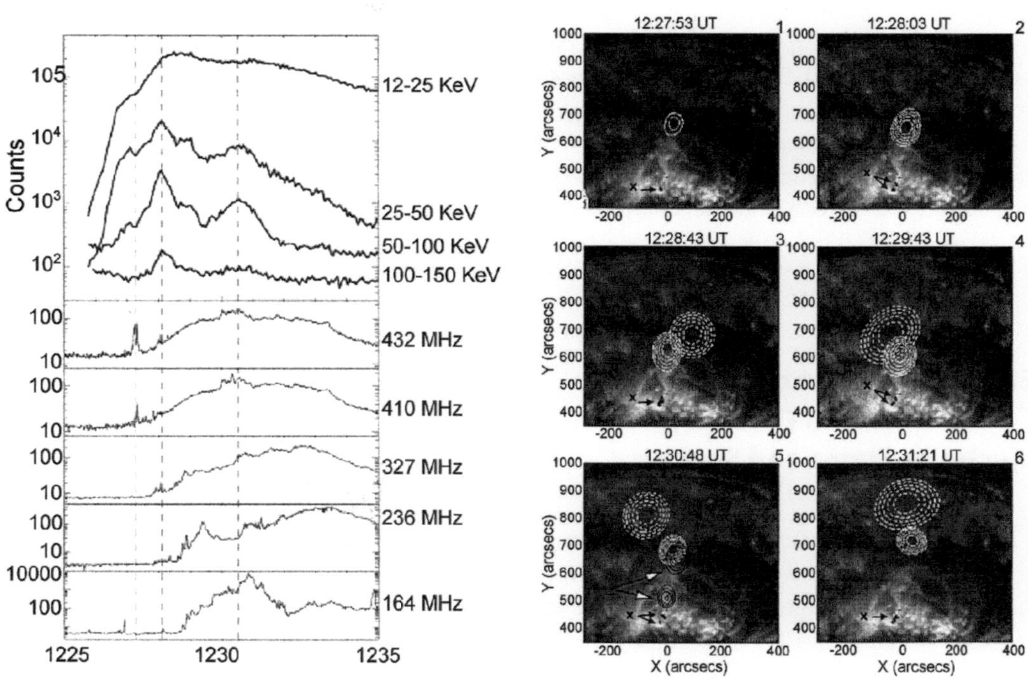

Fig. 22.9 *Left*: from top to bottom: Time evolution of the X-ray RHESSI count-rates in 4 energy bands and time evolution of the radio flux density observed in the flare region by the Nançay Radioheliograph. The *dashed vertical lines* indicate the first radio 432 MHz burst at 12:27:20 UT and the two main HXR peaks above 50 keV around 12:28 UT and after 12:30 UT. Right: RHESSI isocontours (*thick black*) (75, 80, 85, 90, 95% of the maximum) and NRH contours at 327 MHz (*dashed-dotted white*) (75, 80, 85, 90, 95% of the maximum) and 164 MHz (*dotted white*) (75, 80, 85, 90, 95% of the maximum) observed at different times. Note that the 164 MHz emission appears later in the flare (from image 3). The two components at 432 MHz (indicated by *arrows*) are overlaid for comparison (*black contours*) in image 5. The RHESSI and NRH contours are overlaid on the EIT image. The RHESSI contours are obtained in the 40–65 keV range and are indicated by *black arrows* (from Vilmer et al. 2003)

and large scales (>10^5 km) are now involved in the process. These new radio components appear at the time of the extension of strong radio emission below 14 MHz and may trace electron beam injection towards the high corona and the interplanetary medium. The fact that the first peak is associated with very weak radio emission at decimetric/metric wavelengths is consistent with the fact that 10–17% of HXR producing electrons have no detectable emission at decimeter and longer wavelength (e.g., Benz et al. 2005).

Combining HXR images with radio images at different frequencies in the decimetric/metric domains allows to understand the link between the energetic electrons interacting at the sun and the injection of the escaping electrons which produce the radio emissions at the lowest frequencies. The above examples clearly show the interest of large band radio imaging spectroscopy to follow electron paths from the solar surface towards the interplanetary medium. Combined with X-ray images of flares such observations are a crucial support for the future mission Solar Orbiter. One of the key objective of the mission is indeed to trace the magnetic connectivity between the solar surface, the corona and the interplanetary medium.

22.4 Electron Acceleration Sites in Connection with Coronal Mass Ejections, Flares and Shocks

Coronal Mass Ejections (CMEs), i.e. large scale magnetically structured plasmas which are expelled from the sun and propagate to large distances in the heliosphere were discovered in 1973 by (Tousey 1973). Since then, the association and link between flares and CMEs have given rise to a hot and controversial debate. It is now accepted that, even when flares and CMEs can be produced in a same event, there is no causal relationship between these two phenomena; they simply reflect two different ways the corona responds to the same magnetic energy release (e.g. Temmer et al. 2008). When CMEs are associated with flares, strong non thermal emissions are observed in a broad frequency range. Observations as well as models of CMEs show that the developement of a CME involve many loops at different scales. The fully developed CME is also charactized by a flux rope and by a bright and hot front preceded by a shock.

Figure 22.10 shows snapshots of selected field lines from the numerical simulation of a break-out model for Coronal Mass Ejections (MacNeice et al. 2004). Several possible acceleration sites can be associated with the development of the CME: in the bow shock of the CME, in the reconnection sheet formed below the CME or in the interactions regions formed during the evolution of the magnetic structures at different scales. As will be shown below, radio images of metric/decimetric emissions obtained at high temporal cadence prove in fact that efficient electron acceleration in the low and middle corona is rather produced in successive magnetic interactions occuring during the CME development or in the reconnection sheet formed below the CME rather than in the bow shock of the CME.

Fast flare/CMEs have usually a complex development; they are observed to start with a relatively small angular projected dimension and reach their full extent in the low corona (below 2R⊙) in a few minutes. In many cases, radio images obtained at high temporal cadence with the Nançay Radioheliograph show that the radio emission sites are first localized in a coronal region in the vicinity of the flare site and then expand by successive magnetic interactions at progressively larger distances from the flare site. Signatures of these interactions are detected by bursts in the decimetric/metric wavelength domain (Maia et al. 1999). Figure 22.11 shows that there is a good correspondence between the final extent of the radio emitting sites and the extent of the legs of the white light CME structure. In the present case, the CME reached its full size in the low corona within 5 min. This event illustrates a case where electrons are accelerated in successive magnetic interactions occuring during the CME development.

Joint spectral and radio imaging observations at multiple frequencies may also provide evidence for electron acceleration sites associated with magnetic reconnection behind eruptive flux ropes. As an example, Fig. 22.12 Left shows a long duration broad band continuum (type IV burst) drifting towards lower frequencies. This continuum is modulated by successive packets of fast sporadic bursts in close temporal coincidences with hard X-ray peaks (Classen et al. 2003; Pick et al. 2005a). The continuum emission originates

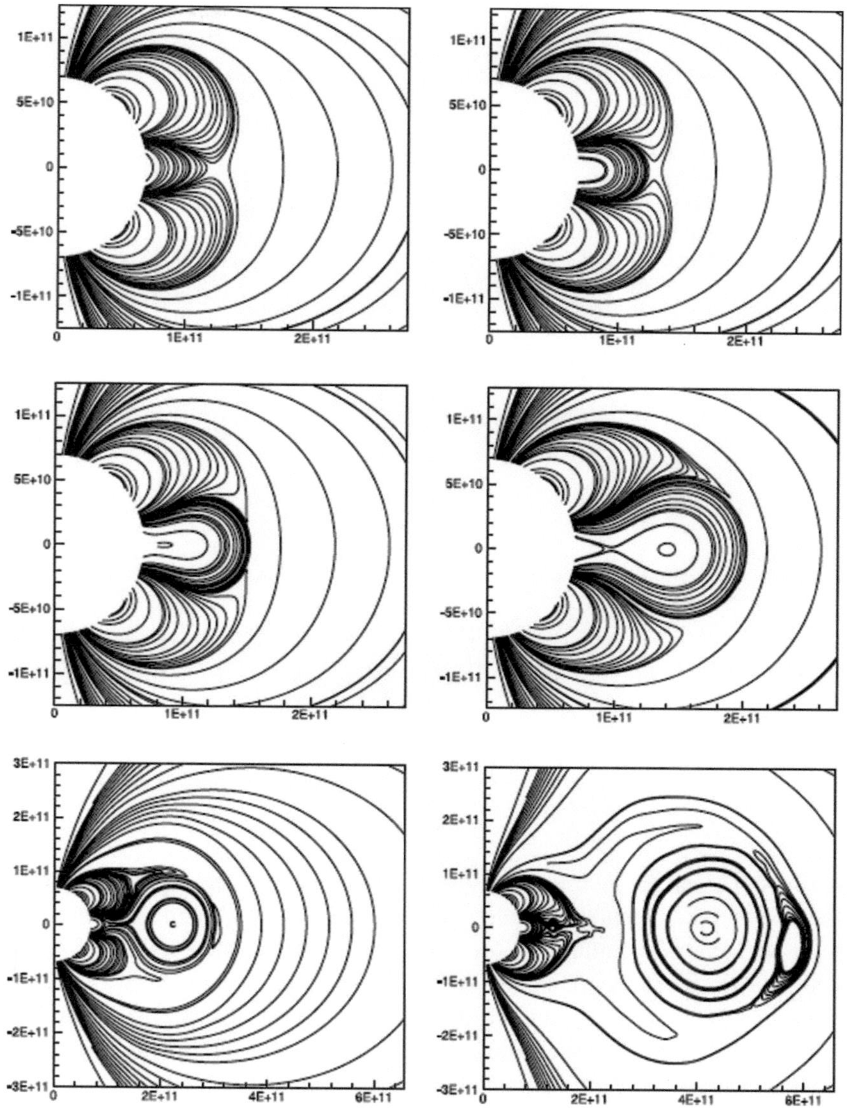

Fig. 22.10 Snapshots of selected field lines in the inner region of the numerical simulation of the break-out model for Coronal Mass Ejections. Snapshot times are, from left to right and top to bottom 0, 13.96, 19.63, 21.95, 23.66, and 26.39 h, respectively. The spatial scale is expanded in the last two frames from 4 × 4 R_s to 8×8 R_s to best show the expanding flux rope (from MacNeice et al. 2004)

from two sources: a fast-moving one with a mean projected speed around 400 km/s and a quasi-stationary one. Both stationary and moving sources are located along the northern edge of an ascending EIT arch overlying the flare (speed of 540 kms^{-1}) and there is no radio emission above the ascending loop. The time coincidence found between the peak flux of the moving and stationary radio sources and of the hard X-ray sources detected by RHESSI implies a causal link: the acceleration site of the radio and X-ray emitting electrons must be the same. These observations are consistent with an electron acceleration site associated with the formation of a current sheet below the erupting twisted flux rope. The accelerated electrons form beams along the newly reconnected field lines and propagate both upward and downward. The moving and the stationary radio sources are located on each side of the current sheet (Pick et al. 2005a). A schematic two-dimensional view of this evolution is outlined in Fig. 22.12 (Right).

A similar conclusion was reached by Dauphin et al. (2005, 2006) for another flare for which broadband modulations were simultaneously observed in X-rays and at radio wavelengths in the late phase of the event (Fig. 22.13). For this event, a rising soft X-ray loop was observed and associated with the onset in the lower corona of a coronal mass ejection. Figures 22.14 and 22.15 show that at the time of the beginning of

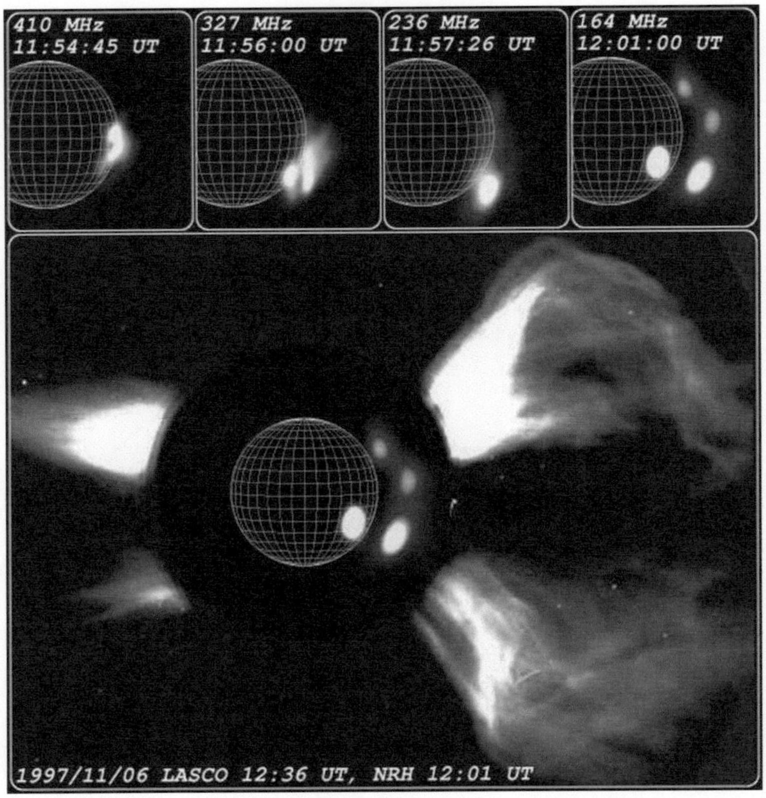

Fig. 22.11 1997 November 6 event. *Upper panel*: Radio images at distinct frequencies which show the complexity of the emitting region: the emission expands toward the Northern hemisphere. *Lower panel*: Composite image of a LASCO/C2 CME image with a radio NRH image at 164 MHz showing that there is a close correspondence between the extent in latitude of the CME seen by LASCO-C2 and the sites of radio emission (adapted from Maia et al. 1999)

the modulations the metric radio sources observed by the NRH are far below the position of the X-ray rising loops and thus of the inferred CME-proxy front. Energetic electrons of a few hundreds of keV produced at the start of the modulations at X-ray and radio wavelengths thus cannot be accelerated by the shock of the CME at decimetric/metric heights but are most probably accelerated in the wake of the CME, i.e. in the reconnecting sites below the CME.

The previous examples illustrate that the most probable site for efficient electron acceleration in the corona is not related to the shock front. This is consistent with other observations in radio and in hard X-rays (see e.g. Klein et al. 2003). This of course does not rule out the possibility of coronal shocks being the accelerators of electrons in the interplanetary medium. This is probably not the most efficient mechanism in the low and middle corona, even if in the corona, the shock may accelerate a small number of electrons, namely the electrons which produce the type II emission. This is illustrated in the same event in Fig. 22.16. The comparison of the positions of the radio sources for the type II burst observed around 09:53 UT and of the image of the X-ray rising loop at the same time clearly shows that the origin of the type II burst is closely associated with the dynamics of the rising loop and that the electrons which radiate the type II burst are accelerated in the bow shock. However, the shock wave driven by the rising loop accelerates only electrons which produce the type II emission but is not efficient at producing hard X-ray emitting electrons at the time of the type II burst as seen in Fig. 22.13. This is consistent with previous observations of HXR and type II bursts.

22.5 Role of the Shock or of Time Extended Coronal Electron Acceleration in the Production of Energetic Particles in the IP Medium?

Solar Energetic Particle events (SEP), which are detected in-situ from GeV down to keV energies, have been divided into two groups, referred to as "impul-

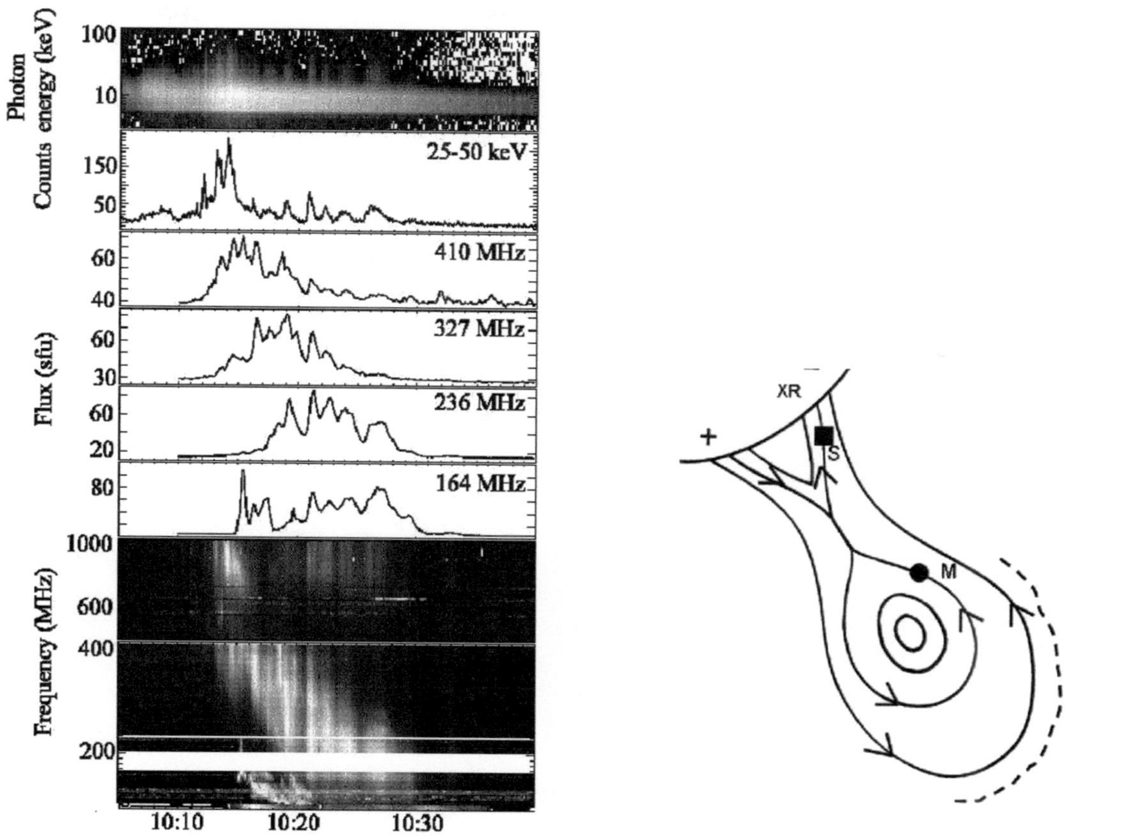

Fig. 22.12 2002 June 02 event. *Left* Comparison between the photon histories measured by RHESSI (*two top panels*), the flux evolution measured at four frequencies by the NRH (*middle four panels*), and the spectral evolution measured by OSRA and by WAVES (*bottom panels*). *Right* Two-dimensional sketch of the magnetic configuration involved in the eruption. A twisted flux rope erupts, driving magnetic reconnection behind it. The particles accelerated in the reconnection region propagate along the reconnected field lines, giving the observed hard X-rays (XR) and the main radio sources (S and M), which correspond to the quasi-stationary sources and moving sources (see text). A shock is propagating at the front edge of the flux rope (*dashed line*) (from Pick et al. 2005a)

sive" and "gradual" events and this classification was commonly accepted until the launch of the Advanced Composition Explorer (ACE) in 1997. In the two-class paradigm for SEP events proposed by Reames (1995, 1999), the flare process accounts for the acceleration of the impulsive events which are not associated with CMEs. On the other hand, for the long duration gradual events the acceleration is dominated by CME driven coronal and interplanetary shocks (not flares). The above two-class paradigm mainly originated from the classification of radio emissions and from the study of their link to interplanetary protons and geomagnetic effects, as introduced by Wild et al. (1963). As discussed in Section 22.2, these authors introduced the concept of two successive phases: an initial impulsive one and a subsequent gradual phase occurring only in large flares. During impulsive flares, the acceleration lasts typically a few minutes, while in the long duration flares, it lasts for tens of minutes, and the acceleration is supposed to take place at extended shocks revealed by type II bursts. The accelerated particles are then either partly trapped in coronal loops, or escape into the interplanetary medium. This model, which assumes a coronal trapping of the accelerated radiating electrons has however difficulties in explaining the fact that the broadband emission of type IV bursts display similar intensity variations at all frequencies, despite the large distance in height of the emitting sources (see Section 22.2). Moreover, the ability of shocks to accelerate coronal particles rapidly up to GeV energies

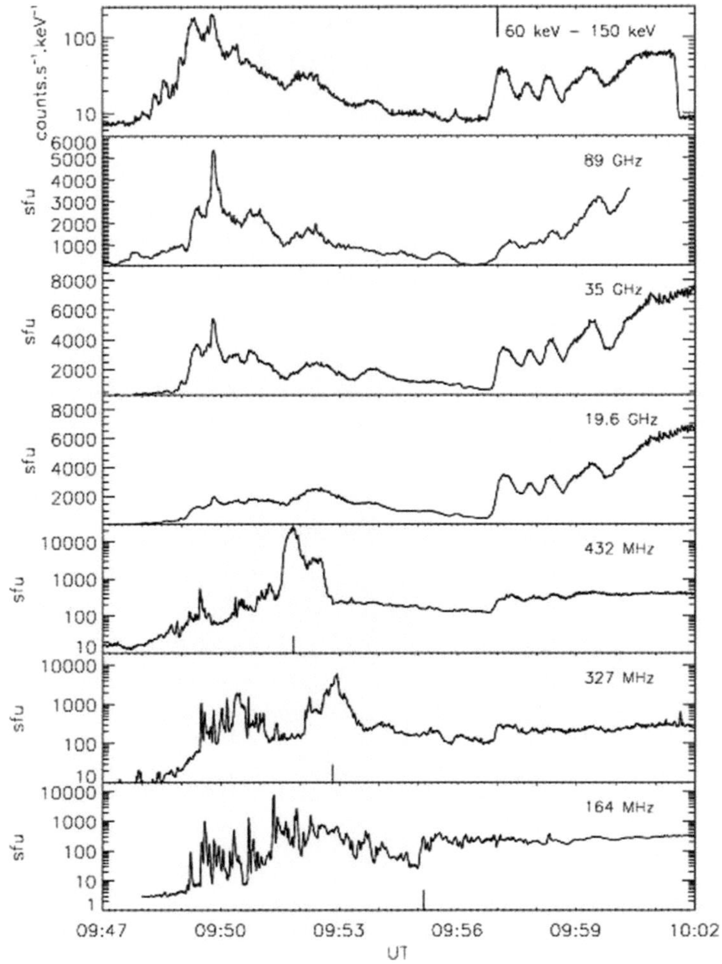

Fig. 22.13 2003 November 3 event. Time evolution of the HXR flux observed above 60 keV with RHESSI and of the millimeter, centimeter and meter radio fluxes observed by Bern instruments and by the NRH. The time of the type II burst at different frequencies below 432 MHz is indicated by *black lines* in the three lower plots and the time of the continuum enhancement is indicated by a *black line* at the top of the figure. After that time, modulations in the centimeter millimeter domains follow the modulations of the hard X-ray flux (from Dauphin et al. 2005)

has stimulated many controversial debates. As already mentioned, it has also been noted that the similarities between HXR and radio signatures of energetic electrons at the sun showed that these electrons could be accelerated quasi continuously for several tens of minutes or hours even in the absence of type II shocks.

In conclusion, it is now generally agreed that the two class paradigm is too simplified. Gradual SEP events can in fact include particles which originate from flares and from CMEs in different ways: particles can be accelerated by CME driven shocks or in the reconnection sites of field lines pulled out by CMEs (see the previous section).

We show here how the radio observations contribute to improve our knowledge on the origin and the propagation of the energetic particles in the interplanetary medium. As radio emissions only give access to electron acceleration, any comparison between radio observations at the sun and the in-situ energetic ions is however an indirect one.

22.5.1 Role of the Shock Wave in the Production of SEP Events

In the 1980s, it was generally believed that large scale shocks in the corona and in the interplanetary medium are the dominant agent for acceleration of energetic electrons (and also ions) in the large gradual events. This idea was based on the classification of radio emissions and of their link to interplanetary protons and geomagnetic effects, recalled above but also on the observations of a new class of kilometric type III bursts in the interplanetary medium associated with metric type II bursts and called Shock Accelerated (SA)

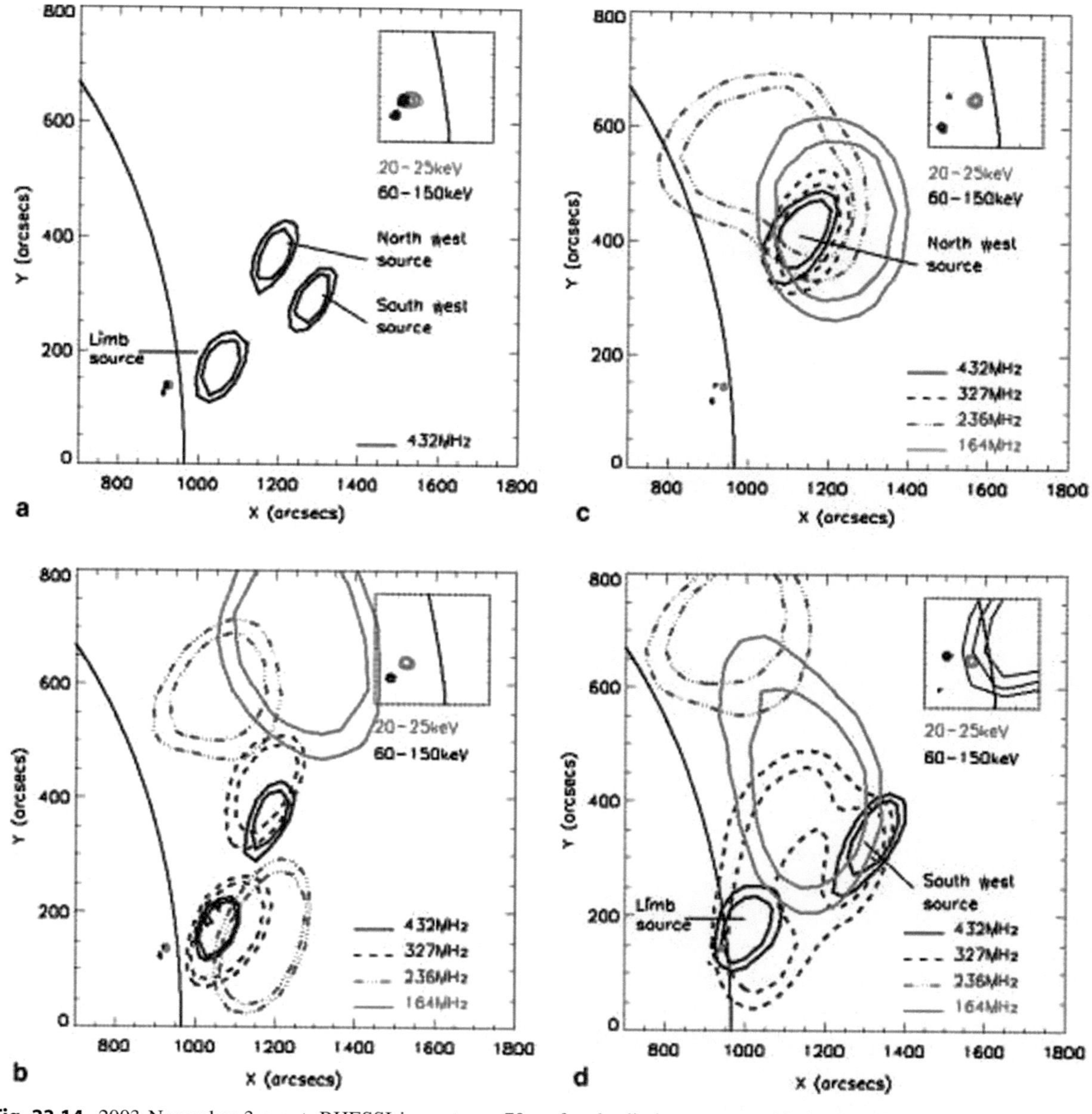

Fig. 22.14 2003 November 3 event. RHESSI isocontours 70, 80, 90% at 20–25 keV (*grey*) and at 60–150 keV (*dark*) and NRH contours (70, 80%) at different frequencies overlaid during different time intervals of the event. The insert panels in each figure show an enlargement of the region close to the X-ray sources. **a** NRH contours at 432 MHz during the first part of the event. The X-ray image is obtained for the 40 s time interval centred at 09:49:25 UT. **b** NRH contours at 432, 327, 236.6 and 164 MHz for the north west bursts taken at respectively 09:49:39:210 UT, 09:49:40:100 UT, 09:49:41:000 UT and 09:49:41:890 UT and for the limb source outside bursts taken at 09:49:18:510 UT, 09:49:23:900 and 09:49:27:500 UT. The X-ray image is taken around 09:49:45 UT. **c** NRH contours at 432, 327, 236.6 and 164 MHz during the first modulation at 09:57:27 UT showing the north-west components of the continuum emission. The X-ray image is taken around 09:57:31 UT. **d** NRH contours at 432, 327, 236.6 and 164 MHz during the fourth modulation at 09:59:26 UT showing the south-west and the limb components of the continuum emission. The X-ray image is taken around 09:59:12 UT (from Dauphin et al. 2005)

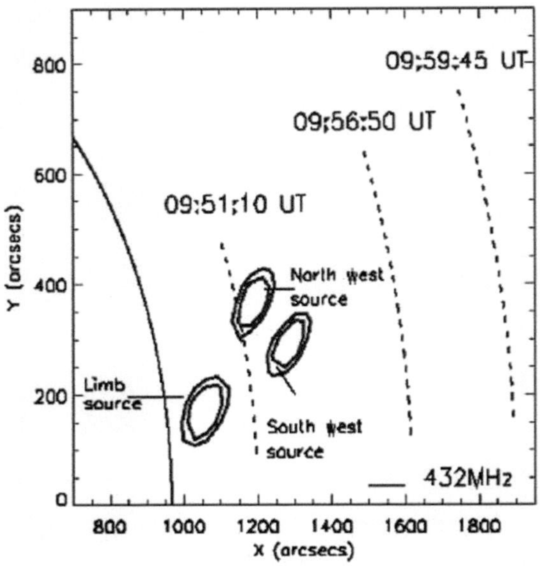

Fig. 22.15 2003 November 3 event. Compound view of the three main sources at 432 MHz at the time of the emission of the type III-like bursts in the first part of the flare. The NRH contours at 70% and 80% shows the limb source at 09:49:18 UT, the north west source at 09:49:41 UT and the south west source at 09:49:13 UT. The *dashed lines* represent the position of the projected height of the front of the X-ray loops and CME-proxy at 09:51:10 UT, 09:56:50 UT and 09:59:45 UT (from Dauphin et al. 2006)

events (Cane et al. 1981). These last events are intense and of long duration (typically over 20 min at 1 MHz) and they display signatures of multiple electron injections (see Fig. 22.17). Because of their association with type II bursts, it was proposed that the SA events are produced by electrons accelerated in the high corona by the same shock that produces the type II bursts. This interpretation was commonly accepted (see e.g., Cane et al. 1984; MacDowall et al. 1987; Kahler et al. 1989). However, MacDowall et al. (1987) underlined that the lack of radio spectral observations in the frequency range 2–20 MHz made difficult the distinction between the regular kilometric type III bursts associated with the prolongation of metric type III activity and the SA events possibly due to shock accelerated electrons. It was also noted by Kundu et al. (1984) that the duration of the SA events was similar to the duration of associated microwave continua. This also argued in favour of a low coronal acceleration process.

More recently, the new data from WIND in the 1–14 MHz window (Bougeret et al. 1995) provided a direct proof of the association between hectometer-kilometric emissions and the radio emissions at higher frequencies. In the example shown in Fig. 22.18, the close similarity of the time profiles of the radio emissions at 3 GHz and 13.82 MHz indicates that the complex type III-like emission detected by WAVES is produced by electrons accelerated in coronal regions as formerly suggested by Kundu et al. (1984). Reiner et al. (2001) also observed that complex type III-like emissions, including the original SA events identified by Cane et al. (1981), are usually associated with major flare/CME events of wide angular extent (see also Pick and Maia 2005).

More recently, Cane et al. (2002) showed that the long duration kilometric type III-events usually start at frequencies higher than the corresponding type II bursts, if present. They thus concluded that, contrary to the original interpretation, these observations argue against a shock-acceleration origin. Cane et al. (2002) also established that $>$20 MeV SEP events are associated with this class of complex type III like events, which also questions the production of SEPs by shock waves. This is consistent with results of Klein et al. (1999) based on the comparison, for two large flares, of gamma-ray, X-ray and radio diagnostics of interacting particles and in-situ detection of $\geq$ 20 MeV protons at 1 AU where they concluded that successive increases of protons fluxes can be traced back to episodes of coronal acceleration.

Finally, Reiner et al. (2007) established, for the very energetic 2002 July 23 γ-ray event, that a good temporal relationship (similar duration and intensity variations) exists not only between the hectometer and decimeter/microwave fluxes but also with the HXR light curves measured by RHESSI. This suggests a single acceleration process for all the particles: while both HXR and microwave emissions are likely produced by a population of downward propagating high-energy electrons ($\sim$100 keV), the low frequency-emissions are generated by a different but linked population of escaping electrons of lower-energy ($<$10 keV) (Fig. 22.19).

These results based on new observations are consistent with the interpretation that during flare/CME events, the particles are accelerated in coronal regions located in the aftermath of CMEs, possibly in the reconnecting cusps appearing below the eruptive flux rope. The opening of the large scale magnetic field due to the passage of the CME allows the escape of the energetic particles to the interplanetary medium.

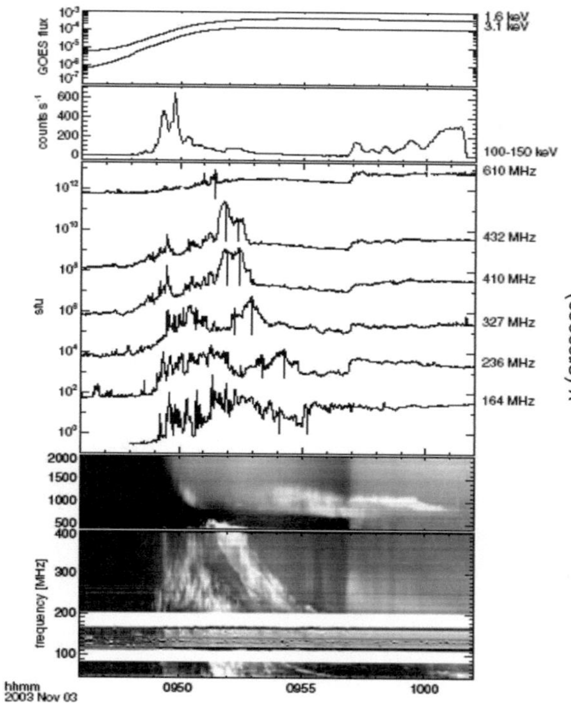

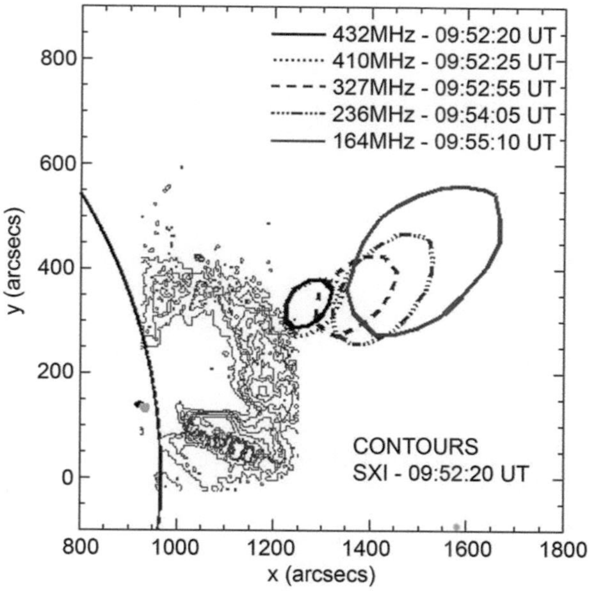

Fig. 22.16 2003 November 3 event. *Left* From top to bottom: time evolution of the GOES X-ray flux (panel 1) and of the X-ray RHESSI counts in the 100–150 keV energy band (background not subtracted) (panel 2); time evolution of the radio flux observed at 610 MHz (from the TRIESTE observatory), 432, 410, 327, 236 and 164 MHz (from the NRH). Radio composite spectrum observed between 2 GHz and 1 MHz by PHOENIX-2 (panel 3), OSRA (AIP Potsdam) (panel 4). Note the continuum enhancement at 09:57 UT corresponding to the second phase of energy release observed in hard X-ray wavelength range. *Right*: Contours of the type II sources (taken at 70% of the maximum) at different frequencies at the times indicated in the figure together with the contours of the GOES/SXI difference image. The X-ray sources from RHESSI are imaged in the 20–25 keV (*grey*) and 60–150 keV (*black*) energy range (adapted from Dauphin et al. 2006)

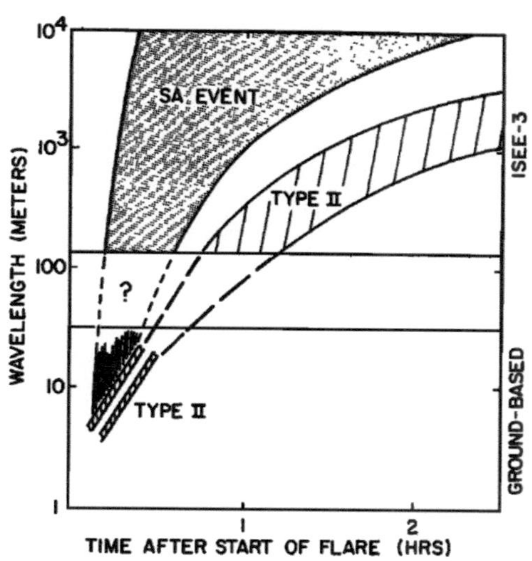

Fig. 22.17 Schematic representation of the relationship between meter wavelength type II activity and the activity observed at kilometric wavelengths (from Cane et al. 1981)

22.5.2 Impulsive Electron Events

Two different kinds of impulsive electron events in the interplanetary medium at energies above 10 keV have been identified by Krucker et al. (1999) and Haggerty et al. (2002) : (i) events released from the sun at the onset of a radio type III burst, which suggests that these electrons are part of the population producing the type III radio emission and ii) events in which the electrons are released up to half an hour later than the onset of the type III burst. Several explanations have been proposed for these delayed events that involve a second acceleration process; the delay acceleration may be due to the coronal counterpart of EIT waves detected by SOHO (Krucker et al. 1999) or to CME-driven shocks (Simnett et al. 2002) or to the change of magnetic field configuration in the corona (Pick et al. 2003). Cane et al. (2003) alternatively proposed

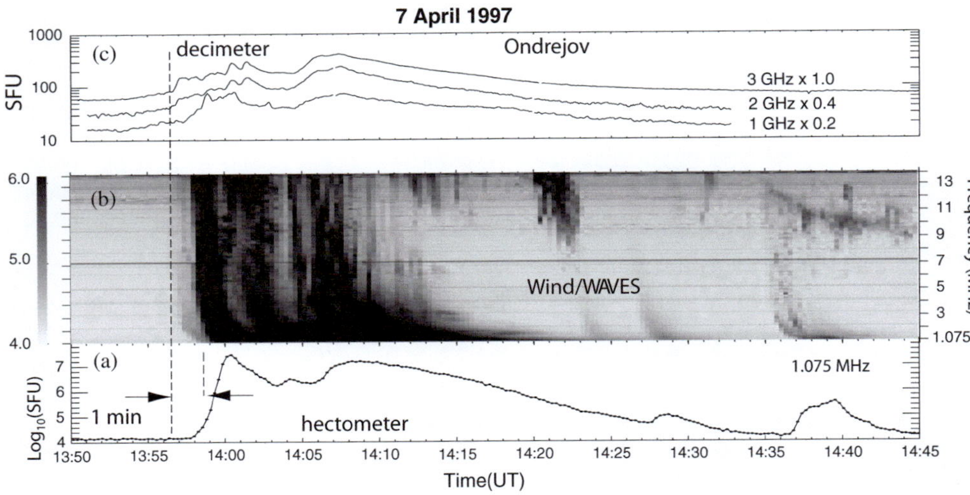

Fig. 22.18 Comparison of the complex type III burst observed on 1997 April 7 by Wind/WAVES in the frequency range from 1 to 14 MHz with the radio emissions observed by Ondrejov observatory from 1–3 GHz. Note the similarity in the shape and duration of the intensity-time profiles in these widely separated frequency regimes (adapted from Reiner et al. 2001)

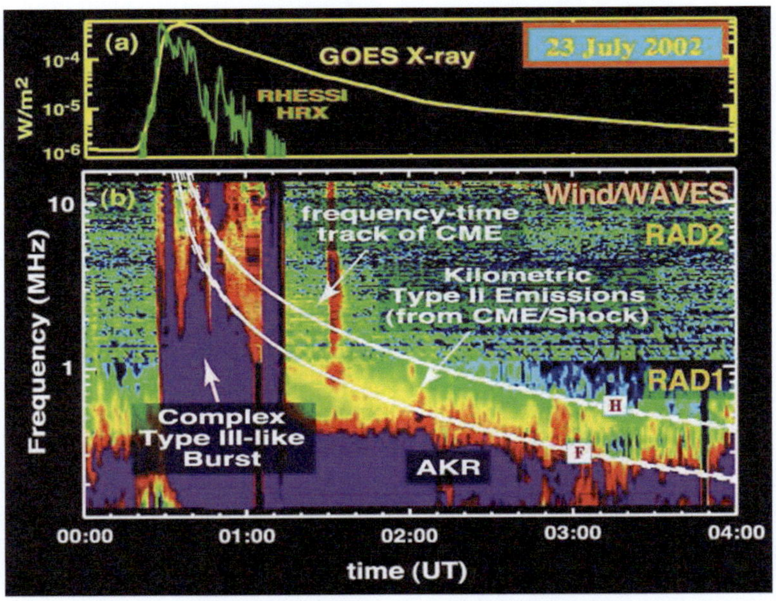

Fig. 22.19 Overview of the X-ray and low-frequency radio emissions associated with the 2002 July 23 X4.8 LDE X-ray event. **a** GOES soft X-ray and RHESSI HXR emissions observed from 00: 00 to 04: 00 UT. **b** Radio dynamic spectrum in the frequency range from 125 kHz to 13.8 MHz, over the same time period, showing the intense, complex type III like emissions (overexposed) associated with the flare and the slowly frequency-drifting type II emissions generated by the propagation of the associated CME through the interplanetary medium. The *curves* on the dynamic spectrum correspond to the frequency-time track of the CME, generating radio emissions at the fundamental and harmonic of the plasma frequency (from Reiner et al. 2007)

that these delays result from the particle transport in the IP medium. From the analysis of the time and places where the energetic particles are released in the corona, Maia et al. (2004) and Klein et al. (2005) showed that the delayed electron events are usually associated with complex and long lasting radio emissions observed in a broad frequency range; the NRH imaging observations show abrupt modifications in the emitting regions

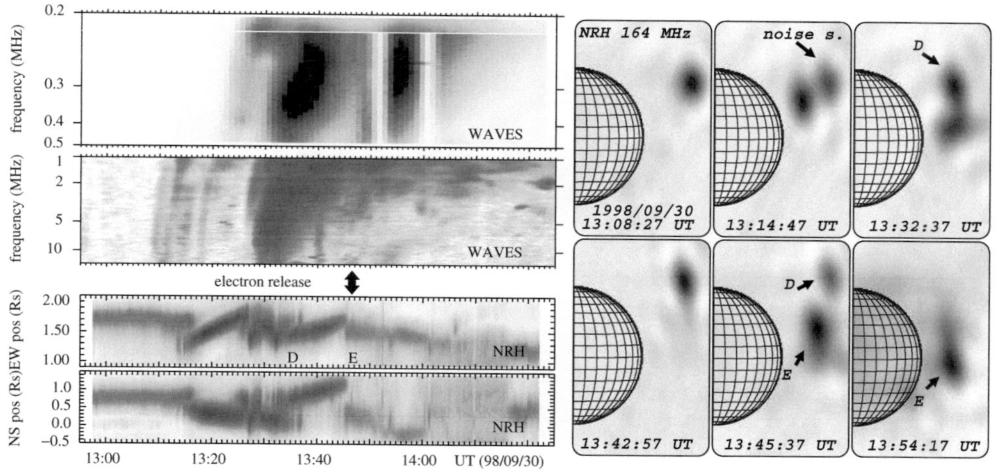

Fig. 22.20 1998 September 30. *Left*: Comparison of the emission observed by the NRH (*bottom*) and WAVES (*top*). The vertical *white bands* seen in WAVES spectrum are due to the saturation. The one-dimensional plot shows a series of radio-sources. D marks a moving continuum, followed by an other one labeled E. The *arrow* indicates the inferred release time at the sun for the electrons with energies above 100 keV. Note that this time coincides with the sudden disappearance of D, the onset of E, and with the low frequency type III burst detected by WAVES. *Right*: Panel of images showing the positions of the sources seen by the NRH (from Maia et al. 2004)

close to the estimated time of the electron release (see Fig. 22.20). These modifications, which are direct signatures of energetic electrons, are related with continua onsets or type II-like features and are also commonly associated with packets of complex type III-like emissions. This association is consistent, as in the previous cases, with an interpretation in which electron acceleration is triggered at different sites by magnetic reconnection in the wake of the CMEs. The electrons are then injected along discrete open magnetic flux structures magnetically connected or not with the spacecraft.

22.5.3 Observations of Energetic Electrons in Radio CMEs and Injection in the Interplanetary Space During a Large SEP Event

Using radioheliograph observations, Bastian et al. (2001) identified expanding loops behind a CME front illuminated by synchrotron radiation from 0.5–1 MeV electrons and called radio CMEs. Another detection of energetic electrons radiating in expanding loops behind CME fronts was performed at higher frequencies (i.e. 432 MHz) by Maia et al. (2007). The progression of the radio loops was followed from a few tenths to more than one solar radius above the solar limb. The expanding loop is associated with a CME which is later detected at higher heights. The radio emission is attributed to incoherent synchrotron radiation due to MeV electrons interacting with the magnetic field of the radio CME loop (0.1 to a few Gauss). Given the brightness distribution of the radio CME, these electrons are most probably accelerated in the aftermath of the rising CME (detected here as a radio loop) in the reconnecting current sheets below the eruptive loop. The event is associated with an in-situ electron event measured by the ACE satellite. The high energy cut-off of the gyrosynchrotron emitting electrons in the loop is found to be of the order of 1 to a few MeV which shows that particles both inside the loops and detected in-situ have comparable energies. Applying a transport model to the electrons detected in-situ showed that not only the inferred onset at the sun but also the duration of the energetic electron release at similar energies are quite comparable for the electrons injected in the radio loop and in the interplanetary medium (Figs. 22.21 and 22.22); electrons are accelerated then released when the CME is still very low in the corona, i.e. a few tenths of radius above the limb. Furthermore, the onset time of relativistic ions, deduced by Bieber et al. (2004) from

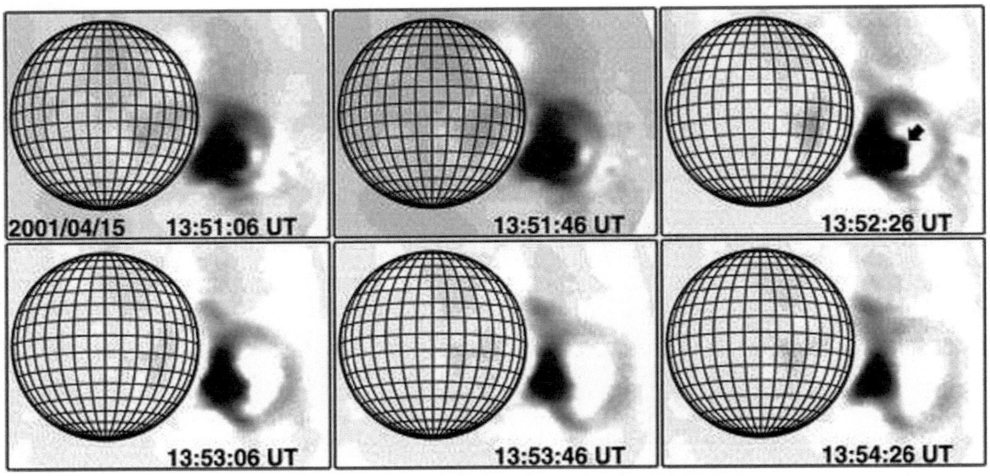

Fig. 22.21 Series of NRH images showing the progression of a radio CME on 2001 April 15. The images are 10 s integrations at an observing frequency of 410 MHz (adapted from Maia et al. 2007)

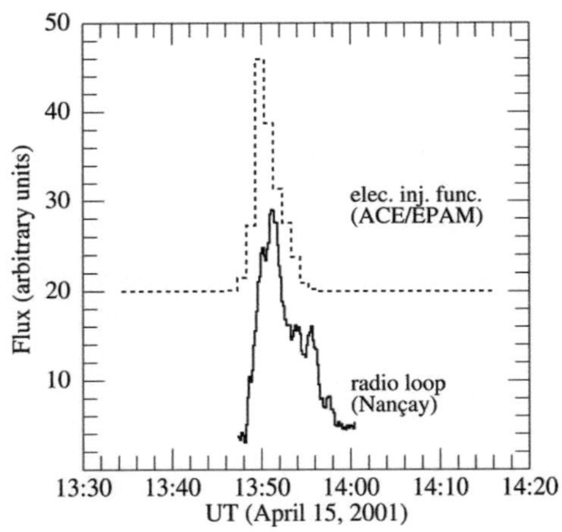

Fig. 22.22 Comparison of the near-relativistic electron injection function and radio loop flux at 432 MHz. The *dashed line* shows the electron injection function that best fits the ACE EPAM electron observations; the *solid line* shows Nançay radioheliograph observations at 432 MHz of the radio loop. Both *curves* agree remarkably well in onset time and duration (from Maia et al. 2007)

ground-based measurements, as well as onset times of ions and electrons at different energies determined by Tylka et al. (2003) are in excellent agreement with the values determined by Maia et al. (2007). These results strongly suggest a similar origin of the electrons injected in the radio loops and of the particles measured in-situ.

22.6 Conclusions

As evidenced in the many observations reviewed in this paper, images of radio bursts performed at high temporal cadence and at several frequencies in the decimetric/metric wavelengths show that the energy release sites and the electron (and more generally particle) acceleration regions are not limited to the flaring active region itself but also imply larger scale structures far away from the flaring site and at different layers of the solar atmosphere. Radio observations provide information about the different possible acceleration sites in the corona: in the flaring active region, but also in successive magnetic interaction sites occurring further and further away from the flare site during the CME development or in the reconnecting current sheets formed below the eruptive flux ropes associated with the CME. While radio observations in the low and middle corona show that the most probable site for efficient electron acceleration in the corona is not related to the shock front, long duration radio emissions (e.g. type IV radiation) rather argue in favor of time and space extended particle acceleration sites in the corona. The opening of the large scale magnetic field due to the passage of a CME allows the escape of the energetic particles to the interplanetary medium. As shown in the previous section, high energy protons and relativistic electrons measured in situ may originate in the low corona and be produced by the same acceleration process as the one producing the relativistic electrons at the origin of the radio CMEs.

As shown in Section 22.3, the combination of HXR images of solar flares with multifrequency images of radio bursts produced in the decimeter to dekameter wavalengths during the propagation of electron beams in the corona allow to trace the magnetic connectivity between the solar surface, the corona and the interplanetary medium. This is one of the key question of the Cosmic Vision Solar Orbiter mission. Such observations are thus crucial complements to the measurements of interplanetary radio emissions and of in situ electron measurements. Together with the existing radioheliographs (such as the NRH) and the future use of non solar dedicated radioheliographs such as LOFAR (de Vos et al. 2009) and MWA (Lonsdale et al. 2009), the development of future ground-based solar dedicated radioheliographs designed to produce images at high spatial resolution and with high-dynamic range over a broad frequency range (e.g. FASR (Frequency Agile Solar Radio Telescope; see e.g. Bastian, in Gary and Keller 2004) and CSRH (Chinese Solar Radioheliograph; see e.g. Yan et al. 2009) projects) is strongly supported in the context of future solar and interplanetary space missions.

Acknowledgements The author acknowledges support from the Centre National d'Etudes Spatiales (CNES) and from the French program on Solar-Terrestrial Physics (PNST) of INSU/CNRS for the participation to the RHESSI project and the support for data analysis. The NRH is funded by the French Ministry of Education, the CNES and the Région Centre.

References

Anderson KA (1958) Phys Rev Lett 1:335
Aschwanden MJ, Benz AO, Dennis BR, Schwartz RA (1995) ApJ 455:347
Aschwanden MJ, Benz AO (1997) ApJ,480:825
Aschwanden MJ, Schwartz RA, Dennis BR (1998) ApJ 502:468
Bastian TS, Pick M, Kerdraon A, Maia D, Vourlidas A (2001) ApJ 558:L65
Benz AO, Barrow CH, Dennis BR, Pick M, Raoult A, Simnett G (1983) Solar Phys 83:267
Benz AO, Grigis PC, Csillaghy A, Saint-Hilaire P (2005) Solar Phys 226:121
Bieber JW, Evenson P, Dröge W, Pyle R, Ruffolo D, Rujiwarodom M, Tooprakai P, Khumlumlert T (2004) ApJ 601:L103
Boischot A, Warwick JW (1959) J Geophys Res 64:683
Bougeret J-L et al (1995) Space Sci Rev 71:231
Cane HV, Stone RG, Fainberg J, Steinberg JL, Hoang S, Stewart RT (1981) Geophys Res Lett 8:1285
Cane HV, Stone RG (1984) ApJ 282:339
Cane HV, Erickson WC, Prestage NP (2002) J Geophys Res (Space Phys) 107:1315
Cane HV, Erickson WC (2003) J Geophys Res (Space Phys) 108:1203
Claßen HT, Mann, G, Klassen A, Aurass H (2003) A&A 409:309
Dauphin C, Vilmer N, Lüthi T, Trottet G, Krucker S, Magun A (2005) Adv Space Res 35:1805
Dauphin C, Vilmer N, Krucker S (2006) A&A 455:339
de Vos M, Gunst AW, Nijboer R (2009) IEEE Proc 97:1431
Frost KJ, Dennis BR (1971) ApJ 165:655
Gary DE, Keller CU (2004) Solar and space weather radiophysics - current status and future developments. In: Gary DE (ed) Center for Solar-Terrestrial Research, New Jersey Institute of Technology, Newark, NJ, USA; Christoph U. Keller, National Solar Observatory, Tucson, AZ, USA. ASTROPHYSICS AND SPACE SCIENCE LIBRARY Volume 314. Kluwer Academic Publishers, Dordrecht
Haggerty DK, Roelof EC (2002) ApJ 579:841
Hey JS (1942) Observatory 64:263
Hey JS (1946) Nature 157:47
Hudson HS (1978) ApJ 224:235
Kahler SW, Cliver EW, Cane HV (1989) Solar Phys 120:393
Klein L, Pick M, Trottet G, Vilmer N, Anderson K, Kane S (1983) Solar Phys 84:295
Klein K-L, Chupp EL, Trottet G, Magun A, Dunphy PP, Rieger E, Urpo S (1999) A&A 348:271
Klein K-L, Schwartz RA, McTiernan JM, Trottet G, Klassen A, Lecacheux A (2003) A&A 409:317
Klein K-L, Krucker S, Trottet G, Hoang S (2005) A&A 431:1047
Klein K-L (2007) In: Bouvier J, Chalabaev A, Charbonnel C (eds) SF2A-2007: Proceedings of the annual meeting of the French society of astronomy and astrophysics held in Grenoble, France, July 2–6, 2007, p 7
Krucker S, Larson DE, Lin RP, Thompson BJ (1999) ApJ 519:864
Kundu MR, Stone RG (1984) Adv Space Res 4:261
Lonsdale CJ et al (2009) IEEE Proc 97:1497
MacDowall RJ, Kundu MR, Stone RG (1987) Solar Phys 111:397
MacNeice P, Antiochos SK, Phillips A, Spicer DS, DeVore CR, Olson K (2004) ApJ 614:1028
Maia D, Vourlidas A, Pick M, Howard R, Schwenn R, Magalhães A (1999) J Geophys Res 104:12507
Maia DJF, Pick M (2004) ApJ 609:1082
Maia DJF, Gama R, Mercier C, Pick M, Kerdraon A, Karlický M (2007) ApJ 660:874
Pick-Gutmann M (1961) Ann Astrophys 24:183
Pick M, Maia D, Wang SJ, Lecacheux A, Hawkins SE (2003) Adv Space Res 32:2527
Pick M, Démoulin P, Krucker S, Malandraki O, & Maia D (2005a) ApJ 625:1019
Pick M, Maia D (2005) Adv Space Res 35:1876
Pick M, Vilmer N (2008) A&A Rev 16:1
Raoult A, Pick M, Dennis BR, Kane SR (1985) ApJ 299:1027
Reames DV (1995) Rev Geophys 33:585
Reames DV (1999) Space Sci Rev 90:413

Reiner MJ, Kaiser ML, Gopalswamy N, Aurass H, Mann G, Vourlidas A, Maksimovic M (2001) J Geophys Res 106:25279
Reiner MJ, Krucker S, Gary DE, Dougherty BL, Kaiser ML, Bougeret J-L (2007) ApJ 657:1107
Roberts (1952) Doctoral thesis, University of Cambridge
Simnett GM, Roelof EC, Haggerty DK (2002) ApJ 579:854
Temmer M, Veronig AM, Vrşnak B, Rybák J, Gömöry P, Stoiser S, Maričić D (2008) ApJ 673:L95
Tousey R (1973) Space Res 13:713
Tylka AJ et al. (2003) Int Cosmic Ray Conf 6:3305
Uchida Y (1960) Pub Astron Soc Jpn 12:376
Vilmer N, Krucker S, Lin RP, The Rhessi Team (2002) Solar Phys 210:261
Vilmer N, Krucker S, Trottet G, Lin RP (2003) Adv Space Res 32:2509
Wild JP, McCready LL (1950) Aust J Phys 3:387
Wild JP, Murray JD, Rowe WC (1954) Aust J Phys 7:439
Wild JP, Sheridan KV, Trent GH (1959a) IAU Symp. 9: URSI Symp. 1: Paris Symposium on Radio Astronomy 9:176
Wild JP, Sheridan KV, Neylan AA (1959b) Aust J Phys 12:369
Wild JP, Smerd SF, Weiss AA (1963) ARA&A 1:291
Yan Y, Zhang J, Wang W, Liu F, Chen Z, Ji G (2009) Earth Moon Planets 104:97

Chapter 23

Coherent Radio Emissions Associated with Solar System Shocks

Iver H. Cairns

Abstract Shock waves are associated with multiple powerful coherent radio emissions within the heliosphere and local interstellar medium. The radio emissions definitely driven by shocks include interplanetary type II (solar radio) bursts,"foreshock" emissions from upstream of Earth's bow shock, and rare emissions from corotating interaction regions (CIRs). Emissions likely driven by shocks, but without definitive observational evidence, include coronal type II bursts, the 2–3 kHz emissions from the outer heliosphere, and drifting pulsating structures from the deep corona. Analogous emissions are also predicted, but not yet observed, for mini-magnetospheres and associated bow shocks on the Moon and for moons like Ganymede, the foreshocks of other planets, particularly Mercury and Jupiter, and supernovae. All these emissions are produced near the electron plasma frequency f_{pe} and/or $2f_{pe}$ via the so-called "plasma emission" mechanism or linear mode conversion, two of the four coherent radio emission mechanisms observed to date. In each case the theoretical interpretation requires coupling of multiple physical processes from microscales to macroscales. Microscale physics includes the time-varying magnetic overshoots of reforming shocks, electron reflection and acceleration at shocks, growth of Langmuir waves in the upstream foreshock, and the linear or nonlinear conversion of Langmuir energy into radio emission at f_{pe} and/or $2f_{pe}$. Intermediate scale physics includes the creation of ripples on the shock on scales of order the decorrelation length of the magnetic field, as well as scattering of the radiation by density irregularities. Macroscale physics includes 3D spatiotemporal variations of the plasma and the shock motion, as well as integration of emission from individual shock ripples over the entire shock. This chapter is a comprehensive review of the field, starting with observations of the emissions definitely and probably driven by shocks. Existing theory is then summarized in some detail, followed by detailed reviews of the observation and theory of type II bursts (both coronal and interplanetary) and the 2–3 kHz emissions from the outer heliosphere, including descriptions of the unresolved issues. The discussion focuses on limitations of the theory and existing observations and ways to address them. The overall conclusions are that the basic theory (electron shock acceleration, development of an electron beam, growth of Langmuir waves, and production of f_{pe} and $2f_{pe}$ radiation for a macroscopic, rippled, shock) appears to explain the primary observations semiquantitatively, that many observational details and theoretical limitations remain unresolved, and that the next ten years ought to be an exciting time that sees theory and observations brought together quantitatively.

23.1 Introduction

Energy releases in plasmas are observed to have multiple effects, including the acceleration or heating of some of the plasma particles and, sometimes, bulk motion of the entire plasma. Via coherent and incoherent processes, accelerated and heated particles can produce radiation across the electromagnetic spectrum, from X-rays to radio waves, that can escape the

I.H. Cairns (✉)
School of Physics, University of Sydney, Sydrey, NSW 2006, Australia
e-mail: i.cairns@physics.usyd.edu.au

M.P. Miralles, J. Sánchez Almeida (eds.), *The Sun, the Solar Wind, and the Heliosphere*, IAGA Special Sopron Book Series 4, DOI 10.1007/978-90-481-9787-3_23, © Springer Science+Business Media B.V. 2011

source plasma and propagate large distances to remote observers. They can also produce plasma waves that do not escape the plasma, such as electrostatic waves near the electron plasma frequency and low-frequency magnetohydrodynamic (MHD) waves, whose energy can be channelled into escaping radiation and into heating and/or accelerating other plasma particles. Shocks are produced if the bulk plasma motion is faster than the group speed of a wave mode subject to nonlinear steepening, waves in this mode are driven by the flowing plasma interacting with another plasma, and the system allows sufficient time for nonlinear steepening into a shock.

Shocks are fundamentally important in laboratory, space, and astrophysical plasmas. They accelerate and reflect some plasma particles, with multiple consequences including: (i) production of high energy particles relevant to space weather, solar flares, and cosmic rays; (ii) particle distributions that drive (iii) high levels of plasma waves via instabilities and (iv) radio emissions, sometimes via coherent processes and sometimes by incoherent processes like gyrosynchrotron emission. Shocks also heat the plasma, in part directly by the steady-state electric field associated with the shock structure (Scudder et al. 1986; Burgess 1995; Hull et al. 1998, 2000; Krasnoselskikh et al. 2002) and in part by damping of waves excited by particles moving through the shock or accelerated by it (Tidman and Krall 1971; Kennel et al. 1985; Burgess 1995; Lembege et al. 2004; Scholer and Matsukiyo 2004; Matsukiyo and Scholer 2006; Hellinger et al. 2007; Yuan et al. 2009; Lembege et al. 2009). Depending on the temperature, heating can produce optical, UV, and X-ray signatures observable remotely. Shocks also introduce spatial and temporal inhomogeneities in the density, velocity, temperature, electric field, and magnetic field in the plasma, often by factors of 4 or more. They also change the plasma's entropy. This chapter focuses on items (i)–(iv) above in connection with coherent radio emissions produced by shocks in our solar system. More details of the physics of shock waves are provided in Section 23.2 below.

Two of the solar system's three most powerful radio emissions are associated with shock waves. The most powerful are the 2–3 kHz radio emissions observed by the Voyager spacecraft: they are believed to be produced beyond the heliopause by a shock wave driven in front of coronal mass ejections (CMEs) and other solar disturbances that have combined into a so-called Global Merged Interaction Region (GMIR) (Kurth et al. 1984; Gurnett et al. 1993; Cairns and Kaiser 2002; Cairns 2004). The heliopause is the plasma discontinuity between the solar wind's "termination shock" and the solar system's "bow shock" (or bow wave) that separates the shocked solar wind plasma from the (likely shocked) plasma of the very local interstellar medium (VLISM) (Zank 1999). Type II solar radio bursts, the weakest of the three, are produced in the Sun's corona and the solar wind (Wild and McCready 1950; Wild et al. 1963; Cane et al. 1982; Nelson and Melrose 1985; Cairns 1986a; Reiner et al. 1998; Reiner 2000; Robinson and Cairns 2000; Cane and Erickson 2005). Interplanetary type IIs are definitely associated with shocks driven ahead of fast CMEs (Bale et al. 1999), while coronal type IIs are believed to be produced by either blast wave shocks (associated with flares) or CME-driven shocks. The last of the three, type III solar radio bursts, are produced in the corona and solar wind by fast streams of electrons accelerated in solar flares and have no known association with shocks (Wild and McCready 1950; Wild et al. 1963; Suzuki et al. 1985; Robinson and Cairns 2000; Cane et al. 2003). At least another five observed or predicted solar system radio emissions are associated with shocks and discussed below. These include the f_p and $2f_p$ radiation associated with Earth's bow shock, formed as a result of the solar wind's interaction with Earth's magnetosphere, which is likely the emission most amenable to definitive observational testing of theory.

The nonthermal radio emissions of interest in this review are produced by coherent mechanisms. One line of evidence for this is provided by the brightness temperature T_b of the radiation being larger than the electron temperature T_e of the emitting plasma. Here

$$T_b = \frac{v_\phi^2}{2k_B f^2} \frac{F(f)}{\Delta\Omega} , \tag{23.1}$$

where v_ϕ is the phase speed of the waves ($\approx c$ except when the wave frequency $f \lesssim 2f_p$), k_B is Boltzmann's constant, $F(f)$ is the flux density (in W m^{-2} Hz^{-1}), and $\Delta\Omega$ is the solid angle of the radio source. The second line of evidence is that the emissions are narrowband, thereby not having the broadband spectrum over orders of magnitude in frequency expected for synchroton and other single-particle emission mechanisms when observed at intensities much larger than the instrumental detection thresholds.

The basic model for coherent radio emissions associated with shocks, which applies to both type II bursts and the 2–3 kHz outer heliospheric radio emissions, is illustrated in Fig. 23.1 in the rest frame of the shock. The plane in Figure 23.1, defined by the upstream plasma flow velocity **u** and magnetic field vector $\mathbf{B}_u$, is a cut of the three-dimensional (3-D) source. The "foreshock" regions are found upstream of the shock but downstream from the magnetic field line tangent to the shock (the so-called tangent line). As explained below, electrons reflected by (or leaked upstream from) the shock are found in the foreshock regions and naturally develop "beam" distributions which are peaked at a non-zero velocity parallel to the magnetic field (more properly the "reduced" distributions, which are integrated over perpendicular velocity space). These beam distributions are unstable to the growth of electrostatic Langmuir waves (Filbert and Kellogg 1979; Cairns 1986a, 1987a, 1987b; Fitzenreiter et al. 1990). Langmuir waves have frequencies close to the electron plasma frequency

$$f_{pe} = \frac{1}{2\pi}\left(\frac{n_e e^2}{m_e \varepsilon_0}\right)^{1/2}, \qquad (23.2)$$

where n_e is the electron number density, e the electron charge, m_e the electron mass, and ε_0 the permittivity of free space. Langmuir waves propagate at relatively low speeds, are reflected by higher density regions, and are subject to significant damping (especially as their wavenumbers increase when entering lower density regions), and so cannot efficiently escape the source region and reach remote observers. Further details on the reflection of electrons from the undisturbed solar wind into the foreshock (and the leakage of heated electrons from downstream of the shock), the development of beam distributions there, and the growth of Langmuir waves are deferred to Section 23.5 below.

The basic model involves coupling of some Langmuir wave energy into radio emission near f_p and $2f_p$, which then propagates to remote observers. The emission processes considered are two of the four known coherent emission processes: so-called "plasma emission" near f_p and $2f_p$ by nonlinear Langmuir wave processes, and so-called "linear mode conversion" (LMC) of Langmuir waves into radio emission near f_p at density gradients. Today the standard processes for the nonlinear plasma emission mechanism are: the electrostatic (ES) decay $L \rightarrow L' + S$ to produce backscattered Langmuir waves L' and ion acoustic waves S from the beam-driven Langmuir waves L; the electromagnetic (EM) decay $L \rightarrow T(f_p) + S'$ to produce radio waves T just above f_p and ion acoustic waves S', stimulated by the ES decay products S; and the coalescence $L + L' \rightarrow T(2f_p)$ of beam-driven L and backscattered L' Langmuir waves to produce radio waves just

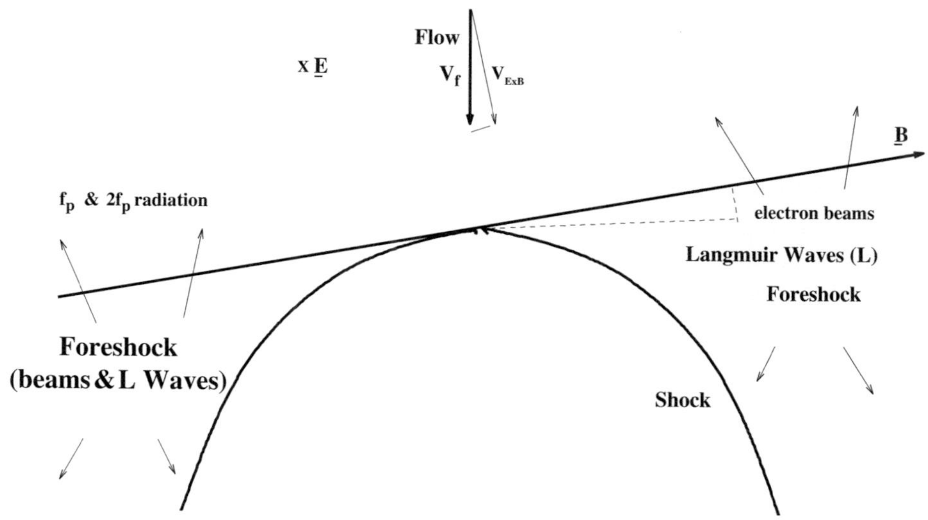

Fig. 23.1 Basic context and model for coherent radio emission associated with shocks, shown in the shock rest frame for a plane defined by the upstream flow velocity **u** and magnetic field vector $\mathbf{B}_u$ that cuts through the 3-D source. Downstream of the magnetic field line tangent to the shock but upstream from the shock, the foreshock regions contain beams of fast electrons (reflected by the shock) which drive high levels of Langmuir waves and radio emissions near f_p and $2f_p$

above $2f_p$ (Cairns and Melrose 1985; Cairns 1988; Robinson and Cairns 1998; Li et al. 2008a, b). These processes conserve frequency (energy) and wavevector (momentum) for the participating waves. In contrast, LMC involves the fact that the L and T modes are not connected in frequency – wavevector ($\omega - \mathbf{k}$) space in a homogeneous plasma, but are connected in $\omega - \mathbf{k}$ space in a density gradient (Budden 1985; Forslund et al. 1975; Yin et al. 1998; Cairns and Willes 2005; Kim et al. 2007, 2008). The connections are restricted to a narrow "radio window" in $\omega - \mathbf{k}$ space as the waves propagate into the density gradient and are subject to damping. For LMC the wave frequency and the component of $\mathbf{k}$ perpendicular to the density gradient are conserved.

Plasma emission is widely favored for f_p and $2f_p$ radiation from Earth's foreshock, type II and III solar bursts, and the 2–3 kHz outer heliospheric emissions (Wild et al. 1963; Kurth et al. 1984; Nelson et al. 1985; Suzuki et al. 1985; Cairns and Melrose 1985; Cairns 1986a, 1988; Gurnett et al. 1993; Robinson and Cairns 1998; Cairns and Robinson 2000; Robinson and Cairns 2000; Knock et al. 2001). Originally discounted because the high observed brightness temperatures were interpreted in terms of nonlinear processes, LMC has enjoyed a renaissance of interest in recent years (Yin et al. 1998; Cairns and Willes 2005; Kim et al. 2007, 2008) and should be considered in detail. Radiation produced at $2f_p$ by the oscillating fields of localized Langmuir wavepackets (Malaspina et al. 2010), proposed in the last year, also needs to be examined fully.

Of course shocks can produce radio emissions by other mechanisms. Incoherent emission of X-ray, UV, optical, and infrared radiation from the regions downstream of shocks, including those associated with stellar winds and bow shocks (which are analogous to the heliosheath region for the 2–3 kHz outer heliospheric emissions) are not considered here. Of interest though is the direct generation of radio emission by "linear" instabilities of the shock-reflected electrons (Wu et al. 1985; Farrell 2001). These mechanisms are very similar to a known fundamental mechanism for coherent radio emissions, the so-called cyclotron maser mechanism (Wu and Lee 1979), but observational evidence for these processes does not yet exist (see Sections 23.3.2 and 23.8 below for more discussion). These mechanisms also appear to require strong tuning of the electron beam parameters to produce radiation near f_p or $2f_p$, rather than over a broadband between these frequencies, and do not appear to be very attractive at this time. In addition, electrons can radiate via the incoherent gyrosynchrotron, synchrotron, or bremsstrahlung mechanisms. An example is the weak gyrosynchrotron emission observed recently from downstream of some CME shocks (Maia et al. 2000; Bastian et al. 2001), while supernova shocks produce radio emission downstream. An aside on this latter context is that the theoretical levels of f_p and $2f_p$ emission produced upstream of supernova shocks should be predicted and compared with observations.

Many of this chapter's topics have been reviewed before, but it appears as though this is the first unified and general review of radio emissions associated with solar system shocks. For instance, previous reviews exist for type II solar radio bursts (Wild et al. 1963; Wild and Smerd 1972; Nelson et al. 1985; Bastian et al. 1998; Reiner 2000; Robinson and Cairns 2000), the 2–3 kHz outer heliospheric radiation (Gurnett and Kurth 1996; Cairns and Zank 2001; Cairns 2004), solar system radio phenomena in general (Wild and Smerd 1972; Gurnett 1995; Cairns and Kaiser 2002), and the radio emission mechanisms (Melrose 1980; Cairns and Robinson 2000). Kellogg (2003) has reviewed Langmuir waves at collisionless shocks. Shock physics and solar system shocks are reviewed elsewhere (Tidman and Krall 1971; Kennel et al. 1985; Russell 1985; Burgess 1995; Lembege et al. 2004; Vrsnak and Cliver 2008).

The foregoing text provides arguments that radio emission associated with shocks involves fundamental plasma physics and is widely important. The chapter proceeds as follows. Section 23.2 summarizes the physics of shocks and identifies unresolved issues that are believed relevant to radio emission. Section 23.3 describes three classes of coherent solar system radio emissions that are definitely observed to be produced by shock waves, including interplanetary type II bursts, focusing on summaries of the observational evidence and the important qualitative theoretical aspects. Section 23.4 describes three classes of observed radio emissions that are believed, but not definitively observed, to be associated with shock waves. Again the focus is on observations and qualitative theoretical aspects. A detailed summary of the standard theory for coherent radio emissions produced by shocks is provided in Section 23.5, together with unresolved issues. Section 23.6 reviews the observations and theory of type II solar radio bursts, focusing on detailed applications of theory to the observations and on unresolved issues. The detailed theory and unresolved observational and theoretical issues for the 2–3 kHz outer heliospheric emissions are reviewed in Section 23.7. The discussion in Section 23.8 focuses on limitations of the theory and on future applications, including to other planetary foreshocks, mini-magnetospheres for moons and unmagnetized planets, and supernova shocks. Section 23.9 contains the chapter's conclusions.

23.2 Relevant Shock Physics

Macroscopic shocks observed thus far in our solar system, whether "travelling" shocks in the solar wind or "bow shocks" that deflect the solar wind or magnetospheric plasma around a planet or moon, respectively, are nonlinearly steepened waves in the fast magnetosonic or whistler modes (Scudder et al. 1986). Since the fast MHD and whistler modes are connected at high frequencies, they are called "fast mode" shocks.

An ideal fast mode shock has discontinuous changes in the plasma parameters across it. Working in the shock's rest frame, upstream of the shock the plasma flows towards the shock at a speed v_u larger than the fast mode speed v_{ms}, which depends on the Alfven speed V_A, sound speed c_s, and angle θ_{bn} between the upstream magnetic field vector $\mathbf{B}_u$ and the (local) normal direction to the shock surface (Russell 1985; Burgess 1995). Downstream of the shock the plasma's flow speed v_d relative to the shock is smaller than v_{ms}. Defining the fast mode or magnetosonic Mach number of a flow with relative speed v_u by $M_{ms} = v_u/v_{ms}$, upstream of the shock $M_{ms} > 1$ exceeds 1 while $M_{ms} < 1$ downstream. Figure 23.2 illustrates the context.

Conservation of mass, momentum, energy, the tangential component of the electric field, and the normal component of the magnetic field across the local shock surface in a fluid (MHD) model leads to the Rankine-Hugoniot or "jump" conditions for the plasma parameters across the shock (Kennel et al. 1985; Melrose 1985; Burgess 1995). For instance, combining the faster flow speed upstream of the shock than downstream (in the shock rest frame) with mass conservation immediately implies that the downstream

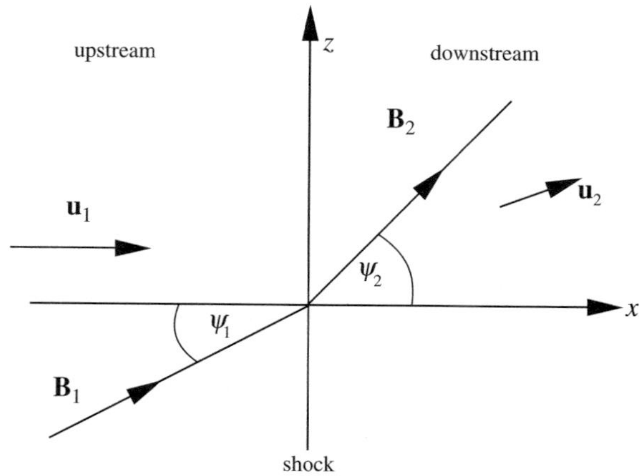

Fig. 23.2 Schematic of the flow velocity and magnetic field vectors upstream and downstream of the shock, in a frame in which the shock is stationary and the upstream flow velocity **u** is oppositely directed to the local shock normal (Burgess 1995). The angle $\psi_1 = 180° - \theta_{bn}$

and upstream densities are related by $n_d/n_u = v_u/v_d$. The increase in density downstream is the counterpart of the flow slowing. It can also be shown that the tangential component and overall strength of the magnetic field and the temperature increase downstream of the shock. The conventional physical picture is that the ram pressure $n_u v_u^2$ of the upstream flow, which is the dominant contribution to the total pressure in the high Alfven and sonic Mach number regime, is primarily balanced by the increased thermal pressure downstream in a time-steady state.

The Rankine-Hugoniot conditions predict that the maximum increase in n_d and B_d (and decrease in v_u) across the shock is a factor of 4 relative to the upstream quantities. In detail these compression ratios (e.g., n_d/n_u or B_d/B_u) increase almost linearly with M_A and M_{ms} for low Mach numbers but asymptote towards 4 once the Mach numbers exceed about 4. However, it is emphasized that the Rankine-Hugoniot analysis does not treat the shock transition region directly but instead assumes that suitable temporal and (macroscopic) spatial averages can be taken such that mass, energy, momentum, the tangential electric field, and the normal magnetic field are conserved across the shock.

The increased strength and more perpendicular orientation of the magnetic field downstream of the shock, compared with upstream, acts like a magnetic mirror. At the simplest level, then, electrons and ions inside the "loss cones" defined by conservation of magnetic moment and energy will go straight through the magnetic mirror and enter the downstream region. Correspondingly, particles outside their loss cone will be reflected back upstream. The mass difference between electrons and ions leads to different gyroradii and it turns out that the thermal electron gyroradius r_{ge} is typically small compared with the shock transition's thickness L, while thermal protons and other ions typically have gyroradii r_{gi} that are comparable to or larger than L. Accordingly, the electrons are well described by standard orbit theory while the ions see the shock as an abrupt discontinuity. This difference in the behavior of electrons and ions leads to a cross-shock electrostatic potential ϕ_{cs}, which affects the electron loss cone and requires a two-fluid (rather than MHD) approach for development of an accurate shock theory. The electron fluid momentum equation can be rearranged to yield

$$\mathbf{E}_{cs} = -\nabla\phi_{cs} = -\frac{1}{en_e}\nabla P_e - \mathbf{v}_e \times \mathbf{B} \,. \qquad (23.3)$$

ignoring the inertia term, where P_e is the electron pressure tensor. In the shock's de Hoffman-Teller frame, in which the electron velocity $\mathbf{v}_e$ is parallel to $\mathbf{B}$ and there is no convection electric field, the $\mathbf{v}_e \times \mathbf{B}$ term vanishes and the cross-shock potential depends only on the electron pressure and density profiles. Under various circumstances (primarily involving the magnetic, density, and perpendicular electron temperature profiles being stepwise continuous and proportional) Eq. (23.3) simplifies to (Hull et al. 1998; Kuncic et al. 2002)

$$\phi_{cs} = 2k_B \Delta T_e = 2k_B T_{e\perp 1}\frac{\Delta B}{B_u} \,, \qquad (23.4)$$

where Δ identifies the change across the shock (from upstream to downstream), k_B is Boltzmann's constant, T_e is the total electron temperature, and $T_{e\perp u}$ is the upstream perpendicular (to $\mathbf{B}$) electron temperature.

The cross-shock potential has fundamental consequences for shocks with high enough M_{ms}. Since the shock heats the electrons (Scudder et al. 1986), ϕ_{cs} must be positive. Accordingly, the cross-shock potential attracts electrons downstream but resists the motion of ions across the shock: indeed, it reflects ions with incident energies less than ϕ_{cs} (Leroy et al. 1982). These reflected ions are then specularly reflected, with their initial reflected normal velocity the opposite of their incoming normal velocity. This reflection represents a significant energization, since the ions start off at a speed $\approx 2u$ relative to the upstream ions, and leads to the ions having a ring in perpendicular velocity space. The ion gyromotion allows the ions to gyrate upstream approximately half a gyroperiod before coming back towards the shock with sufficient energy to surmount the potential and move downstream. The ring is unstable to the growth of waves both upstream and downstream, leading to additional thermalization.

Crucially, the reflected ions have a considerable current, leading to an increase in magnetic field in the "foot" region upstream, an overshoot just downstream of the main ramp, and then a periodic undershoot-overshoot pattern in the magnetic field profile. Figure 23.3 illustrates the characteristic magnetic and potential profiles across a "supercritical" shock for which ion reflection is important.

Electrons moving through the spatially varying magnetic and electric fields of the shock can

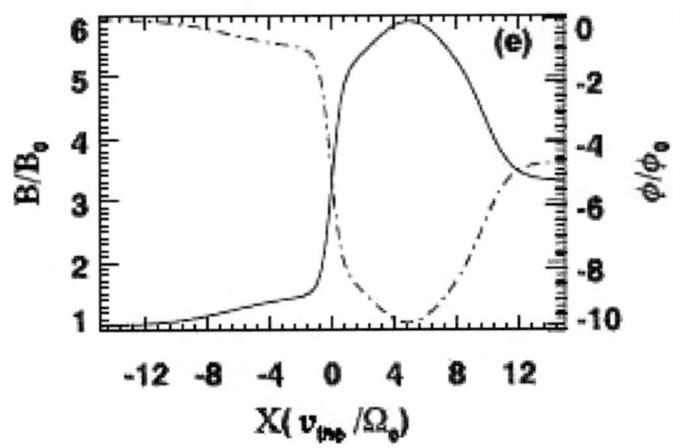

Fig. 23.3 Model spatial profiles for the normalized magnetic field (*solid line*) and cross-shock potential (*dash-dot line*) (Yuan et al. 2008a). The upstream is to the left, with the foot located where $-12 < X < 0$, the ramp centered at $X = 0$, and the overshoot peaking at $X = -5$

be reflected or transmitted, because the increasing magnetic field from upstream to downstream acts as a magnetic mirror that is partially counteracted by the cross-shock potential pulling electrons downstream. Accordingly, as detailed in Section 23.5 below, these profiles are vital in this paper because they determine (1) which (upstream) electrons are reflected back upstream into the foreshock and (2) which (downstream) electrons are able to leak upstream into the foreshock. The cross-shock potential also has a role in heating the downstream electrons and driving waves inside the shock structure, since it creates a void at low energy in the electron distribution function $f(v_{\|},v_{\perp})$ of upstream electrons moving downstream and reflects downstream electrons that attempt to move upstream but have insufficient energy (Scudder et al. 1986; Hull et al. 1998). Waves should then grow to fill in the void and remove unstable features in $f(v_{\|},v_{\perp})$, thereby resulting in at least parallel electron heating. Strong levels of electrostatic waves, presumably driven by these electron-ion and electron-electron drifts, are indeed observed in the ramp and downstream region of the shock (Rodriguez and Gurnett 1975; Onsager et al. 1989). It is recognized that the void feature caused by the cross-shock potential can, when partially relaxed, resemble an electron beam (Scudder et al. 1986): these beams have been observed by Feldman et al. (1983) and Fitzenreiter et al. (2003) in the near vicinity of the shock ramp. However, this beam is rapidly removed by wave growth and associated quasilinear relaxation to produce the "flat top" electron distributions found throughout the macroscopic regions downstream of shocks (Feldman et al. 1983; Scudder et al. 1986; Fitzenreiter et al. 2003). Only quasithermal levels of Langmuir waves are found within the ramps and downstream regions of shocks (Rodriguez and Gurnett 1975; Onsager et al. 1989).

With observable electron beams not produced in the macroscopic regions downstream of shocks (but only in the very near vicinity of the shock ramp), the foreshock region is the natural place to look for electron beams and associated Langmuir waves. These are indeed widely observed throughout the macroscopic foreshock regions. The natural mechanism to develop and recreate beam features in $f(v_{\|},v_{\perp})$ for (upstream) electrons reflected back upstream into the foreshock is detailed in Section 23.4, as is the beam instability which often drives intense Langmuir waves there.

Above it is assumed that the shock structure remains constant in time. Here we focus on the facts that (1) the magnetic overshoot in Fig. 23.2 can significantly exceed the maximum (of 4 relative to B_u) predicted by the Rankine-Hugoniot conditions and (2) the temporal and spatial averages required for the Rankine-Hugoniot analysis do not require a time-steady shock structure. Indeed, for a long time it has been predicted based on simulations that shocks become unsteady at high enough M_A and M_{ms}, undergoing continuous self-reformation processes rather than being constant in time (Leroy et al. 1982; Lembege et al. 1987; Krasnoselskikh et al. 2002; Hellinger et al. 2002, 2007; Yuan et al. 2009; Lembege et al. 2009). The reformation appears to be correlated with the specularly reflected ions as they gyrate first upstream and then downstream (Leroy et al. 1982; Hellinger et al. 2007; Yuan et al. 2009; Lembege et al. 2009), with reformation corresponding to a group of gyrating ions driving waves which steepen into a new shock front upstream of the old one, which then decays. Both the magnetic overshoot and reformation thus appear to be associated with the gyrating ions. Thus far observational evidence of reformation remains elusive and indirect (Lobzin et al. 2007).

An important aspect of reformation is that the magnetic compression ratios need not be limited by the Rankine-Hugoniot conditions but can instead be larger at some stages of the reformation cycle: this would lead to cyclic variations in the properties and number of electrons being reflected back (and leaked) into the foreshock. Such cyclic variations of the reflected electrons have been demonstrated for reforming shocks and shown to be material (e.g., factors of 2–4 in the reflected fraction, depending on the shock parameters) and likely important to type II bursts (Yuan et al. 2007, 2008b).

The foregoing review of shock physics has only touched on some aspects of active research. Other current research addresses the physics of reformation for fast mode shocks and the nature of the dissipation and wave growth processes at them (Lembege et al. 2004; Scholer and Matsukiyo 2004; Matsukiyo and Scholer 2006; Hellinger et al. 2007; Yuan et al. 2009; Lembege et al. 2009), but is not reviewed here.

23.3 Definite Shock-Driven Radio Emissions

This section presents observational data and associated theoretical interpretations for three classes of solar system radio emissions that have been definitively associated with shocks: interplanetary type II bursts, radiation from Earth's foreshock, and radiation from some shocks bounding corotating interaction regions (CIRs).

23.3.1 Interplanetary Type II Bursts

Interplanetary type II bursts were discovered by Cane et al. (1982) and immediately interpreted in terms of f_p and $2f_p$ radiation associated with a shock. Figure 23.4 provides an example of an interplanetary type II and illustrates several important aspects. First, interplanetary type IIs are often very intermittent and fragmented, rather than continuous. Second, these bursts often have a fundamental (or F) band (f_p radiation) and a harmonic (H) band ($2f_p$ radiation) during the event as a whole, but often only one band or indeed no emission is observable at a given time. Figure 23.4 shows a clear H band only around 1800 on 25–26 August, with evidence for both F and H emission only prior to about 0600 on 25 August.

The third, and perhaps most important, aspect is that the type II emission is well organized by the lines labelled F and H in this $1/f - t$ dynamic spectrum (Reiner et al. 1998; Reiner 2000; Lobzin et al. 2008). These lines are defined by assuming that: (i) the solar wind density obeys $n_e(r) \propto r^{-2}$, as expected for a constant speed, time-steady, spherically symmetric wind; (ii) the shock speed v_{sh} is constant, and (iii) the emission is at $f_p(r(t))$ and $2f_p(r(t))$ upstream of the shock. Then the shock location is $r(t) = r_0 + v_{sh}(t - t_0)$, where r_0 and t_0 are the heliocentric distance and time when the emission starts, and it is easy to show using Eq. (23.1) that (Reiner et al. 1998; Reiner 2000)

$$\frac{1}{f(t)} = \frac{1}{mf_p(r(t))} = Av_{sh}(t - t_0)\,. \qquad (23.5)$$

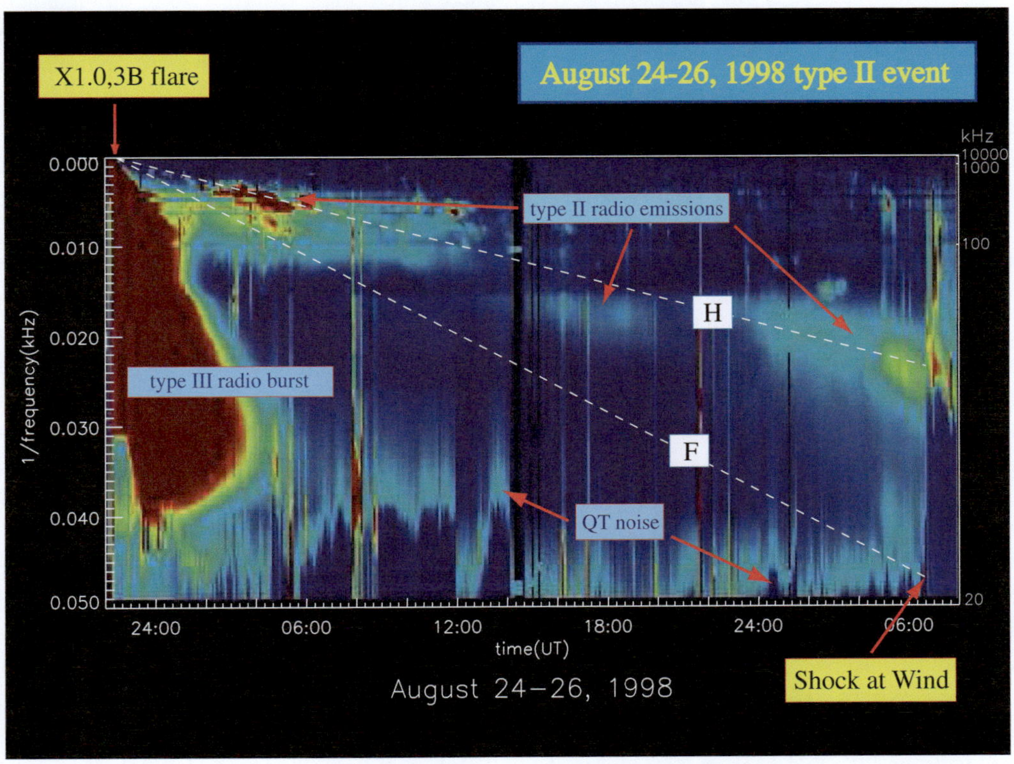

Fig. 23.4 Dynamic spectrum in $1/f - t$ space of an interplanetary type II burst, as well as an interplanetary type III burst and quasithermal plasma noise at f_p (Reiner 2000). The type II radiation is well organized by the two lines F and H discussed in the text. The type II shock arrived near 06:40 on 26 August

Here $m = 1$ or 2 for F or H radiation, respectively, and the constant A is determined by the density at 1 AU when the shock arrives at Earth: the only free parameter then is v_{sh}. Thus, the organization of the type II radiation along two straight lines in Fig. 23.4's $1/f - t$ dynamic spectrum provides a very strong argument that the radiation is indeed produced at f_p and/or $2f_p$ upstream of the shock wave. Moreover, the slope of the line provides a way to estimate v_{sh}.

Definitive evidence that interplanetary type II bursts are produced upstream of a shock, in foreshock regions filled with electrons reflected and accelerated at the shock and with Langmuir waves driven by the electrons, is provided by Fig. 23.5 (Bale et al. 1999). The bottom two panels show clearly the shock's magnetic and density signatures, with the downstream region entered about 0640:30 UT. The top panel is a $f - t$ dynamic spectrum of the electric field fluctuations detected by the Wind spacecraft's WAVES instrument: it shows strong levels of radio waves just above the local f_p and near $2f_p$ (prior to 0638:30 UT), intense Langmuir waves near f_p from about 0638:30 UT until

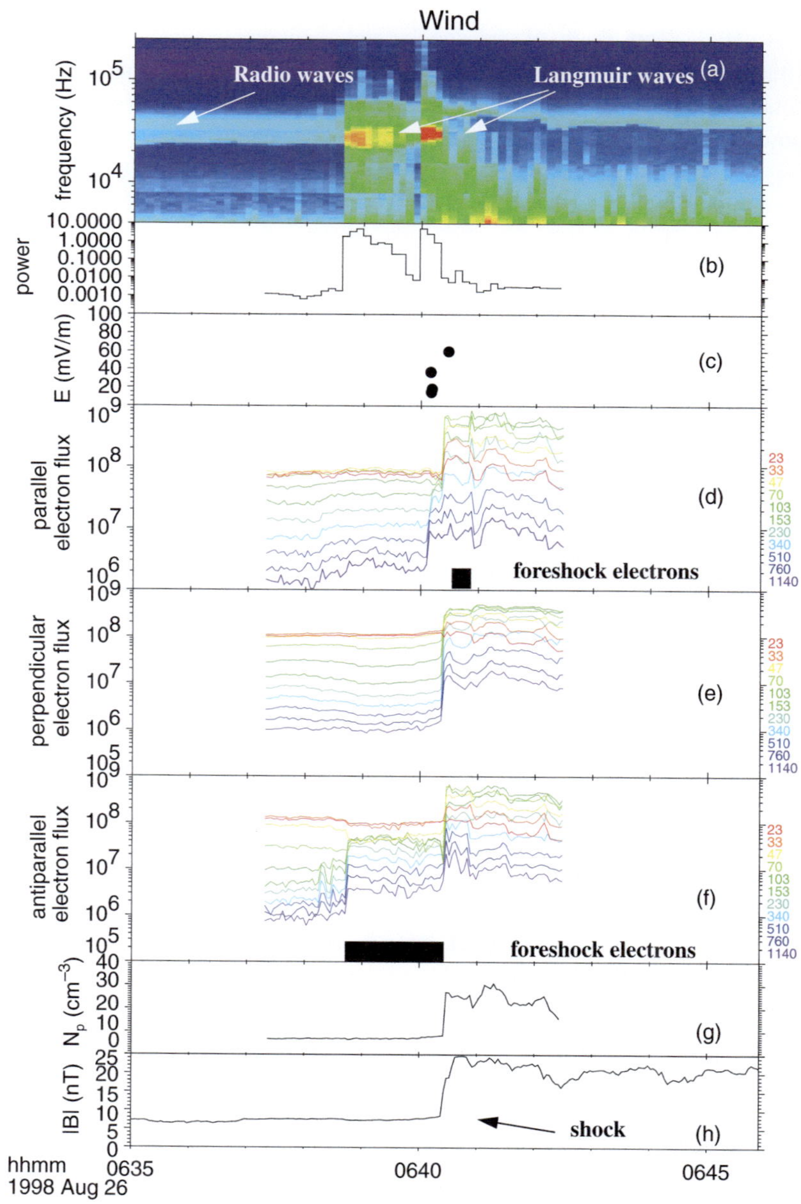

Fig. 23.5 Detailed radio and in situ observations of the source region of an interplanetary type II burst on 26 August 1998 (Bale et al. 1999). Described more in the text, the panels from top to bottom contain **a** an electric field spectrogram, **b** the power in Langmuir waves, **c** the times of TDS capture events, **d** –**f** the fluxes in different energy bands (colors) of electrons moving parallel, perpendicular, and anti-parallel, respectively, to **B**, and **g** and **h** n_e and B, respectively, as functions of time

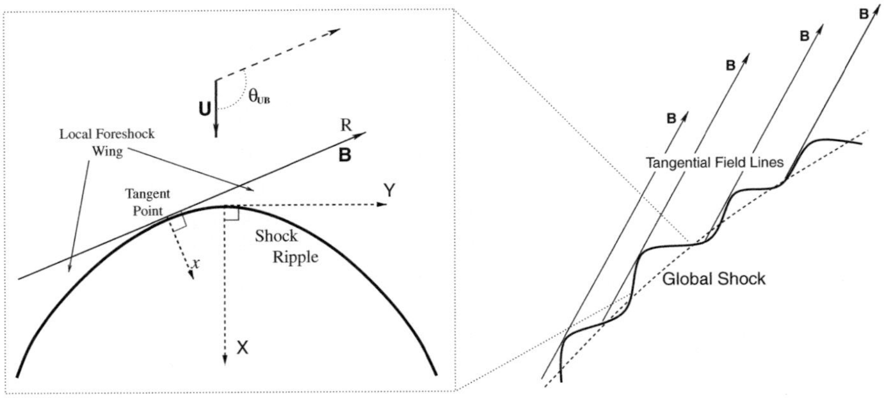

Fig. 23.6 Schematic (Knock et al. 2003b) that (*left*) defines the parameters of a ripple and (*right*) shows how multiple ripples on the macroscopic shock can explain qualitatively the electron and Langmuir observations in Fig. 23.5

0640:30 UT upstream of the shock, and then quasithermal plasma noise at the downstream f_p after the shock. The next two panels show the Langmuir wave power and the times of particularly intense Langmuir waves captured by the Time Domain Sampler subinstrument, respectively. The fourth to sixth panels show the fluxes of electrons with different energies (colored lines) moving parallel, perpendicular, and anti-parallel to **B**, respectively. Large increases in the flux of electrons moving parallel and anti-parallel to **B** are observed in the foreshock, consistent with reflection back upstream of accelerated electrons, while the simultaneous increase in perpendicular electron fluxes at all energies marks the shock transition (and the observation of the denser, shock-heated downstream electrons). During the upstream period **B** remained essentially constant in direction.

In Fig. 23.5 it is crucial to note the absence of three vital signatures in the downstream region. First, there are no strong (or even significantly nonthermal) Langmuir waves downstream of the shock (e.g., after 0640:30 UT). Second, no radio emissions at twice the downstream f_p (at about four times the upstream f_p) are observed. Third, changes in the downstream electron fluxes tend to occur simultaneously in all three panels, thereby providing no evidence for any processes that might produce beams of electrons in the downstream region rather than just the shock-heated downstream electrons. Thus, for this event at least, there is no evidence for any processes that might produce f_p or $2f_p$ radiation downstream of the shock.

It should be questioned how the shock can produce (Fig. 23.5) enhanced levels of upstream electrons moving anti-parallel to **B** for several minutes at the spacecraft and then simultaneously have electrons reach the spacecraft moving parallel to **B**, with **B** essentially constant in direction for the whole period. A planar shock cannot produce this signature, so Bale et al. (1999) interpreted the signature in terms of the shock having a ripple or indentation (Fig. 23.6), so that as the shock moved outward the spacecraft observed electrons first from one side of the ripple and then from both sides just before the shock crossing.

This ripple is evidently at intermediate scales between the scale of the macroscopic shock (of order 0.2 – 1 AU) and those for the microphysics of the shock and the electron – Langmuir wave – radiation interactions. Bale et al. (1999) estimated that the ripple had a characteristic height (or distance along the radial direction) in the directions parallel and antiparallel to **B** of $\approx 1.4 \times 10^8$ m and $\approx 2.5 \times 10^7$ m, respectively. More accurate analyses of the Bale et al. shock and two others yield (Fitzenreiter et al. 2003; Pulupa and Bale 2008) heights and distances along $\pm$**B** in the ranges $(1-7) \times 10^7$ m and $(8-14) \times 10^7$ m, respectively. A recent theoretical model for type IIs (Knock et al. 2003b; Knock and Cairns 2005; Cairns and Knock 2006) assumes that the entire shock has ripples with characteristic scales of order the decorrelation length l_b of the magnetic field in the solar wind, $\approx 10^9$ m or 0.01 AU at the orbit of Earth. Intuitively the shock should have ripples on this scale since the shock surface will depend sensitively on the spatially-varying fast mode speed, which will certainly vary on scales of order l_b. Observations show such ripples (Neugebauer and Giacalone 2005).

Other inhomogeneities and structures exist in the solar wind, including shocks associated with previous CMEs or outbursts and those associated with corotating interaction regions. Interactions between a fast CME-driven shock driving a type II burst and other shocks that are travelling more slowly, or more likely the regions with increased numbers of accelerated and heated electrons associated with these earlier shocks, might then be expected to lead to more intense type II emission and fine structures. Gopalswamy et al. (2001, 2002) have interpreted the fine structures observed in interplanetary type IIs with interactions of a faster CME-driven shock with a previous, slower, CME-driven shock ("CME cannibalism"), while Reiner et al. (1997, 2001) and Hoang et al. (2007) have interpreted observed fine structures in terms of a CME-driven shock interacting with CIR shocks. Theoretical calculations also demonstrate that fine structure should be produced by the shock moving through such interaction regions and the spatially inhomogeneous corona and solar wind (Knock and Cairns 2005; Cairns and Knock 2006), as shown in Section 23.6 below. Returning to the observations, the type II – CIR associations of Reiner et al. (1997, 2001) and Hoang et al. (2007) are based on "direction finding": as a spacecraft rotates, an antenna in the spin plane detects a time-varying, periodic, signal, which maximises when the antenna is perpendicular to the source direction (in the spin plane) for a radio signal. Direction finding with two widely separated spacecraft allows the (assumed common) radio source to be triangulated and then associated with the predicted location of the CIR.

More generally theory predicts that the properties of the accelerated electrons, Langmuir waves, and radio emissions should depend on the shock speed and 3-D locus, the spatially varying **B** field, and the properties of the spatially varying upstream plasma, e.g., n_e, $\mathbf{u}$, T_e, and T_i (Holman and Pesses 1983; Cairns 1986a; Knock et al. 2003a, b). Thus, solar wind inhomogeneities should induce fine structures into the type II burst (Knock et al. 2003b; Knock and Cairns 2005). The degrees to which inhomogeneities and other structures in the solar wind, as opposed to ripples and 3-D inhomogeneities in the macroscopic shock, induce the observed fine structures and variabilities in type II bursts will be addressed partially in Section 23.6 below and in future work.

Until quite recently the interplanetary type II in Fig. 23.5 and Bale et al. (1999) was the only one whose source region had been definitively observed by spacecraft and analysed in detail. Recently, however, Fitzenreiter et al. (2003) and Pulupa and Pulupa and Bale (2008) presented observations of upstream Langmuir waves and reflected electrons for five additional events, at least two with local type II emission. Fitzenreiter et al. (2003) found beams in the upstream reduced electron distribution for 4 events. Pulupa and Bale (2008) showed for three events (2 plus the August 1998 type II) that the reflected electrons have weak loss cone features and bumps in the reduced electron distribution function (integrated over perpendicular velocity space), and obey inverse velocity dispersion (i.e., faster electrons arrive before slower electrons). Furthermore, in all three cases the electron observations were consistent with the shock surface being rippled, with electrons being detected both parallel and anti-parallel to **B** in the original Bale et al. event and one other. While the three observed events may be anomalous, this would be most unlucky and the observed properties are consistent with those for Earth's foreshock, described next. Accordingly the simplest interpretation is adopted here: that the type II source observations to date are typical. Moreover, to the best of this reviewer's knowledge, significantly nonthermal Langmuir waves are never observed downstream of shocks (except perhaps sometimes in the shock ramp itself as the beam induced by the cross-shock potential is relaxed into a flat-top distribution – see Section 23.2 for more detail), but instead only upstream.

23.3.2 f_p and $2f_p$ Radiation from Earth's Foreshock

Radio signals at twice the solar wind plasma frequency near Earth were first observed in 1974 and have been studied extensively since (Dunckel 1974; Gurnett 1975; Hoang et al. 1981; Cairns and Melrose 1985; Cairns 1986b; Burgess et al. 1987; Reiner et al. 1997; Kasaba et al. 2000). The radiation was interpreted almost immediately by analogy with type II and III bursts in terms of electrons accelerated at the bow shock and then associated Langmuir waves and radio emissions at $2f_p$. Cairns (1986b) first claimed the existence of f_p radiation and showed definitively that the $2f_p$ signals are transverse electromagnetic radiation in the free-space modes, using (1) the simultaneous observation of harmonic radiation from regions with different densities present simultaneously in the foreshock, (2) the persistence of these signals into the undisturbed solar wind and other regions of the foreshock, (3) the inability of electrostatic modes to propagate significant distances, and (4) the free-space mode being the only essentially undamped mode significantly above f_p in a stable plasma like the solar wind (Fig. 23.7). He also showed that occasionally radiation at $3f_p$, $4f_p$, and $5f_p$ is observable, as subsequently claimed for a type II solar radio burst (Kliem et al. 1992). Burgess et al. (1987) showed definitively that f_p radiation is also produced in Earth's foreshock (Fig. 23.8), using the same arguments as (1)–(5) above but for f_p radiation. At times, then, f_p and $2f_p$ radiation from different parts of the foreshock are observable.

Another argument for the radiation source being the (3-D) foreshock is based on "direction finding", as explained above for interplanetary type IIs. Early workers (Hoang et al. 1981; Cairns 1986b) used a single spacecraft to establish consistency with the foreshock, while Reiner et al. (1997) used direction-finding with two spacecraft to triangulate the radio source and found it to be in the foreshock.

A final argument for $2f_p$ radiation being generated in the foreshock comes from the observed spatial variations in intensity. Specifically, Lacombe et al. (1988) and Kasaba et al. (2000) found that the $2f_p$ radiation has its largest fields relatively close to the upstream boundary of the foreshock, in approximate coincidence with the largest fields of Langmuir waves, and not in the solar wind or the deep foreshock. This is expected if the radio source lies in the foreshock and is associated with the Langmuir waves or their driving electron beams.

The foreshock regions of Earth's bow shock are observed routinely and found to contain high levels of Langmuir waves and energetic electrons streaming away from the bow shock (Filbert and Kellogg 1979; Anderson et al. 1981; Bale et al. 1997; Cairns et al. 1997; Bale et al. 2000). When the solar wind properties and shock location are almost con-

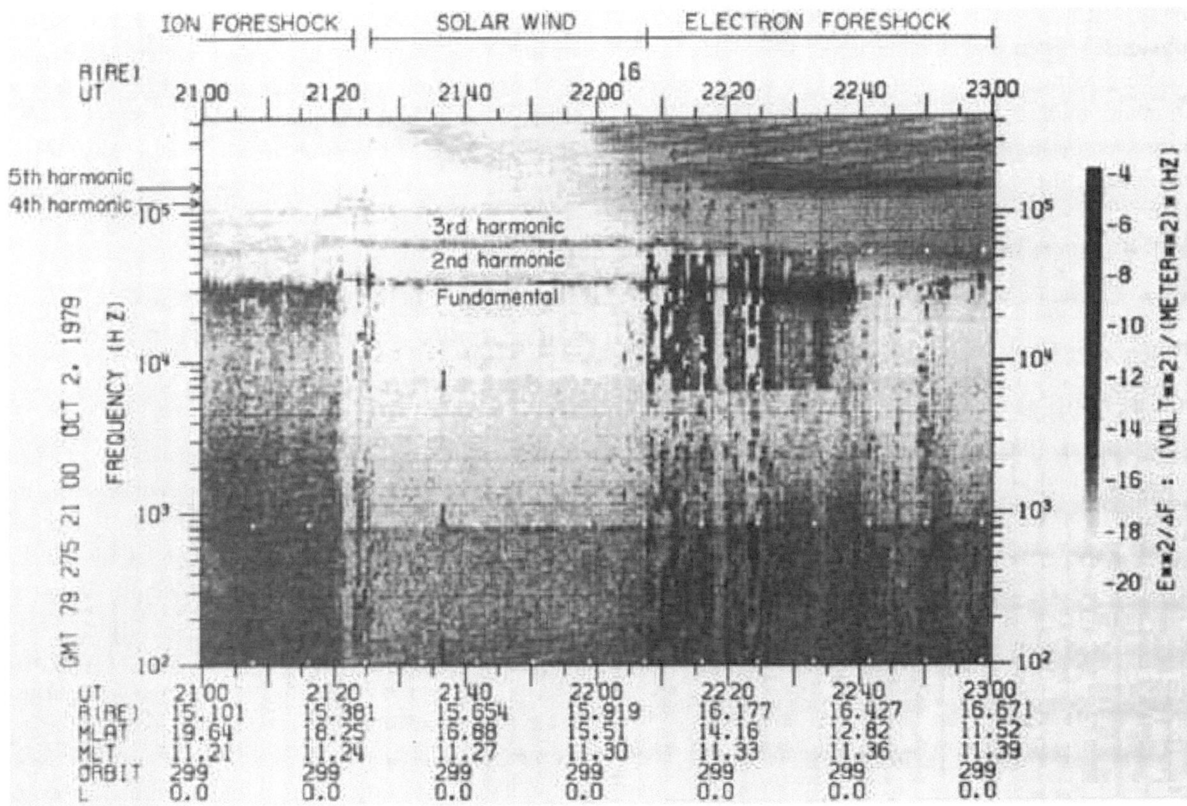

Fig. 23.7 Electric field spectrogram of ISEE-1 spacecraft data clearly showing f_p, $2f_p$, and $3f_p$ radiation in the solar wind (2 October 1979) and the foreshock (Cairns 1986b). Fine structures in the $2f_p$ radiation are associated with regions of different density being simultaneously present in the foreshock

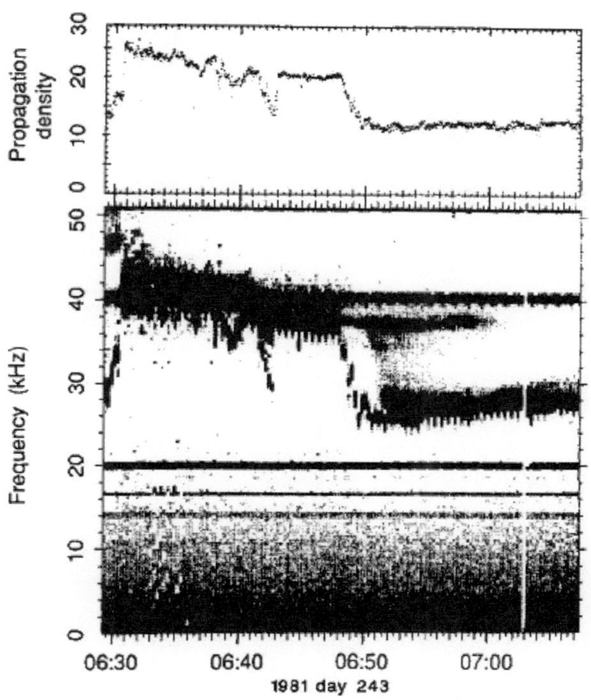

Fig. 23.8 Electric field spectrogram of ISEE-1 spacecraft data showing fine structures in f_p and $2f_p$ radiation associated with regions of different density being simultaneously present in the foreshock (Burgess et al. 1987). This establishes the electromagnetic free-space mode character of the f_p signals

stant in time a spacecraft entering the foreshock from the solar wind first observes small numbers of very energetic electrons and then larger numbers of slower electrons. This "inverse" velocity dispersion, meaning fast electrons closest to the upstream foreshock boundary and slow ones well downstream from the boundary, is characteristic. It leads naturally to beam distributions of electrons (Fig. 23.9), as observed in foreshock regions where the relative beam number density is highest and the beam speed varies slowly with position (Fitzenreiter et al. 1984; Fitzenreiter et al. 1990). The velocity dispersion and development of electron beams is explained physically in terms of "cutoff" and "time-of-flight" effects in Section 23.5 below, based on the research of Filbert and Kellogg (1979) and Cairns (1987a). The electron beams show clear evidence of loss cone effects, consistent with conservation of the magnetic moment in the reflection process at the shock, which is sometimes called fast Fermi acceleration (Wu 1984).

The properties of the Langmuir waves also vary with position relative to the foreshock boundary: the waves are quasithermal in the solar wind but their average fields first increase with increasing penetration into the foreshock, rising to a peak downstream of but relatively close to the foreshock boundary, before decreasing with increasing distance downstream from the boundary. The waves are intrinsically bursty, with fields that vary rapidly in time at a given location and also vary rapidly with position. Some wavepackets have very strong fields of order 40 mV/m although fields of order $0.01 - 10$ mV/m appear more typical (Anderson et al. 1981; Bale et al. 1997; Cairns et al. 1997; Bale et al. 2000; Kasaba et al. 2000; Henri et al. 2009; Malaspina et al. 2009, 2010). Very recent observations from the STEREO spacecraft suggest that some Langmuir wavepackets are actually eigenstates of density depressions (Ergun et al. 2008; Malaspina et al. 2009, 2010), while others are beam-driven wavepackets. The threshold field for the ES decay $L \rightarrow L' + S$ depends on the beam speed v_b and its spread, as well as on other properties, thereby varying with location throughout the foreshock. Values between 0.1 and 20 mV m^{-1} have been calculated for various locations in the foreshock (Robinson and Cairns 1995; Cairns et al. 1997), thereby suggesting that many observed wavepackets are above the ES decay threshold and that the ES decay should proceed. Very recent observations and associated simulations for type III bursts, complementing earlier observations and theory (Cairns and Robinson 1995), provide an independent argument by analogy that ES decay pro-

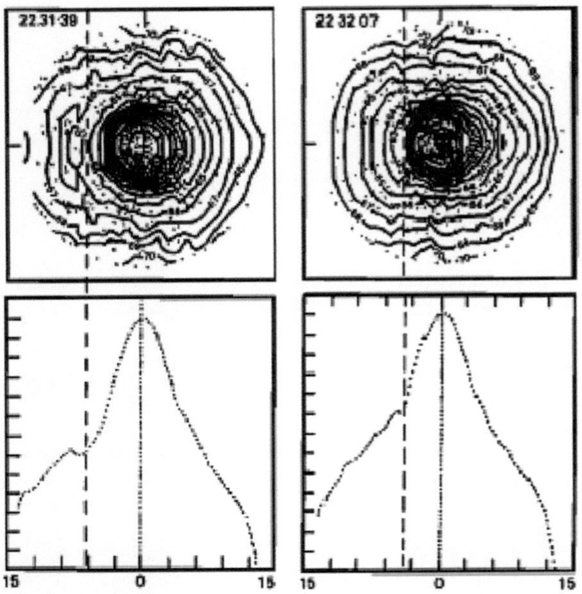

Fig. 23.9 Examples of electron beams observed in Earth's foreshock by the ISEE-1 spacecraft (Fitzenreiter et al. 1990): the *top panels* show the gyrotropic 2D distributions $f(v_{\parallel}, v_{\perp})$ and the *bottom panels* the reduced distributions $f_r(v_{\parallel})$. *Dot symbols* show the locations of the phase space regions sampled and contours are spaced by 1 in units of $\ln f$. Note the clear evidence for loss cone features in $f(v_{\parallel}, v_{\perp})$ and for beams in $f_r(v_{\parallel})$ for these well-chosen examples

ceeds in Earth's foreshock for wavepackets with large enough fields (Henri et al. 2009).

The waves have field statistics consistent with quasithermal plasma noise in the solar wind and thermal waves subject to an instability just inside the foreshock boundary (Cairns et al. 2000), before then developing lognormal statistics consistent with stochastic growth theory (SGT) deeper in the foreshock (Cairns and Robinson 1997, 1999; Sigsbee et al. 2004; Krasnoselskikh et al. 2007). When ES decay is important, the probability distribution of wave (envelope) fields should be reduced below the lognormal prediction at fields above the threshold. While this has been observed for Langmuir waves in the source regions of type III bursts it has not been observed in the foreshock statistics. Nevertheless, since an SGT state is very close to marginal stability, the observed field statistics (Fig. 23.10) provide strong justification for assuming that the electron distribution is close to marginal stability, that SGT applies for fields below the ES decay threshold, and that the ES decay proceeds as needed for the radio emission processes.

Plotted in Fig. 23.10 (left) are the electric fields of bursty Langmuir waves in Earth's foreshock as a function of the distance D_f of the spacecraft downstream from the tangent magnetic field line measured along the solar wind velocity, otherwise known as the DIFF parameter of Filbert and Kellogg (1979). The intrinsic burstiness of the waves is clear. On the right is the probability distribution $P(X)$ of the scaled wave electric field variable $X = (\log E - \mu(D_f))/\sigma(D_f)$, where each wave field sample E is compared with the mean value $\mu(D_f)$ and standard deviation $\sigma(D_f)$ of the logarithm of the field at the location D_f for the sample. SGT predicts that E should be lognormally distributed, so $\log E$ should be Gaussian distributed and X should be Gaussian distributed with zero mean and unit standard deviation. Both the wave fields and distribution $P(X)$ are for a period when the solar wind was unusually stable and the spacecraft's time-varying location in the foreshock could be calculated with great certainty (Cairns et al. 1997). Power-law mod-

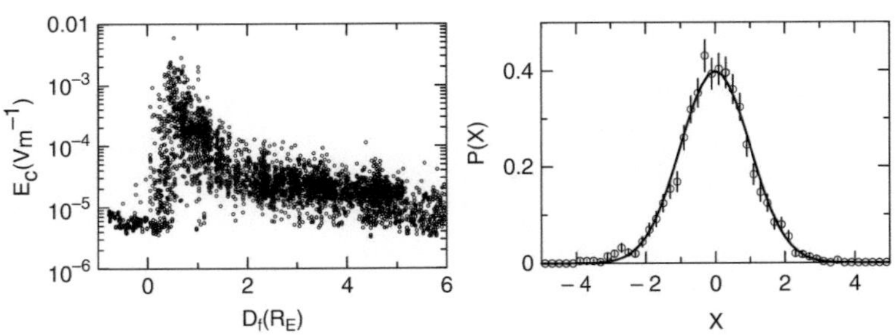

Fig. 23.10 (*Left*) Bursty Langmuir wave fields measured as a function of the distance D_f in Earth's foreshock (Cairns and Robinson 1999). (*Right*) Corresponding field statistics for Langmuir waves in Earth's foreshock (Cairns and Robinson 1999), presented in terms of the probability distribution $P(X)$ described in the text. The *solid line* shows the SGT prediction. Excellent agreement is apparent

els for $\mu(D_f)$ and $\sigma(D_f)$ were obtained, consistent with the raw field data, by minimizing χ^2 between the predicted and observed distributions $P(X)$ and so placing strong constraints simultaneously across most of the foreshock. Excellent agreement with the SGT prediction is apparent in Fig. 23.10 (right). Subsequent work shows similar results (Sigsbee et al. 2004; Krasnoselskikh et al. 2007). The small differences (Krasnoselskikh et al. 2007) are attributable to inadequate consideration of spatial variations in $\mu(D_f)$ and $\sigma(D_f)$ and to sampling on fast timescales commensurate with or smaller than the timescales of the stochastic fluctuations in the growth rate considered explicitly by SGT (Robinson et al. 1993; Cairns and Robinson 1997). SGT appears to apply widely in space and astrophysical wave phenomena (Robinson et al. 2006).

Other mechanisms than plasma emission have been proposed for foreshock f_p and $2f_p$ radiation. For instance, Wu, Gaffey, Farrell, and others (Wu et al. 1985; Farrell 2001) proposed cyclotron maser emission from a relativistic ring-beam distribution of electrons leaving the shock. This mechanism is predicted to produce radiation with fine structures spaced by close to the electron cyclotron frequency f_{ce}. The mechanism also requires careful tuning of the beam parameters (including rather large perpendicular speeds for the ring) for the radiation to be close to $2f_p$ rather than near $1.5f_p$ or other frequencies. This tuning and the typical absence of radiation except very close to f_p and $2f_p$ are strong arguments against such cyclotron maser mechanisms being relevant for foreshock radiation, as is the absence of a detailed theoretical mechanism for the ring-beam. The reason for making this last point is that while mirror reflection indeed yields a loss-cone, it appears that the conditions for significant emission requires the ring-beam to be centered near pitch-angles of about 45° (Farrell 2001) whereas mirror reflection for typical shocks yields loss-cone angles less than 20° (Yuan et al. 2008a).

When the electron cyclotron frequency f_{ce} is unusually large ($\geq$ 400 Hz) fine structures regularly spaced by $f_{ce}/2$ are sometimes observed in f_p radiation and the corresponding electrostatic waves (Cairns 1994). Intuitively these phenomena might lead to $2f_p$ radiation with fine structures spaced by f_{ce}, but this has not yet been observed for the foreshock's $2f_p$ radiation: if so, then this splitting may be relevant to split-band fine structures in coronal type II bursts (Cairns 1994), as discussed in Section 23.4.1.1 below. It is noted that the loss cone features expected on the reflected electron distribution (Fitzenreiter et al. 1990), due to conservation of the magnetic moment in the reflection process, can lead theoretically to fine structure near f_{ce} in electrostatic waves below f_p when the electron beam speed is very slow (Lobzin et al. 2005). These theoretical calculations have not been extended yet to the faster electron beams with $v_b/V_e \geq 3 - 20$ of primary interest here. This should be done, focusing on frequency fine structures in the Langmuir waves caused by loss cone features that might survive into radio emissions near f_p and $2f_p$ produced by nonlinear processes, as well as linear instabilities for the radio waves. Similarly, more observations are encouraged of fine structures in f_p and $2f_p$ radiation and associated electrostatic waves near f_p for Earth's foreshock and other radio emissions associated with shocks. The reason is that these observations may place strong constraints on the mechanisms responsible for the radiation. It is probably worthwhile stating here, though, that multiple mechanisms may exist for f_p and $2f_p$ radiation, with some occurring under some conditions but not others. Accordingly, different radio emissions may have different detailed mechanisms, while a single source might have different mechanisms active simultaneously, e.g., in distinct spatial regions of the source.

Finally, very recently Malaspina et al. (2010) calculated the $2f_p$ emission resulting from the nonlinear currents of Langmuir eigenstates, finding that it may be significant and even dominate the nonlinear Langmuir processes considered above. This novel mechanism needs to be evaluated in detail.

In summary, strong evidence exists that Earth's f_p and $2f_p$ radiation are associated with high levels of Langmuir waves driven by electron beams (perhaps with loss cone features) produced at Earth's bow shock. Direct generation of the radiation appears implausible at this time, leaving nonlinear Langmuir processes and linear mode conversion as the favored mechanisms.

23.3.3 Radio Emission from Corotating Interaction Regions

Hoang et al. (1992) presented Ulysses observations of Langmuir waves and radio emissions associated with

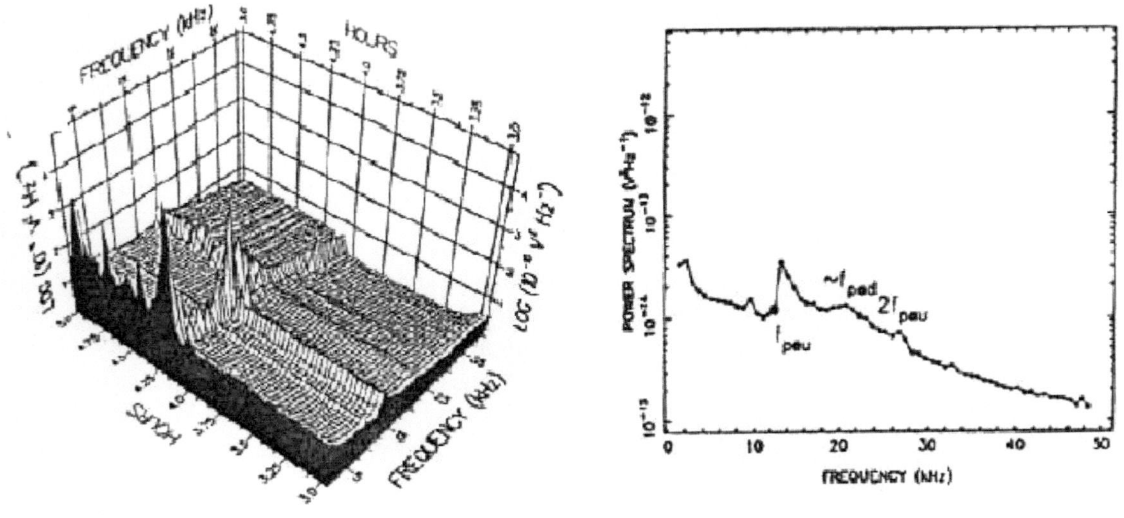

Fig. 23.11 Ulysses electric-field observations before and after the forward CIR shock of 27 May 1991 reported by Hoang et al. (1992): **a** Time-series of electric field spectra; **b** Spectrum near 03:56 UT in the upstream region. Here f_{ped} and f_{peu} refer to the values of f_p downstream and upstream, respectively, of the shock.

corotating interaction regions (CIRs) beyond 1 AU. They found Langmuir waves upstream of three of four forward CIR shocks, including the two with the largest simultaneous values of M_A $(2.9 - 3.6)$ and θ_{bn} $(\geq 65°)$, and noted that the period observable with Langmuir waves increased with θ_{bn}. Figure 23.11 shows the wave and radio data for the strongest shock, on 27 May 1991. Hoang et al. did not report nonthermal Langmuir waves downstream of the forward shocks or in the vicinity of their three reverse shocks, neither upstream nor downstream.

For the two strongest shocks described above Hoang et al. (1992) observed radiation at twice the local upstream value of f_p, when upstream of the shock; e.g., the signal near 27 ± 2 kHz in Fig. 23.11. The signal near $f_p \approx 14 \pm 1$ kHz is the combination of quasithermal plasma noise, nonthermal Langmuir waves, and perhaps foreshock f_p radiation, dominated by the first. This represents the first detection of $2f_p$ radiation generated upstream of a CIR shock.

Controversially, Hoang et al. (1992) suggested that the broad signal near 20 kHz might be fundamental radiation from downstream of the shock. This "downstream" source model is the opposite of the standard "upstream foreshock" model and would require a comprehensive reevaluation of the standard theory. However, strong arguments against this interpretation are advanced here based on the same observations reported by Hoang et al. (1992). First, the authors state that the signal was observed for more than 1 day before the shock crossing, as opposed to less than 30 min for the upstream $2f_p$ radiation, and had a much smaller frequency drift than typical type II bursts. Second, significantly nonthermal Langmuir waves were observed upstream but not downstream of the shock and, third, the shock does not appear to be unusual. Accordingly, much simpler interpretations of the data are that the long-lived signal near 20 kHz is either (i) coming from remote upstream source regions where f_p or $2f_p$ is close to 20 kHz or (ii) not related to the CIR but is instead coming from a distant and unrelated source. It is not known whether so-called planetary continuum radiation from Jupiter, Saturn, and Earth, have been ruled out as a possible source of the 20 kHz signal.

Thus, Hoang et al. (1992) observations and analyses establish that CIR shocks can stimulate $2f_p$ radiation from upstream foreshocks with Langmuir waves, thereby generalizing the phenomenon beyond CME shocks and bow shocks. However, their work does not establish the generation of radio signals near f_p and $2f_p$ in the region downstream of CIR shocks. Instead, the one event reported in which a radio signal was observed upstream of a CIR shock at a frequency near the downstream value of f_p is much more simply interpreted in terms of f_p or $2f_p$ emission from a remote upstream foreshock source, that happens to be near f_p downstream of the shock crossing, or an unrelated background source (e.g., Jovian continuum radiation). Generation of coherent radiation in the regions downstream of shocks is therefore not considered further below.

23.4 Possible Shock-Driven Radio Emissions

This section addresses solar system radio emissions that are believed to be associated with shocks, but for which definitive evidence of the shocks does not exist. The sources range from the solar corona to the outer boundaries of the solar system.

23.4.1 Coronal Shocks: Type II Bursts

Coronal type II bursts were discovered almost 50 years ago at metric wavelengths as the slow drift counterpart to the fast-drift emissions categorized as type III bursts (Wild 1950; Wild and McCready 1950). Figures 23.12 and 23.13 show modern examples of coronal type II bursts. Type IIs were rapidly interpreted in terms of f_p and $2f_p$ emission associated with a shock, since the "exciter" speeds inferred from the frequency drift rates and standard coronal density models obtained from eclipse data (Baumbach 1937; Allen 1947) were of order 500–2,000 km s^{-1} and so commensurate with a low multiple of the Alfven speed (Wild 1950). In contrast, type IIIs have exciter speeds of order $0.1-1$ times the speed of light. Shocks are therefore the most plausible exciters of type II bursts.

Identification of type II bursts is not always easy, or even consistent between people interpreting the same data set or the same event observed by different instruments (Cane and Erickson 2005). Many reasons exist, including (1) the frequent superposition of multiple radio bursts in large events (see the schematic of solar activity in Fig. 23.14), especially type IIIs onto a type II, (2) multiple type IIs sometimes occur simultaneously during an event, with overlapping signatures in frequency-time space, including fine structures, (3) the frequency domains of groups of weak type IIIs sometimes drift at a similar rate to a metric type II and can be misidentified as a type II event (Cane et al. 2003), and (4) it is very difficult to identify weak intermittent emissions, events with large intensities that saturate the standard data displays, and weaker emissions during intense events, all of which can apply to type IIs. Moreover, while the existence of multiple classes of fine structures on metric type II bursts sometimes makes event identification simpler, these fine structures can sometimes complicate event interpretation instead. Figure 23.13 shows a coronal type II event with several overlapping sets of bands, changing drift rates, fine structures, and other radio bursts.

Cane and Erickson (2005) state that they definitely identify metric type IIs by requiring harmonic structure for an event with a suitably slow drift rate and duration (e.g., 2–10 min in the frequency range ≈ 300–30 MHz), supplemented if required by the fine structures defined below. This rests on the work of Roberts (1959) and Prestage (1994). Roberts stated "Harmonic structure is clearly visible in 60 percent of the bursts recorded and in an even higher percentage

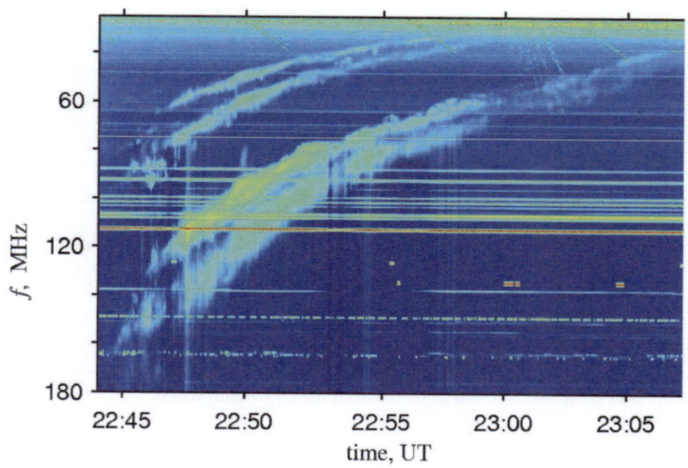

Fig. 23.12 A metric split-band type II solar radio burst (Lobzin et al. 2008) generated in the solar corona. Both fundamental and harmonic bands are clearly evident, each split in two, and the emission in a given band is significantly time variable and blobby

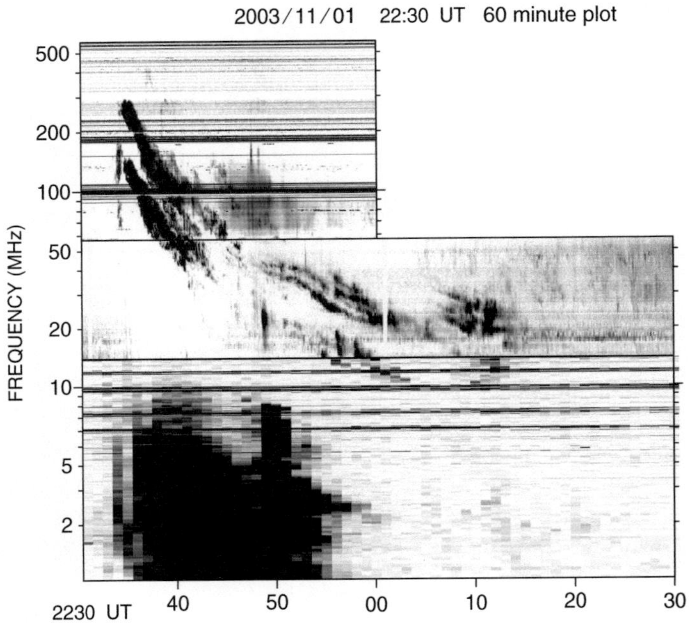

Fig. 23.13 A type II event with two main sets of bands (note the second set starting near 2310 UT), split-band fine structure, and significant changes in drift-rate (Cane and Erickson 2005). The emission in a given band is significantly time variable (blobby). Type III solar bursts are visible near 2235 UT for 80–180 MHz and for the period 2235 – 2300 below 10 MHz

of those bursts which are of at least moderate intensity and consist of well defined ridges". Prestage (1994) characterized all 69 type II bursts observed by the Culgoora digital spectrograph from June 1992 to June 1994, whence Cane and Erickson (2005) concluded: (1) 83% of the events had both fundamental and harmonic bands and (2) none of the remaining events were very intense and more than half occurred at low enough frequencies that the fundamental band was likely below Culgoora's observing limit. Accordingly it appears that harmonic structure is a crucial criterion for identifying an event as a metric type II burst.

While metric type IIs were rapidly interpreted in terms of f_p and $2f_p$ radiation produced by the plasma emission mechanism by electrons accelerated at a shock wave (Wild 1950; Wild and Smerd 1972; Nelson et al. 1985), more detailed models were developed slowly. The first detailed qualitative model for why the shock-accelerated electrons should form beam distributions that drive Langmuir waves and so f_p and $2f_p$ radiation is the foreshock model of Cairns (1986a). It involves "cutoff" distributions produced by time-of-flight effects, exactly as for Earth's foreshock (Filbert and Kellogg 1979; Cairns 1987a) and interplanetary type II bursts (Knock et al. 2001). Exten-

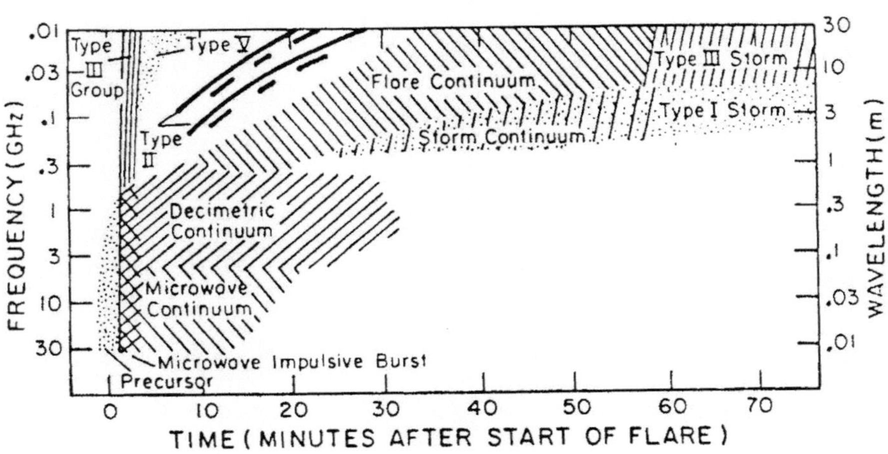

Fig. 23.14 Schematic of the classes of solar radio bursts (Suzuki et al. 1985)

sions of the foreshock theory are presented in detail in Section 23.6, for both coronal and interplanetary type IIs, and are not discussed further in this subsection. Instead, the remainder of this subsection addresses the fine structures of metric type IIs, the origin of the shock waves, the connections between coronal and interplanetary type IIs, and recent work on metric type IIs.

23.4.1.1 Fine Structures of Metric Type IIs

There are several accepted fine structures for metric type IIs. First, "split-bands" in which the fundamental and harmonic bands of a type II are each split in 2, with a frequency difference of order 10–20% (meaning $\Delta f/f \approx 0.1$–0.2) and close mimicking of variations in the corresponding split fundamental and harmonic band. Figure 23.12 shows a classic split-band event. Second, "multiple lane" events in which multiple sets of type II-like bands exist with different frequency separations, drift rates, and start and end times. Third, "herringbone bursts" are sometimes superposed onto the type II band (or "backbone"), comprising bursts that start on the type II band and drift rapidly towards higher and/or lower frequencies with drift rates commensurate with type III bursts (Roberts 1959; Cane et al. 1981). Herringbone bursts have a different frequency-time shape than type IIIs, with decreasing duration at increasing frequency difference from the type II band, whereas the duration of a type III burst increases with increasing frequency offset from its starting frequency (Cairns and Robinson 1997).

Split-bands are sometimes interpreted in terms of plasma emission from both upstream and downstream of the shock, with the frequency difference allowing the shock's Alfven Mach number to be constrained (Smerd et al. 1974; Vrsnak et al. 2002). However, as pointed out above, there is no accepted theoretical model or observational evidence for plasma emission from the downstream region. Another simple interpretation is that the bands arise from distinct spatial regions with slightly different plasma density encountered by the shock (McLean 1967). Explicit calculations show that the foreshock theory of type II bursts can indeed account for some split-band features in this way (Knock and Cairns 2005). However, finding that the splitting is commonly $\Delta f/f \approx 0.1$–0.2 and does not appear to have a larger range of observed values is a difficulty for this interpretation due to the corona being strongly inhomogeneous.

Another interpretation, based on observations of fine-structures regularly spaced at $f_{ce}/2$ in f_p radiation upstream of Earth's foreshock (Cairns 1994), is that split-bands involve splitting by $f_{ce}/2$ for the fundamental band (e.g., the split bands are close to f_p and $f_p + f_{ce}/2$) and f_{ce} for the harmonic band (e.g., bands at $2f_p$ and $2f_p + f_{ce}$) (Cairns 1994). If correct, then the frequencies of split-band type IIs contain information on the spatial profiles of both $n_e(\mathbf{r})$ and $B(\mathbf{r})$ in the corona. The standard 10–20% splitting of type II bursts then implies $f_{ce}/f_p \approx 0.05$–0.2 in the source region, which appears reasonable and might provide a qualitative explanation for the limited range of values $\Delta f/f$ for the splitting. This interpretation is discussed further in Section 23.6 below.

Multiple-lane events are simply interpreted in terms of emission from distinct source regions on the macroscopic shock (McLean 1967; Cairns and Knock 2006). These source regions then can have different plasma densities, propagation speeds, and Alfven Mach numbers for the shock, allowing the emissions to sometimes appear separate and drift at different rates and sometimes drift at the same rate but with an offset in frequency. In addition, several distinct shocks could be produced in an event, including a blast-wave shock and one driven in front of a CME, and produce multiple sets (or lanes) of emission features.

A widely-accepted and detailed theoretical interpretation for herringbone bursts does not exist. Qualitatively, however, they are accepted to involve fast streams of electrons accelerated at the shock and radiating via the plasma emission mechanism. One model involves unusual values of θ_{bn} (Holman and Pesses 1983) while another involves interaction of the shock with a localized coronal structure (e.g., a current sheet) which leads to impulsive, short-duration, acceleration of unusually large numbers of electrons by a localized region of the shock (Cairns and Robinson 1997).

In addition to these fine structures, type IIs are sometimes interpreted in terms of multiple shocks moving through the corona. These can produce emissions that sometimes overlap and cross in frequency-time space but sometimes follow one another in time along similar tracks (see Fig. 23.13). Usually these are not regarded as a fine structure but instead simply as separate type II bursts.

Another fine structure that is obvious in metric type IIs, but rarely discussed as a fine structure, is the intermittent and blobby nature of the emission. Sometimes type IIs appear "wispy", with intermittent broadband emission in multiple separated time periods that form the envelope of the type II band (e.g., Klassen et al. 1999; Mel'nik et al. 2004). This fine structure is clear in Figs. 23.12 and 23.13. It requires explanation.

The last fine structure sometimes considered is emission at the third and higher harmonics, as observed rarely for Earth's foreshock radiation (Cairns 1986b). Although this is controversial, Kliem et al. (1992) make a good case for third harmonic structure (in conjunction with fundamental and second harmonic bands) in one very strong type II burst. Arguably, however, definitive evidence for this fine structure does not yet exist.

Finally, consider the observations that (1) groups of weak type IIIs sometimes drift at a similar rate to a metric type II and can be misidentified as a type II event (Cane et al. 2003) (2) some low-frequency events identified as type IIs appear to contain fast-drift fine structures reminiscent of type IIIs (Mel'nik et al. 2004), and (3) some type IIIs show significant changes in intensity when their frequency-time tracks cross a type II burst and the type III electrons cross the type II shock (an increase-decrease-increase profile with decreasing frequency) (MacDowall 1989). New qualitative interpretation for these three phenomena are: (i) Some type IIIs become detectable only when the beam electrons reach the near vicinity of a radio-quiet shock that is unable to produce type II emission, so that repeated beam injections lead to repeated fast drift bursts whose observable envelope drifts in $f - t$ space at the rate corresponding to the shock, thereby mimicking a type II burst but being composed of fast-drift elements; (ii) Other type IIIs will be detectable before the electrons encounter the shock, but will have changes in intensity as the electrons move from the downstream region into the upstream plasma, being directly relevant to MacDowall's events (MacDowall 1989). Recent simulations of type III bursts (with quasilinear electron-Langmuir wave physics and the standard nonlinear processes discussed in Section 23.5) show strong intensity variations when beam electrons enter localised increases or decreases in T_e or T_i (Li et al. 2010). Further work is required to confirm or reject these interpretations.

23.4.1.2 Origin of Shocks and Connections to Interplanetary Type IIs

Two classes of shocks have been proposed to produce observable metric type II bursts: (1) blast wave shocks associated with flares and other explosive events, and (2) "piston-driven" shocks ahead of CMEs. It is now accepted that essentially all interplanetary type IIs are associated with CMEs (Cane and Stone 1984; Cane et al. 1987; Reiner et al. 1998; Reiner 2000; Gopalswamy et al. 2000; Gopalswamy 2006; Vrsnak and Cliver 2008). However, the situation for metric type IIs is not clear.

An important issue is that the term CME is usually defined in terms of coronagraph observations, which often do not cover the height range $< 2 \times 10^8$ m $\approx 0.3 R_S$ (where $R_S = 7 \times 10^8$ m is the solar radius) of flare sites or metric type II bursts. Note that $f_p > 100$ MHz for $r < 1.5 R_S$ for the Baumbach-Allen density model, for instance. Moreover, shocks take time to develop and become observable (Vrsnak and Cliver 2008). A blast-wave shock is also expected to weaken and eventually become unobservable, as it propagates and loses energy by heating and accelerating the downstream plasma.

Clear evidence exists for large amplitude waves with speeds $\approx 1{,}000$ km s^{-1} excited by flares, in the form of Moreton waves in the chromosphere (Moreton 1960) and their counterparts in the corona at EUV, X-ray, and other wavelengths (Moses et al. 1997; Thompson et al. 1998; Narukage et al. 2002; Khan and Aurass 2002; Vrsnak and Cliver 2008). Case studies of multiple sets of events observed at multiple wavelengths (Warmuth et al. 2004a, b) argue for a common flare origin. Moreover, the classic Uchida "sweeping skirt" model accounts for Moreton waves as the chromospheric signature of a fast-mode blast-wave MHD shock that produces the associated metric type II burst (Uchida 1974).

Arguments exist for and against type IIs being associated with blast wave shocks or CME-driven shocks. These arguments are reviewed in detail elsewhere (Cane and Erickson 2005; Cliver et al. 1999, 2004; Mancuso and Raymond 2004; Vrsnak and Cliver 2008) and no attempt is made here to review or justify them. For example, though, arguments against all metric type IIs having CME-driven shocks include the metric emission often apparently coming from behind the leading edges of CMEs, the exciter speeds inferred from the drift rates of metric type IIs differing greatly from the speeds of associated CMEs, and the absence sometimes of observable CMEs for a given type II (Cane and Erickson 2005; Vrsnak and Cliver 2008). Similarly, arguments against all metric type IIs having a blast-wave shock include the high correlations between metric type IIs and fast CMEs and the poor correlation between flare size and type II occurrence (Cliver et al. 1999; Cane and Erickson 2005). It is noted that there are significant difficulties in removing projection effects, viewing, and obtaining speeds for Earth-directed CMEs (Mancuso and Raymond 2004), while the shock speeds extracted from radio data depend heavily on the density model adopted. In addition, the regions of the shock emitting observable radiation need not be at the nose of the shock, but instead near the flanks (Stewart and Magun 1980), since it is regions where **B** is close to perpendicular to the local shock normal that are favoured (Stewart and Magun 1980; Holman and Pesses 1983; Cairns 1986a; Knock et al. 2001). Coupling the foreshock type II theory with simulations of shocks evolving in data-driven models of the inhomogeneous solar corona, and then comparing the predicted dynamic spectra with observations, will help resolve this issue. Progress on such simulations is described in Sections 23.5 and 23.6 below.

Perhaps the clearest argument as to whether all metric type IIs are produced by CME-driven shocks, as interplanetary type IIs are accepted to be, would be if it was routine for the shock to continuously produce radio emission into the interplanetary medium so that the metric type II burst can be followed continuously through the decametric band (e.g., 100 – 1 MHz) into an interplanetary type II burst. Multiple observational searches for this have occurred and it is now clear that this must be exceedingly rare, if it ever happens. Figure 23.15 shows a metric type II band from $\approx$ 180 MHz near 2248 UT that extends down to 2259 UT near 38 MHz and an interplanetary type II that starts near 9 MHz at 2252 UT (Cane and Erickson 2005). It is evident that the coronal and interplanetary type IIs do not connect in this event, due to simultaneous emission at widely different and non-harmonically related frequencies. This event therefore requires two separate shocks for the coronal and interplanetary emissions and argues against a CME-driven explanation for both phenomena (Cane and Erickson 2005).

As a counterpart to Fig. 23.15, consider Fig. 23.16: it shows a metric type II that continues from $\approx$ 80 MHz (not shown) near 0515 UT down to at least $\approx$ 1.8 MHz near 0720 UT (Cane and Erickson 2005), also studied by Cliver et al. (2004). An emission interpretable

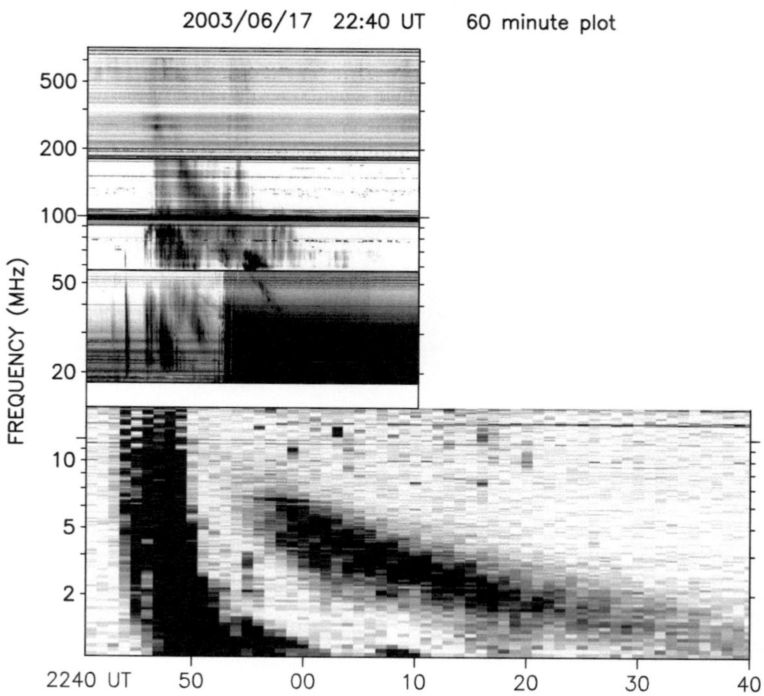

Fig. 23.15 A coronal type II and a interplanetary type II that do not connect (Cane and Erickson 2005). See the text for more details

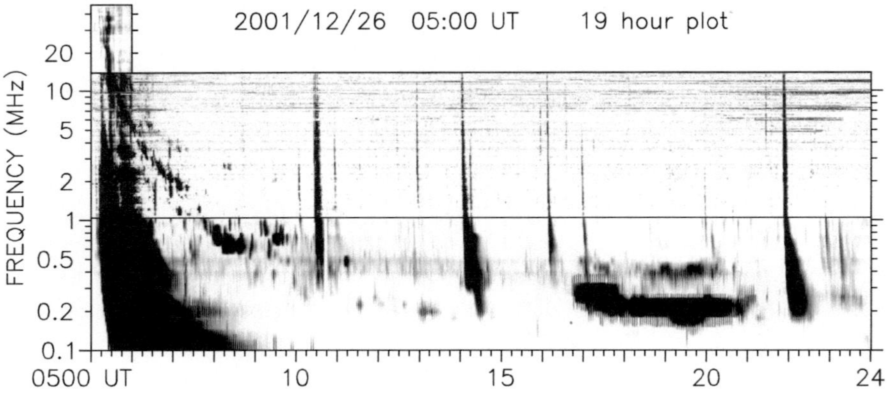

Fig. 23.16 Emissions that are interpretable as a coronal type II and a interplanetary type II and which are at least close to overlapping in frequency and time (Cane and Erickson 2005). The connection and interpretation of the emissions are discussed in the text

as an interplanetary type II is present from $\approx$ 0730 UT at a frequency just above 1 MHz, continuing intermittently until about 1000 UT, then between 1130 and 1400 UT and then between 1700 and 2130 UT. Cane and Erickson (2005) believe that the "interplanetary" emission after about 0720 is not connected to the metric emission, because the interplanetary event clearly starts at a measurably lower frequency than the last remnant of metric emission. Here the difference in frequency is less than a factor of 2, arguably being about a factor of 1.5 in Cane and Erickson's Fig. 12 (as opposed to their Fig. 13 shown here as Fig. 23.16), but they interpret the initial interplanetary emission as harmonic rather than fundamental as for the metric type II. On the other hand Cliver et al. (2004) interpret the same event as clear evidence of continuation of a metric type II into an interplanetary type II.

Detailed inspection of Fig. 23.15 shows many similarities between the metric and interplanetary type II emission present, including their intermittent and blobby nature and their organization into fundamental and harmonic bands. In this reviewer's opinion, however, the detailed relationship between the two emissions is not clear. The two emissions arguably overlap in time. However, if these data were clearly in the metric domain then the obvious interpretations would be in terms of either multiple lanes or two contemporaneous shocks. It would be good to perform detailed analyses of these data in inverse frequency – time space (i.e., $1/f - t$ space), as in Reiner et al. (1998), Reiner (2000), and Lobzin et al. (2008), so as to test these two interpretations. However, this reviewer's opinion is that the single (CME-driven) and two shock (CME-driven and blast wave) interpretations both remain viable for this event and cannot be ruled out, contrary to both earlier claims.

In summary, at this time no consensus exists on how to resolve these issues and the apparently contradictory arguments as to whether any metric type IIs continue into the solar wind, whether all type IIs (coronal and interplanetary) are produced by CME-driven shocks, and whether both. Perhaps, however, it is time to dispense with the either/or approach, favored though it is by Occam's Razor, and consider whether both types of shocks can occur and produce observable emission (i.e., be "radio-loud") in events, usually one at a time but occasionally both simultaneously. Put another way, might some metric type IIs be produced by blast-wave shocks, some by CME-driven shocks, and some involve multiple nearly simultaneous bands associated with one (or more) shock of each type? Perhaps the strongest arguments for this approach are that on one hand it appears certain that only very rarely, if ever, does a metric burst continue directly into an interplanetary type II burst, while on the other hand all interplanetary type IIs are accepted to be driven by CMEs. Even if one argues that the conditions for production of observable type II radiation are rarely met in the corona and solar wind, consistent with theory (Holman and Pesses 1983; Nelson et al. 1985; Knock et al. 2001, 2003a, b; Knock and Cairns 2005; Cairns and Knock 2006) summarized in Sections 23.5 and 23.6, the evident rarity of type IIs continuing from the corona into the solar wind is prima facie evidence for two different classes of shocks in most events, while the strong correlation with CMEs is prima facie evidence for CME-driven shocks playing a strong role.

23.4.2 Drifting Pulsating Structures

High-frequency observations (600 – 2,000 MHz) reveal intense solar radio emissions in association with flares and associated magnetic reconnection events, including the ejection of plasmoids (Karlicky 2003; Karlicky and Barta 2004). Many additional emissions are observed from 1 to 20 GHz, e.g., Wang et al. (2001) and Yan et al. (2001). Figure 23.17 shows multiple radio and microwave events covering the domain 4.5 GHz – 40 MHz, in association with a soft X-ray event that started near 0939 UT, peaked near 1,028 UT, and ended near 1049 UT (Karlicky 2003). A strong metric type II burst is evident below about 300 MHz from about 1016 – 1023 UT, implying a shock is present. Multiple fast-drift type III-like emissions are observed across the domain 200 – 4,500 MHz from about 1014 – 1023 UT, but attention is focused here on two so-called drifting pulsating structures: the first from 1016:00 – 1017:30 UT for 450 – 1,000 MHz and the second stronger event from 1017:20 – 1022:00 UT in the range 450 – 1,500 MHz (Karlicky 2003; Karlicky and Barta 2004). The intense portion of the second event has a clear drift in frequency, albeit with several weaker broadband bursts that extended to higher frequencies and appear to be superposed. The intense portion of this event is strongly reminiscent of the contemporaneous metric type II burst, only at about 6 times the frequency. Significant circular polarizations were sometimes observed, particularly at higher frequencies.

At a glance it is attractive to interpret the intense portions of the drifting pulsating structures, especially the second which strongly resembles the contemporaneous type II burst, in terms of a travelling shock. The interpretation adopted (Fig. 23.18) involves magnetic

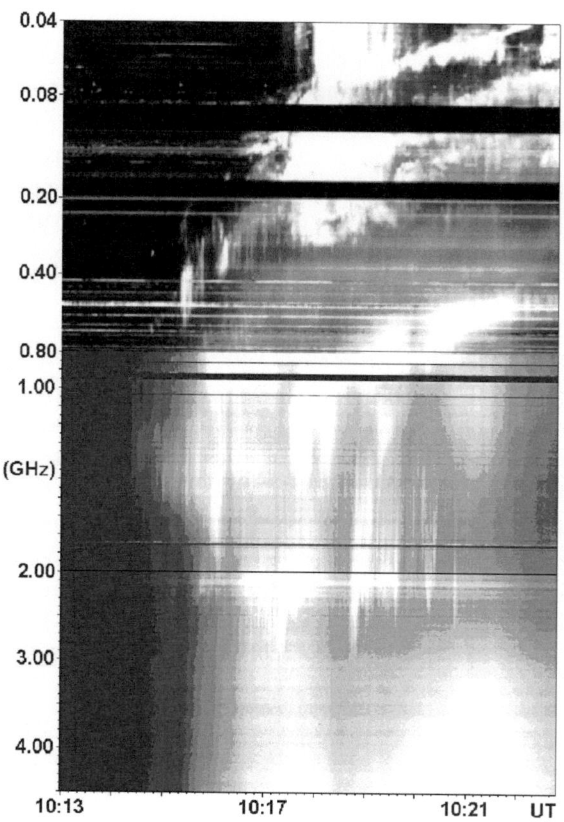

Fig. 23.17 Metric and decimetric spectra of the 12 April 2001 event of Refs. Karlicky (2003); Karlicky and Barta (2004). A type II burst is observed above 400 MHz and two drifting pulsating structures are observed in the range 450 – 1,500 MHz, the second of which strongly resembles the type II burst. See text for details

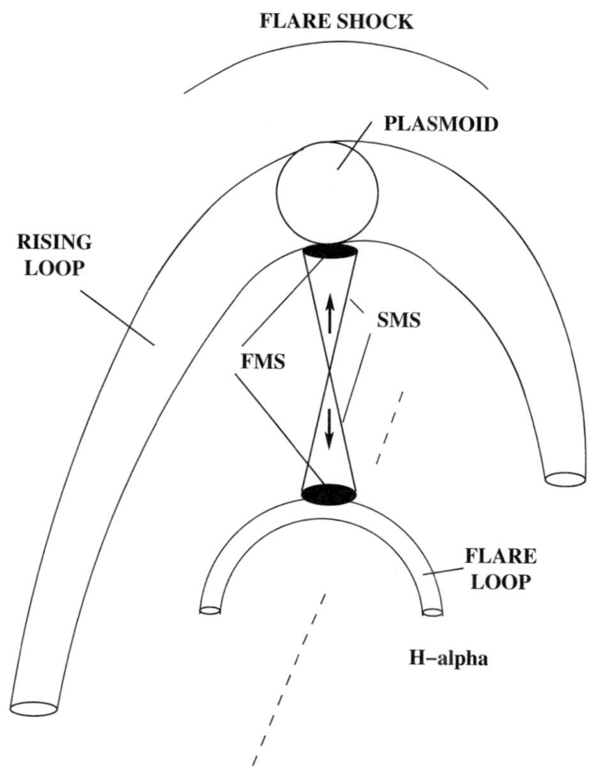

Fig. 23.18 Model of Karlicky and Barta (2004) for interpreting drifting pulsating structures. Fast outflows from magnetic reconnection regions, which develop at current sheets between rising and flare loops, are shown with arrows. SMS refers to a slow mode MHD shock and FMS to a standard fast-mode MHD shock

reconnection at multiple current sheets between an underlying flare loop and higher loops, leading to fast outflows (Karlicky and Barta 2004). Slow mode shocks are predicted to be formed as part of the standard structure of the reconnection outflow region, along the "side" boundaries, while fast mode shocks develop where the reconnection outflows impact the rising and flare loops (called "termination" shocks). A plasmoid may develop as part of the reconnected rising loop. The frequency drift of the decimetric emissions is interpreted in terms of one or more loops or the plasmoid rising, with associated motions of the shocks and the reconnection sites. With loops participating in the reconnection process the structure can continue to rise and produce drifting radio emission, with fine structures related to the time variations in the multiple-loop system. Karlicky and Barta (2004) specifically point out that one event (their Fig. 18) had fine structures that strongly resemble herringbone fine structures on type II bursts, a further argument that a shock is involved.

Karlicky and Barta (2004) state that the radiation is produced by the plasma emission mechanisms as a result of fast electron beams accelerated at these shocks. They then used the radiation frequency to constrain the plasma density in the multiple-loop system. These authors did consider several models for radio fine structures, most involving upper hybrid waves driven by temperature anisotropies and then linear mode conversion, rather than the standard "plasma emission" mechanisms involving electron beam-driven Langmuir waves and nonlinear radiation processes involving Langmuir waves. Extension of the Knock et al. model for type II bursts (Knock et al. 2001; Knock and Cairns 2005) to these emissions would be useful in establishing the need for non-standard emission mechanisms and in quantitatively modelling drifting pulsating structures and other decimetric emissions.

23.4.3 Outer Heliospheric Radio Emissions

In 1983 the two Voyager spacecraft were beyond the orbit of Saturn and separated by several astronomical units (AU) when they observed radio emissions at frequencies $f \approx 2$–3 kHz with very similar dynamic spectra (Kurth et al. 1984). Figure 23.19 is a recent dynamic spectrum of these emissions from 1982 until July 2009, showing the emission frequency versus time with the intensity color coded. It was obtained from the Voyager Plasma Wave Subsystem (PWS) site http://www-pw.physics.uiowa.edu/w̃sk/vgr/recent.html, courtesy of D.A. Gurnett and W.S. Kurth, and is similar to figures in other publications (Kurth et al. 1987; Gurnett et al. 1993, 1998). The intense and continuous red band near 2.4 kHz is interference from Voyager 1's power supply system. The signals below 1 kHz at all times, and the diffuse light blue signals below 2 kHz from 1982 to mid-1992, are dominated by interference from other instruments and spacecraft and systems. The signals of interest here are the three relatively intense, longlasting episodes of emission from ≈ 1.8–3.6 kHz (1983–1984, 1992–1995, and 2003–2004), together with weaker events (e.g., early 1986, late 1989, and late 1991) in the same frequency range. Note that these major episodes are separated by an approximately 9–10 year period, reminiscent of the solar cycle, and that the third outburst started in mid-2003 and was relatively short-lived.

Figure 23.20 summarizes the plasma regions and discontinuities expected in the outer heliosphere (Zank 1999; Izmodenov et al. 2004). The superalfvenic, supersonic solar wind plasma undergoes a shock transition at the termination shock, as recently observed by Voyagers 1 and 2 at heliocentric distances of 91 and 84 AU, respectively (Stone et al. 2005, 2008). The inner heliosheath contains shocked solar wind plasma, which is slowed to speeds ≈ 100 km s^{-1}, compressed by a factor ≈ 2–4, heated to temperatures $\approx 10^6$ K, deflected in direction, and the magnetic field amplified and rotated, at the termination shock. Analogous to the terrestrial magnetopause, the heliopause is a rotational discontinuity between the inner and outer heliosheaths that separates the shocked solar wind plasma from interstellar plasma. Finally, if the solar system moves superalfvenically or supersonically, then the VLISM plasma will be shocked at a bow shock, and modified similarly to the solar wind at the termination shock. The outer heliosheath contains the VLISM plasma processed by the bow shock (or a bow wave if the flow is subsonic and subalfvenic).

Kurth et al. (1984) interpreted the radio waves as originating in the outer heliosphere and, most likely, a signature of the solar wind's interaction with the very

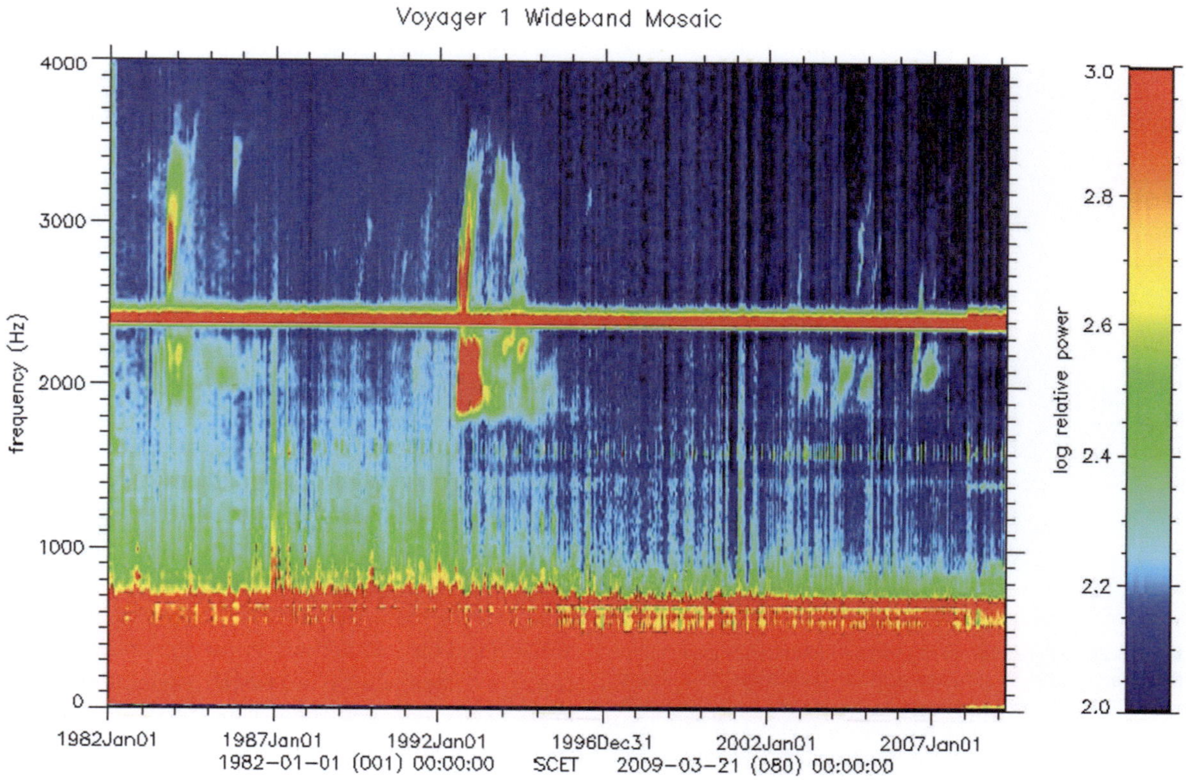

Fig. 23.19 Voyager 1 dynamic spectrum for the period 1982–2009. Major episodic radiation events are visible, as well as weak drifting events. See the text for more details

local interstellar medium (VLISM). Now the favored interpretation is that the radiation is produced beyond the heliopause by shocks moving away from the Sun in front of global merged interaction regions (GMIRs) resulting from solar activity (Gurnett et al. 1993; Gurnett and Kurth 1996; Cairns and Zank 2001, 2002; Mitchell et al. 2004).

The Voyager emissions can be categorized into two classes (Kurth et al. 1987; Cairns et al. 1992; Gurnett et al. 1993), as illustrated by Figures 23.19 and 23.21: (1) The "2 kHz component", which remains in the frequency range 1.8–2.6 kHz, is longer lasting ($\approx$ 3 years), and does not drift significantly in frequency. (2) "Transient" or "drifting" emissions which drift up in frequency, have a range of starting and ending frequencies within the domain 1.8–3.6 kHz, frequency drift rates in the range $\approx$ 1–3 kHz/year, and last for $\approx$ 100–300 days.

Clear evidence exists for frequency fine structures in both the 2 kHz component and the transient emissions, particularly for the latter. For instance, Fig. 23.19 and 23.21 show that the transient emissions during 1994 often occurred as pairs of signals with very similar drifts that are offset in frequency. This "pairing" characteristic is not understood but is reminiscent of split-band and multiple-lane type II solar radio bursts (Wild et al. 1963; Nelson and Melrose 1985) discussed in Section 23.4.1 above. Possible interpretations include the shock moving across 2 regions with slightly different densities or the splitting being an intrinsic feature of the emission process. It is worth pointing out that the density profiles in the heliosheath and heliopause are predicted to be asymmetric about the relative VLISM-Sun velocity vector $\mathbf{v}_{VLISM}$, due to the influence of the interstellar magnetic field (Pogorelov et al. 2009; Opher et al. 2009a). These density asymmetries may be directly relevant to the observed pairing. If, however, the splitting is assumed to be intrinsic and at half the electron cyclotron frequency, as for some rare events of f_p emission in Earth's foreshock (Cairns 1994), then the inferred magnetic field is $\approx$ 20 nT. This appears to be unrealistically large.

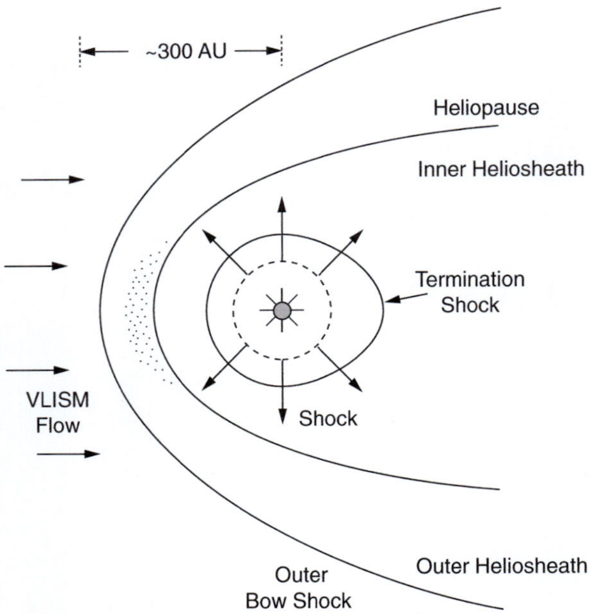

Fig. 23.20 Major plasma regions and boundaries expected for the solar wind – VLISM interaction. The *dashed line* denotes a GMIR shock moving outwards. The *dotted region* shows the source region predicted for the Voyager emissions by the GMIR model and Priming/GMIR theory (Cairns and Zank 2002)

Following earlier work on event triggers (McNutt 1988; Grzedzielski and Lazarus 1993), Gurnett et al. (1993) postulated that the episodic radio bursts are produced when global merged interaction regions (GMIRs) cross the heliopause. GMIRs are formed by the merging of multiple interacting CMEs and other fast plasma flows produced by solar activity into a global disturbance of the plasma density, magnetic field and flow speed that propagates outwards faster than the ambient solar wind. Figure 23.22 shows when multiple spacecraft observed the shock driven by the GMIR associated with the 1992–1994 radiation event, as well as the Forbush decreases in cosmic ray flux caused by cosmic rays being reflected and scattered by the enhanced and disturbed magnetic field of the GMIR material. Detection of these signatures by the widely separated Voyager and Pioneer spacecraft confirms that the disturbance was truly global.

Figure 23.23 confirms the association between large GMIRs and the major radio outbursts (Gurnett and Kurth 1995): the 1983–1984 and 1992–1994 outbursts are associated with the two largest Forbush decreases observed in the Deep River Neutron Monitor data. Moreover, the time-lag between the Forbush decreases at Earth and the radio onsets at Voyager are consistent, at 415 ± 4 days, and the two GMIR propagation speeds to Voyager were consistent at $\approx 830 \pm 20$ km s^{-1}. Using these speeds and time-lags, together with plausible estimates for shock slowing, Gurnett et al. (1993) estimated the source to be at a radial distance $R \approx$ 140–190 AU. These are plausible distances for the heliopause.

Recently Kurth and Gurnett (2003) combined GMIR time-of-flight effects with direction-finding and an amplitude-dependent triangulation technique to constrain further the distance and direction to

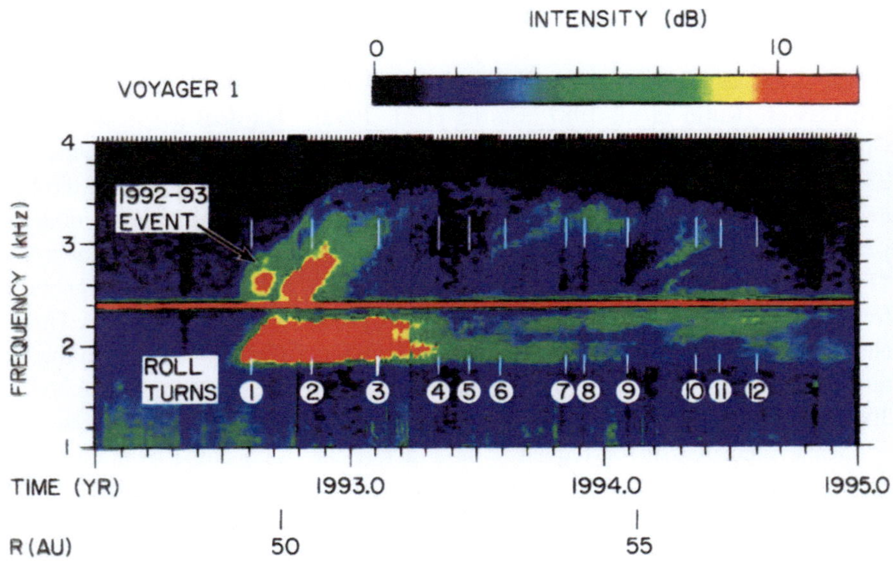

Fig. 23.21 Two classes of radiation event are demonstrated here for the 1992–1994 event (Gurnett et al. 1998): the "2 kHz component" and drifting "transient emissions". *Numbers* show the times of spacecraft rolls and associated direction-finding analyses discussed below

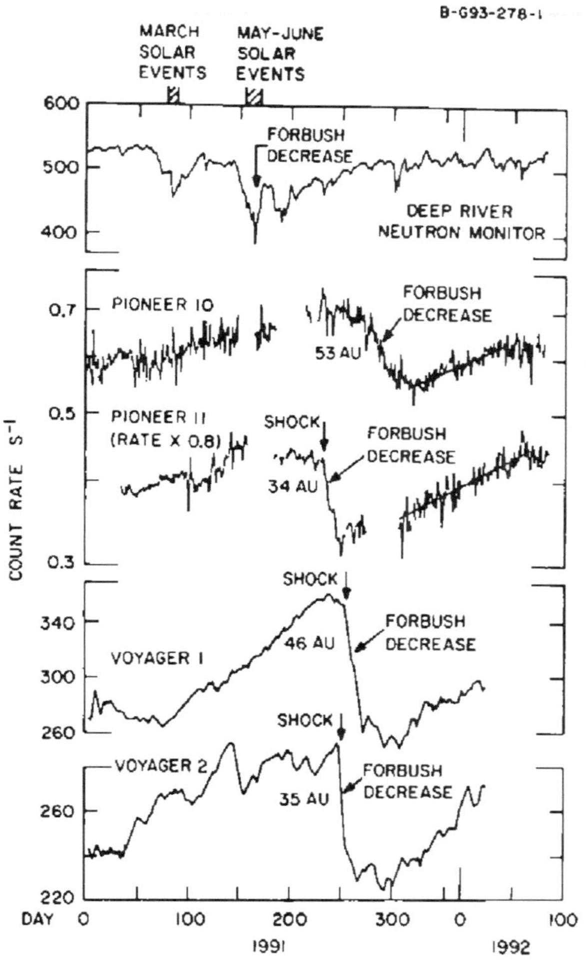

Fig. 23.22 Forbush decreases in cosmic rays caused by the GMIR and its shock at Earth and various widely separated spacecraft in the outer heliosphere for the 1992–1994 radio event (Gurnett et al. 1993)

the radiation source, extending earlier work (Gurnett et al. 1993, 1998). The three techniques yield simultaneous solutions for multiple observations of the transient emissions during the 1992–94 event, as shown in Fig. 23.24 as a function of galactic latitude and longitude (Kurth and Gurnett 2003). The results are clear. First, the source of the transient emissions is initially very close to the nose direction for the heliopause. Second, the source direction changes with time but lies along a line closely parallel to the galactic plane. Third, the source generally moves away from the nose direction and the modulus of its ecliptic latitude typically increases with time. The first and third of these results confirmed earlier analyses (Gurnett et al. 1993, 1998). A detailed physical explanation for these results was not attempted. However, it was speculated that the second result arises because the apparent source axis is aligned with **B** in the outer heliosheath, implying that **B** is parallel to the galactic plane. More recent work suggests a different interpretation (Gurnett et al. 2006; Cairns et al. 2006).

The radio emission is widely accepted to be f_p or $2f_p$ radiation produced upstream of a shock wave (Kurth et al. 1984; Gurnett et al. 1993; Cairns and Zank 2001, 2002). This means that the observed radiation frequency is a remote measure of the source density, thereby also constraining the source location. It is widely accepted that the observed radiation frequencies are inconsistent with f_p or $2f_p$ emission from the solar wind or inner heliosheath, but that a source in the outer heliosheath or VLISM is viable.

Before proceeding it is emphasized that a GMIR shock is not a sufficient condition by itself to obtain observable radio emission. The reason is that the Voyager spacecraft do not observe any local radio emission when the GMIR shocks pass over the Voyager spacecraft. Instead two conditions (or triggers) are required, first that a suitable GMIR shock exists, and second that the shock enters a plasma region where it is able to excite observable radio emission (Cairns and Zank 2001, 2002).

In summary, although in situ observations of the shock in the radio source do not yet exist, the observational evidence for the GMIR shock stimulating the emission beyond the heliopause is very strong. Current theory and interpretations of the 2–3 kHz emissions all involve this model, due originally to Gurnett et al. (1993). A review of theoretical research is provided in Section 23.7 below.

23.5 Theory

This section summarizes the standard theory for radio emission at f_p and $2f_p$ produced by shocks. It proceeds first by discussing the reflection and acceleration of electrons, then the formation of beam distributions, then the growth of and power flux into the Langmuir waves, and finally the production of f_p and $2f_p$ radio emission by specific nonlinear Langmuir wave processes. Each of these steps is illustrated with examples from either Earth's foreshock or interplanetary type II bursts.

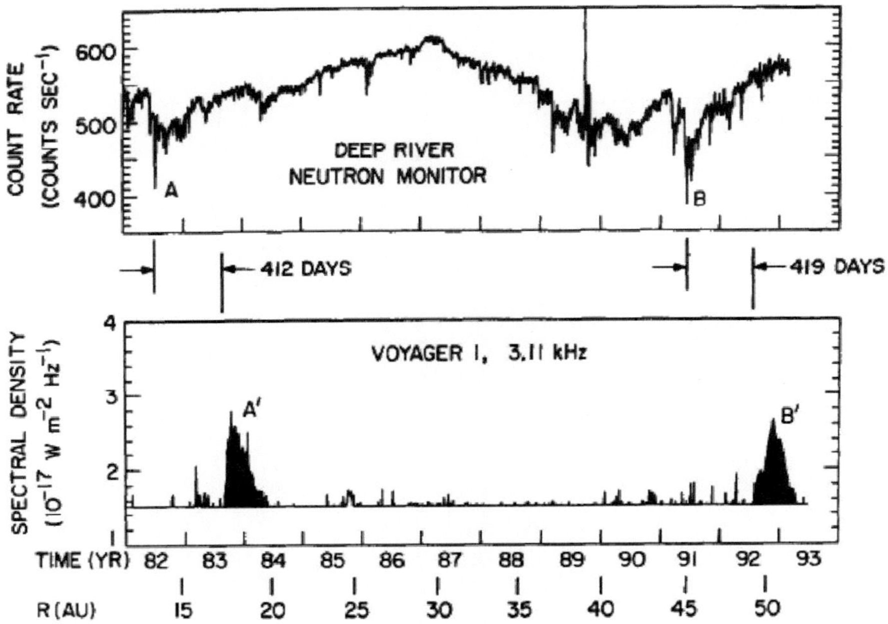

Fig. 23.23 Correlation between the largest Forbush decreases at Earth (A and B) and the major 2–3 kHz radio events (A′ and B′) (Gurnett and Kurth 1995)

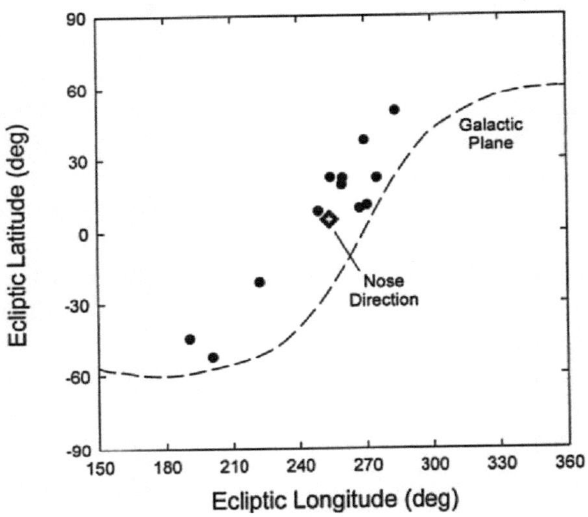

Fig. 23.24 Source directions of transient emissions as functions of [1950] ecliptic latitude and longitude derived from analyses of roll modulations, relative amplitudes between Voyagers 1 and 2, and GMIR-radio timelags (Kurth and Gurnett 2003). Sources start near the heliopause nose and typically move away with increasing time

23.5.1 Electron Motion, Reflection and Acceleration

In Figs. 23.1 and 23.6 the shock is stationary and the upstream plasma moves towards the shock at velocity $\mathbf{u}$, carrying magnetic field lines across an observer fixed relative to the shock. Accordingly, a convection electric field $\mathbf{E}_u = -\mathbf{u} \times \mathbf{B}_u$ exists in general (except when $\mathbf{u}$ and $\mathbf{B}$ are parallel): it acts on all charged particles and causes them to move with a mass and charge-independent drift velocity $\mathbf{v}_d = \mathbf{E}_u \times \mathbf{B}_u/|B_u|^2$ with $v_d = u|\sin\theta|$ where θ is the angle between $\mathbf{B}_u$ and $\mathbf{u}$. Put another way, $\mathbf{v}_d$ is the component of $\mathbf{u}$ perpendicular to $\mathbf{B}_u$.

Electron reflection is best described in the de Hoffman-Teller frame (Toptyghin 1980; Wu 1984; Ball and Melrose 2001), where the convection electric field vanishes. This frame moves along the shock front with speed $v_d \tan\theta_{bn}$, where θ_{bn} is the angle between the local shock normal and $\mathbf{B}_u$. Conservation of the magnetic moment and energy are assumed, the latter subject to the cross-shock potential $\phi_{cs}(\mathbf{r})$:

$$\frac{v_\perp^2(\mathbf{r}_1)}{B(\mathbf{r}_1)} = \frac{v_\perp^2(\mathbf{r}_2)}{B(\mathbf{r}_2)}, \tag{23.6}$$

$$m_e\left(v_\parallel^2(\mathbf{r}_1) + v_\perp^2(\mathbf{r}_1)\right) - 2e\phi_{cs}(\mathbf{r}_1) = m_e\left(v_\parallel^2(\mathbf{r}_2) + v_\perp^2(\mathbf{r}_2)\right) - 2e\phi_{cs}(\mathbf{r}_2), \tag{23.7}$$

respectively, where $\mathbf{r}_1$ and $\mathbf{r}_2$ are two locations. The magnetic field $\mathbf{B}(\mathbf{r})$ and potential ϕ_{cs} are then specified by a model, e.g., Eq. (23.4). Conservation of the magnetic moment leads to a loss cone in the reflected electron distribution, which depends on the magnetic compression at the shock, as predicted by Wu (1984) and Fitzenreiter et al. (1990), among others.

The shock model used in applications to type II bursts, Earth's foreshock, and the 2–3 kHz emissions ignores the shock foot and overshoot/undershoot structures, instead assuming a simple linear ramp between the upstream and downstream states (Knock et al. 2001; Kuncic et al. 2002; Mitchell et al. 2004) specified by the Rankine-Hugoniot conditions (Kennel et al. 1985; Melrose 1985; Burgess 1995). Of particular relevance then are the magnetic mirror ratio B_d/B_u and corresponding change in ϕ_{cs} via Eq. (23.4), which depend upon the Alfven Mach number M_A, sonic Mach number M_S, and θ_{bn}. The potential modifies the shock's loss cone, determined by B_d/B_u, and makes it more difficult to reflect electrons with low $v_\parallel$ since it attracts all electrons downstream.

An important future extension of existing models is to include the effects of the magnetic and potential overshoots, which increase the magnetic compression ratio to a ratio $B_m/B_u > B_d/B_u$, decrease the loss cone, and result in significantly more reflected electrons. Yuan et al. (2007, 2008a) found factors of 2–4 increase in the number of reflected electrons for shocks with overshoots compared with non-overshoot shocks.

In the de Hoffman-Teller frame electrons with initial speed $v_\parallel^i$ upstream are reflected with parallel speed $v_\parallel^r = -v_\parallel^i$. Then, moving into the shock rest frame of Figs. 23.1 and 23.2 one finds

$$v_\parallel^r = 2v_d \tan\theta_{bn} - v_\parallel^i \,. \tag{23.8}$$

The associated change in energy is large when θ_{bn} is close to 90°. This increase in energy can be understood in terms of the shock's magnetic mirror reflecting the electron analogously to a ping-pong bat accelerating a ball, so the acceleration is sometimes called mirror reflection or fast Fermi acceleration (Wu 1984). Alternatively, in the shock's rest frame it is understood in terms of the electrons undergoing a ∇B plasma drift in the direction anti-parallel to the convection electric field, leading to the name shock drift acceleration (cf. Toptyghin 1980; Ball and Melrose 2001).

23.5.2 Formation of Electron Beams

Electron beams are formed in the foreshock by two effects (Cairns 1987a). The first is determined by the shock's effective speed along the magnetic field lines, leading to a minimum "escape cutoff" speed for which electrons can enter the foreshock (Cairns 1987a). The second is a time-of-flight effect that relates to motion of charged particles in the foreshock (Filbert and Kellogg 1979).

The escape cutoff forms beams because the shock moves along the magnetic field lines at an effective speed $v_\parallel = v_d \tan\theta_{bn}$, in the shock rest frame of Fig. 23.1. Accordingly only particles with

$$v_\parallel > v_c = v_d \tan\theta_{bn} \tag{23.9}$$

can outrun the shock and escape into the foreshock. This constraint determines a minimum cutoff speed $v_\parallel = v_c$, equal to the de Hoffman-Teller speed, for electrons to be able to leave the shock: this therefore naturally forms a beam of electrons with $v_\parallel > v_c$ upstream of any shock (Cairns 1987a), provided only that any electrons are able to escape upstream (which requires $v_c < c$).

The time-of-flight mechanism relies instead on the motion of charged particles in the foreshock. The kinematic motion of a charged particle in a locally homogeneous magnetized plasma can be described by

$$\mathbf{v}(t) = \mathbf{v}_\parallel + \mathbf{v}_g(t) + \mathbf{v}_d \,. \tag{23.10}$$

This particle moves with constant speed $v_\parallel$ along $\pm\mathbf{B}_u$, has the standard gyromotion $\mathbf{v}_g(t)$ with gyro speed v_g, and is subject to a constant drift velocity $\mathbf{v}_d$ perpendicular to $\mathbf{B}_u$ due to the convection electric field. Each particle then moves in a plane defined by $\mathbf{u}$ and $\mathbf{B}_u$, so the full 3-D problem of a set of particles moving in the 3-D foreshock can be broken into a set of calculations in a stack of 2-D planes, with each particle moving in a 2-D plane defined by $\mathbf{u}$ and $\mathbf{B}_u$.

Put another way, the particle gyrocenter moves along the line defined by $\mathbf{v}_\parallel + \mathbf{v}_d$. The slope of this line is defined by the ratio of $v_\parallel$ to v_d, with lines of larger $v_\parallel/v_d$ being more closely parallel to $\mathbf{B}$. Now consider a point (R, x) in the foreshock, where R is the distance along the tangent field line and x is the perpendicular distance along $\mathbf{v}_d$. The fastest electrons reaching that point from the shock move approximately along a line parallel to **B**, while the slowest ones move along the line tangential to the shock that passes through (R, x). Even slower electrons cannot come from the shock, so the particle kinematics directly impose a a minimum $v_{\parallel}$ for shock-accelerated electrons reaching (R, x), thereby again imposing a cutoff on the electron distribution function and implying that a beam feature exists. For fast beams relatively close to the foreshock boundary the equation for the line directly gives (Filbert and Kellogg 1979; Cairns 1987a)

$$v_{\parallel,min} \approx v_d \frac{R}{x} = v_d \tan\theta_{bn} = v_c \,, \qquad (23.11)$$

consistent with Eq. (23.9). Alternatively, this can be seen from the tangential nature of the path, which means that the particle path is parallel to the local shock surface and so $v_{\parallel} = v_c$.

Thus, electrons reaching a location in the foreshock from the shock, whether reflected upstream electrons or electrons leaked from the shock's downstream region, naturally have a beam distribution. This is imposed by two effects, the escape cutoff at the shock itself and the cutoff imposed by particle kinematics (time-of-flight effects) in the foreshock itself. Both yield the same cutoff in $v_{\parallel}$, as argued above. It should be recognized, however, that the electron reflection can cause the peak in the reduced electron distribution function $f_r(v_{\parallel}) = \int dv_{\perp} v_{\perp} f(v_{\perp}, v_{\parallel})$ to be well above $v_{\parallel} = v_c$: the two cutoff effects impose a minimum $v_{\parallel}$ and beam character on $f_r(v_{\parallel})$ but multiple beams are possible.

The variation of v_c along the shock surface and the spatial variation in v_c with R and x in Eq. (23.11) results in a significant variation in the foreshock beam speed. This speed is large near the foreshock boundary but $< c$ (there is actually a region downstream of the foreshock boundary that contains no reflected electrons, since $v_c > c$ there, if gyromotion effects are ignored) and decreases monotonically with increasing x and depth $DIFF = x/\sin\theta$. Thus the fastest beams are expected near the foreshock boundary, with the slowest beams deep inside the foreshock.

The kinematic cutoff effect can lead to electron beams rebuilding in the foreshock, particularly when faster electrons nearer the foreshock boundary are slowed by wave-particle interactions. Consider a group of such electrons: they leave their effective source region near the foreshock boundary with a reduced $v_{\parallel} = v_1$ and now move on particle paths (lines) that are more highly inclined relative to **B** than their original path, crossing the trajectories of some electrons with higher $v_{\parallel}$ that are coming directly from the shock and forming a beam at lower $v_{\parallel}$. This movement of electrons to foreshock regions they cannot access without energy loss corresponds to an effective rebuilding of the electron beam at speeds close to but below v_c (Cairns and Fung 1988) and might lead to multiple unstable beams at a given location.

As well as having beam-like character, foreshock electron distributions also should have loss cone features in the reflected electrons (imposed by the conservation of magnetic moment and energy in the de Hoffman-Teller frame, as discussed in Section 23.5.1), while leaked electrons will be found inside the loss cone (Fitzenreiter et al. 1990). Figure 23.9 shows the electron distributions in Earth's foreshock to have both beam and loss cone features [Fitzenreiter et al. 1984], as discussed in Section 23.3.2. The combination of a loss cone and a cutoff in $v_{\parallel}$ can form a well-defined ring-beam distribution, as found explicitly by Yuan et al. (2007, 2008a) using test-particle simulations.

The electron distribution functions $f(v_{\perp}, v_{\parallel})$ and $f_r(v_{\parallel})$ can be calculated using Liouville's Theorem (Cairns 1987a; Fitzenreiter et al. 1990; Knock et al. 2001; Kuncic et al. 2002). This involves tracing particle paths back to the shock, calculating B_d/B_u and ϕ_{cs} (without or with overshoots) on the spatially varying shock, unfolding the effects of the shock acceleration using Eq. (23.8), equating $f(v_{\parallel}^r, v_{\perp})$ to the assumed upstream distribution $f_{in}(v_{\parallel}^i, v_{\perp})$ (or the downstream distribution for leaked electrons), and then integrating over $v_{\perp}$ to obtain the f_r.

Figure 23.25 illustrates the reduced electron distributions $f_r(v_{\parallel}, R, x)$ at 4 locations upstream of Earth's bow shock, as well as bow shock and the particle paths to the 4 locations for the cutoff velocity and the maximum $v_{\parallel}$ calculated. Spatial coordinates (R, x) are used, with $\mathbf{B}_u$ oriented along the $+R$ axis and $\mathbf{v}_d$ oriented along the $+x$ axis. The top and bottom panels also state the average beam speed v_b, effective thermal width Δv_b, and number density n_b of the beam relative to the background electron number density. These are calculated by taking appropriate moments of f_r. Spatial

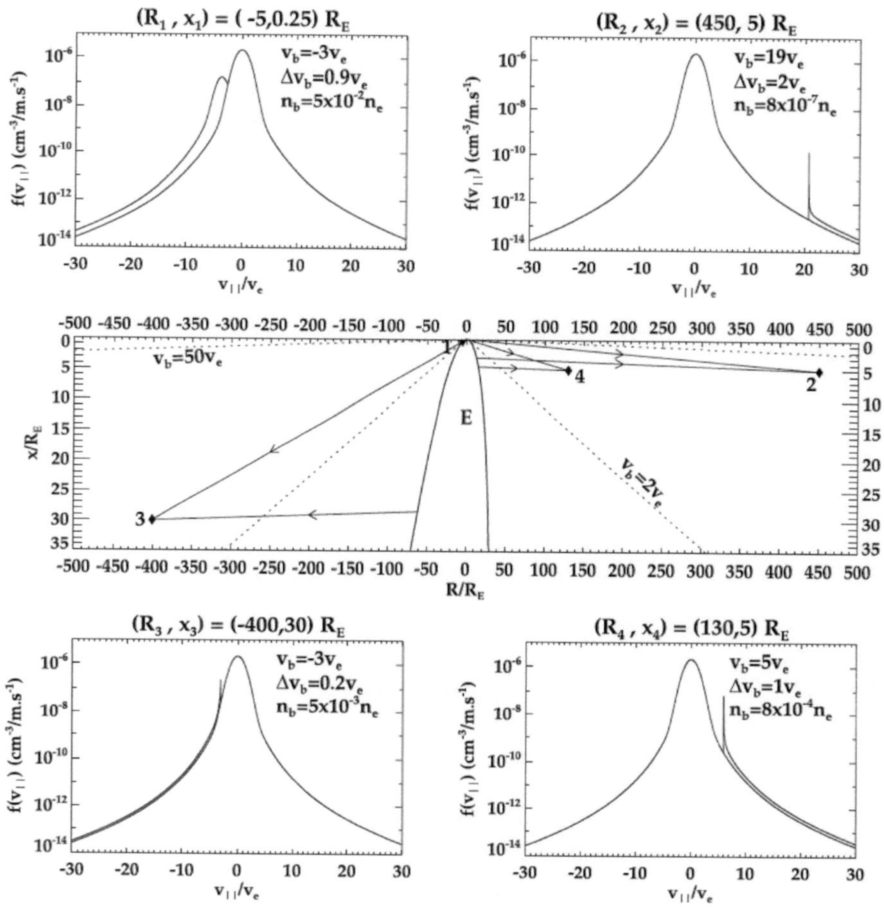

Fig. 23.25 Reduced electron distributions (*top* and *bottom panels*) at four locations in Earth's foreshock (Kuncic et al. 2004). The *middle panel* shows the Earth, bow shock, and particle paths corresponding to the cutoff speed and maximum $v_{\|}$ calculated, in (R, x) coordinates. The beam quantities v_b, Δv_b, and n_b quoted in the *top* and *bottom panels* are defined in the text

variations in v_c are evident, increasing with increasing R and decreasing x as predicted by Eq. (23.11), while n_b varies dramatically, being largest closer to the shock and for $v_b \leq 5V_e$, where V_e is the electron thermal speed. Beams are clearly evident in all 4 cases: the cutoff is dominant in determining the beam's character in all but the top left case, where variations in reflection efficiency along the shock determine the location of the peak in f_r.

Spatial variations in the $n_b(R, x)$, $v_b(R, x)$, and $\Delta v_b(R, x)$ have been calculated based on Liouville's equation and the above electron reflection and leakage physics (Fitzenreiter et al. 1990; Cairns et al. 1997). These are also used in Sections 23.5.3 and 23.5.4 below and existing predictions for type II bursts, Earth's foreshock, and the 2–3 kHz emissions (Knock et al. 2001; Knock and Cairns 2005; Cairns and Knock 2006; Florens et al. 2007; Kuncic et al. 2002, 2004; Cairns et al. 2004; Mitchell et al. 2004, 2005, 2009). In brief, $n_b(R, x)$ varies with v_b, R, and x, being zero at the upstream foreshock boundary, then increasing to a peak where $v_b \approx 3V_e$ for Earth's foreshock, and then decreasing again as v_b decreases and x increases for a given R. This is because the maximum energy gain for mirror reflection is limited to about a factor of 10 (Ball et al. 2001), so there is a balance between the energy gain factor via Eq. (23.8), its variation over the shock surface, and the fraction of incoming upstream electrons able to be accelerated to speeds of v_b or greater. In addition, n_b decreases as R increases, due to dilution of the total number of reflected electrons into an increasingly large foreshock volume. The fraction of the foreshock filled with beams fast enough and dense enough to drive significant Langmuir waves and radio emission, is therefore relatively small and typically found close to the tangent field lines, a $v_b \geq 3V_e$, and relatively close to the shock.

23.5.3 Growth of Langmuir Waves

Cutoff electron distributions thus naturally have a beam-like enhancement at high $v_\parallel$ and can drive Langmuir waves via the conventional electron beam or "bump-on-tail" instability (Filbert and Kellogg 1979; Cairns 1987b). The instability is driven by a positive gradient $\partial f_r(v_\parallel)/\partial v_\parallel$ of the reduced electron distribution function f_r and the growth rate is proportional to $\partial f_r(v_\parallel)/\partial v_\parallel$ evaluated at the wave phase speed. The characteristic growth rate γ_K of Langmuir waves resonant with the beam (meaning those with parallel phase speeds near v_b) is given by (Melrose 1985)

$$\gamma_K = \frac{n_b}{n_e}\left(\frac{v_b}{\Delta v_b}\right)^2 \omega_p. \qquad (23.12)$$

Here n_e is the background electron number density (assumed $>> n_b$), $\omega_p = 2\pi f_p$, and small constants of order unity are neglected. Thus large growth rates are predicted for beams that are relatively dense, fast, and cold.

The energy for the Langmuir waves to grow comes from the electron beam, causing the electrons to move towards lower kinetic energy and $v_\parallel$. Quasilinear theory (e.g., reviews by Grognard 1985 and Melrose 1985) treats quantitatively the relaxation of the electron distribution function and the growth of the waves. Homogeneous quasilinear theory for a delta function beam (Melrose 1985) predicts that the system evolves to a state in which 2/3 of the initial beam kinetic energy $1/2 m_e n_b v_b^2$ goes to the waves and 1/3 is retained by the electrons, whose distribution function is flattened into a plateau in $v_\parallel$ space. This plateau has $\partial f_r/\partial v_\parallel = 0$ and so has zero growth rate for Langmuir waves. Warm beams have less energy available for wave growth, now $1/2 m_e n_b v_b \Delta v_b$ (Melrose 1985).

Spatial inhomogeneities significantly affect quasilinear relaxation. One effect is due to reabsorption (or damping) of Langmuir waves by slower electrons. Qualitatively, it arises as follows for a spatiotemporally localized pulse of electrons: faster electrons outrun slower electrons to a given location, forming a beam in $f_r(v_\parallel)$ and driving Langmuir waves at relatively high phase speeds v_ϕ, which are subsequently reabsorbed by slower late-arriving electrons (since then $\partial f/\partial v_\parallel < 0$ at $v_\parallel = v_\phi$), thereby moving energy back into the beam and reducing its energy loss from the homogeneous prediction. It was proposed that inhomogeneous effects limit quasilinear relaxation and the beam instability in Earth's foreshock for example, allowing high levels of Langmuir waves to persist with electron distributions that are only partially quasilinearly relaxed and sometimes still have beam features present (Cairns 1987b). This proposal is supported by some numerical simulations for Earth's foreshock (Klimas and Fitzenreiter 1988).

Stochastic growth theory (SGT) takes the ideas of inhomogeneities and incomplete quasilinear relaxation several steps further by assuming that a wave-particle system is very close to a state of marginal stability (averaged over time and volume) and that the wave gain $G(t) = \ln(E(t)/E_0) = \int_{-\infty}^{t} \gamma(t)$ is a stochastic variable. Here $E(t)$ is the wave electric field at time t, E_0 is a reference field, and $\gamma(t)$ is the growth rate. The stochastic nature of G then predicts, by itself, the crucial qualitative point that the waves should be intrinsically bursty. Similarly, the logarithmic dependence of G on the wave field enables SGT to explain qualitatively the production of waves with a wide range of wave fields, from quasithermal to those approaching or exceeding the thresholds for nonlinear processes. Provided that many fluctuations in γ occur during some characteristic time for the waves then the Central Limit Theorem predicts that the probability distribution of E will be lognormal (meaning that $\ln E$ is Gaussian distributed). Figure 23.10 provides strong evidence that Langmuir waves in Earth's foreshock are in a SGT state. Indeed, SGT appears to apply widely, accounting for the wave properties in over 14 applications to date (Robinson et al. 2006).

One model for how an SGT state is achieved for electron beam-driven Langmuir waves involves the beam moving through an inhomogeneous plasma which has multiple evolving sites where wave growth is favored, leading to enhanced localized growth and modification of the particle distribution inside the sites, while the beam rebuilds between growth sites due to faster electrons outrunning slow ones and increasing $\partial f_r/\partial v_\parallel$ (Robinson et al. 1993; Cairns and Robinson 1997). Indicative calculations suggest that the model is viable for Earth's foreshock (Cairns and Robinson 1997) but have not yet been performed for other foreshock sources.

As well as providing an explanation of the bursty nature of the Langmuir waves, SGT provides a rationale for assuming that the waves are near marginal stability (suitably averaged over time and space) and for making a quantitative prediction for the amount of energy entering the Langmuir waves irrespective of the details of the wave growth and saturation. Specifically, the power flux entering the Langmuir waves equals the total time derivative of the free energy available from quasilinear relaxation of the electron beam, yielding in steady-state (Knock et al. 2001)

$$\frac{d}{dt}W_L = \mathbf{v}.\frac{\partial}{\partial \mathbf{r}}\left(\frac{n_b v_b \Delta v_b}{3}\right) \approx \frac{n_b v_b^2 \Delta v_b}{3l} . \qquad (23.13)$$

Here $W_L = \epsilon_0 E_L^2/2$ and the factor of 3 comes from multiplying the quasilinear prediction that 2/3 of the initial available kinetic energy reaches the waves with the 1/2 for the definition of kinetic energy. The derivative is now usually approximated by v_b/l, as in the rightmost form of Eq. (23.13), where $l = (R^2 + x^2)^{1/2}$ is the distance from the shock to the observer location along the trajectory for $v_{\parallel} = v_b$ (Knock et al. 2001).

Figure 23.26 illustrates the spatial variations in the beam quantities that enter Eq. (23.13) for a single ripple of an interplanetary shock with properties similar to the Bale et al. (1999). The beam properties furnished by the Liouville calculation of electrons reflected from the shock are subjected to quasilinear flattening, resulting in a plateau in the range $v_- \leq v_{\parallel} \leq v_+$ that connects the background thermal distribution to the accelerated electron component with no positive slope regions. Here $v_b = (v_+ + v_-)/2$, $\Delta v_b = (v_+ - v_-)/2$, and n_b is defined by conserving electron number. The combination of Eq. (23.13) and calculations like those in Fig. 23.26 predict robustly that the power inflow into Langmuir waves varies substantially with

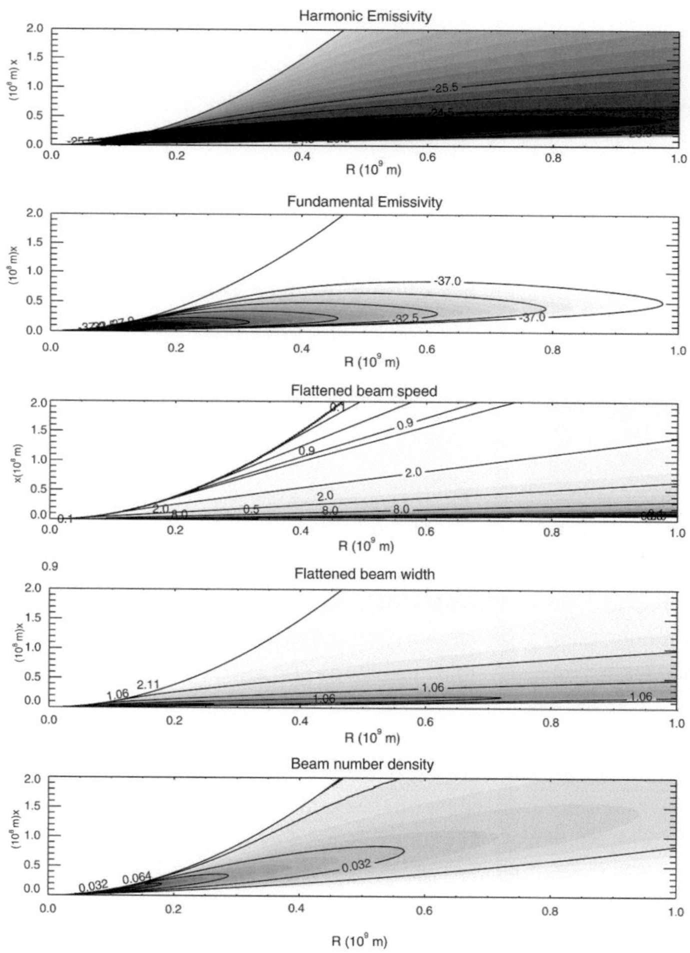

Fig. 23.26 Prediction of the quasilinearly-relaxed electron beam properties n_b/n_e, $\Delta v_b/v_b$, and v_b/V_e as a function of R and x in the top three panels, similar to Fig. 23.4 of Knock et al. (2001), for a ripple on an interplanetary type II shock with properties similar to the Bale et al. (1999) event. These allow prediction of the power flux into the Langmuir waves, via Eq. (23.13), and into radio waves (bottom two panels), as $\log_{10}(j_M)$ with j_M in units of W m^{-3} via Eqs. (23.16) and (23.18)

position in the foreshock, rising from negligible values for the fastest beams (with very low n_b) very close to the foreshock boundary, to a peak for beams with moderate speed but much larger n_b, before decreasing again for the very slow but dense and relatively narrow beams in the deep foreshock.

It is pointed out that retaining the vector nature of the derivative in Eq. (23.13) leads to spatial gradients with respect to both R and x individually, rather than just l, and that these gradients can compete and change sign (Kuncic et al. 2002). Specifically, Kuncic et al. (2002, 2004) found that the full derivative leads to the power flux being negative in regions where $v_b \lesssim 3V_e$ and the gradient in the available free energy with respect to x becomes negative as the shock become less efficient in producing accelerated electrons. That is, while wave growth still occurs for $v_b \leq 3V_e$ due to SGT effects, the net effect is of damping in the deep foreshock. More complete evaluation of these effects is needed.

It is possible to predict the average Langmuir field $\langle E_L(\mathbf{r})\rangle$ as a function of position using Eq. (23.13) as a starting point and using the standard wave growth equation

$$\frac{d}{dt}W_L = \alpha - \gamma W_L\,, \qquad (23.14)$$

where α is the energy input from spontaneous emission and other effects not dependent on W_L, and γ is the net damping rate. Combining this equation and Eq. (23.13) yields

$$\alpha_{spon} + \Lambda_{beam} = (\Gamma_L + \Gamma_{scat} + \Gamma_{ES})\langle W_L\rangle\,. \quad (23.15)$$

Here $\langle W_L \rangle = \epsilon_0 \langle E_L \rangle^2/2$, α_{spon} is the average rate for spontaneous emission (Melrose 1985), Λ_{beam} is the average rate of power input into the beam given by the right hand side of Eq. (23.13), Γ_L is the average net Landau damping rate (set to zero because the system is near marginal stability on average), Γ_{scat} is the nonlinear rate for diffusive scattering of the Langmuir waves out of resonance with the beam due to refraction by ambient density fluctuations, and Γ_{ES} is the nonlinear rate for the electrostatic decay process $L \to L' + S$. These quantities can all be specified using analytic theory, the Liouville and reflection analysis (for Λ_{beam}), experimental quantities, and reasonable assumptions for the levels and wavenumbers of the ambient density fluctuations (Kuncic et al. 2004; Malaspina et al. 2009). Note that the nonlinear damping associated with the coalescence $L + L' \to T$ is not included in Eq. (23.15) since it is much smaller than Γ_{ES}.

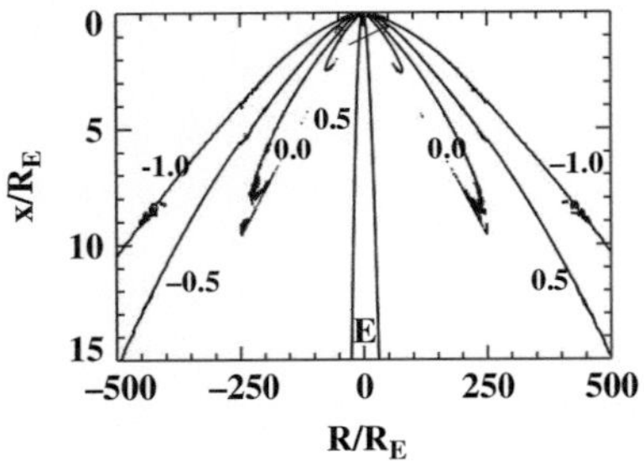

Fig. 23.27 Predictions for $\langle E_L(\mathbf{r})\rangle$ as a function of R and x for Earth's foreshock, in contours of $\log_{10}[\langle E_L\rangle/(1\text{mV m}^{-1})]$ (Malaspina et al. 2009)

Using refined expressions for Γ_{ES} and the vector approach to the convective derivative in Eqs. (23.13) and (23.14), Fig. 23.27 predicts $\langle E_L(\mathbf{r})\rangle$ in the foreshock for a set of reasonable foreshock parameters applicable to a specific data period (Malaspina et al. 2009). The maximum $\langle E_L\rangle$ is now 10 mV m^{-1} and occurs for beams with $v_b/V_e \approx$ 5–10, not right at the foreshock's upstream edge. These values are not inconsistent with observations (Anderson et al. 1981; Bale et al. 1997; Cairns et al. 1997; Bale et al. 2000; Kasaba et al. 2000; Malaspina et al. 2009) and are predicted by Kuncic et al. (2004) and Malaspina et al. (2009) to vary substantially with the solar wind parameters as the shock's reflection ability and the wave damping terms vary. It is emphasized that these predictions allow the fall-off with r to be estimated, yielding a power-law dependence $\langle E_L\rangle \propto r^p$ with index $p = -1.0 \pm 0.2$ that agrees very well with the index -1.01 ± 0.12 obtained from recent STEREO spacecraft observations (Malaspina et al. 2009).

23.5.4 Radiation Processes

The standard nonlinear processes considered to produce f_p and $2f_p$ radio emission are the following: The electrostatic (ES) decay $L \to L' + S$ to produce backscattered Langmuir waves L' and ion acoustic waves S from the beam-driven Langmuir waves L; the electromagnetic (EM) decay $L \to$

$T(f_p) + S'$ to produce radio waves T just above f_p and ion acoustic waves S', stimulated by the ES decay products S; and the coalescence $L + L' \rightarrow T(2f_p)$ of beam-driven L and backscattered L' Langmuir waves to produce radio waves just above $2f_p$(Cairns and Melrose 1985; Cairns 1988; Robinson and Cairns 1998; Li et al. 2008a, b). A general reason for favoring decay processes over coalescence processes is that decay processes require only a single population of nonthermal waves to produce nonthermal product waves, while coalescence processes require both participating populations to be nonthermal (Melrose 1985; Cairns and Melrose 1985).

The coalescence $L + L' \rightarrow T(2f_p)$ is the simplest known emission process for $2f_p$ radiation and previous analyses show that it can produce levels of emission comparable to those observed (Cairns 1988; Knock et al. 2001; Kuncic et al. 2002; Mitchell et al. 2004). It requires nonthermal levels of both L and L' waves to produce nonthermal $2f_p$ radiation (Cairns and Melrose 1985; Melrose 1985) and the ES decay is the fastest and most robust known nonlinear Langmuir process that can produce the required backscattered Langmuir waves. Specifically ES decay dominates scattering off thermal ions in almost all circumstances (Cairns 2000). (The exception is when the exponential growth time becomes smaller than the inverse of the ion sound wave frequency, a situation not expected in Earth's foreshock or realistic solar system environments; Cairns 2000.) Strong evidence also exists that ES decay proceeds for Langmuir waves in type III source regions and Earth's foreshock (Anderson et al. 1981; Cairns 1988; Cairns and Robinson 1995; Robinson and Cairns 1995; Henri et al. 2009), and it is directly relevant that the S waves produced by ES decay can stimulate the EM decay that produces f_p radiation (Robinson et al. 1994). Finally, EM decay is faster than scattering off thermal ions (with a similar proviso expected to that above for ES decay) and is the fastest known three-wave nonlinear Langmuir process for producing f_p radiation.

Standard analytic plasma theory yields the efficiencies with which energy is converted from beam-driven Langmuir waves into the L' waves ($\phi_{L'}$), f_p radiation (ϕ_F), and $2f_p$ radiation (ϕ_H). These are then combined with the power flux into the Langmuir waves to yield the volume emissivities j_M of radiation (the power output per unit volume and solid angle) throughout the foreshock (Robinson et al. 1994; Robinson and Cairns 1998; Dulk et al. 1998; Knock et al. 2001; Mitchell et al. 2004):

$$j_M = \frac{\phi_M}{\Delta\Omega_M} \frac{m_e n_b v_b^3}{3r} \frac{\Delta v_b}{v_b}, \tag{23.16}$$

$$\phi_F = 72\sqrt{3}\frac{\gamma_{L'}}{\gamma_S}\left(\frac{V_e}{c}\right)^3 \frac{v_b}{\Delta v_b} \frac{e^{-u_c^2}}{u_c\sqrt{\pi}}\zeta_F, \tag{23.17}$$

$$\phi_H = \frac{18\sqrt{3}}{5\gamma}\sqrt{\frac{m_i}{\gamma m_e}} \frac{v_b^2 V_e^3}{c^5} \frac{v_b}{\Delta v_b}\zeta_H . \tag{23.18}$$

Here $M = F$ or H, $\Delta\Omega_F = 2\pi$, $\Delta\Omega_H = 4\pi$, and the radiation is produced into bandwidths $\Delta f_F = 3(V_e/v_b)^2(\Delta v_b/v_b)$ and $\Delta f_H = 12(V_e/v_b)^2(\Delta v_b/v_b)$ that are dependent on the local beam parameters. The quantities $\gamma_{L'}$ and γ_S are the damping rates for the L' and S waves, respectively, $\gamma = 1 + \eta T_i/T_e$ is specified by Cairns (2000) as

$$\eta = \frac{T_e}{2T_i}\left(\sqrt{1 + \frac{12T_i}{T_e}} - 1\right) \tag{23.19}$$

and relates to the phase speed of ion acoustic waves, T_e and T_i are the electron and ion (proton) temperatures, and m_e and m_i are the electron and ion masses, respectively. The quantities ζ_F and ζ_H are the overlap fractions of Langmuir waves that are able to contribute to fundamental and harmonic emission, respectively, and are defined by (Robinson et al. 1994; Dulk et al. 2008) as

$$\zeta_F = \exp{-}\left[\frac{4\gamma m_e}{45 m_i}\left(\frac{v_b}{\beta\Delta v_b}\right)^2\left(\frac{3}{2}\sqrt{\frac{m_i}{\gamma m_e}} - \frac{v_b}{V_e}\right)^2\right] \tag{23.20}$$

$$\zeta_H = \frac{c}{2v_b}\sqrt{\frac{\pi}{6}}\frac{\beta\Delta v_b}{v_b}\left[\operatorname{erf}\left(\frac{\frac{\sqrt{3}V_e}{c} + \frac{2}{3}\sqrt{\frac{\gamma m_e}{m_i}}}{\frac{V_e}{c}\frac{\sqrt{2}\beta\Delta v_b}{v_b}}\right) + \operatorname{erf}\left(\frac{\frac{\sqrt{3}V_e}{c} - \frac{2}{3}\sqrt{\frac{\gamma m_e}{m_i}}}{\frac{V_e}{c}\frac{\sqrt{2}\beta\Delta v_b}{v_b}}\right)\right]. \tag{23.21}$$

Fundamental radiation is produced very close to the local f_p and so is strongly scattered and diffused by density irregularities in the source. It is also subject to loss by linear mode conversion, this time from electromagnetic radiation to Langmuir waves (most of whose energy is subsequently damped and not re-emitted as radiation). The expressions involving u_c in Eq. (23.17) are the escape factor for fundamental radiation (Robinson and Cairns 1998), with the value $u_c \approx 2$ found to be typical.

The flux density of radiation into the mode M is given by integrating Eqs. (23.17) and (23.18) over the source and accounting for propagation of the radiation from the source to the observer at $\mathbf{r}_0$:

$$F_M(f,t,\mathbf{r}_o) = \Sigma_{t'} \frac{\Delta\Omega_M}{\Delta f} \int d^3 V \frac{j_M(f,\mathbf{r},t')}{|\mathbf{r}-\mathbf{r}_o|^2}, \quad (23.22)$$

where

$$t = t' + \int d\tau / v_g(\mathbf{r}) \approx t' + |\mathbf{r}-\mathbf{r}_o|/c. \quad (23.23)$$

These forms slightly generalize those in Knock et al. (2001, 2003b) and Mitchell et al. (2004). Qualitatively the flux is calculated by integrating the volume emissivity over the source, taking into account the inverse distance squared fall-off of the radiation and the time required to reach the observer, and summing over all possible emission times t'. The first form in Eq. (23.23) assumes integration along the ray path for radiation with spatially-dependent group speed v_g, while the second form assumes straight-line propagation at the speed of light. Refined versions reject contributions from straight-line paths that lie through regions where the local f_p exceeds the radiation frequency f, since this radiation would be reflected and not reach the observer.

The bottom two panels of Fig. 23.26 shows j_F and j_H as functions of foreshock position for parameters appropriate to Bale et al.'s (1999) shock near 1 AU (Knock et al. 2001). Other examples are in Knock et al. (2001), Kuncic et al. (2002), Cairns et al. (2004) and Mitchell et al. (2004). Fundamental radiation is primarily produced where v_b is large, near the tangent field line, due to the strong dependences of ϕ_F, but particularly ζ_F, on v_b. Harmonic radiation is produced over a larger area, but with lower peak emissivity, for these parameters.

How efficient need the radiation processes be? Figure. 23.28 answers this question for Earth's foreshock. The peak values for ϕ_F and ϕ_H are similar and $\approx 10^{-8}$ for nominal conditions. However, ϕ_F is strongly peaked near the upstream edge of the foreshock, where fast beams are found, while ϕ_H varies slowly over a much broader volume albeit peaked near the tangent point. The sharply peaked behavior for fundamental emission is due to the rapid variations of ζ_F in Eq. (23.21) with v_b, which also varies rapidly with x near the foreshock boundary. The peak values for ϕ_F and ϕ_H depend strongly on V_e, v_b/V_e, $\Delta v_b/v_b$, and T_i/T_e and so vary appreciably within the heliosphere's many environments where shocks and associated foreshock emission are expected.

It is pointed out explicitly now that ϕ_F and ϕ_H as specified are averages over the angular spectrum, thereby not retaining the explicit angular dependences of the underlying emission processes. The EM decay, for instance, has an intrinsic dipolar directivity peaking perpendicular to $\mathbf{B}$, while the $2f_p$ coalescence has a quadrupolar directivity with peaks at 45° to $\pm\mathbf{B}$. Both dependences are included when calculating the averaged ϕ_M above. Reasons for not including these dependences are that F radiation is produced only just above f_p and so heavily scattered by density inhomogeneities in the source, while both F and H radiation are scattered in angle by density irregularities and subject to refraction by large scale density inhomogeneities along the myriad paths between source and observer (Riddle 1974; Steinberg et al. 1985; Robinson and Cairns 1998; Thejappa et al. 2007). These effects are believed to modify the intrinsic directivities significantly, most likely dominating them, with scattering tending to isotropize both the F and H radiation. This provides some justification for assuming isotropic emission at a level corresponding to the average over the angular emission rate. However, detailed descriptions of scattering and the effects of large-scale refraction are not included yet in the theory.

Other emission processes exist and were introduced in Sections 23.1 and 23.3.2: linear mode conversion (LMC), antenna radiation from Langmuir eigenstates, and direct emission by ring-beams. None of these have yet been included in the standard foreshock theory, although LMC is poised to be included. The current status of these mechanisms, including issues with extending the foreshock theory to include them, is discussed in detail in Section 23.8 below.

Initial attempts to compare the foregoing theory quantitatively with observational data are reviewed in the next two sections for type II bursts and the 2–3 kHz outer heliospheric radiation. Earth's foreshock is the arguably the optimum source for such

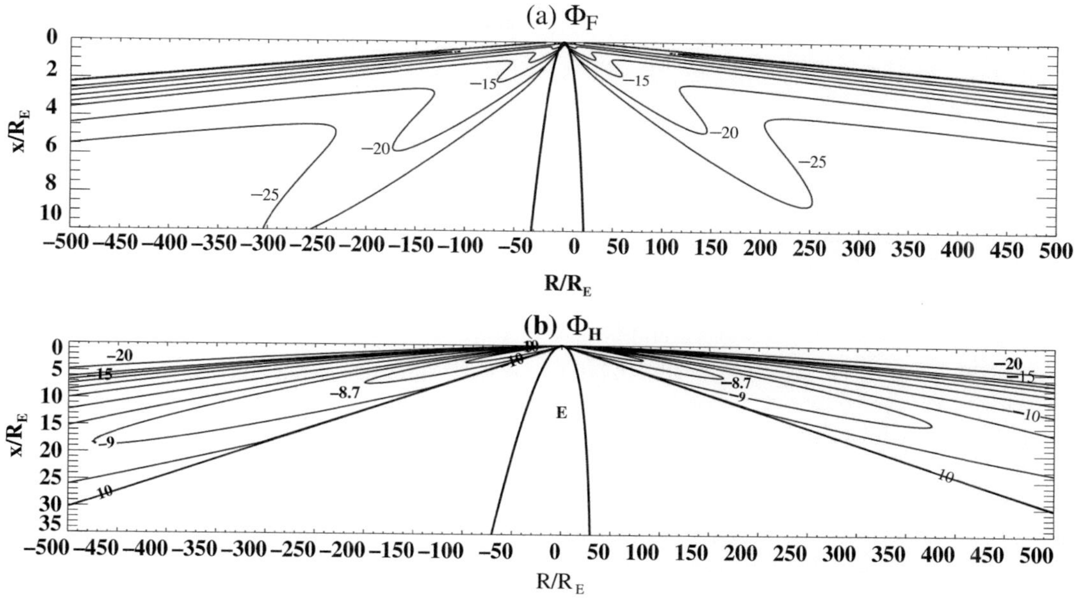

Fig. 23.28 Efficiencies ϕ_F and ϕ_H predicted by Eqs. (23.17) and (23.18) for fundamental and harmonic emission by the standard nonlinear processes, respectively, for Earth's foreshock under nominal conditions (Kuncic et al. 2004)

theory-data comparisons, due to the readier availability of radio data and the detailed plasma parameters in the source and at the shock. While comprehensive studies are still required, initial attempts are very encouraging. Kuncic et al. (2002) find that the theory predicts $2f_p$ fluxes within a factor of 2 of the Geotail spacecraft observations of Kasaba et al. (2000) for the same solar wind conditions. They also reported agreement to within a factor of 3–6 in flux with Lacombe et al. (1988) ISEE-1 observations of $2f_p$ radiation very near the upstream foreshock boundary. Finally, Kuncic et al. (2002) found agreement within a factor of 2 for the peak fluxes of $2f_p$ radiation observed by Cairns (1986b), with the observed radiation sometimes being much weaker, while the predicted flux 4×10^{-17} W m^{-2} Hz^{-1} of f_p radiation lay within the range $\approx 10^{-18} - 10^{-16}$ W m^{-2} Hz^{-1} observed.

23.6 Type II Radio Bursts

This section reviews detailed theoretical predictions for type II bursts, both interplanetary and coronal, and existing attempts to compare theory and observations. The observations and associated context are provided in Section 23.4.1 above. This section proceeds by considering the emission from an individual shock ripple (or equivalently a single, unrippled, macroscopic shock), then dynamic spectra from rippled shocks passing through purely model and then data-driven models for the corona and solar wind. Results and problems are also presented of a recent attempt to combine the type II theory with a global MHD simulation of a coronal shock. The section ends by summarizing outstanding issues and current research activities.

23.6.1 Flux Predictions for a Single Ripple or Unrippled Macroscopic Shock

The radio flux predicted by the theory in Section 23.5 depends sensitively on the shock and plasma parameters, as well as on observer location. This subsection combines Eqs. (23.10)–(23.23) to predict the radio flux at a single time from a single shock ripple, or equivalently an unrippled macroscopic shock with the same parameters, corresponding to a snapshot of the emission from a time-varying shock moving through a plasma. The background plasma is modelled simply in terms of a single electron component described as a gyrotropic kappa distribution with

$$f_\kappa(v_\parallel,v_\perp) = \frac{n_e \Gamma(\kappa+1)}{\Gamma(\kappa-1/2)} \pi^{-3/2} V_e^{-3} \left(1 + \frac{v_\parallel^2 + v_\perp^2}{V_e^2}\right)^{-(\kappa+1)}, \tag{23.24}$$

so that $f_\kappa(v) \propto v^{-2(\kappa+1)}$ for large $v^2 = v_\parallel^2 + v_\perp^2$. For Fig. 23.29 the plasma and shock parameters are appropriate to 1 AU: the paraboloidal ripple has a radius of curvature of 10^9 m at its nose, while the upstream plasma parameters are $T_e = 3T_i = 1.5 \times 10^5$ K, $n_e = 7$ cm^{-3}, $u = U = 300$ km s^{-1}, $B = 6$ nT and is oriented at an angle $\theta_{UB} = 85°$ relative to the ripple's velocity vector, and the observer is located 10^9 m upstream from the ripple's nose along its velocity vector (see the left hand image of Fig. 23.6 for more details).

Figure 23.29 shows that faster shocks are predicted to produce more intense type II bursts while sufficiently slow shocks should not produce observable radio emission (Knock et al. 2003a). This is not inconsistent with observational findings that faster (and larger) CMEs tend to produce brighter type II bursts (Cane and Stone 1984; Cane et al. 1987; Gopalswamy et al. 2001; Cairns et al. 2003). Not unexpectedly, the emission level decreases as the fraction of superthermal background solar wind electrons decreases (as the κ parameter increases from 2 to 5), since the shock-reflected electrons are initially preferentially superthermal due to Eqs. (23.7) and (23.9). Qualitatively, then, shocks moving through regions with enhanced populations of superthermal electrons (e.g., in the vicinity of CIRs or after previous flares or CMEs) are predicted to produce larger levels of radio emission for otherwise identical shock parameters. This is directly relevant to the localized emissions observed when CME shocks and CIRs interact (Reiner et al. 1998; Hoang et al. 2007; Cairns et al. 2003) and to "cannibalization" events and others in which a second CME moving through approximately the same spatial volume produces an observable radio burst whereas the first CME did not (Gopalswamy et al. 2001, 2002; Gopalswamy 2006).

The fluxes in Fig. 23.29 and similar predictions below are only observable if they exceed the background imposed by the galactic background radiation and observing instrument. The flux of the galactic background radiation varies significantly with observing frequency (Dulk et al. 2001; Hillan et al. 2010), being of order $\approx 10^{-19}$, $10^{-20.6\pm0.2}$, and $< 10^{-21.3}$ Wm^{-2}Hz^{-1} at frequencies 1–10 MHz, 300 kHz, and 100 kHz, respectively. In comparison, the Wind spacecraft's noise level is $\approx 10^{-21.4}$ Wm^{-2}Hz^{-1} in the range 100–400 kHz and $10^{-20.7}$ Wm^{-2}Hz^{-1} at 1 MHz (Dulk et al. 2001). Clearly the shock should be fast (high u and M_A) and the plasma have large numbers of superthermal electrons for the type II emission to be observable.

Figure 23.30 shows that shocks for which θ_{UB} is within $\approx 50°$ of being perpendicular are predicted to have higher levels of radio emission (Knock et al. 2003a), with quasiparallel shocks predicted to have weak emission (especially for large κ). While this has been inferred remotely for some coronal type II bursts (e.g., Stewart and Magun, 1980), detailed observational testing of this prediction for in situ type II bursts has not yet been performed. Fig-

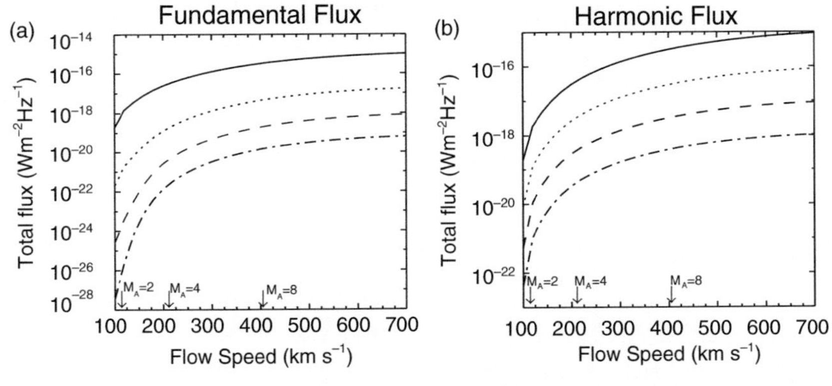

Fig. 23.29 Predicted (**a**) fundamental and (**b**) harmonic flux for a ripple on a type II shock as a function of the shock speed $U = v_{sh} - v_{sw}$ relative to the upstream plasma flow (Knock et al. 2003a). Each line is for a different κ parameter, ranging from 2 (*solid*) to 5 (*dot-dash*) as the relative fraction of nonthermal solar wind electrons decreases. The other shock and plasma parameters are listed in the text

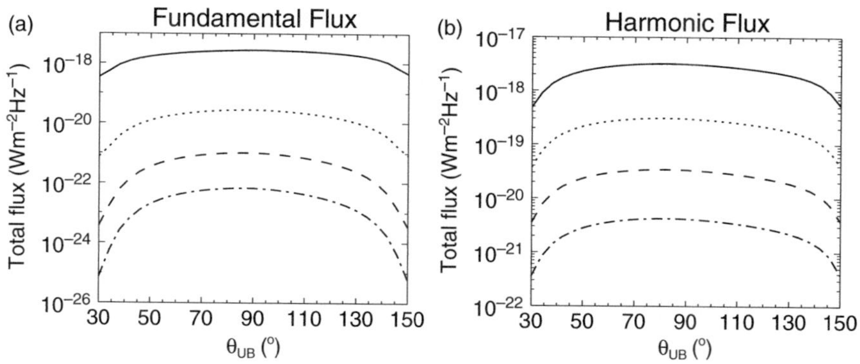

Fig. 23.30 Predicted (**a**) fundamental and (**b**) harmonic flux for a ripple with $U = 300$ km s^{-1} as a function of the angle θ_{UB} between the magnetic field vector and the ripple's relative velocity vector (Knock et al. 2003a). The line styles for κ and the shock and plasma parameters are as in Fig. 23.29

ure 23.30 shows that the dependence is weak, though, for quasiperpendicular orientations of $\mathbf{B}_u$ to the shock velocity vector. The reason is that Section 23.5's theory explicitly includes the radio emission from electrons leaving the shock at all possible local values of θ_{bn}. Accordingly all the shocks considered in Figure 23.30 have a tangent point where $\theta_{bn} = 90°$: what changes with θ_{UB} is the location of the tangent point on the ripple (see Fig. 23.6) and the local shock strength there (since u and so M_A and the shock jumps decrease with increasing distance away from the shock's nose).

The facts that the type II theory includes all reflected electrons and shows no significant variations near $\theta_{UB} = 90°$ in Fig. 23.30 (consistent with the above reasoning), and the absence of two emission regions in Fig. 23.26, provide direct contrary arguments to the elegant ideas of Holman and Pesses (1983) for herringbone bursts and the ordinary type II (backbone) emission. Specifically, instead of shock drift acceleration and the associated electron beams directly producing plasma radiation with the appearance of herringbones and the type II backbone as θ_{bn} and/or θ_{UB} are varied, the calculation naturally includes all θ_{bn} and finds no significant difference in beam or emission character or emission level as θ_{bn} is varied (or θ_{UB} is varied for quasiperpendicular values). Instead, it appears that intrinsic differences are required in the shock character as θ_{bn} is varied for the Holman and Pesses model to survive. An alternative interpretation is that interactions of the type II shock with localized spatial structures (e.g., current sheets) yields enhanced time-localized electron acceleration events and associated radio emission that are observed as herringbone bursts. Both these interpretations require detailed examination.

Figure 23.31 shows that the predicted radio flux scales with $b^{-2} \propto R_c^2$, where b is the curvature of the 3-D paraboloid $X = -b(Y^2 + Z^2)$ modelling the ripple in Figure 23.6 and R_c is the shock's radius of curvature at the nose (Knock et al. 2003a). This can be understood from the combination of $j_M \propto b$ due to the $1/r^2$ dependence in Eq. (23.22) and the volume of emission varying as b^{-3} for a distant observer. Thus, larger shocks should produce more intense radio emission and, moreover, the flux scales with the surface area of the shock. This is consistent with the results of Cane and Stone (1984), Cane et al. (1987) and Gopalswamy et al. (2001) that larger (and faster) shocks are more likely to produce observable type II bursts.

Other trends exist as functions of n_e, T_e, and B_u but are relatively weak compared with those described above (Knock et al. 2003a). In summary, Figs. 23.29, 23.30, and 23.31 demonstrate that stringent conditions exist on the shock and plasma conditions for observable type II emission to be observed. Specifically, the shock should have sufficiently large u, M_A, θ_{UB}, size (small b), and numbers of nonthermal electrons (small κ) and the observer should be sufficiently close. Since the shock and plasma properties should vary significantly in the temporally- and spatially-varying corona and solar wind, type IIs should be intrinsically bursty and time-variable in the corona and solar wind.

23.6.2 Dynamic Spectra and Macroscopic Predictions

This subsection addresses the prediction of dynamic spectra for macroscopic rippled shocks moving through the corona and solar wind. This requires

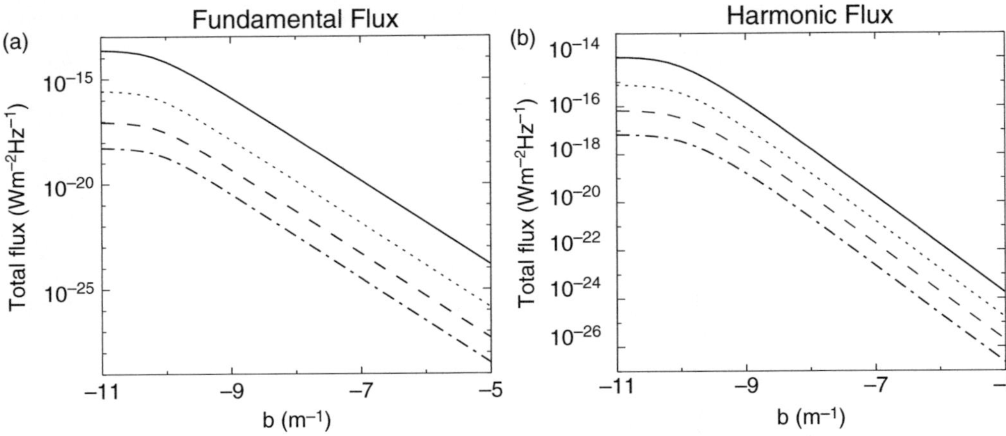

Fig. 23.31 Predicted (**a**) fundamental and (**b**) harmonic flux for a ripple with $U = 300$ km s^{-1} as a function of the curvature parameter b (Knock et al. 2003a). The *line styles* for κ and the shock and plasma parameters are as in Fig. 23.29

the calculation of radio emission for multiple ripples across the macroscopic shock, as well as modelling of the inhomogeneous corona and solar wind and the time-varying 3-D locus of the shock.

When multiple ripples are present on the shock some points in the upstream plasma can be connected to multiple ripples (for electrons with different $v_{\|}$, leading to multiple beams being present at these points, while other ripples may obstruct the particle paths leading to particular regions, thereby "shadowing" other ripples. These shadowing and multiple beam effects by neighboring ripples therefore directly affect the particle paths and so the electron distributions predicted upstream of macroscopic shocks (Knock et al. 2003b), in principle potentially modifying the predicted levels of Langmuir and radio waves. Importantly, calculations in Fig. 23.32(Left) show that ripples are independent to a good approximation (Knock et al. 2003b): the flux predicted for multiple realizations of the same seven 2-D ripples randomly located within a spatial interval, when multiple beam and shadowing effects are included, is within $\approx 30\%$ of that predicted assuming the ripples to be independent. Accordingly, it can be assumed that rip-

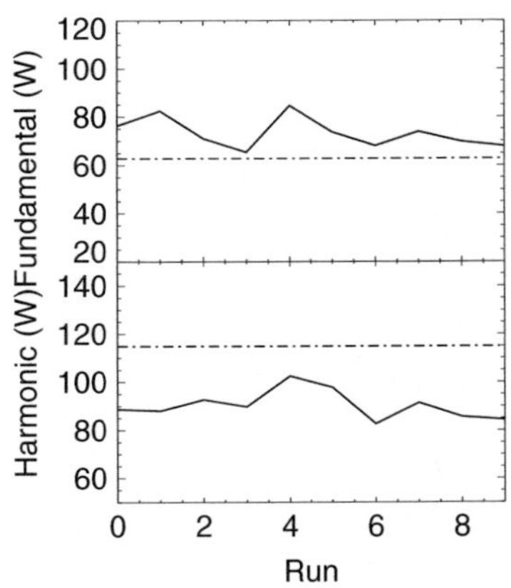

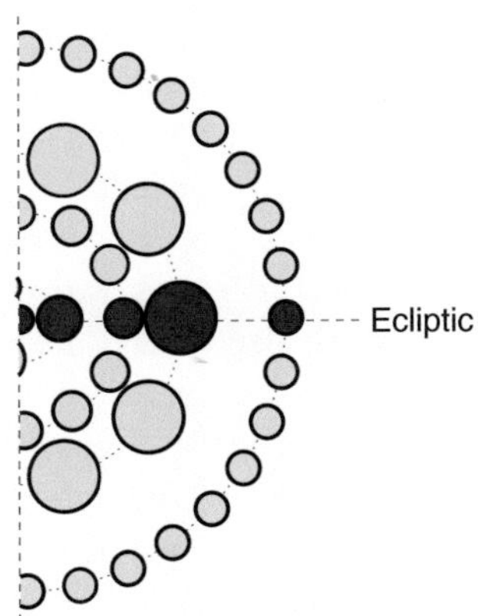

Fig. 23.32 (*Left*) Fundamental and harmonic fluxes predicted for multiple realizations of 7 randomly-located ripples, including multiple beam and shadowing effects on the electron distribution functions (*solid line*), compared with the summed flux from the same 7 ripples when calculated in isolation (*dashed line*) (Knock et al. 2003b). (*Right*) Illustration of how ripples are packed with half-hemispherical symmetry onto the macroscopic shock (Knock and Cairns 2005). See text for more details

ples are independent to within the level of accuracy currently considered necessary, greatly simplifying the calculations and length of the simulation runs. Put another way these calculations show that having multiple simultaneous ripples on the macroscopic shock for a constant density medium does not qualitatively or semiquantitatively affect the overall flux and dynamic spectrum predicted. The effects of an inhomogeneous medium on the dynamic spectrum are considered next.

Prediction of dynamic spectra for rippled shocks moving through the inhomogeneous corona and solar wind requires specification of (1) the properties of ripples considered on the macroscopic shock, (2) the time-evolving radius of curvature, 3-D location, and average velocity of the macroscopic parabolic shock, and (3) the properties of the inhomogeneous solar wind plasma. Two approaches exist to date. The first is to include ripples, use an analytic model for the macroscopic shock's velocity, and prescribe the plasma environment using an analytic or data-driven model (Knock et al. 2003b; Knock and Cairns 2005; Cairns and Knock 2006; Florens et al. 2007). The other is to ignore ripples, prescribe the shock motion and plasma environment using an MHD simulation, and develop approximate analytic formulae to predict the emission associated with the moving macroscopic shock (Schmidt and Gopalswamy 2008). Both are attractive. Most likely the optimum future approach is to include ripples but use an MHD simulation to prescribe the shock motion through an initial plasma model driven by data, thereby containing elements of both existing approaches (Cairns and Knock 2006). Since this optimum approach does not yet exist, the results of the two existing approaches are described next.

Ripples are important because they lead directly to fine structure in the dynamic spectrum, associated both with the intrinsic ripple lifetime and variations in plasma parameters across the macroscopic shock (Knock et al. 2003b; Knock and Cairns 2005; Cairns and Knock 2006), they are observed (Bale et al. 1999; Pulupa and Bale 2008), and they allow emission over the macroscopic shock to be calculated relatively efficiently. Ripples are assumed to be paraboloidal perturbations that evolve (i.e., appearing and disappearing) on a time scale $\tau_{rip} = R_c/V_A$, where R_c is the ripple's radius of curvature and V_A is the Alfven speed. This ripple lifetime provides a direct physical interpretation (with associated predictions) for the intrinsically time- and frequency-localized bursts of emission (sometimes called wisps) that make up coronal and interplanetary type II bursts. Specifically, the burst timescale should be τ_{rip} and the frequency extent $\Delta f \approx R_c f d/dr[\ln n_e(r)]$; their observed variations with f (and so r) should therefore constrain the radial profiles of $R_c(r)$, $V_A(r)$, and $n_e(r)$.

In analyses to date the ripple properties are indeed assumed to vary with r, with R_c Gaussian distributed around the decorrelation length of the magnetic field (Collier et al. 2000; Knock et al. 2003b; Knock and Cairns 2005; Neugebauer and Giacalone 2005). Computational limitations currently prevent the ripples being randomly packed onto the macroscopic shock and the contribution to the dynamic spectrum being calculated exactly. Instead, the ripples are closely packed with modified azimuthal symmetry about the Sunward direction, as shown in Fig. 23.32(Right): looking Sunward with the ecliptic plane horizontal, the eastern and western hemispheres of the macroscopic shock are packed independently and in an azimuthally symmetric fashion with ripples. The ripples are closely packed, with R_c equal to their separation distance, and their properties are chosen in the ecliptic plane. To include solar wind variability on ripple scales, the plasma density, velocity, temperatures, and magnetic field are sometimes perturbed with Gaussian-distributed fluctuations about the plasma model for the ecliptic plane. Then the radiation produced by a given ripple in the ecliptic plane is calculated, assuming no interactions with neighboring ripples. Computational limitations presently require (see Schmidt and Gopalswamy (2008), for an alternative approach) the assumption of azimuthal symmetry about the shock's average velocity vector but with different ripples in the eastern and western hemisphere. Nevertheless the falloff in the radiation flux with distance between the observer and each ripple is calculated exactly along straight line propagation paths. Moreover, if the plasma frequency along the path to the observer for a given ripple exceeds the radiation frequency, then the radiation is "blocked" and is not detected by the observer. As discussed in Section 23.5, isotropic emission patterns are assumed and neither scattering nor large-scale refraction are not included.

The analytic model developed thus far for the shock's motion, 3D locus, and properties is straightforward (Knock et al. 2003b). It assumes the macroscopic shock to be a paraboloid about the shock's average velocity vector and can include a constant acceleration. The shock's radius of curvature can be constant or can vary with r, thereby allowing the shock to propagate ballistically or to evolve.

The inhomogeneous plasma environment is specified either (1) analytically, including the desired radial and azimuthal variations in the plasma quantities, CIRs, magnetic clouds, and other desired inhomogeneities (Knock et al. 2003b; Knock and Cairns 2005), or (2) using the data-driven model of Florens et al. (2007), itself based on an earlier model of Reiner et al. (1998) for the electron number density only. The data-driven model assumes that the solar wind is constant over a solar rotation, converts the temporal variations in plasma quantities measured by a spacecraft near Earth at 1 AU into azimuthal variations of these quantities, and then uses the monthly-averaged solar wind speed, the Parker solar wind model, and assumed power-law variations for $T_e(r)$ and $T_i(r)$ to obtain models for $n_e(r,\phi)$, $v_{sw}(r,\phi)$, $\mathbf{B}(r,\phi)$, $T_e(r,\phi)$, and $T_i(r,\phi)$ as functions of r and azimuthal angle ϕ relative to Earth's location on a given day. This data-driven model naturally has fast and slow solar wind streams and realistic solar wind structures corresponding to particular type II events. However, it has limitations, particularly related to the assumption of the Parker spiral magnetic field, which means that the macroscopic variations in the direction of $\mathbf{B}$ are not realistic. Finally, the Rankine-Hugoniot conditions are used to specify M_A, θ_{bn}, etc. over the macroscopic shock and/or each individual ripple. Note that this prescription ignores the magnetic overshoot and reformation. Both these effects are predicted to be quantitatively important at the level of factors of 2–4 (Yuan et al. 2007, 2008a, b).

Figure 23.33 demonstrates the prediction of the dynamic spectrum of a type II burst from 200 MHz to 30 kHz (right) for a model wind (left) that contains two CIRs and two magnetic clouds (Knock and Cairns 2005). The shock is moving directly towards Earth with initial shock height, speed, and acceleration of $1.1R_S$, 1,100 km s^{-1}, and -3.2 m s^{-1}, respectively. Full descriptions of the plasma model are available elsewhere (Fig. 8 in Ref. Knock and Cairns 2005), but the CIRs and magnetics cloud are apparent in maps of $T_e(r,\phi)$ (shown), $v_{sw}(r,\phi)$, $\mathbf{B}(r,\phi)$, $T_i(r,\phi)$, and $\kappa(r,\phi)$, as well as the schematic (shown). Based on Figs. 23.29 and 23.30 and analogs in. Knock et al. (2003a), one predicts theoretically that these structures and spatial variations in plasma parameters should lead directly to frequency fine structures in the dynamic spectrum. This prediction is verified in Fig. 23.33(right), which identifies the spatial structure responsible for each frequency fine structure. The figure also shows a strong burst of metric emission, followed by a substantial gap in frequency and time before interplanetary emission starts below about 2 MHz. This is associated with $V_A(r)$ peaking at a height corresponding to $f_p \approx 4$ MHz, with $M_A \lesssim 2$ for 12 MHz $\leq f_p \leq$ 10 MHz and so not expected to produce observable emission (cf. Figure 23.29).

Figure 23.33 therefore provides strong qualitative support for the interpretations that (i) variations in u/V_A due to a peak in $V_A(r)$ cause gaps between metric and decametric emission for a shock/disturbance (Mann et al. 1999; Gopalswamy et al. 2001; Mann et al. 2003), (ii) multiple lanes and even split-bands can be associated with emission from multiple regions of a shock (McLean 1967; Knock et al. 2003b; Knock and Cairns 2005), and (iii) spatial variations in the plasma and interactions with CIRs, magnetic clouds, previous CMEs, and other coronal and interplanetary structures can give rise to hotspots (or null emission regions) on the shock and cause observable fine structures in the dynamic spectrum (Reiner et al. 1998; Gopalswamy et al. 2001, 2002; Cairns et al. 2004; Knock and Cairns 2005; Cairns and Knock 2006; Florens et al. 2007; Hoang et al. 2007).

It is expected that observers at different locations will observe different dynamic spectra, due to different source-observer distances and frequency blocking for example, and different source locations on the sky. Figures 23.34 and 23.35 show radio dynamic spectra, some information on the solar wind inhomogeneities, and radio source locations (projected into the plane of the sky) for the same shock and plasma model as in Fig. 23.33 for two distinct observers (Cairns and Knock 2006). One observer is well off to the eastern side of the Earth at solar-ecliptic coordinates (100,–100,0) Gm, potentially the STEREO-B spacecraft, and the second is close to the Earth at location (148,1,0) Gm, for instance the Wind spacecraft. The source location corresponds to the direction finding information an ideal observing instrument would have available. Of course, extraction of source information on type II bursts from the dynamic spectra and direction-finding data for two or more widely separated

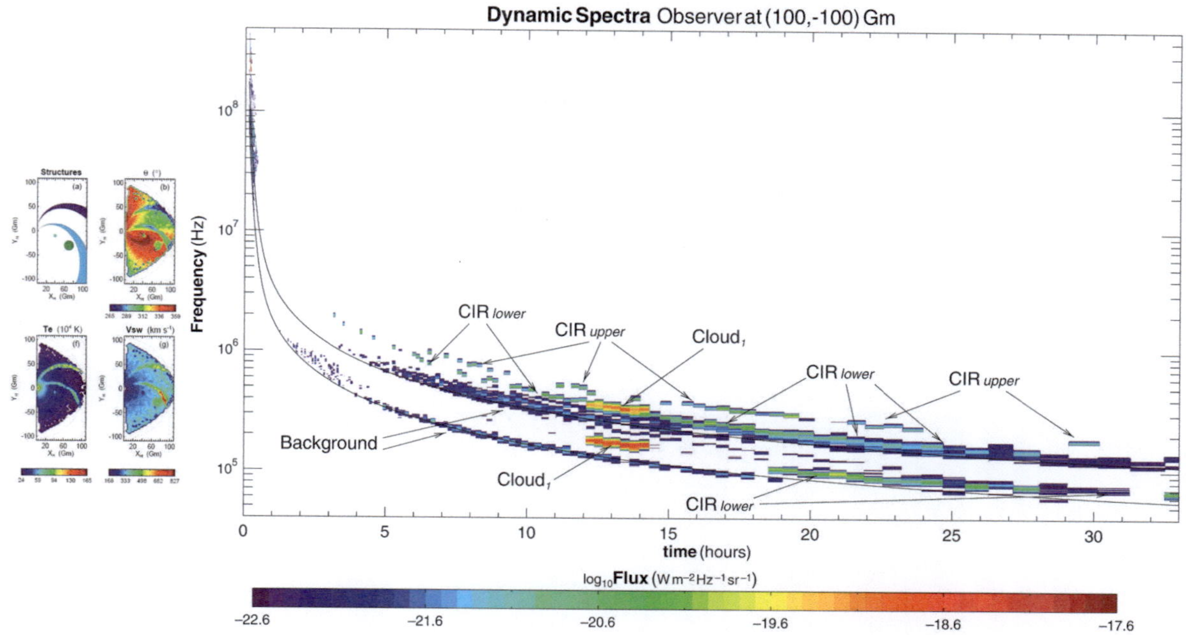

Fig. 23.33 (*Left*) Schematic and $\theta = \theta_{UB}$, T_e and v_{sw} maps of the interplanetary medium, with the Earth at $(X_H, Y_H) = (1.5,0)$ Gm and (*Right*) dynamic spectrum predicted for a shock moving through the coronal and interplanetary plasma environments summarized in (*Left*) (Knock and Cairns 2005)

observers is one major goal of NASA's STEREO mission (e.g., Bougeret et al. 2008).

Clear differences are indeed visible in the dynamic spectra (top panels) for the two observers in Figs. 23.34 and 23.35. These are due to the different relative distances between observers and elements (ripples) of the macroscopic source, as well as frequency blocking effects. Thus, the figures support the theoretical prediction that dynamic spectra observed in multiple locations indeed contain information on relative source-observer locations and the inhomogeneous plasma environment. As in Figure 23.33, some features in the dynamic spectra relate specifically to the interaction of the shock with macroscopic solar wind features. For instance, the intense (red) short-lived features at constant frequency at the times of the vertical and parabolic white lines in the top and leftmost bottom panel, respectively, correspond to the macroscopic shock crossing a magnetic cloud while the multiple long-lived curving features relate to the shock's interaction with CIRs (Reiner et al. 1998; Knock and Cairns 2005; Cairns and Knock 2006; Hoang et al. 2007).

Direction finding, however, may be required to constrain this information, as discussed next. Moreover, other physics related to scattering and directivity patterns may need to be added to better explain the detailed dynamic spectra: angular broadening and time delays due to scattering by density irregularities are likely to smooth fine structure in the dynamic spectrum, while anisotropic intrinsic directivity patterns for either radiation component would further modify the predictions for different observers.

The two rightmost bottom panels of Figs. 23.34 and 23.35 show the source location, projected onto the plane of the sky, for fundamental and harmonic radiation at the time when the macroscopic shock is crossing the first magnetic cloud (vertical and parabolic white lines in the figures' top and leftmost bottom panels). Emission from individual ripples is clearly visible. In addition, Fig. 23.34, for the western observer, shows the 3-D macroscopic source shape to be a paraboloid that is not seen exactly perpendicular to the direction of the source centroid's motion (e.g., ripples at a constant polar angle but not identical azimuthal angles do not project onto a straight line). The theory's predictions therefore suggest that direction-finding with STEREO and other spacecraft might permit the source's 3-D shape, including asymmetries, and direction of motion to be inferred remotely. If achieved

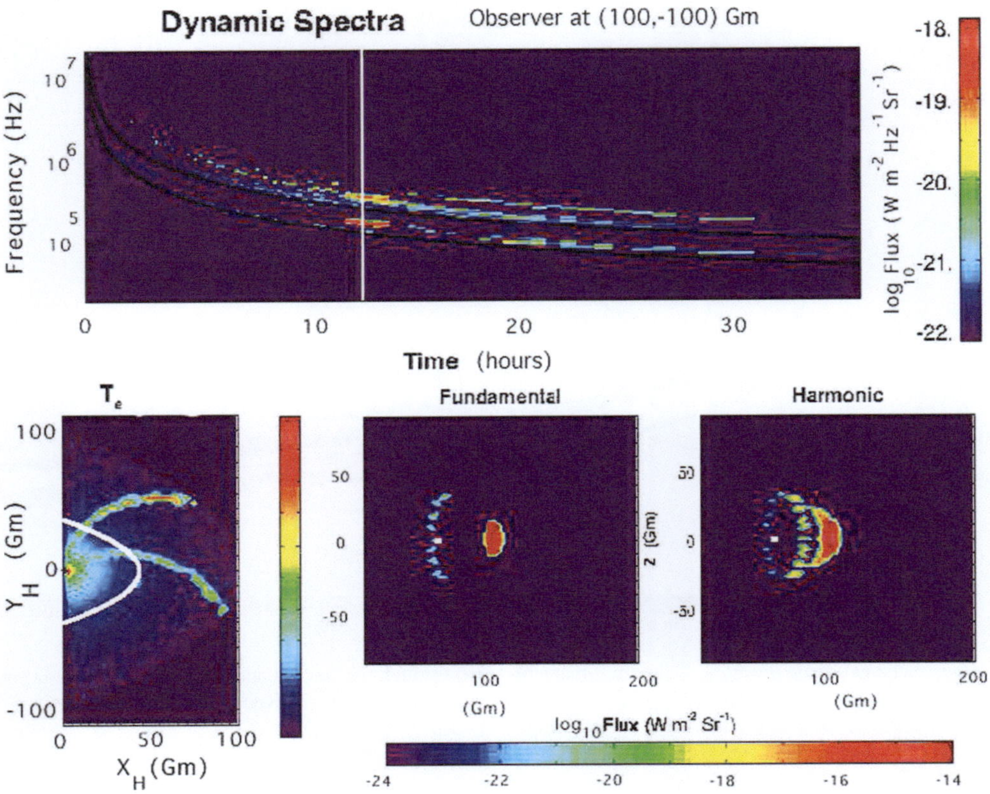

Fig. 23.34 (*Top*) Predicted dynamic spectrum, (*bottom left*) model electron temperature as a function of position in the ecliptic plane, and (*bottom center* and *right*) snapshots of the predicted source locations of fundamental and harmonic radiation, projected into the plane of the sky, for an observer at location (100,–100,0) Gm in solar-ecliptic coordinates (Cairns and Knock 2006). The snapshots are taken for the shock location and time shown by the parabolic and vertical white lines, respectively, in the leftmost and top panels

observationally, this would be extremely useful in predicting the arrival or not of space weather events at Earth.

The source seen by the head-on observer is shown in Fig. 23.35. Complementary information on the shock's 3-D structure from the eastern observer is evident. In particular, the azimuthal ripple-packing symmetries assumed in this theoretical implementation is clear. Despite the symmetry being unrealistic for a real type II shock it does elucidate the role of the macroscopic magnetic field direction: note that the western (right-side) ripples are on average much more intense than the eastern ripples, consistent with the angles between the macroscopic shock normal and Parker spiral field being closest to 90° and so with Fig. 23.30 predicting larger emission for otherwise identical ripple parameters. Put another way, the western hemisphere of the shock (on the right in Fig. 23.35) is quasiperpendicular while the eastern hemisphere is quasiparallel, so that the dominant emission is predicted from the western hemisphere. This appears to be consistent with the bias of radio-loud type IIs in the western hemisphere observed by Gopalswamy et al. (2008).

The foregoing examples have involved analytic models of the corona and solar wind. Since the plasma properties affect the predicted radio emission, it is clearly vital to realistically model the plasma environment for a given event if an accurate prediction of the dynamic spectrum is desired. The data-driven model of Florens et al. (2007), based on solar wind data for the solar rotation before the radio event, appears to be the most advanced available for all the required plasma properties. Figure 23.36 illustrates this model for the 24–26 August 1998 type II event observed by Bale et al. (1999), clearly demonstrating the complex solar wind structure for this event and the need to have

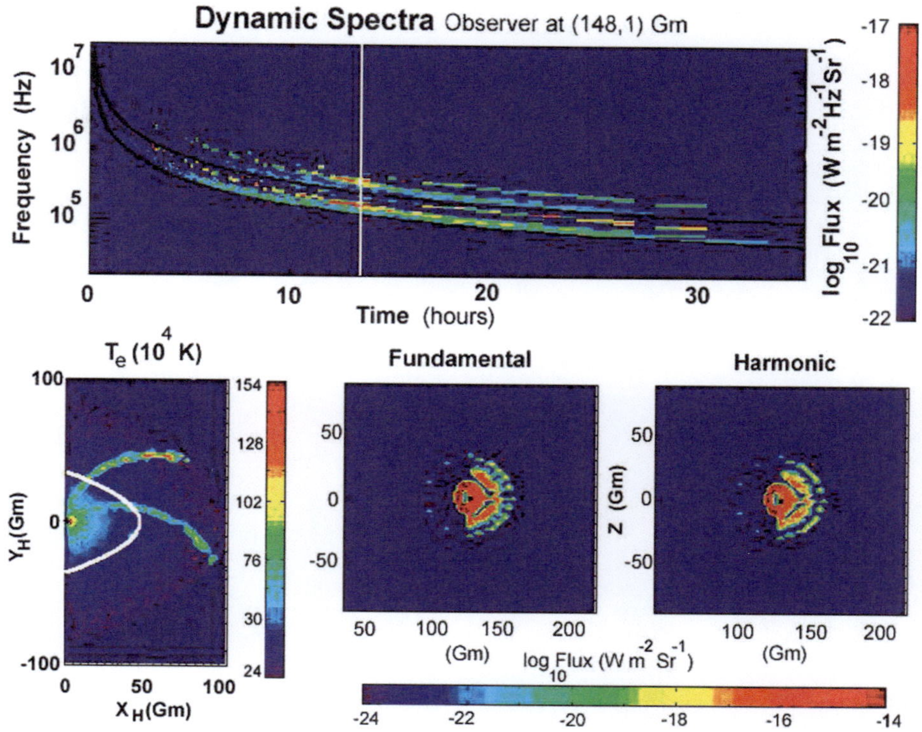

Fig. 23.35 Dynamic spectrum, electron temperature, and snapshots of the source location in Fig. 23.24's format for an observer very close to the Earth, at location (148,1,0) Gm (Cairns and Knock 2006)

event-specific solar wind (and coronal) models. It is pointed out that the more recent data-driven model of Opitz et al. (2009), uses both STEREO spacecraft to better model $n_e(r, \phi)$ and $v_{sw}(r, \phi)$ but does not predict **B**, T_e, and T_i. The Florens et al. (2007) model should be extended beyond the Opitz et al. (2009) model in due course using data from two or more spacecraft (e.g., both STEREOs and one or both of ACE and Wind), also allowing the basic assumption that the coronal sources are time-invariant to be tested.

Figure 23.37 compares the dynamic spectrum observed by the Wind spacecraft near Earth (top panel) with the following theoretical prediction: the dynamic spectrum is predicted using the foreshock type II theory for a shock with the properties identified by Bale et al. (1999) that moves through the inhomogeneous 2D solar wind plasma (Fig. 23.36) calculated using Wind spacecraft data and the Florens et al. (2007) model. Reasonable qualitative agreement with the timing and frequencies of bursts of enhanced emission is apparent: for instance the bursts near 2330, 0100, 0400, and 0600 on 24–25 August and the relatively continuous emission below ≈ 60 kHz after 1300 on 25 August (Florens et al. 2007). The upper panel is in dB relative to the spaceraft background while the lower panel is in absolute units, thereby complicating the task of testing quantitative agreement and requiring the spacecraft background to be accurately known for detailed comparisons. In their preliminary analysis Florens et al. (2007) find agreement typically to within a factor of 10 above 200 kHz, where the background is caused by galactic synchrotron emission (Dulk et al. 2001). They note that varying the shock direction significantly can alter the predicted flux by orders of magnitude. Nevertheless the degree of agreement found is very encouraging.

Finally, consider the novel work of Schmidt and Gopalswamy (2008), who used MHD simulations to predict the shock motion and properties through the model corona and combined these with analytic expressions based on the foreshock type II theory of Knock and collaborators (Knock et al. 2001, 2003a, b; Knock and Cairns 2005) to predict the dynamic spectrum and source locations of coronal type II bursts. Figure 23.38 illustrates the dynamic spectrum. It

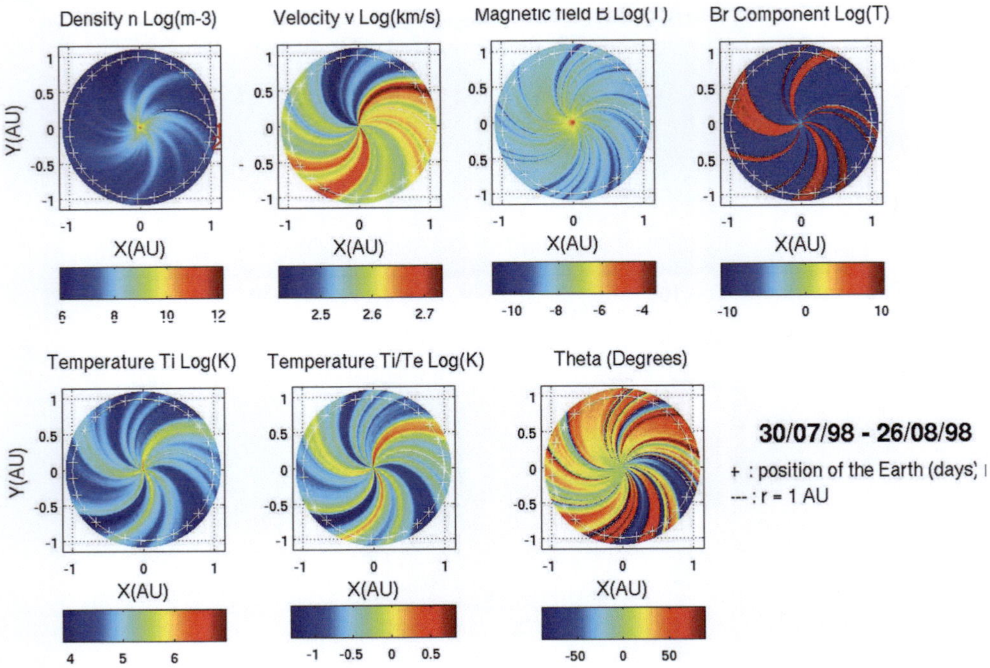

Fig. 23.36 Two-dimensional solar wind model of Florens et al. (2007) for the period 30 July to 26 August 1998 based on Wind spacecraft data (D.S. Hillan, personal communication 2010). The Sun is at $(X,Y) = (0,0)$ and the Earth at (1,0) on 30 July and then moves clockwise one plus sign per day around the white circle through the wind pattern. The *top* panels are (*left to right*) the solar wind density, speed, magnetic field strength $|\mathbf{B}|$, and inwards/outwards sense of the radial component of **B**, while the *bottom* panels are the ion temperature, ratio of ion to electron temperature, and angle θ between **B** and the radial direction. Note that $\theta \approx +45°$ for the Parker solar wind model

appears attractive, with multiple emissions resulting from different areas of the shock interacting with model loops and other structures, and fluxes that correspond with the observed ranges. The problem is that the authors find that both the peak emission and the bulk of the emission are produced behind the shock (e.g., their Figs. 2, 3, and 4 and Sections 5.3, 5.6, and 5.7). This is a contradiction, since the foreshock theory can only produce electron beams and associated radio emission upstream of the shock. It is also inconsistent with all the available data for Earth's foreshock and type II bursts where we have *in situ* data, with the possible exception of the Hoang et al. (1992) interplanetary events (which are also arguably best explained in terms of remote or upstream emission – see Section 23.3.2 above).

Thus, while combining analytic approximations to the emission model with MHD simulations that model the shock motion and plasma structures is very attractive and almost certainly a viable way forward, it appears that the execution of this idea is flawed in Schmidt and Gopalswamy (2008). A necessary improvement is to require that the emission is zero downstream of shocks. It is possible that this condition was not imposed because otherwise the predicted emission was weak, plausibly because the authors assumed the plasma electrons to have a Maxwellian rather than a kappa distribution function. This severely reduces the number of fast electrons and so the levels of emission predicted in the simulation: see the strong dependence of the flux in Figs. 23.29 and 23.30 on κ, where strongly nonthermal distributions with $\kappa = 2$ and 5 yield fluxes that differ by 4 orders of magnitude (Knock et al. 2003a).

23.6.3 Outstanding Issues and Future Work

The fundamental limitation of work to date on type II bursts, both coronal and interplanetary, is the lack of detailed quantitative comparisons between observa-

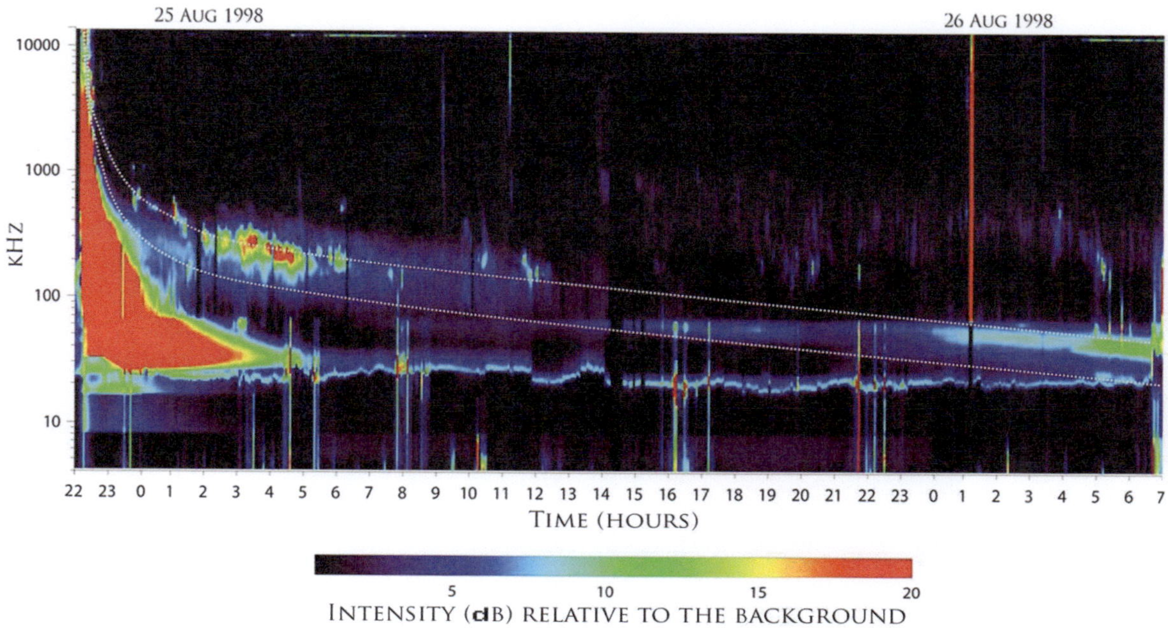

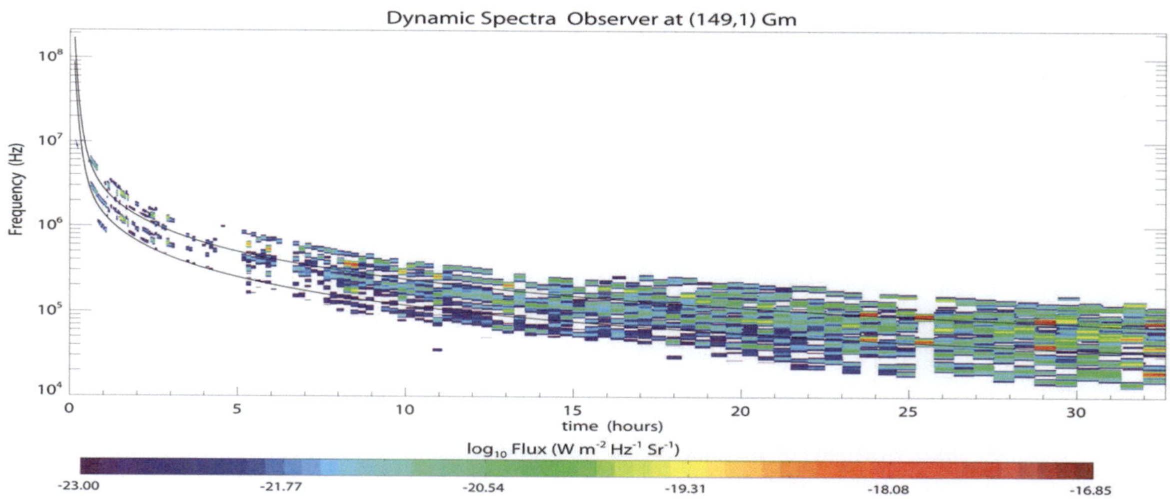

Fig. 23.37 Dynamic spectra for the 24–26 August 1998 type II burst of Bale et al. (1999): (*top*) observed and (*bottom*) predicted (Florens et al. 2007)

tions and theory for well-observed events. These can only be performed using a quantitative type II theory that uses reasonable models of the actual coronal and solar wind plasma properties, only recently available (Knock and Cairns 2005; Cairns and Knock 2006; Florens et al. 2007), the instrumental background from galactic radiation and thermal plasma waves (Dulk et al. 2001; Hillan et al. 2010), and accurate information and constraints on the shock's 3D time-varying locus (e.g., velocity, acceleration, shape, size, and expansion) from either simulations or an analytic model driven by coronal and interplanetary observations. This information is only available for a few events.

Nevertheless the time is now ripe for such detailed testing of available theories for type II bursts, especially because the direction-finding and triangulation capabilities of the two STEREO spacecraft and the

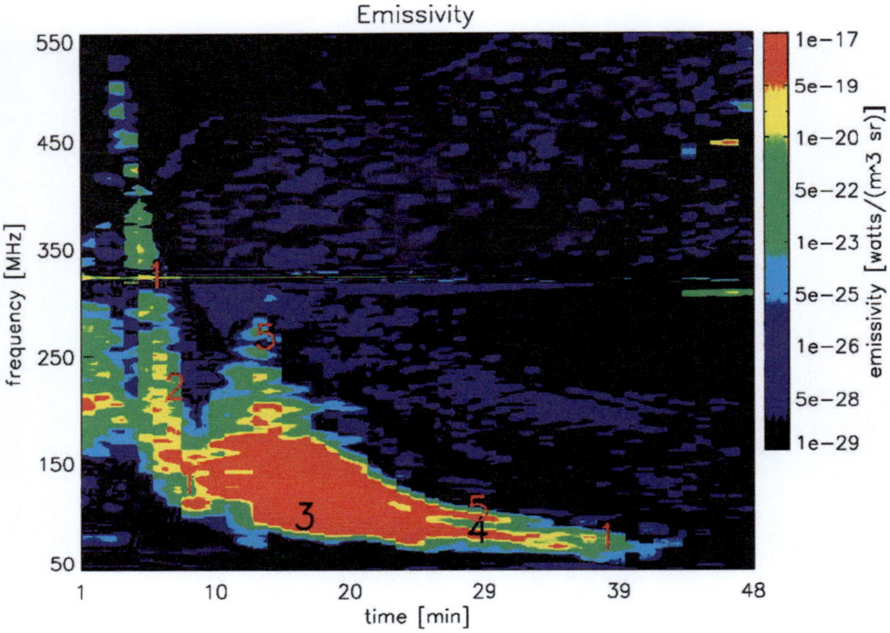

Fig. 23.38 Simulated type II burst of Schmidt and Gopalswamy (2008), using the combination of analytic approximations to the emission theory with MHD simulations of the shock motion and plasma structures. The numerical labels identify fine structures cause by the interaction of the shock with different coronal structures

Wind spacecraft (Bougeret et al. 2008) should enable the 3D location of the source and its emission hotspots to be determined better than ever before, while the coronagraph and other solar instruments on STEREO, TRACE, Solar Dynamics Observatory, and on the ground should permit CMEs, flares, and other disturbances to be characterized better than before. As a preliminary example of what is to come, Fig. 23.39 compares Wind data with the type II theory (including the data-driven solar wind model for this event) for the type II of 3 December 2004 (D.S. Hillan, personal communications, 2009, 2010). Visually there is reasonable semiquantitative agreement, with good agreement in morphology – there are two main bands, with intensifications at similar frequencies and times) – and in magnitude – observations and theory are plotted on the same intensity scale, in dB relative to the instrumental background, and typically the agreement is within 5–10 dB. Quantitatively, a cross-correlation analysis of this event yields a coefficient of order 50% with small offsets in frequency and time. Clearly detailed data-theory comparisons are still required, but efforts to date suggest that the existing theory is attractive and able to account broadly for some type II phenomena.

Once an accurate data-tested theory becomes available for type II bursts it will allow us to unlock the great potential that exists for predicting space weather events (at Earth and elsewhere in the solar system) that are driven by CMEs, their shocks, and the associated changes in magnetic field, plasma flow speed, and energetic particles based on solar and interplanetary radio data. In particular, iteratively comparing theory and data for the radio dynamic spectrum (as well coronagraph and energetic particle data) should allow extraction of the 3-D time-varying shock locus, including its velocity and acceleration, and so prediction of whether and when the shock and CME will impact Earth's magnetosphere. As an adjunct to this, robust automatic identification systems exist for type II and III bursts in coronal (metric) radio data (Lobzin et al. 2009, 2010) and could be usefully extended to STEREO and other interplanetary spacecraft datasets.

It should now be clear that the interaction of a shock with the inhomogeneous coronal and solar wind plasma is demonstrated to produce radio emission that is strongly reminiscent of coronal and interplanetary type II bursts, including the intensity and qualitative patterns of dynamic spectrum from $\approx$ 300 MHz to 10 kHz and many of the observed fine structures.

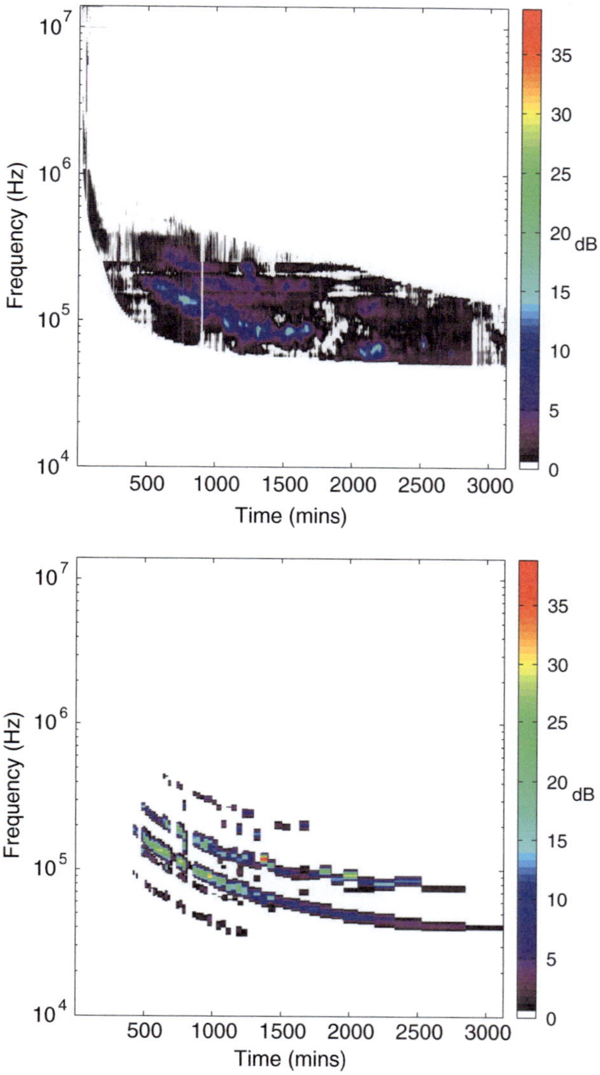

Fig. 23.39 Dynamic spectra for the interplanetary type II burst of 3 December 2004: (*above*) Wind spacecraft observations, restricted to events of interest, and (*below*) theoretical predictions based on the type II foreshock theory with the data-driven model for the solar wind (Florens et al. 2007) and shock parameters from LASCO (D.S. Hillan, personal communications, 2009, 2010)

However, equally evident is the fact that a significant number of outstanding issues exist, both in terms of theory/simulation and observations. The observational issues include the following:

- Are ripples at intermediate scales on the macroscopic shock vital to type II bursts or not, and do they cause the intrinsic blobs/burstiness/wispiness of type IIs due to their finite lifetime and spatial (frequency) extent? If not, then what causes the intrinsic blobs/burstiness/wispiness?
- Are metric type IIs all driven by blast-wave shocks, with interplanetary type IIs all produced by CME-driven shocks, or are some metric type IIs produced by CME-driven shocks and some by blast-wave shocks? At one level this is almost irrelevant for the type II theory, since all that is required is a shock wave and an upstream plasma environment that can be modelled. However, in more detail, the radio dynamic spectrum will depend on the shock's time-varying strength, velocity, and 3-D location, which all depend on whether the shock is a blast-wave (e.g., presumably weakening and slowing with time) or piston-driven (e.g., the CME provides a kinetic energy and momentum reservoir for the shock and may be accelerated outwards). Future comparisons of type II radio data with predictions based on coupling the type II theory with MHD simulations of shock evolution, cf. Schmidt and Gopalswamy (2008), may allow useful constraints to be set. These questions are also clearly important for observers and for understanding the correlations with CMEs and flares. See Section 23.4.1.2 for more discussion.
- Are herringbone fine structures caused by the shock interacting with (upstream) current sheets or other coronal structures, by special conditions for a part of the shock that are currently unknown, or are they caused by fast electrons from the flare site or CME that impulsively follow magnetic field lines (temporarily connected, for instance, by magnetic reconnection or instabilities of the CME's contact discontinuity) through the shock and into the foreshock where they produce bursts of radio emission via the type II theory? In this latter possibility the electrons would not have loss cone features in their distribution function, while in the first and third intuitively there would be much larger numbers of electrons with speeds above $c/3$ by analogy with type III bursts and their likely origin in magnetic reconnection sites. These possibilities can all be modelled quantitatively with small modifications of the existing type II theory.
- Are split-band type IIs best understood in terms of emission from two regions upstream of the shock with different f_p (and other plasma parameters), as can be explained simply and naturally by the

foreshock type II theory (McLean 1967; Knock and Cairns 2005; Cairns and Knock 2006; Florens et al. 2007) or are some actually due to simultaneous emission upstream and downstream of the shock (Smerd et al. 1974; Vrsnak et al. 2002)? Given the current absence of any theoretical justification for plasma emission downstream of the shock, the simplest test would be to ascertain whether the type II theory can predict suitable split-band-like structures for the events considered by Vrsnak et al. (2002). If so, then the upstream-downstream interpretation for type II bursts should be considered disproven unless future theoretical work provides a viable theoretical model or definitive observational evidence becomes available.

- Another interpretion for split-band type IIs involves fine structures spaced by $f_{ce}/2$ and f_{ce} for the fundamental and harmonic bands, respectively, based on observations of such splitting in f_p radiation produced in Earth's foreshock (Cairns 1994). Given that the observations in Earth's foreshock sometimes show 3 or more regularly-spaced bands in f_p radiation (Cairns 1994), coronal type II data should be examined for evidence of 3 or more bands in split-band events. In addition, the profile $B(r)$ obtained from fitting the observed bands should be compared with magnetic profiles obtained from photospheric measurements and associated modeling.
- Observations of slow-drift type II-like features composed of fast-drift bursts (Cane and Erickson 2005; Mel'nik et al. 2004) and type III bursts whose intensity changes as their electrons cross type II shocks (MacDowall 1989) need to be compared with theoretical predictions (Li et al. 2010) for type III electrons encountering localized temperature increases behind shock waves. It needs to be established whether the qualitative variations in intensity with frequency are consistent between theory and observations. If not, then alternative explanations for intensification of weak type III bursts near shocks need to be developed.
- Are the Langmuir waves in type II foreshocks well described by SGT, is Eq. (23.13) a good approximation, and what proportion of Langmuir energy with fields above the ES decay threshold is found in ordinary wavepackets versus the Langmuir eigenstates (Ergun et al. 2008) sometimes called Intense Localized Structures or ILSs (Nulsen et al. 2007)?

Significant theoretical issues exist and should be examined; many can be considered minor extensions and generalizations of the basic foreshock model for type II bursts described above. They relate to the fundamental question that started this subsection, as to whether the type II theory can go beyond its qualitative and even semiquantitative successes and actually quantitatively explain the observations. Issues with the basic theory include the following, the first three involving microphysics and the next three shock ripples and intermediate scale physics (issues with data-driven models for the solar wind and corona are discussed separately):

- The current theory calculates the electron reflection and acceleration under the assumption that the shock's magnetic mirror and potential jump are specified by the Rankine-Hugoniot predictions, thereby ignoring both the existence of significant overshoots in B and ϕ_{cs} and the effects of shock reformation (which also appears to increase the maximum values of B and ϕ_{cs}). Test-particle calculations suggest that inclusion of these effects will increase the energy flux into the foreshock electron beams and radio emissions by factors of 2–4 (Yuan et al. 2007, 2008a, b).
- Theoretical calculations of the evolution of electron beam - Langmuir wave systems with inhomogeneous plasma backgrounds should study the evolution to an SGT state, the fraction of Langmuir energy found in Langmuir eigenstates (ILS), and quantitatively justify the use of Eq. (23.13).
- Other emission mechanisms should be considered quantitatively and included in the theory, including linear mode conversion, emission from Langmuir eigenstates, direct linear processes, and frequency fine structures near $f_{ce}/2$ and f_{ce}. These options are discussed in detail in Section 23.8.1 below.
- The current numerical implementation of the theory packs ripples on the macroscopic shock assuming azimuthal symmetry about the shock's velocity vector within each of the eastern and western hemispheres, with the ripple parameters determined in the ecliptic plane. The azimuthal symmetry then means that ripples out of the ecliptic plane have parameters different than would be predicted directly from the current 2-D (cylindrical) data-driven solar wind model. Removing this symmetry would make the predictions more consistent

with the 2-D solar wind model currently used and eventually allow a 3-D solar wind model to be used.

- Detailed studies of shock ripples on intermediate scales should be performed to test how ripples develop and whether the decorrelation length of the magnetic field is the most appropriate predictor for the ripple scale, extending initial studies (Neugebauer and Giacalone 2005). While the current model is intuitive and attractive, more complete theoretical justifications should be sought. In addition, at present shock ripples are evolved statistically, not deterministically, representing another approximation.
- For a rippled curved shock electrons reflected by one ripple may be able to move upstream and cross another region of the shock, thereby potentially entering the downstream region and even producing Langmuir waves and radio emission downstream. This process has not been investigated theoretically yet. Difficulties anticipated include scattering by the downstream magnetic turbulence and the reduced contrast of the beam electrons compared with the heated downstream electron distribution. Overall this option appears unattractive due to the lack of evidence for downstream Langmuir waves and radio emission at Earth's foreshock or, especially, interplanetary type II bursts (Bale et al. 1999; Pulupa and Bale 2008).
- The effects of intrinsic directivity patterns and scattering of radiation by density irregularities, and refraction by large-scale variations in plasma density, can reduce the observed flux significantly and need consideration. This is discussed further in Section 23.5.4.
- Accurate mixed analytic-numerical implementations should be developed for the type II theory that can be "bolted-on" to global MHD (and other) simulations of the corona and solar wind. The work of Schmidt and Gopalswamy (2008) is an important first step in this direction that needs to be corrected and extended (see Section 23.6.2 for details). Such future implementations will likely need to include nonthermal particle distributions, ripple physics, intermediate scale turbulence, and the basic type II foreshock theory. The global simulation code will need to be initialized with coronal and solar wind structures provided by data-driven models like that of Florens et al. (2007).

Data-driven models for the corona and solar wind are of primary importance and need further development. While currently essentially state of the art, the model of Florens et al. (2007) needs to be extended along the following lines.

- The magnetic field remains Parker-like in the wind model and type II theory, despite observations (e.g., Fig. 23.36) often showing this to be a poor approximation to **B** and despite the type II predictions depending significantly on the angle between **B** and the shock ripple's normal (see Fig. 23.30) (Knock et al. 2003a; Knock and Cairns 2005). More realistic data-driven modelling of **B** is thus important, for instance by modelling the radial and tangential components of **B** separately and so allowing non-Parker magnetic field directions across the macroscopic shock.
- Presently the average solar wind speed (over a solar rotation) is used to relate longitude and time, potentially leading to significant errors in the positions of wind structures and so in the frequency, timing, and flux of radio emission. The 2-spacecraft approach of Opitz et al. (2009) can be used to directly test the persistence of structures (in time and longitude) between spacecraft, and so to better locate structures in (r,ϕ) space.
- The assumed stationarity of the solar wind parameters over 27 days is not always reasonable. The combination of Wind and STEREO will allow three-point assessment of this assumption and the development of more accurate plasma models for the range of longitude surveyed by the STEREO spacecraft.
- Currently solar wind turbulence and other short-spatial scale variations in the solar wind parameters are smoothed in the data-driven model, yet are likely to be especially important close to the Sun because random variations in ripple parameters (e.g., due to turbulence) can modify the frequency and flux of bursts by factors of at least two (Knock and Cairns 2005). Turbulence in the important wind parameters on intermediate and macroscopic scales could be identified observationally using the 2- or 3-spacecraft approach of Opitz et al. (2009), and then used to better model turbulence and plasma inhomogeneities in the theory.
- Last but not least, coronal parameters are necessary to make better predictions above 10 MHz, espe-

cially to understand the herringbone and other fine structures on type II bursts and to evolve our understanding of type II bursts towards predictive capability for (i) the corona's plasma environment and structures and (ii) the properties and motion of coronal shocks.

At first sight the lists of substantial issues above may appear daunting. However, they actually represent great progress over the last 20 years since they are almost all detailed questions that had to wait until a broadly viable quantitative theory for type II bursts was developed.

23.7 Outer Heliospheric Emissions

As reviewed in Section 23.4.3, the Voyager spacecraft have observed episodic bursts of radio emissions at 2–3 kHz in association with global merged interaction regions (GMIRs) reaching the vicinity of the heliopause. This section reviews the existing "GMIR/Priming" theory for the emissions, including comparisons with the observations. Earlier theories involving emission from the foreshock Sunwards of the termination shock (Macek et al. 1991; Cairns et al. 1992) or the inner heliosheath are reviewed elsewhere (Cairns and Zank 2001).

The GMIR/Priming theory combines Gurnett et al.'s (1993) GMIR model for the radiation events with the theory reviewed in Sections 23.5 and 23.6 for radio emission upstream of a shock and a theory for priming the outer heliosheath for subsequent triggering of a radiation event by the GMIR shock (Cairns and Zank 2001, 2002). The combined theory (Cairns and Zank 2002; Cairns et al. 2004; Mitchell et al. 2004) provides a quantitative theoretical basis for Gurnett et al.'s GMIR model. It involves the following primary concepts:

1. The observed radio emission is f_p radiation produced in foreshock regions upstream of a rippled GMIR shock (Cairns and Zank 2001, 2002; Cairns et al. 2004; Mitchell et al. 2004).
2. The radiation turns on (or is triggered) when the GMIR shock enters a region primed with an enhanced superthermal electron tail just beyond and near (within ≈ 50 AU of) the heliopause nose (Cairns and Zank 2001, 2002; Cairns et al. 2004; Mitchell et al. 2004).
3. The priming mechanism involves pickup ions that generate lower hybrid (LH) waves which then resonantly accelerate the electron tail by a process called lower hybrid drive (LHD) (Cairns and Zank 2001, 2002).
4. The pickup ions result from charge-exchange in the outer heliosheath of Region 3 neutrals produced originally in the solar wind (Mitchell et al. 2009) (formerly the neutrals were from the inner heliosheath (Cairns and Zank 2002)).
5. Constraints on LHD localize the priming to the outer heliosheath near the magnetic draping region (Cairns and Zank 2001, 2002; Mitchell et al. 2009).

As summarized in Section 23.4.3, the basic reason why a priming mechanism is required is that the GMIR shock does not produce observable emission in the solar wind, as evidenced by Voyager spacecraft observations when the GMIR shocks pass them and by the lack of observable radiation so far at frequencies consistent with f_p or $2f_p$ in the solar wind at distances beyond about 10 AU. Instead, the emission turns on when the GMIR shock reaches a suitably primed region beyond the heliopause. Questions that must be resolved by the GMIR/Priming theory include why and where does the radiation turn on, why does the GMIR shock not produce detectable radiation in the solar wind and inner heliosheath, and what region primed to emit radiation, and what is the priming mechanism?

Recently the IBEX spacecraft observed a ribbon of energetic neutral atoms (ENAs) that snakes across the sky between the locations of the two Voyager spacecraft and encloses (but off-center) the direction to the nose of the heliopause (McComas et al. 2009). One interpretation of this ribbon is that the ENAs result from pickup ions in the outer heliosheath (with energies of characteristic solar wind-VLISM interactions) and that the ribbon's location determines the direction of **B** in the outer helioseath (McComas et al. 2009; Heerikhuisen et al. 2010). These data and interpretation provide an independent line of evidence for pickup ions in the outer heliosheath, supporting previous strong theoretical arguments (e.g., review by Zank 1999, and references therein).

This section proceeds by reviewing the priming mechanism and the predictions for the radio flux and dynamic spectrum, before summarizing recent work on the strength and direction of the VLISM magnetic field (associated with the priming mechanism and the observed source locations). Outstanding observational and theoretical issues are then identified.

23.7.1 The Priming Mechanism

The priming mechanism is a result of charge-exchange interactions in the outer heliosheath between protons from the VLISM (Very Local Interstellar Medium) and neutral hydrogen atoms ("neutrals") coming from either the inner heliosheath (Cairns and Zank 2001, 2002) or the solar wind (Mitchell et al. 2009), sometimes called Region 3 and 2 neutrals (Zank 1999), respectively. These two sets of neutrals both result from charge-exchange of VLISM neutrals but involve protons from regions with very different properties (Zank 1999; Izmodenov et al. 2004): either cool protons ($T \approx 10^4$ K) with large bulk speed $\approx$ 400–800 km s^{-1} from the undisturbed solar wind or else hot protons ($T \approx 10^5 - 10^6$ K) with relatively small bulk speeds $\approx$ 100 km s^{-1} from the inner heliosheath (due to the termination shock). The importance here of charge-exchange is that it changes the distribution function (and momentum) of the plasma's protons, potentially leading to the generation of waves, and the acceleration and heating of particles, as explained next.

Charge-exchange involves an electron instantaneously moving from a hydrogen atom to a proton, ideally with no change in velocity of the proton or atom, i.e.,

$$H(\mathbf{v}_n) + p^+(\mathbf{v}_p) \rightarrow p^+(\mathbf{v}_n) + H(\mathbf{v}_p)\,. \qquad (23.25)$$

The newly charge-exchanged protons experience the plasma's convection electric field **E** and magnetic field **B** and are said to be "picked up" by the plasma, changing their previous straight-line motion into a gyromotion and developing an $\mathbf{E} \times \mathbf{B}$ plasma drift at the plasma's bulk velocity perpendicular to **B**. If $\mathbf{v}_n \cdot \mathbf{B} \approx 0$ and the distribution of $\mathbf{v}_n$ is narrow then the pickup ions have a ring distribution in velocity space perpendicular to **B** with ring speed $v_r \approx v_n$. (If $|\mathbf{v}_n \cdot \mathbf{B}| \neq 0$, then the proton distribution is more properly a ring-beam distribution, but this complication is ignored below.)

The proton distribution $f_{pu}(\mathbf{v}, \mathbf{r}, t)$ obeys a Boltzmann (or Liouville) equation with source and loss terms corresponding to charge-exchange. It can be solved as an integral along particle paths that reach the observation point **r** at time t, including the $\mathbf{E} \times \mathbf{B}$ drift, magnetic mirroring, draping of **B** across the heliopause surface, and propagation along $\pm\mathbf{B}$ (Zank 1999; Mitchell et al. 2009). Figure 23.40 presents a contour plot of $f_{pu}(v_\perp, v_\parallel)$ and a reduced distribution $f_r(v_\perp) = \int dv_\parallel\, f_{pu}(v_\perp, v_\parallel)$ at a point in the outer heliosheath about 30 AU from the heliopause nose. Both the contour plot and reduced distribution show that the charge-exchanged solar wind neutrals form a clear ring-beam with $v_r \approx v_\perp \approx v_{sw} = 500$ km s^{-1} and $v_\parallel \lesssim 100$ km s^{-1}. However, neutrals from the inner heliosheath only form a broad shoulder for $v_\perp \approx$ 50–200 km s^{-1} and not a ring-beam. Thus while it is confirmed that pickup ions should produce a strong ring-beam in the outer heliosheath (Cairns and Zank 2001, 2002), contrary to earlier expectations it is the solar wind (Region 3) neutrals that form the ring-beam and not inner heliosheath (Region 2) neutrals (Mitchell et al. 2009).

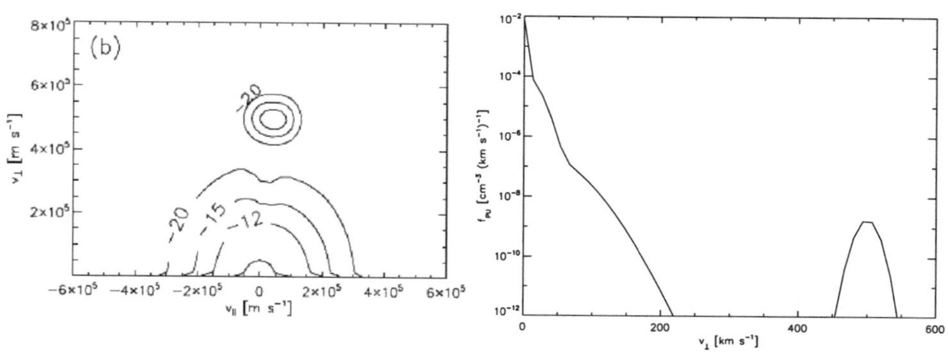

Fig. 23.40 (*Left*) Contour plot of $f_{pu}(v_\perp, v_\parallel)$ and (*Right*) the reduced distribution $f_r(v_\perp)$ for pickup protons in the outer heliosheath about 30 AU from the heliopause nose (Mitchell et al. 2009). Charge-exchanged solar wind neutrals form a clear ring-beam with $v_r \approx v_\perp \approx v_{sw} = 500$ km s^{-1} and $v_\parallel \lesssim 100$ km s^{-1} but neutrals from the inner heliosheath do not

et al. 2009). Thus, the theory proceeds as before, but with a change in the detailed source of pickup ions. This change turns out to be beneficial on balance, since although v_r is a factor ≈ 4 larger and the ring number density n_r a factor ≈ 10 lower, the kinetic energy available from the pickup ions goes up by a factor of ≈ 1.6. The full consequences of this change in the source of pickup protons have not yet been explored.

The reason that pickup ions are important in this context is that when the ring is sufficiently narrow and slow in velocity space and the plasma β (equal to the ratio of the thermal pressure to the magnetic pressure) is sufficiently low, then the ring is unstable to the growth of LH waves which can then accelerate electrons and heat the pickup ions via a process sometimes called lower-hybrid drive (LHD) (McBride et al. 1972; Omelchenko et al. 1989; Cairns and Zank 2002). LH waves have low frequencies $f \approx f_{LH} \approx (f_{ce} f_{ci})^{1/2}$), are primarily electrostatic, and propagate almost perpendicular to $\mathbf{B}$ with $k_\perp/k_\| \approx (m_i/m_e)^{1/2}$. Their phase fronts then move very fast along $\mathbf{B}$, and the associated parallel electric fields can accelerate electrons to large speeds via the Cherenkov resonance. Ring distributions can effectively drive LH waves that resonate with both the ion and electron distributions, with $\omega \approx \omega_{LH} \approx k_\perp v_r \approx k_\| v_{\|,e}$. Under these conditions the LH waves accelerate a plateau-like superthermal tail out of the thermal electron distribution (McBride et al. 1972; Omelchenko et al. 1989; McClements et al. 1993; Shapiro et al. 1998; Cairns and Zank 2001, 2002), with maximum speed

$$v_m = (m_i/m_e)^{1/2} v_r \,. \qquad (23.26)$$

For the ring-beam in Fig. 23.40 LHD in the outer heliosheath should lead to $v_m \approx 15 V_e \approx 1.6 \times 10^7$ m s^{-1}, where $V_e \approx 10^6$ m s^{-1} is the electron thermal speed, and a total tail fraction $\approx 10^{-6}$ of the background electron number density (Mitchell et al. 2009).

Constraints on LHD localize the priming mechanism to the outer heliosheath but near the heliopause, where interstellar magnetic field lines are draped over the heliopause (Cairns and Zank 2002). (Recent suggestions that $\mathbf{B}_{VLISM}$ is closely aligned with $\mathbf{v}_{VLISM}$ mean that the heliopause nose need not be close to the draping region (Opher et al. 2009a; Pogorelov et al. 2009)). Specific reasons include: (1) Simulations show that LHD is only efficient when

$$v_r/V_A \lesssim 5 \qquad (23.27)$$

(Omelchenko et al. 1989; Shapiro et al. 1998); (2) the LH waves must have minimal damping by thermal ions and electrons. Figure 23.41 shows estimates of the ratio v_r/V_A (Cairns and Zank 2002) along the Sun-heliopause nose axis, calculated assuming charge-exchange of inner heliosheath neutrals ($v_r = 100$ km s^{-1}), n_e obtained from a plasma-neutral simulation (Zank et al. 1996), and values for B calculated using the convected field approximation for $\mathbf{B}_{VLISM}$ (assumed perpendicular to $\mathbf{v}_{VLISM}$ with $B_{VLISM} = 0.15$ nT). Magnetic draping and flow stagnation at the heliopause lead to v_r/V_A decreasing from values ≈ 10 in the VLISM to values $\lesssim 5$ close to the heliopause nose, before increasing again in the solar wind. Enhanced lower hybrid damping precludes effective LHD occurring in the inner heliosheath. The reason is that $T_i \gtrsim 10^6$ K and $T_e \gtrsim 10^5$ K there, due to heating at the termination shock, whence $v_r \lesssim V_i$ and $v_m \lesssim 2V_e$ so that growth of LH waves is quenched. Thus Fig. 23.41 predicts that the LH instability and enhanced superthermal electron tail are limited to the magnetic field draping region of the outer heliosheath, presumably within $\lesssim 50$ AU of the heliopause nose.

The calculation in Fig. 23.41 is explicitly for neutrals from the inner heliosheath and not the solar wind, despite the new results in Fig. 23.40, but can be generalized by increasing v_r to suitable values. Indeed, exactly the same figure would follow for solar wind neutrals with $v_r = 400$ km s^{-1} if B_{VLISM} were

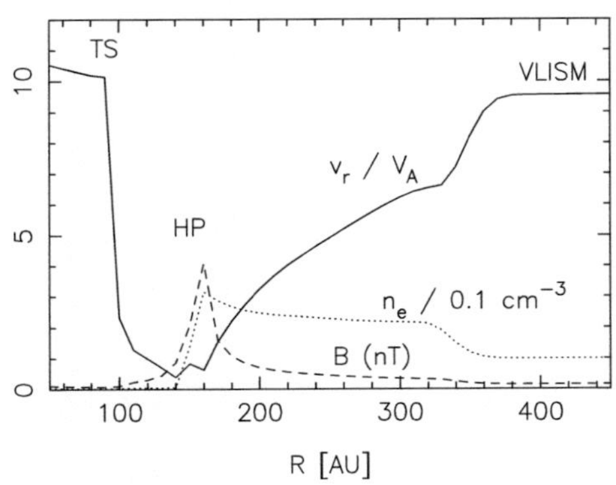

Fig. 23.41 Spatial variations of v_r/V_A, B, and n_e along the Sun-heliopause nose axis based on plasma-neutral simulations and the convected field approximation for Region 2 neutrals (Cairns and Zank 2002). The locations of the termination shock and heliopause are marked by symbols TS and HP, respectively

increased by a factor of 4 to 0.6 nT. The basic result is that Eq. (23.27) can still be satisfied for solar wind neutrals but requires higher values of B_{VLISM}. Recent global heliospheric simulations are independently coming to similar values $\gtrsim 0.4$ nT by considering where the Voyager spacecraft crossed the termination shock, the directions of the downstream flow velocities, and energetic particle observations upstream of the termination shock (Opher et al. 2009a; Pogorelov et al. 2009). Better calculations of the magnetic draping also allow such figures to be refined. An appropriate conclusion is that analogs of Fig. 23.41 imply that the GMIR-Priming theory remains viable, and able to explain the localization of the priming to the outer heliosheath near the heliopause, provided B_{VLISM} is $\gtrsim 0.4$ nT (Cairns et al. 2006; Mitchell et al. 2009; Pogorelov et al. 2009).

23.7.2 Predicted Radio Fluxes and Dynamic Spectra

Predictions for the properties of electron beams, Langmuir waves and radio waves produced upstream of the GMIR shock can now be obtained (Cairns et al. 2004; Mitchell et al. 2004, 2009) by combining Section 23.7.1's priming mechanism with the foreshock radio theory in Sections 23.5 and 23.6 for a plasma environment specified by independent global simulations of the solar wind – VLISM interaction (e.g., Zank et al. 1996; Zank 1999; Izmodenov et al. 2004; Opher et al. 2009a; Pogorelov et al. 2009). Put another way, the priming theory predicts the existence (or not) and properties of the superthermal electron tail produced by LHD that is superposed onto the background electron distribution for a given location of the shock, and the shock then reflects and accelerates the local electron distribution to form electron beams that drive Langmuir waves and radio emission. The theory presently predicts 6 quantities for a single paraboloidal ripple on the global shock at a given location: the reduced electron distribution $f_r(v_\parallel, \mathbf{r})$, gyrotropic electron distribution $f_r(v_\parallel, v_\perp, \mathbf{r})$ and volume emissivities $j_F(\mathbf{r})$ and $j_H(\mathbf{r})$ of radiation throughout the 3D foreshock, and the fundamental and harmonic radiation fluxes $F_F(\mathbf{r}_{ob})$ and $F_H(\mathbf{r}_{ob})$ at an observer location $\mathbf{r}_{ob}$, respectively. The fluxes are calculated by integrating the volume emissivities throughout the 3-D source while taking into account the $|\mathbf{r} - \mathbf{r}_{ob}|^{-2}$ falloff and straight-line propagation from each source element to the observer. Of course, the dynamic spectrum at a given location can be calculated by allowing the shock to move into regions with different $n_e(\mathbf{r})$ and analysing the time-varying frequency and flux.

Before proceeding it is stated that the calculations below are all for priming using the parameters appropriate to inner heliosheath neutrals, rather than the solar wind neutrals that should be used (Mitchell et al. 2009). These calculations therefore need to be redone for solar wind neutrals. The qualitative implications of these changes are expected to be small, but are summarized at the end of this subsection.

The background electron distribution before priming is assumed to be a generalized kappa distribution given by Eq. (23.24). Values for κ, n_e, T_e, and B are specified in Table 23.1. The axis of the ripple is assumed parallel to $\mathbf{U}$ and perpendicular to $\mathbf{B}$. The properties of the LHD tail are given by Eq. (23.26) and the ratio $n_T/n_e = 10^{-5}$ predicted assuming the pickup ions have number density $n_{ce} = 10^{-4}$ cm^{-3} appropriate to the inner heliosheath. Finally, it is relevant that $f_p = 2.6$ kHz for $n_e = 0.1$ cm^{-3}.

Table 23.1 Nominal shock and plasma parameters for the outer heliosheath

U	R_c	T_e	n_e	κ	B
600 km s^{-1}	0.42 AU	8,000 K	0.1 cm^{-3}	5	0.1 nT

The importance of priming is demonstrated first. Figure 23.42 compares $f_r(v_\parallel)$ at two foreshock locations for the following situations: shock acceleration of background electrons with no LH priming (dotted curves), LH priming but no shock acceleration (dashed curves), and both shock acceleration and LH priming (solid curves). The calculations are due to J.J. Mitchell (personal communication, 2004) and assume inner heliosheath neutrals (Cairns 2004). The foreshock locations are $(R,x) = (100,15)$ Gm and (50,3) Gm for the left and right panels, in terms of the usual foreshock coordinate system, with the latter case having larger $v_c \approx v_d R/x$. Comparing the solid and dotted curves it is evident that priming increases the number of reflected electrons by many orders of magnitude, while differences between the solid and dashed curves

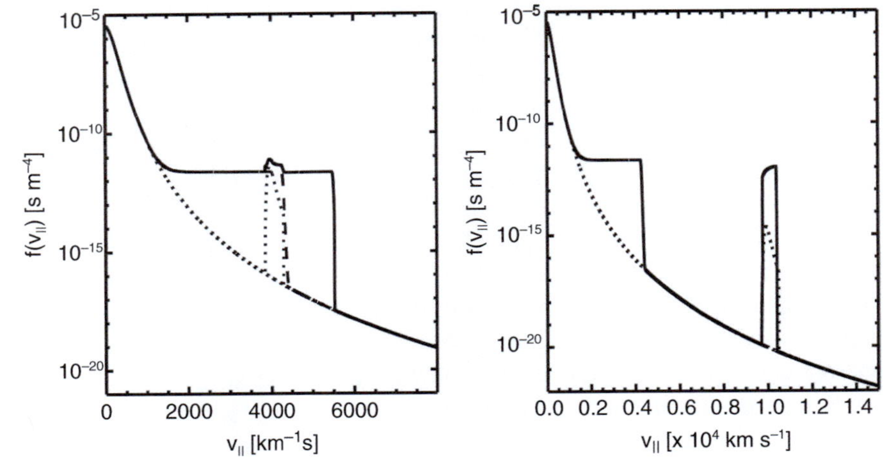

Fig. 23.42 Distributions $f(v_{\|})$ predicted at two foreshock locations upstream of the GMIR shock with both LH priming and shock acceleration (*solid curves*), LH priming but no shock acceleration (*dotted curves*), and shock acceleration by no LH priming (*dashed curves*): [*left*] $(x,R) = (15{,}100)$ Gm and [*right*] (3,50) Gm. (Cairns 2004)

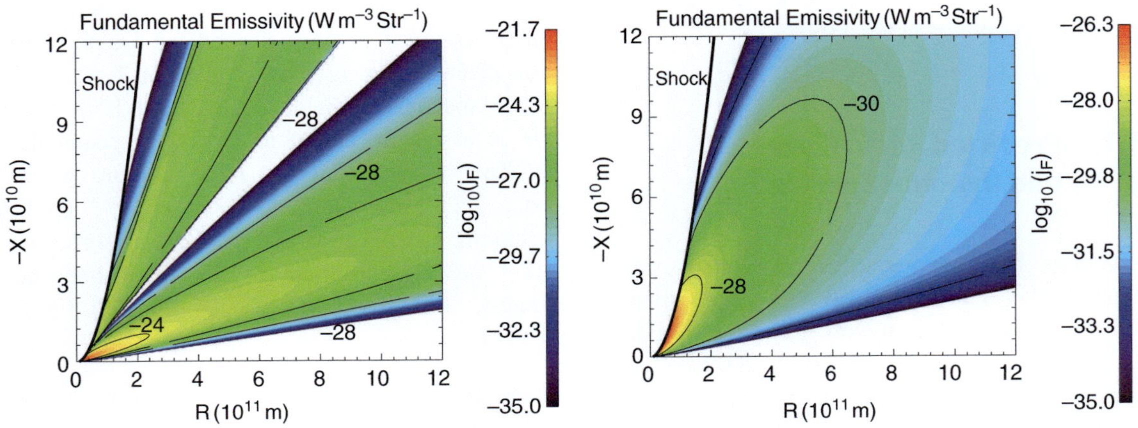

Fig. 23.43 Volume emissivity $j_F(R,x)$ for f_p radiation (*left*) with and (*right*) without LHD priming for the plasma parameters in Table 23.1 (Cairns et al. 2004). The color bar shows $\log_{10} j_F(R,x)$. The shock is solid black line and the calculation is for the plane defined by **U** and **B**

demonstrate the importance of the shock in accelerating electrons and forming beam distributions that drive Langmuir waves. Moreover, the larger number of fast seed electrons available in the primed case at large $v_{\|}$ means that priming increases the relative beam number density proportionately more at the larger $v_{\|}$ found closer to the foreshock boundary than at smaller $v_{\|}$. Eq. (23.20) then predicts that priming will favor f_p radiation over $2f_p$ radiation (Cairns and Zank 2001, 2002), as shown below.

The importance of LH priming to the production of radiation is demonstrated in Fig. 23.43, by plotting $j_F(\mathbf{r})$ with and without LH priming for the central plane of the 3D foreshock. Priming increases the maximum values of j_F by a factor $\approx 10^4$ and causes two regions with significant j_F to develop (Cairns et al. 2004; Mitchell et al. 2004; Cairns 2004): the region with large v_b near the foreshock boundary $(x = 0)$ corresponds to shock-accelerated tail electrons while the other corresponds to accelerated background electrons.

Priming is vital for $2f_p$ radiation, as shown in Fig. 23.44. However, the primary reason for Fig. 23.44 is to compare the volume emissivities of f_p and $2f_p$ radiation: the maximum value of j_H is 3–4 orders of magnitude less than for j_F, providing a very strong argument that fundamental radiation should dominate harmonic radiation by 3–4 orders of magnitude for the 2–3 kHz radiation. This conclusion follows in both the primed and unprimed cases, with the enhancement factor due to priming larger for fundamental radiation due to Eq. (23.20) and the larger number of fast electrons

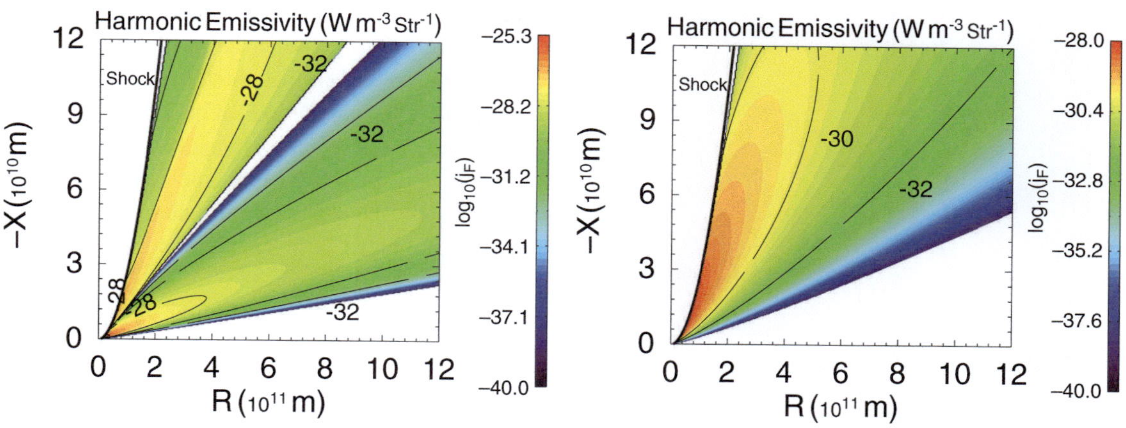

Fig. 23.44 Volume emissivity $j_H(R, x)$ for $2f_p$ radiation (*left*) with and (*right*) without LHD priming, in the same format and with the same shock and plasma parameters as Fig. 23.43 (Mitchell et al. 2004)

in the primed case. Analyses for other plasma parameters shows that the lack of $2f_p$ radiation is primarily due to the low value of $T_e = 8{,}000$ K assumed for the outer heliosheath (Mitchell et al. 2004). Thus, the theory provides a strong argument that the 2–3 kHz radiation should be exclusively f_p radiation (Cairns and Zank 2001, 2002; Cairns et al. 2004; Mitchell et al. 2004), consistent with the interpretation of the abrupt lower frequency edge of the 2 kHz component in Figure 23.21 as a cutoff at f_p for the outer heliosheath (Gurnett et al. 1993; Cairns and Zank 2002).

The fluxes predicted for a remote observer can now be calculated by integrating the predictions for $j_F(R, x)$ and $j_H(R, x)$ over the 3-D foreshock as in Eq. (23.22). The results are summarized in Table 23.2 and the important qualitative implications are: (1) The flux of f_p radiation should dominate the $2f_p$ flux by over 4 orders of magnitude. Accordingly harmonic structure is unlikely to be observed for the 2–3 kHz radiation and the observed radiation is almost certainly f_p radiation. (2) Priming is critical, since it increases the predicted flux of f_p radiation by a factor $\approx 10^4$ for these parameters. This can plausibly account for the radiation turning on beyond the heliopause once the GMIR shock enters the primed region. (3) The predicted f_p flux for this single ripple of characteristic size 1 AU is of order that observed by the Voyager spacecraft, $\approx 1.8 \times 10^{-17}$ Wm^{-2}Hz^{-1} (Gurnett et al. 1993), for these parameters. These results confirm the theoretical predictions of Cairns and Zank (2001, 2002).

Table 23.2 Fluxes predicted for an observer in the ecliptic plane 50 AU from the Sun along the Sun - heliopause nose axis for the shock and plasma parameters in Table 23.1

	f_p flux (Wm^{-2}Hz^{-1})	$2f_p$ flux (Wm^{-2}Hz^{-1})
Tail	3×10^{-17}	1×10^{-22}
No Tail	8×10^{-22}	7×10^{-25}

Consider now the dynamic spectrum predicted (Fig. 23.45) as the GMIR shock travels through the solar wind and inner heliosheath before entering the primed region and eventually the VLISM (Mitchell et al. 2004). The calculation assumes the spatial profiles in density, flow speed, and ion and electron temperature given by the two-shock, cylindrically-symmetric, plasma-neutral, 4-fluid (3 neutral fluids and 1 plasma fluid coupled by charge-exchange and ordinary collisions) simulation of Zank et al. (1996). For simplicity, the shock is assumed to have constant speed and shape (U and R_c are specified in Table 23.1), **U** is directed along the Sun – heliopause nose line which is the symmetry axis of the shock and system, and **B** is always perpendicular to **U**. Priming is assumed to occur in the outer heliosheath only, consistent with the theory above.

In Fig. 23.45 from early times until about day 300 the frequency bands drifting downwards from 8 kHz to about 200 Hz are f_p (primarily) and $2f_p$ radiation produced when the GMIR is in the solar wind. The

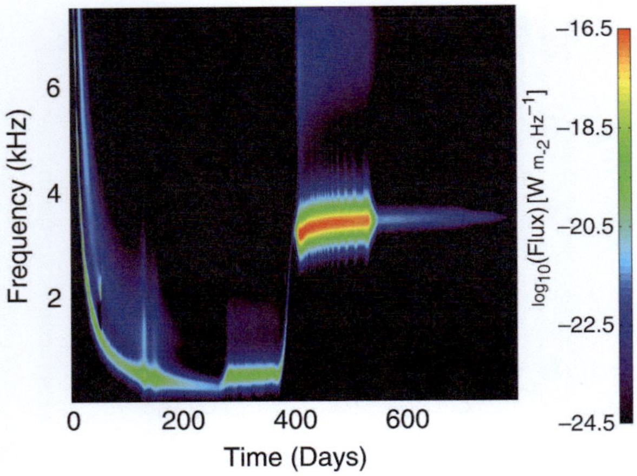

Fig. 23.45 Dynamic spectrum predicted for the GMIR shock described in the text, with the logarithm of the flux color-coded for an observer located 50 AU from the Sun and just off the Sun-heliopause nose axis (Mitchell et al. 2004). The vertical ribbing from days 400–550 is a plotting artifact

time origin corresponds to the GMIR leaving the Sun. (The enhancement near day 150 is when the GMIR passes near the observer.) The radiation produced in the solar wind is predicted to be very weak, typically well below about $10^{-18.5}$ Wm^{-2}Hz^{-1}, whereas the Voyager threshold is near 10^{-17} Wm^{-2}Hz^{-1}. The GMIR is in the inner heliosheath from about day 250 to day 400, primarily producing f_p radiation, after which the rapidly rising tone corresponds to f_p emission from the heliopause density ramp. When the GMIR enters the outer heliosheath the f_p radiation intensity is predicted to increase by ≈ 4 orders of magnitude, to values in excess of the Voyager threshold. Moreover, this emission remains almost constant in frequency and lasts for approximately 150 days, starting shortly after day 400. The GMIR enters the VLISM near day 550 and the emission then becomes very weak.

The following predictions and results follow from Fig. 23.45. First, the radiation should be too weak to be observable by the Voyager spacecraft when the GMIR is in the solar wind, the inner heliosheath, and the VLISM. Second, the radiation should turn on when the GMIR enters the primed outer heliosheath, because the radiation flux is predicted to increase above the Voyager threshold. Third, the f_p radiation predicted for the primed outer heliosheath closely resembles the 2 kHz component (compare Figs. 23.21 and 23.45). Fourth, the GMIR shock is not predicted to produce $2f_p$ radiation in the outer heliosheath or any other region that is observable above the Voyager threshold. Fifth, the timing and frequencies of the radio event predicted above the Voyager threshold are semiquantitatively consistent with the observations. (The densities, and so radiation frequencies, are slightly high in the outer heliosheath and VLISM for this simulation.)

The flux, frequency, and timing of the radio emissions are predicted to vary with the properties of the GMIR, neutral population leading to LHD, and the source plasmas (particularly the outer heliosheath plasma). Specifically the flux increases with increasing U, tail speed v_m, n_T, n_e, and R_c, and with decreasing κ (Mitchell et al. 2004), analogous to Figs. 23.29, 23.30, and 23.31 above for type II bursts. Changing the pickup ion source to solar wind neutrals from inner heliosheath neutrals is predicted to increase v_m and decrease n_T and initial calculations predict that the radio fluxes should increase by a factor ≈ 30 (Mitchell et al. 2009) (primarily due to the increased number of fast tail electrons available for the GMIR shock to accelerate), although detailed calculations remain to be done.

In summary, the combination of the priming mechanism, the foreshock emission theory, and the GMIR shock described in Fig. 23.45 provide an underlying and semiquantitative theoretical basis for Gurnett et al.'s (1993) GMIR model for the radiation (Cairns et al. 2004; Mitchell et al. 2004). Specifically, this GMIR/Priming theory can account semiquantitatively for (Cairns and Zank 2002; Cairns et al. 2004; Mitchell et al. 2004, 2009): the radiation's turn-on in the outer heliosheath; the apparent lack of emission when the GMIR is in the solar wind, inner heliosheath, and the VLISM; the lack of harmonic structure in the observed radio emissions; the characteristic flux of the observed emission; and the characteristics of the 2 kHz component. Nevertheless the theory is not complete, with several issues requiring resolution, including the lack of strong drifting emissions analogous to the observed transient emissions in Fig. 23.45, the propagation of radiation into the inner heliosphere, and a fully quantitative study of the radio flux (Cairns et al. 2004; Cairns 2004). These issues are discussed in more detail in Section 23.7.4 below.

23.7.3 Orientation and Strength of the VLISM Magnetic Field

Figure 23.24 shows that the source regions inferred for the 2–3 kHz emissions (specifically the transient emissions) lie on an approximately linear band in the plane of the sky that is almost parallel to the galactic plane and contains the direction to the heliopause nose (Kurth and Gurnett 2003). Kurth and Gurnett (2003) then interpreted the linear band in terms of the direction of the VLISM magnetic field $\mathbf{B}_{VLISM}$, with $\mathbf{B}_{VLISM}$ then being parallel to the galactic plane. This finding appears to be consistent with results for the large-scale magnetic field in the Milky Way (Frisch 2003).

Subsequently draping of the interstellar magnetic field over the heliopause was studied to see whether the priming mechanism would strongly localize the emission source near the heliopause, for instance via Eq. (23.27) and spatial variations in the Alfven speed $V_A(\mathbf{r}) \propto B(\mathbf{r})/n_e(\mathbf{r})$ (Cairns 2004; Mitchell et al. 2008). While draping does indeed lead to a linear band of enhanced B and so reduced v_r/V_A on the plane of the sky that is parallel to $\mathbf{B}_{VLISM}$, the enhancement is typically only about 10% compared with a surrounding roughly circular region (Cairns 2004; Cairns et al. 2006; Mitchell et al. 2008). Accordingly, draping alone does not appear to explain the observed band of source locations.

A very attractive idea is that the emission region is a band on the sky where $\mathbf{B} \cdot \mathbf{n} \approx 0$ where $\mathbf{n}$ is the normal vector to the GMIR shock and $\mathbf{B}$ is the magnetic field just upstream of the shock (Gurnett et al. 2006). The foreshock theory naturally incorporates this constraint since the constraint $\mathbf{B} \cdot \mathbf{n} = 0$ just identifies the location of the tangent point on the shock (e.g., Fig. 23.1) and the foreshock electrons with large values of $v_\parallel$ are connected to regions of the shock where $\mathbf{B} \cdot \mathbf{n} \approx 0$. Moreover, since each plane defined by $\mathbf{U}$ and $\mathbf{B}$ that passes through the shock has a point where $\mathbf{B} \cdot \mathbf{n} = 0$ there is a band of locations on the 3D shock where $\mathbf{B} \cdot \mathbf{n} = 0$ that is approximately perpendicular to $\mathbf{B}$ in the plane of the sky (Gurnett et al. 2006). Gurnett et al. (2006) therefore interpreted the observed band as being perpendicular to $\mathbf{B}_{VLISM}$.

The $\mathbf{B} \cdot \mathbf{n} \approx 0$ idea can actually lead to the source being elongated parallel to $\mathbf{B}_{VLISM}$ or perpendicular to $\mathbf{B}_{VLISM}$ (Mitchell et al. 2008). The reason is that the relevant question is whether the region where quasilinear relaxation of foreshock electron beams and significant emission occurs has a size parallel to $\mathbf{B}$ that is small or large compared with the size of the perpendicular domain on the shock where $\mathbf{B} \cdot \mathbf{n} = 0$ (Mitchell et al. 2008). The calculations in Figs. 23.43 and 23.44, and analogs elsewhere (Cairns et al. 2004; Mitchell et al. 2004) show that the volume emissivity of radiation is only large for distances $\lesssim 10^{11}$ m ≈ 1 AU parallel to $\mathbf{B}$. Viewed from the Sun this is a distance of about 0.4°. In comparison, the domain on the macroscopic GMIR shock where $\mathbf{B} \cdot \mathbf{n} \approx 0$ extends at least 30° from the heliopause nose (Mitchell et al. 2008). Thus, recognizing the importance of the $\mathbf{B.n} \approx 0$ idea, the GMIR/Priming theory quantitatively predicts that the observed band is perpendicular to $\mathbf{B}_{VLISM}$ in the plane of the sky.

An independent means to obtain the direction of $\mathbf{B}$ in the outer heliosheath, and thence in the VLISM, is provided by the IBEX spacecraft's observations (McComas et al. 2009) of a ribbon of ENAs on the sky with energies and directions characteristic of the solar wind-VLISM interaction. In one interpretation the ribbon shows where ENAs are produced from pickup ions that have a ring distribution in the outer heliosheath and resulted from charge-exchange of solar wind neutrals, exactly as in the revised GMIR/Priming theory (Mitchell et al. 2009). Moreover, the simulations of Heerikhuisen, McComas and colleagues (McComas et al. 2009; Heerikhuisen et al. 2010) show that the ring must not be scattered into a shell if the ribbon is to be observable. In this interpretation the ring should be where $\mathbf{B}_{OS} \cdot \mathbf{r} \approx 0$ (Heerikhuisen et al. 2010), essentially identical to the $\mathbf{B} \cdot \mathbf{n}$ interpretation for the radio sources (Gurnett et al. 2006). Heerikhuisen et al. (2010) have estimated the direction of $\mathbf{B}_{OS}$ from the IBEX ENA data (it originates from close to ecliptic coordinates (224,41)), but this direction has not yet been compared with other estimates (Gurnett et al. 2006; Frisch 2003; Pogorelov et al. 2009; Opher et al. 2009a). It remains to be seen whether these are all consistent.

The IBEX data and their pickup ion interpretation also offer a way to constrain the magnitudes of B_{OS} and B_{VLISM}. This is because pickup ion rings subject to LHD and the constraint $v_r/V_A \lesssim 5$ of Eq. (23.27) are not expected to scatter into a shell but instead to fill the ring to lower $v_\perp$ (Omelchenko et al. 1989;

McClements et al. 1993; Shapiro et al. 1998; Cairns and Zank 2001, 2002), whereas pickup rings that do not satisfy Eq. (23.27) are expected to drive MHD waves and isotropize the ring into a shell distribution (see Zank 1999, and references therein). Importantly, a ring filled towards lower $v_{\perp}$ still satisfies the geometric constraint $\mathbf{B}_{OS} \cdot \mathbf{r}$ for the ribbon's creation, whereas a shell does not. Accordingly, the existence of the IBEX ribbon and associated ENA simulations (McComas et al. 2009; Heerikhuisen et al. 2010) provides an independent argument that the conditions for LHD, the priming mechanism, and the GMIR/Priming theory are met, whence $V_A \gtrsim v_{sw}/5$ in the outer heliosheath.

The magnitudes of B_{OS} and B_{VLISM} are thus constrained by Eq. (23.27). This leads to $B_{OS} \approx B_{VLISM} > 0.2$ nT for solar wind neutrals (Mitchell et al. 2009). (The constraint is a factor ≈ 4 weaker for inner heliosheath neutrals; Cairns 2004, Mitchell et al. 2008.) Another constraint is that the plasma beta be less than 1 for LH wave growth to be favored, yielding $B_{VLISM} > 0.03$ nT (Cairns 2004). These large values of B_{VLISM} make it likely that M_A in the VLISM is less than 1 so that an outer bow shock appears increasingly unlikely.

Recent global heliospheric simulations are independently yielding $B_{VLISM} \approx 0.4$–0.6 nT by considering where the Voyager spacecraft crossed the termination shock, the directions of the downstream flow velocities, and energetic particle observations upstream of the termination shock (Opher et al. 2009a; Pogorelov et al. 2009). These analyses should be considered more direct and stronger arguments for large values of B_{VLISM} than the arguments above based on the GMIR/Priming model. Perhaps more importantly, the large values of $B_{VLISM} > 0.4$ nT obtained by comparing the global simulations with non-radio data provide a strong argument that the constraint of Eq. (23.27) is satisfied for the outer heliosheath and so that the proposed priming mechanism and GMIR/Priming theory are viable. Another argument is provided above based on the existence and properties of the IBEX ENA ribbon.

23.7.4 Issues and Future Research Directions

The theoretical predictions for the 2–3 kHz radiation are not in as advanced a state as for type II bursts. A number of important issues exist, all of which can be resolved with further research.

The primary issue can be considered the lack of detailed quantitative predictions for the dynamic spectrum for a macroscopic GMIR shock moving through realistic asymmetric models for the inhomogenous plasma structures of the heliopause and outer heliosheath. Recent global neutral-plasma simulations show that the plasma and magnetic fields become strongly asymmetric when highly tilted and strong $\mathbf{B}_{VLISM}$ are considered (Opher et al. 2009a; Pogorelov et al. 2009). By analogy with theoretical predictions for type II bursts (McLean 1967; Knock and Cairns 2005; 2006; Florens et al. 2007), these asymmetries can be expected to produce emissions reminiscent of split-band and multiple-lane type II bursts that might explain the existence of both the 2 kHz component and transient emissions, the multiplicity of transient emissions, and fine structures in the 2 kHz component. Specifically, the hope is that some portions of the shock will encounter regions of almost constant plasma density while others will encounter regions with positive density gradients that have appropriate length scales (several to 20 AU) to yield the upwards drifting transient emissions as the shock moves up the density ramp. Slowing of the shock as it moves from the solar wind to the outer heliosheath is also expected to be vital but is yet to be included.

The next issue to do with making the theory fully quantitative is that existing calculations (Cairns et al. 2004; Mitchell et al. 2004) only consider a single ripple with $R_c = 0.42$ AU (and projected length ≈ 1 AU in the plane of the sky) whereas the macroscopic GMIR shock has a characteristic scale of over 100 AU near the heliopause, if approximately spherical, and so should have over $100^2 = 10^4$ active ripples of this size that might produce radio emission (Cairns et al. 2004). Restricting the active ripples to the region of the outer heliosheath where the plasma is primed and strong magnetic draping occurs will restrict the number of ripples over the above estimate, but not by more than an order of magnitude. Accordingly, scaling up the existing flux predictions to the macroscopic shock means that the flux may be underestimated by a factor $\approx 10^4$ (since the flux is proportional to R_c^2 and the number of ripples (Knock et al. 2003a; Cairns et al. 2004)). Since the existing calculations yield fluxes of order those observed, this is a significant quantitative problem that requires an effective loss mechanism to bring the scaled up predictions into balance with the observations. One such mechanism is discussed next.

Propagation of the radiation into the inner heliosheath and solar wind is a major issue for the GMIR/Priming theory: the radiation is predicted to be produced near f_p upstream of a shock that is moving away from the Sun, thereby being prevented from immediately propagating Sunwards by the density increase behind the shock (Cairns and Zank 2001, 2002). This has a qualitative explanation that has not been quantified. The qualitative explanation (Cairns and Zank 2001, 2002) is that scattering by density irregularities diffuses the f_p radiation around the sides of the GMIR shock until it reaches locations where the GMIR shock has not crossed the heliopause, where the radiation frequency is now much greater than the local plasma frequency and so the radiation can propagate directly across the shock and into the inner heliosphere. Existing calculations predict that scattering should be important in the solar wind and inner heliosheath (Cairns 1995, 1996; Armstrong et al. 2000).

Quantitative ray tracing calculations are needed to calculate the fraction of the emitted radiation that can reach the inner heliosphere by diffusing around the GMIR shock. Intuitively, this fraction is expected to be much less than 1% (Cairns and Zank 2001), meaning that the majority of the radiation moves out into the VLISM; this allows and allowing the flux predictions for multiple active ripples on the GMIR shock to possibly lead to fluxes of order those observed (Cairns et al. 2004). Accordingly, radiation from GMIR-like shocks leaving the myriad active stars in the Galaxy like the Sun may well be a major contributor to the galactic background below about 100 kHz. Another point is that the apparent necessity for the radiation to leak around the sides of the GMIR shock before entering the inner heliosheath, rather than moving directly across near the nose of the shock, means that there must be a geometric "shadow zone" immediately behind the portion of the GMIR shock that has crossed the heliopause. Spacecraft inside the shadow zone will observe only weak radiation that has been scattered into the region. This shadow zone concept may explain why the radio events observed after 2002 (see Fig. 23.19) have been weaker than the earlier events despite the Voyager spacecraft being closer to the heliopause nose and so the predicted source regions.

Detailed simulations of the LHD process and the constraint of Eq. (23.27) are required for parameters relevant to the outer heliosheath, including the effects of both charge-exchange and ordinary collisions. While Eq. (23.27) is consistent with previous simulations (e.g., of Omelchenko et al. 1989; McClements et al. 1993; Shapiro et al. 1998), this constraint needs to be confirmed and to be developed as a function of β.

Solar cycle variability may also affect the likelihood of producing radio emissions and may produce density structures in the outer heliosheath relevant to transient emissions. Specifically, the 11-year solar cycle injects periodic variations in the numbers of solar wind and inner heliospheric neutrals plus density waves that propagate into the outer heliosheath (Zank and Muller 2003; Scherer and Fahr 2003). Mitchell et al. (2005) showed that there should be a solar cycle dependence on the priming, with maximum priming 2–3 years after solar maximum, and so optimum conditions to produce radio emissions. While this is consistent with the radio data, the predictions need to be redone for solar wind neutrals and the consequences explored of the density waves injected into the outer heliosheath. Qualitatively, these density structures provide natural density ramps over a restricted range of heliolatitudes and heliolongitudes for the shock to move up and produce upwards-drifting radio emissions like the transient emissions.

The strong dependence of the predicted radio emission on the shock speed is another issue that requires more careful consideration. This is relevant in at least 2 ways. The first is that shocks are predicted to slow markedly in the inner and outer heliosheaths (e.g., Gurnett et al. 1993; Zank 1999), leading theoretically to substantially lower amounts of radio emission via analogs of Fig. 23.29, but this effect has not been considered in the dynamic spectra and other quantitative calculations of the radio emission (Cairns et al. 2004; Mitchell et al. 2004). Secondly, this may explain why the 2003–2004 emission event is much weaker than the 1983–1984 and 1992–1994 events (Cairns 2004): the lower GMIR speed inferred for the 2003–2004 radiation event (≈ 560 km s^{-1}) compared with the 1983-1984 and 1992-1994 events (≈ 850 km s^{-1}) should lead to a smaller radiation intensity by a factor ≈ 2, thereby potentially moving much of the radio emission produced below the detection thresholds of the Voyager plasma wave instruments. (Note that the most intense radio emissions observed to date have intensities only a factor ≈ 3 above the Voyager thresholds.)

Attractive future modifications to the foreshock theory were identified in Sections 23.5 and 23.6.3 in the contexts of Earth's foreshock radiation and type II bursts. Some of these apply also to the 2–3 kHz emissions. Further discussion of these is deferred to the next section.

23.8 Discussion

Detailed applications of the foreshock theory to Earth's foreshock, type II bursts (both coronal and interplanetary), and the 2–3 kHz radio emissions are shown in Sections 23.5, 23.6, and 23.7 to broadly be in good qualitative and even semiquantitative agreement with available observations. Nevertheless, these same sections identified a number of areas of improvement for the basic theory itself as well as for the detailed applications of the theory. Rather than repeat those here, this section instead focuses on issues involving the emission mechanisms for the radiation and on future applications of the theory.

23.8.1 Improvements to the Foreshock Theory

Section 23.6.3 describes a number of important revisions proposed for the foreshock theory, including the reflection of electrons at the shock (principally the inclusion of overshoots in the magnetic field and cross-shock potential), inclusion of other emission processes, the modelling of ripples on the macroscopic shock, better modeling the directivity patterns, propagation, and scattering of the radiation, and the development of "bolt-on" numerical implementations of the theory that can be combined easily with advanced numerical, data-driven models for the background plasma and evolution of the shock. Attention is focused here on emission processes other than the standard ES decay, EM decay, and $2f_p$ coalescence processes. These include linear mode conversion (LMC) (Forslund et al. 1975; Budden 1985; Yin et al. 1998; Cairns and Willes 2005; Kim et al. 2007, 2008), radiation from localized Langmuir eigenstates (Malaspina et al. 2010), direct radiation via electron cyclotron maser emission (Farrell 2001), and fine structures at $f_{ce}/2$ in f_p radiation from Earth's foreshock (Cairns 1994). These are not included in the basic foreshock theory, but are now discussed in turn.

Before starting, however, an important point is made: the current foreshock theory yields predictions in reasonable semiquantitative agreement (typically a factor of order 3–10) with available observations, implying that these other processes are not likely to be crucial unless a basic building block of the theory (like the power flux into the Langmuir waves) is not estimated accurately. Moreover, strong arguments exist for favoring the standard processes (Sections 23.3 and 23.5). Nevertheless, it is important to thoroughly explore and develop these non-standard emission processes. Reasons include the fundamental nature of the physics, the need to extend the foreshock theory so that it can quantitatively explain the observed radiation to better than a factor of 2, and the possibility that these processes will apply to other radio emissions.

Recent work has almost placed LMC in a suitable state for insertion into Eq (23.17) and consideration on an equal basis to the standard nonlinear processes. The reasons are as follows. Recent simulations (Kim et al. 2007, 2008) have established the energy and power conversion efficiencies for a specified incoming wavevector (shown to differ by the ratio of the group speeds of the electrostatic and electromagnetic waves, thereby being a quantitatively important difference) and reconciled previous analytic and simulation work in (Forslund et al. 1975; Yin et al. 1998), while analytic and numerical calculations have shown how to average the power (or energy) conversion efficiencies over the angular and length scale distributions of incoming Langmuir waves and density irregularities (Cairns and Willes 2005). The averaged efficiencies are commensurate with those for the nonlinear processes above, sometimes being larger and sometimes smaller depending on the beam parameters, and are smaller than the unaveraged efficiencies by factors of order $(V_e/c)^2$ (Cairns and Willes 2005). This provides a robust argument that LMC needs to be fully considered (Cairns and Willes 2005) – as argued previously on the less secure grounds of the unaveraged power conversion efficiencies being $\approx 50\%$ and so orders of magnitude larger than required to explain the observed fluxes. Future work should therefore involve the addition of the averaged LMC efficiencies into Eq. (23.17) on an equal basis to the nonlinear efficiencies.

The newly discovered Langmuir eigenstates (Ergun et al. 2008; Malaspina and Ergun 2008), sometimes called Intense Localized Structures or ILSs (Nulsen et al. 2007), have significant nonlinear currents j_{NL} and electric fields E_{NL} at both f_p and $2f_p$ that are associated with the time varying Langmuir fields at f_p and their spatial gradients. These currents can drive radio waves at f_p and $2f_p$ due to coupling between the current and wave fields leading to non-zero power input $\mathbf{j}_{NL} \cdot \mathbf{E}_{NL}$. Effectively the eigenstate radiates as an antenna. A similar mechanism was tried earlier (Papadopoulos et al. 1978; Goldman et al. 1980) for Langmuir wave packets subject to the process of wave collapse in strong turbulence, in which nonlinear self-focusing dominates wave dispersion and wavepackets intensify and collapse to spatial scales of order $10\lambda_D$ (Zakharov 1972; Robinson 1997), but found to be unimportant. The critical difference in the new calculations is that the eigenstate (or ILS) has a length scale larger than or commensurate with the wavelength of $2f_p$ radiation ($\lambda \approx c/2f_p$), so that the source cannot be assumed to be small compared with the wavelength: this results in survival of the dipole contribution to the radiated fields (Malaspina et al. 2010), unlike the antennas for collapsing wavepackets considered previously (Papadopoulos et al. 1978; Goldman et al. 1980).

Malaspina et al. (2009) calculated the $2f_p$ power radiated by an ILS (Fig. 23.46 and found it to be sufficient to contribute significantly to the radiation observed in Earth's foreshock, perhaps even dominating the standard nonlinear processes. The predicted power depends sensitively, though, on the number, characteristic fields, and length scales of the ILS in the foreshock, which are not yet known observationally or predictable theoretically. Further work is required on this antenna mechanism for ILS, which needs to be incorporated into the standard foreshock theory and its applications to solar system shocks.

Now consider cyclotron maser mechanisms for direct generation of f_p and $2f_p$ radiation (Wu et al. 1985; Farrell 2001). This requires careful tuning of parameters to obtain emission near the observed frequencies of $\approx 1.9f_p - 2.1f_p$ rather than $1.4f_p$, $2.5f_p$ or any other frequency, since $f_{ce} \ll f_p$ and the observed emission is at high harmonics

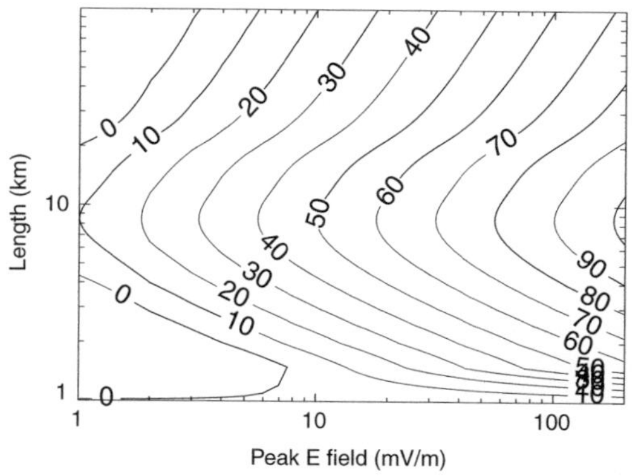

Fig. 23.46 Prediction of the $2f_p$ power in dB radiated by a Langmuir eigenstate (ILS) in Earth's foreshock as a function of the ILS peak electric field and length scale (Malaspina et al. 2010). The contour labelled 0 corresponds to a power of 2×10^{-11} W

($\approx$ 50–200) of f_{ce} in most source regions of interest in this paper (and specifically for Earth's foreshock). Furthermore, the process requires an energetic ring-beam distribution, with characteristic perpendicular speeds at least $0.1c$. While such ring-beam distributions can be produced by mirror reflection (Yuan et al. 2007, 2008a), they have very low number densities ($n_b/n_e \leq 10^{-6}$) at such large $v_\perp \approx v_\parallel$ (see Fig. 23.25 and the papers of Fitzenreiter et al. (1990) and Cairns et al. 1997). These two problems must be resolved before direct cyclotron maser emission can be considered a viable competitor to the standard nonlinear processes or the other alternative processes above.

Finally, no detailed or accepted theoretical explanation exists for splitting at $f_{ce}/2$ for f_p radiation in Earth's foreshock (Cairns 1994), which may also be relevant to split-band type II bursts and fine structures in the transient component of the 2–3 kHz outer heliospheric radiation. Mechanisms involving linear growth of oblique Langmuir waves due to loss cone or ring-beam features in the reflected electron distribution (Lobzin et al. 2005), followed by LMC or nonlinear conversion processes to radiation, or direct cyclotron maser emission may be attractive. Alternatively, magnetization effects on the standard nonlinear processes for f_p and $2f_p$ radiation need to be investigated for frequency fine structures, as opposed to the polarization analyses performed prevously.

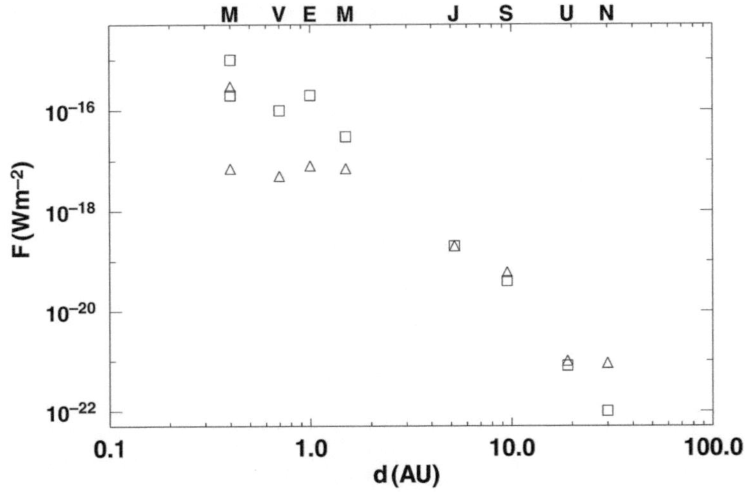

Fig. 23.47 Total fluxes (W m^{-2}) of f_p (*triangles*) and $2f_p$ (*squares*) radiation from planetary foreshocks for an observer O located a distance $2a_s$ upstream from each planet along the Sun–planet line, where a_s is the bow shock's standoff distance (Kuncic and Cairns 2005). The radio fluxes are calculated for nominal solar wind conditions, except for Mercury, where the fluxes are calculated for $\theta_{BU} = 90°$ and 130° (lower values)

23.8.2 Other Applications

The foreshock theory for radio emissions at f_p and $2f_p$ has numerous other applications in the solar system and in astrophysics. These include the foreshocks of other planets, mini-magnetospheres on the Moon, Mars, and the moons of other planets, the radio background of the interstellar medium, and emission from supernova shocks.

It is evident that the theory developed for Earth's foreshock can be generalized to any planet or moon with a bow shock. Kuncic and Cairns (2005) have performed these calculations for the planets, finding (Fig. 23.47) that Mercury is a particularly attractive target for observing foreshock radiation (e.g., with Bepi-Colombo). Similar figures for spacecraft at a fixed absolute distance from the planet's bow shock suggest that Jupiter should produce the largest flux, followed by Earth.

Localized and strong magnetic fields on moons can form a mini-magnetosphere and directly reflect the solar wind (or superalfvenic corotation flow within some planetary magnetospheres) – see Harnett and Winglee (2003) and references therein – and form a bow shock that then reflects electrons and give rise to radio emissions via the standard foreshock model (Kuncic and Cairns 2004). Figure 23.48 shows the foreshocks predicted for various orientations of the mini-magnetosphere to the solar wind for the Moon, with the shading proportional to the volume emissivity of radiation (Kuncic and Cairns 2004). Observable fluxes of radio emission are predicted for the Moon (Kuncic and Cairns 2004), perhaps accounting for some of the signals observed near f_p when the Wind spacecraft traversed the Moon's wake (Kellogg et al. 1996).

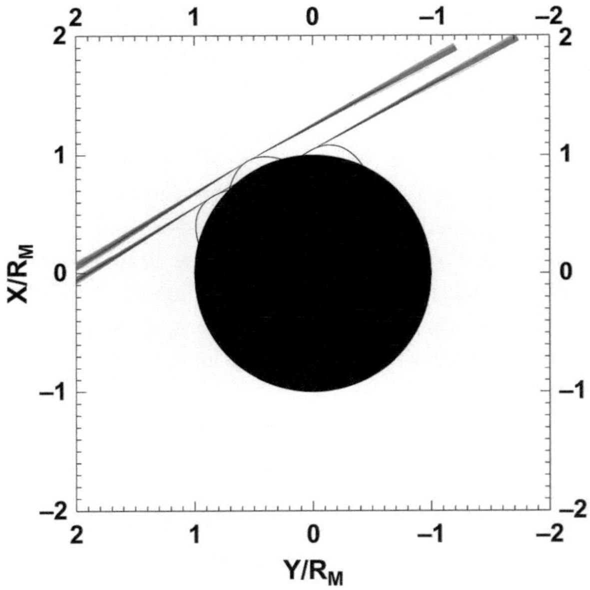

Fig. 23.48 Predicted volume emissivities and source regions for $2f_p$ radiation for three different locations of mini-magnetospheres (or magnetic anomalies) on the surface of the Moon when in the solar wind (Kuncic and Cairns 2004)

Similar situations may be applicable to moons in the Jovian or Saturnian magnetospheres (e.g., Ganymede) and to perturbations of the bow shocks of Mars and

Venus due to remnant magnetic fields. Further research is needed to quantify these ideas.

Finally, supernova shocks are a very attractive target for prediction of f_p and $2f_p$ radiation in a classic astrophysical context. Radiation over a broad frequency range can be expected when the shock overtakes ejecta from earlier periods of the star's evolution or reaches the system's termination shock and heliopause, since a broad range of densities is expected. It remains to be seen whether coherent foreshock emission will be predicted to have observable levels and whether it will be a significant contributor to the galactic background radiation at low frequencies $\lesssim 100$ kHz (Dulk et al. 2001; Hillan et al. 2010). Of course, emission from GMIR shocks for other stars, like the 2–3 kHz radiation for our Sun, may also be large contributors to the galactic background radiation, as commented in Section 23.7.4.

23.9 Concluding Remarks

Many powerful coherent radio emissions are associated with shocks in our solar system, from the deep corona to the solar wind to the outer heliosheath and perhaps the very local interstellar medium. All are produced near the electron plasma frequency f_{pe} and/or $2f_{pe}$. Emissions definitely driven by shocks include interplanetary type II bursts, radiation from Earth's foreshock, and rare emissions from CIRs. Emissions likely driven by shocks, but without definitive observational evidence, include drifting pulsating structures at GHz frequencies from the deep corona, coronal type II bursts, and the 2–3 kHz emissions from the outer heliosphere. Analogous emissions are also predicted, but not yet observed, from the foreshocks of the other planets (particularly Mercury and Jupiter, which are predicted to have the most easily observable emissions after Earth), mini-magnetospheres and associated bow shocks on the Moon and other moons that can move into the solar wind or exist in superalfvenic, corotating, flows in planetary magnetospheres (e.g., Ganymede). Foreshock f_p and $2f_p$ radiation is also expected upstream of supernova shocks and also for the equivalent of GMIRs from other active stars, perhaps being a major contributor to the galactic background radiation at low frequencies $\lesssim 1$ MHz.

A detailed theory exists for foreshock f_p and $2f_p$ radiation, based on reflection and acceleration of electrons into the foreshock by the shock's magnetic mirror, the formation of electron beams (with loss cone features that can have many attributes of ring-beam distributions) in the foreshock due to imposition of a minimum parallel velocity to reach a given foreshock location by two effects (one imposed at the shock and one by time-of-flight effects) that typically coincide, the generation of intense electrostatic Langmuir waves by the electron beams, and the conversion of Langmuir energy into radiation by standard nonlinear processes involving Langmuir waves. This theory couples multiple physical processes from microscales to macroscales, with the foregoing processes being microscale physics, the creation of ripples on the shock with sizes of order the decorrelation length of the magnetic field and scattering of radiation by density irregularities being intermediate scale physics, while the macroscale physics includes 3D spatiotemporal variations of the plasma and the shock motion, as well as integration of emission from individual shock ripples over the entire shock.

This theory has been implemented analytically and numerically and applied in some detail to Earth's foreshock radiation, type II bursts, and the 2–3 kHz outer heliospheric radiation. The predictions appear to be in good qualitative and even semiquantitative agreement with available observations, explaining the electron properties well and typically accounting for the observed Langmuir fields and radiation fluxes to within a factor of 3–10.

A number of improvements and tests have been identified for the microphysics of the theory, such as including overshoots in the magnetic field and cross-shock potential for the electron reflection and beam formation, observational testing of the assumption of marginal stability and SGT for the waves, simulation-based testing of the energy transfer rate into the waves from the beams based on marginal stability, and incorporation of additional radiation mechanisms such as linear mode conversion and emission from Langmuir eigenstates and ILSs. The physics of ripple formation and evolution needs to be understood better, so as to better model the characteristic sizes and packing of ripples onto the macroscopic shock. Scattering and intrinsic directivity effects for the radiation are also acknowledged to be important and require inclusion. Another area requiring improvement is the modeling of inhomogeneities of the background plasma (density, magnetic field, flow speed, temperatures, and the fraction and distribution function of nonthermal particles) and the 3D motion and location of the shock. Initial 2D models for the solar wind and high corona, driven by spacecraft data at 1 AU, exist but improvements have been identified. These types of data-driven models and accurate shock models are increasingly important for the modelling of type II bursts and the 2–3 kHz radiation, since the levels and frequency-time fine structures of the radiation are predicted to depends strongly on the local properties of the shock and plasma. Within the next few years it is believed that theory-data comparisons for type II bursts will be useful in constraining the properties of CME-driven shocks in the solar wind: these will allow prediction of whether, when, and with what properties a CME will impact Earth's magnetosphere and produce space weather events. Similarly, modelling a large-scale 3D GMIR shock interacting with realistic asymmetric models for the solar wind – LISM interaction may naturally explain the two classes and detailed properties of the 2–3 kHz outer heliospheric emission. It is recognized that major progress in testing the theory, and making associated refinements, will occur with the advent of reliable bolt-on versions of the foreshock theory that can be combined with advanced, data-driven, global simulations of both the background plasma and shock properties.

Unresolved theoretical and observational issues for type IIs, Earth's foreshock radiation, and the 2–3 kHz radiation are described in detail in Sections 23.5, 23.6.4, and 23.7.4. These include the one- versus two-shock debate for coronal and interplanetary type IIs, regarded here as of minor relevance (because the foreshock theory requires a shock but is agnostic as to its origin) although the evidence appears to favor a blast-shock for most metric type IIs and a CME-driven shock for almost all interplanetary type IIs. Others are the nature of the fine structures on type II bursts and the 2–3 kHz emissions, and the importance of scattering and propagation effects. Avenues to resolve these issues appear to exist and should be pursued. Resolving them may well lead to progress in understanding coronal structures and the interaction between the solar wind and VLISM, as well as the orientation and strength of the magnetic field in the VLISM.

Advanced ground-based instruments like the Frequency Agile Solar Radiotelescope (FASR), LOw Frequency ARray (LOFAR), and Murchison Widefield Array (MWA) will produce high dynamic range and high time- and frequency-resolution dynamic spectra, as well as images with higher time and angular resolution than before. In addition, the new Solar Dynamics Observatory spacecraft (SDO) and the existing Solar and Heliospheric Orbiting (SOHO), STEREO, and Wind spacecraft have excellent capabilities for observing many solar and interplanetary phenomena, including those related to type II bursts, CIR shocks, and Earth's foreshock radiation. The Voyager spacecraft continue to operate well and move towards the heliopause and the outer heliosheath, thereby being well poised to answer questions related to the origin of the 2–3 kHz radiation. Novel, high quality data will therefore be available to answer many unresolved issues raised in this review.

A major conclusion of this review is that the time is now ripe for making major progress on coherent radio emissions from shocks, since an attractive, apparently viable, and widely applicable theory exists, as do abundant high quality observational data. Put another way, a primary focus of current and future research should be on developing quantitative predictions of the theory, comparing these with observational data, and refining the theory as required. In this regard, quantitative testing and refining of the theory for Earth's foreshock should perhaps be of highest priority, due to the wealth of observational data on the electrons, Langmuir waves, radio emission and the solar wind properties, plus well-tested models for the location and basic properties of the bow shock.

In conclusion, it appears that the basic foreshock theory for radio emission from shocks (electron reflection and acceleration, development of electron beams, growth of Langmuir waves, and production of f_{pe} and $2f_{pe}$ radiation for a macroscopic, rippled, shock) appears likely to explain semiquantitatively the primary observations for the three best-observed and modelled applications, that many solar system radio emissions appear to be associated with shocks and qualitatively consistent with the basic theory, and that many observational details and theoretical limitations remain but do not appear likely to fundamentally alter the theory. The next 10 years ought to be an exciting time that sees theory and observations brought together quantitatively, and type II bursts and the 2–3 kHz radiation become important in predicting space weather at Earth and the impacts of solar activity on the local interstellar medium.

Acknowledgements The author thanks the Referee, colleagues D.S. Hillan, B. Li, V.V. Lobzin, and P.A. Robinson for constructive comments, Editors M.P. Miralles and J. Sanchez Almeida for patience and encouragement, and D.S. Hillan, S.A. Knock, and J.J. Mitchell for providing unpublished figures. Funding from the Australian Research Council is gratefully acknowledged.

References

Allen CW (1947) Interpretation of electron densities from corona brightness. Mon Not R Soc 107:426

Anderson RR, Parks GK, Eastman TE, Gurnett DA, Frank LA (1981) Plasma waves associated with energetic particles streaming into the solar wind from the Earth's bow shock. J Geophys Res 86:4493

Armstrong JW, Coles WA, Rickett BJ (2000) Radio wave scattering in the outer heliosphere. J Geophys Res 105:5149

Bale SD, Burgess D, Kellogg PJ, Goetz K, Monson SJ (1997) On the amplitude of intense Langmuir waves in the terrestrial electron foreshock. J Geophys Res 102:11281

Bale SD, Reiner MJ, Bougeret J-L (1999) The source region of an interplanetary type II radio burst. Geophys Res Lett 26:1573

Bale SD, Larson DE, Lin RP, Kellogg PJ, Goetz K, Monson SJ (2000) On the beam speed and wavenumber of intense electron plasma waves near the foreshock edge. J Geophys Res 105:27353

Ball L, Melrose DB (2001) Shock drift acceleration of electrons. Publ Astron Soc Aust 18:361

Bastian TS, Benz AO, Gary DE (1998) Radio emission from solar flares. Annu Rev Astron Astrophys 36:131

Bastian TS, Pick M, Kerdraon A, Maia D, Voulidas A (2001) The coronal mass ejection of 1998 April 20: direct imaging at radio wavelengths. Astrophys J 558:L65

Baumbach S (1937) Strahlung, ergiebigkeit und elektronendichte der sonnenkorena. Astron Nachr 263:131

Bougeret J-L, Goetz K et al (2008) S/Waves: the radio and plasma wave investigation on the STEREO mission. Space Sci. Rev 136:487

Budden KG (1985) The propagation of radio waves. Cambridge University Press, New York, NY

Burgess D (1995) Collisionless shocks. In: Kivelson MG, Russell CT (eds) Introduction to space physics. Cambridge University Press, New York, NY, p 129

Burgess D, Harvey CC, Steinberg J-L, Lacombe C (1987) Simultaneous observation of fundamental and second harmonic radio emission from the terrestrial foreshock. Nature 330:732

Cairns IH (1986a) The source of free energy for Type II solar radio bursts. Proc. Astron. Soc. Aust 6:444

Cairns IH (1986b) New waves at multiples of the plasma frequency upstreamof the Earth's bow shock. J Geophys Res 91:2975

Cairns IH (1987a) The electron distribution upstream from the Earth's bow shock. J Geophys Res 92:2315

Cairns IH (1987b) A theory for the Langmuir waves in the Earth's foreshock. J Geophys Res 92:2329

Cairns IH (1988) A semiquantitative theory for the $2f_p$ radiation observed upstreamfrom the Earth's bow shock. J Geophys Res 93:3958

Cairns IH (1994) Fine structure in plasma waves and radiation near the plasma frequency in Earth's foreshock. J Geophys Res 99:2975

Cairns IH (1995) Radio wave scattering in the outer heliosphere: Preliminary calculations. Geophys Res Lett 22:23433

Cairns IH (1996) On radio wave scattering in the outer heliosphere. In: Winterhalter D et al (eds) "Solar wind eight", AIP Conference Proceedings 382, p 582

Cairns IH (2000) Role of collective effects in dominance of scattering off thermal ions over Langmuir wave decay: analysis, simulations, and space applications. Phys Plasmas 7:4901

Cairns IH (2004) Radiation from the outer heliosphere and beyond. In: Poletto G, Suess S (eds) The sun and the heliosphere as an integrated system, Astrophysics and Space Science Library, vol 317. Kluwer, Dordrecht, p 65

Cairns IH, Fung SF (1988) Growth of electron plasma waves above and below the electron plasma frequency in the foreshock. J Geophys Res 93:7307

Cairns IH, Kaiser ML (2002) Solar system radio emissions. In Russ Stone W (ed) The review of radio science 1999–2002. IEEE Press, Piscataway, p 749

Cairns IH Knock SA (2006) Predictions for dynamic spectra and source regions of type II radio bursts in the inhomogeneous corona and solar wind. In: Rucker HO, Kurth WS, Mann G (eds) Planetary radio emissions VI proceedings. Verlag. Ost. Akad. Wissenschaften, p 419

Cairns IH, Melrose DB (1985) A theory for the $2\ f_p$ radiation upstream of the Earth's bow shock. J Geophys Res 90:6637

Cairns IH, Robinson RD (1987) Herringbone bursts associated with type II solar radio emission. Solar Phys 111:365

Cairns IH, Robinson PA (1995) Ion acoustic wave frequencies and onset times during type III solar radio bursts. Astrophys J 453:959

Cairns IH Robinson PA (1997) First test of stochastic growth theory for Langmuir waves in Earth's foreshock. Geophys Res Lett 24:369

Cairns IH, Robinson PA (1999) Strong evidence for stochastic growth theory of Langmuir-like waves in Earth's foreshock. Phys Res Lett 82:3066

Cairns IH, Robinson PA (2000) Roles played by electrostatic waves in producing radio emissions. In: Stone RG, Weiler KW et al (eds) Radio astronomy at long wavelengths, Geophysics Monograph Series 119. American Geophysical Union, Washington, DC, p 37

Cairns IH, Willes AJ (2005) Angle-averaged efficiencies for linear mode conversion between Langmuir waves and radiation in an unmagnetized plasma. Phys Plasmas 12:052315

Cairns IH Zank GP (2001) Theories for radio emissions from the outer heliosphere. In: Scherer K, Fichtner H et al (eds) The outer heliosphere: the next frontiers. Pergamon, Amsterdam, p 253

Cairns IH, Zank GP (2002) Turn-on of 2–3 kHz radiation beyond the heliopause. Geophys Res Lett 29, 2001GL014112

Cairns IH, Robinson PA, Anderson RR (2000) Thermal and driven stochastic growth of Langmuir-likewaves in the solar wind and Earth's foreshock. Geophys Res Lett 27:61

Cairns IH, Kurth WS, Gurnett DA (1992) The outer heliospheric radio emissions: (2) Foreshock source models. J Geophys Res 97:6245
Cairns IH, Knock SA, Robinson PA, Kuncic Z (2003) Type II solar radio bursts: theory and space weather implications. Space Sci Rev 107:27
Cairns IH, Mitchell JJ, Knock SA, Robinson PA (2004) Towards a quantitative theory for 2–3 kHz emission from beyond the heliopause. Adv Space Res 34(1):88
Cairns IH, Mitchell JJ, Pogorelov N, Zank GP (2006) Magnetic draping, 2–3 kHz radio emissions, and constraints on the interstellar magnetic field. In: AIP Conference Proceedings 858, p 329
Cairns IH, Robinson PA, Anderson RR, Strangeway RG (1997) Foreshock Langmuir waves for unusually constant solar wind conditions: data and implications for foreshock structure. J Geophys Res 102:24249
Cane HV, Erickson WC (2005) Solar type II radio bursts and IP type II events. Astrophys J 623:1180
Cane HV, Stone RG (1984) Type II solar radio bursts, interplanetary shocks, and energetic particle events. Astrophys J 282:339
Cane HV, Sheeley NR Jr, Howard RA (1987) Energetic interplanetary shocks, radio emission, and coronal mass ejections. J Geophys Res 92:9869
Cane HV, Stone RG, Fainberg J, Stewart RT, Steinberg J-L, Hoang S (1981) Radio evidence for shock acceleration of electrons in the solar corona. Geophys Res Lett 8:1285
Cane HV, Stone RG, Fainberg J, Steinberg J-L, Hoang, S (1982) Type II solar radio events observed in the interplanetary medium. I - General characteristics. Solar Phys 78:187
Cane HV, Erickson WC, Prestage NP (2003) Solar flares, type III radio bursts, coronal mass ejections, and energetic particles. J Geophys Res 107:2001JA000320
Cliver EW, Webb DF, Howard RA (1999) On the origin of solar metric type II bursts. Solar Phys 187:89
Cliver EW, Kahler SW, Reames DV (2004) Coronal shocks and solar energetic proton events Astrophys J 605:902
Collier MR, Szabo A, Slavin JA, Lepping RP (2000) IMF length scales and predictability: the two length scale medium. Int J Geopmag Aeron 2:3
Dulk GA, Erickson WC, Manning R, Bougeret J-L (2001) Calibration of lowfrequency radio telescopes using the galactic background radiation. Astron Astrophys 365:294
Dulk GA, Leblanc Y, Robinson PA, Bougeret J-L, Lin RP (1998) Electron beams and radio waves of solar type III bursts. J Geophys Res 103:17223
Dunckel N (1974) Low frequency radio emissions from the Earth and Sun. PhD thesis, Stanford University, Palo Alto
Ergun RE, Malaspina DM, Cairns IH et al (2008) Eigenmode structure in solarwind Langmuir waves. Phys Rev Lett 101:051101
Farrell WM (2001) Direct generation of o-mode emission in a dense, warm plasma: applications to interplanetary type II emissions and others in its class. J Geophys Res 106:15701
Feldman WC, Anderson RC, Bame SJ, Gary SP, Gosling JT et al (1983) Electron velocity distributions near the Earth's bow shock. J Geophys Res 88:96
Filbert PC, Kellogg PJ (1979) Electrostatic noise at the plasma frequency beyond the Earth's bow shock. J Geophys Res 84:1369
Fitzenreiter RJ, Scudder JD, Klimas AJ (1984) Detection of bump-on-tail electron reduced electron velocity distributions at the electron foreshock boundary. Geophys Res Lett 11:496
Fitzenreiter RJ, Scudder JD, Klimas AJ (1990) Three dimensional analyticmodel for the spatial variation of the foreshock electron distribution: Systematics and comparisons with ISEE observations. J Geophys Res 95:4155
Fitzenreiter RJ, Ogilvie KW, Bale SD, Vinas AF (2003) Modification of the solar wind electron velocity distribution at interplanetary shocks. J Geophys Res 108:1415
Florens MSL, Cairns IH, Knock SA, Robinson PA (2007) Data-driven solar wind model and prediction of type II bursts. Geophys Res Lett 34:L04104
Forslund DW, Kindel JM, Lee K, Lindman EL, Morse RL (1975) Theory and simulation of resonant absorption in a hot plasma. Phys Rev A 11:679
Frisch PC (2003) Boundary conditions of the heliosphere. J Geophys Res 108(10):LIS 11-1
Goldman MV, Reiter GF, Nicholson DR (1980) Radiation from a strongly turbulent plasma: Application to electron beam-excited solar emissions. Phys Fluids 23:388
Gopalswamy, N (2006) Coronal mass ejections and Type II radio bursts. In: Gopalswamy N, Mewaldt R, Torsti J (eds) Solar eruptions and energetic particles, AGU Monograph Series 165, p 207
Gopalswamy N, Kaiser ML, Thompson BJ, Burlaga LF et al (2000) Radio-rich solar eruptive events. Geophys Res Lett 27:1427
Gopalswamy N, Yashiro S, Kaiser ML., Howard RA, Bougeret J-L (2001) Radio signatures of coronal mass ejection interaction: coronal mass ejection cannibalism? Astrophys J 548:L91
Gopalswamy N, Yashiro S, Kaiser ML, Howard RA, Bougeret J-L (2001) Characteristics of coronalmass ejections associated with long-wavelength type II radio bursts. J Geophys Res 106:29219
Gopalswamy N, Yashiro S, Kaiser ML., Howard RA, Bougeret J-L (2002) Interplanetary radio emission due to interaction between two coronal mass ejections. Geophys Res Lett 29:106
Gopalswamy N, Yashiro S, Xie H, Akiyama S, Aguilar-Rodriguez E et al (2008) Radio-quiet fast and wide coronal mass ejections. Astrophys J 674:560
Grognard RJ-M (1985) Propagation of electron streams. In: McLean DJ, Labrum NR (eds) Solar radiophysics, Cambridge University Press, Cambridge, New York, p 173
Grzedzielski S, Lazarus AJ (1993) 2- to 3-kHz continuum emissions as possible indications of global heliospheric breathing. J Geophys Res 98:5551
Gurnett DA (1975) The Earth as a radio source: the non-thermal continuum. J Geophys Res 80:2751
Gurnett DA (1995) Heliospheric radio emissions. Space Sci Rev 72:243
Gurnett DA, Kurth WS (1995) Heliospheric 2–3 kHz radio emissions and their relationship to large Forbush decreases. Adv Space Res 16(9):279

Gurnett DA, Kurth WS (1996) Radio emissions from the outer heliosphere. Space Sci Rev 78:53

Gurnett DA, Allendorf SC, Kurth, WS (1998) Direction-findingmeasurements of heliospheric 2–3 kHz radio emissions. Geophys Res Lett 25:4433

Gurnett DA, Kurth WS, Allendorf SC, Poynter RL (1993) Radio emission from the heliopause triggered by an interplanetary shock. Science 262:199

Gurnett DA, Kurth WS, Cairns IH, Mitchell JJ (2006) The local interstellar magnetic field direction from direction-fnding measurements of heliospheric 2–3 kHz radio emissions. In: AIP conference proceedings 858, 129

Harnett EM, Winglee, R (2003) 2.5D fluid simulations of the solar wind interacting with multiple dipoles on the surface of the Moon. J Geophys Res 108(A2):1088

Heerikhuisen J, Pogorelov NV, Zank GP, Crew GB, Frisch PC et al (2010) Pickup ions in the outer heliosheath: a possible mechanism for the Intestellar Boundary Explorer ribbon. Astrophys J Lett 708:L126 [Erratum 710, L172 (2010)]

Hellinger P, Travnicek P, Matsumoto, H (2002) Reformation of perpendicular shocks: hybrid simulations. Geophys Res Lett 29:2234

Hellinger P, Travnicek P, Lembege B, Savoini P (2007) Emission of nonlinear whistler waves at the front of perpendicular supercritical shocks: hybrid versus full particle simulations. Geophys Res Lett 34:L14109

Henri P, Briand C, Mangeney A, Bale SD, Califano F, Goetz K, and Kaiser ML (2009) Evidence for wave coupling in type III emissions. J Geophys Res 114:A03103

Hillan DS, Cairns IH, Robinson PA, Mohamed A (2010) Prediction of background levels for the Wind/WAVES instrument and implications for the galactic background radiation. J Geophys Res 115:A061012

Hoang S, Fainberg J, Steinberg J-L, Stone RG, Zwickl RH (1981) The $2f_p$ circumterrestrial radio emission as seen from ISEE 3. J Geophys Res 86:4531

Hoang S, Pantellini F, Harvey CC, Lacombe C, Mangeney A et al (1992) Interplanetary fast shock diagnosis with the radio receiver on Ulysses. In: Marsch E, Schwenn R (eds) Solar wind seven, 465. COSPAR Colloquium series 3

Hoang S, Lacombe C, MacDowall RJ, Thejappa, G (2007) Radio tracking of the interplanetary coronal mass ejection driven shock crossed by Ulysses on 10 May 2001. J Geophys Res 112:A09102

Holman GD, Pesses ME (1983) Solar type II radio emission and the shock drift acceleration of electrons. Astrophys J 267:837

Hull AJ, Scudder JD, Frank LA, Paterson WR, Kivelson MG (1998) Electron heating and phase space signatures at strong and weak quasi-perpendicular shocks. J Geophys Res 103:2041

Hull AJ, Scudder JD, Fitzenreiter RJ, Ogilvie KW, Newbury JA et al (2000) Electron temperature and de Hoffman - Teller potential change across the Earth's bow shock: new results from ISEE 1. J Geophys Res 105:20,957

Izmodenov VV (2004) The heliospheric interface: models and observations. In: Poletto G, Suess S (eds) The sun and the heliosphere as an integrated system, Astrophysics and Space Science Library, vol 317. Kluwer, Dordrecht, p 23

Karlicky M (2003) High-frequency radio signatures of solar-eruptive flares. Space Sci Rev 107:81

Karlicky M, Barta M (2004) Diagnostics of solar flare reconnection. Nonlin Proc Geophys 11:471

Kasaba Y, Matsumoto H, Omura Y, Anderson RR, Mukai T et al (2000) Statistical studies of plasma waves and Backstreaming electrons in the terrestrial electron foreshock observed by Geotail. J Geophys Res 105:79

Kellogg PJ (2003) Langmuir waves associated with collisionless shocks: a review. Planet Space Sci 51:681

Kellogg PJ, Goetz K, Monson SJ, Bougeret J-L, Manning R, Kaiser ML (1996) Observations of plasma waves during a traversal of the Moon's wake. Geophys Res Lett 23:1267

Kennel CF, Edmiston JP, Hada, T (1985) A quarter century of collisionless shock research. In: Stone RG, Tsurutani BT (eds) Collisionless shocks in the heliosphere: a tutorial review, , Geophysical Monograph 34, p 1

Khan J, Aurass H (2002) X-ray observations of a large-scale solar coronal shock wave. Astron Astrophys 383:1018

Kim, E-H, Cairns IH, Robinson PA (2007) Extraordinary-mode radiation produced by linear-mode conversion of Langmuir-waves. Phys Rev Lett 99:015003

Kim, E-H, Cairns IH, Robinson PA (2008) Mode conversion of Langmuir to electromagnetic waves at magnetic field-aligned density inhomogeneities: simulations, theory, and applications to the solar wind and corona. Phys Plasmas 15:102110

Klassen A Aurass H, Klein K-L, Hofmann A, Mann G (1999) Radio evidence on shock wave formation in the solar corona. Astron Astrophys 343:287

Kliem B, Krueger A, Treumann RA (1992) Third plasma harmonic radiation in type II bursts. Solar Phys 140:149

Klimas AJ, Fitzenreiter RJ (1988) On the persistence of unstable bump-on-tail electron velocity distributions in the Earth's foreshock. J Geophys Res 93:9628

Knock SA, Cairns IH (2005) Type II radio emission predictions: Sources of coronal and interplanetary spectral structure. J Geophys Res 110:A01101

Knock SA, Cairns IH, Robinson PA, Kuncic Z (2001) Theory of type II solar radio emission from the foreshock region of an interplanetary shock. J Geophys Res 106:25041

Knock SA, Cairns IH, Robinson PA, and Kuncic,Z Theoretically predicted properties of type II radio emission from an interplanetary foreshock. J. Geophys. Res 108, 2002JA009508 (2003a)

Knock SA, Cairns IH, Robinson PA (2003b) Type II radio emission predictions: multiple shock ripples and dynamic spectra. J Geophys Res 108, 2003JA009960

Krasnoselskikh VV, Lembege B, Savoini P, Lobzin VV (2002) Nonstationarity of strong collisionless shocks: theory and full particle numerical simulations. Phys Plasmas 9:1192

Krasnoselskikh VV Lobzin VV, Musatenko K, Soucek J, Pickett JS, Cairns IH (2007) Beam-plasma interaction in randomly inhomogeneous plasmas and statistical properties of small-amplitude Langmuir waves in the solar wind and electron foreshock. J Geophys Res 112:A10109.

Kuncic Z, Cairns IH (2004) Radio emission from mini-magnetospheres on the Moon. Geophys Res Lett 31:L11809

Kuncic Z, Cairns IH (2005) Planetary foreshock radio emissions. J Geophys Res 110:A07107

Kuncic Z, Cairns IH, Knock S, Robinson PA (2002) A quantitative theory for terrestrial foreshock radio emissions. Geophys Res Lett 29:1161

Kuncic Z, Cairns IH, Knock S (2002) Analytic model for the electrostatic potential jump across collisionless shocks, with application to Earth's bow shock. J Geophys Res 107:1218

Kuncic Z, Cairns IH, Knock S (2004) A quantitative theory for terrestrial foreshock radio emissions: 1. Predicted properties. J Geophys Res Lett 109:A02108

Kurth WS, Gurnett DA (2003) On the source location of low-frequency heliospheric radio emissions. J Geophys Res 108(10):LIS 2-1

Kurth WS, Gurnett DA, Scarf FL, Poynter RL (1984) Detection of a radio emission at 3 kHz in the outer heliosphere. Nature 312:27

Kurth WS, Gurnett DA, Scarf FL, Poynter RL (1987) Long-period dynamic spectrograms of low-frequency interplanetary radio emissions. Geophys Res Lett 14:49

Lacombe C, Harvey CC, Hoang S, Manganey A, Steinberg J-L, Burgess D (1988) ISEE observations of radiation at twice the solar wind plasma frequency, Ann. Geophys 6:113

Leblanc Y, Dulk GA, Cairns IH, Bougeret J-L (2000) Type II-flare continuum in the corona and solar wind. J Geophys Res 105:18215

Lembege B, Dawson JM (1987) Self-consistent study of a perpendicular collisionless and nonresistive shock. Phys Fluids 30:1767

Lembege B et al (2004) Selected problems in collisionless shock physics. Space Sci Rev 110:161

Lembege B, Savoini P, Hellinger P, Travnicek PM (2009) Nonstationarity of a two-dimensional perpendicular shock: competingmechanisms. J Geophys Res 114:A03217

Leroy M, Winske D, Goodrich CC (1982) The structure of perpendicular bow shocks. J Geophys Res 87:5081

Li B, Cairns IH, Robinson PA (2008a) Simulations of coronal type III solar radio bursts: 1. Simulation model. J Geophys Res 113:A06104

Li B, Cairns IH, Robinson PA (2008b) Simulations of coronal type III solar radio bursts: 2. Dynamic spectrum for typeical parameters. J Geophys Res 113:A06105

Li B, Cairns IH, Robinson PA (2010) Imprints of coronal temperature disturbances on type III bursts. Astron Astrophys 510:L6

Lobzin VV, Cairns IH, Robinson PA (2008) Evidence for windlike regions, acceleration of shocks in the deep corona, and relevance of 1/f dynamic spectra to coronal type II bursts. Astrophys J Letts 677:L129

Lobzin VV, Krasnoselskikh VV, Schwartz S, Cairns IH, Lefbvre B, Decreau P (2005) Generation of downshifted oscillations in the electron foreshock: a loss-cone instability. Geophys Res Lett 32(18):L18101

Lobzin VV, Krasnoselskikh VV, Bosqued JM et al (2007) Nonstationarity and reformation of high-Mach-number Quasiperpendicular shocks: Cluster observations. Geophys Res Lett 34:L05107

Lobzin VV, Cairns IH, Robinson PA, Steward G, Patterson G (2009) Automatic recognition of type III solar radio bursts: themethod and first observations. Space Weather 7:S04002

Lobzin VV, Cairns IH, Robinson PA, Steward G, Patterson G (2010) Automatic recognition of coronal type II radio bursts: the automated radio burst identification system method and first observations. Astrophys J Lett 710:L58

Macek WM, Cairns IH, Kurth WS, Gurnett DA (1991) Plasma wave generation near the inner heliospheric shock. Geophys Res Lett 18:357

MacDowall RJ (1989) Effects of interplanetary shocks on kilometric type III radio bursts. Geophys Res Lett 16:923

Maia D, Pick M, Vourlidas A, Howard RA (2000) Development of coronal mass ejections: radio shock signatures. Astrophys J 528:L49

Malaspina DM, Ergun RE (2008) Observations of three-dimensional Langmuir wave structure. J Geophys Res 113:A12108

Malaspina DM, Li B, Cairns IH, Robinson PA, Ergun RE (2009) Terrestrial foreshock Langmuir waves: STEREO observations, theoretical modelling, and quasi-linear simulations. J Geophys Res 114:A12101

Malaspina DM, Cairns IH, Ergun RE (2010) The 2fp radiation from localized Langmuir waves. J Geophys Res 115:A01101

Mancuso S (2007) Coronal transients and metric type II radio bursts II. Accelerations at low coronal heights. Astron Astrophys 463:1137

Mancuso S, Raymond JC (2004) Coronal transients and metric type II radio bursts I. Effects of geometry. Astron Astrophys 413:363

Mann G, Klassen A, Aurass H, Classen HT (2003) Development of shock waves in the solar corona and the interplanetary space. Astron Astrophys 400:329

Mann G, Aurass H, Klassen A, Estel C, Thompson BJ (1999) Coronal transient waves and coronal shock waves. In: Vial J-C, Kaldeich-Schmann B (eds) 8th SOHO workshop: plasma dynamics and diagnostics in the solar transition region and corona, ESA Special Publication 446, p 477

Matsukiyo S, Scholer M (2006) On microinstabilities in the foot of high Mach number perpendicular shocks. J Geophys Res 111:A06104

McBride JB, Ott E, Boris JP, Orens JH (1972) Theory and simulation of turbulent heating by the modified two-stream instability. Phys Fluids 15:2367

McComas DJ, Allegrini F, Bochsler P, Bzowski M, Christian ER et al (2009) Global observations of the interstellar interaction from the interstellar boundary explorer (IBEX). Science 326:959

McClements KG, Bingham R, Su JJ, Dawson JM, Spicer DS (1993) Lower hybrid resonance acceleration of electrons and ions in solar flares and the associated microwave emission. Astrophys J 409:465

McLean DJ (1967) Band splitting in type II solar radio bursts. Proc Astron Soc Aust 12:401

McNutt RL Jr (1988) A solar-wind 'trigger' for the outer heliosphere radio emissions and the distance to the terminal shock. Geophys Res Lett 15:1307

Mel'nik VN, Konovalenko AA, Rucker HO, Stanislavsky AA, Abranin EP et al (2004) Observations of solar Type II bursts at frequencies 10–30 MHz. Solar Phys 222:151

Melrose DB (1980) The emission mechanisms for solar radio bursts. Space Sci Rev 26:3

Melrose DB (1985) Instabilities in space and laboratory plasmas. Cambridge University Press, Cambridge, New York

Mitchell JJ, Cairns IH, Robinson, PA (2004) Theory for 2–3 kHz radiation from the outer heliosphere. J Geophys Res 109:A06108

Mitchell JJ, Cairns IH, Heerikhuisen, J (2009) Pick-up ions and the 2–3 kHz radio emissions, Geophys Res Lett 36: L12109

Mitchell JJ, Cairns IH, Muller H-R, Zank GP (2005) Prediction of the timing of the 2–3 kHz radio emission within the solar cycle. Geophys Res Lett 32(7):L07101

Mitchell JJ, Cairns IH, Pogorelov, N, Zank GP (2008) Draping of the local interstellar mediummagnetic field over the heliopause. J Geophys Res 113:A04102

Moreton GE (1960) Hα observations of flare-initiated disturbances with velocities 1000 km/sec. Astron J 65:494

Moses D, Clette F, Delaboudineiere JP, Artzner GE et al (1997) EIT observations of the extreme ultraviolet Sun. Solar Phys 175:571

Narukage N, Hudson HS, Morimoto T, Akiyama S et al (2002) Simultaneous observation of aMoreton wave on 3 November 1997 in Hα and soft X-rays. Astrophys J 572:L109

Nelson GJ, Melrose DB (1985) Type II bursts. In: McLean, Labrum NR (eds) Solar radiophysics, Cambridge University Press, Cambridge, New York, p 333

Neugebauer M, Giacalone J (2005) Multispacecraft observations of interplanetary shocks: nonplanarity and energetic particles. J Geophys Res 110:A12106

Nulsen AL, Cairns IH, Robinson PA (2007) Statistics and shape of Langmuir wave packets observed by Ulysses in an interplanetary type III radio source. J Geophys Res 112:A05107

Omelchenko YA, Sagdeev RA, Shapiro VD, Shevchenko VI (1989) Numerical simulation of quasilinear relaxation of an ion ring and production of superthermal electrons. Sov J Plasma Phys 15:427

Onsager TG, Holzworth RH, Koons HC, Bauer OH, and Gurnett DA (1989) High-frequency electrostatic waves near Earth's bow shock. J Geophys Res 94:13397

Opher M, Alouani Bibi F, Toth G, Richardson JD, Izmodenov VV, Gombosi TI (2009a) A strong, highly-tilted interstellar magnetic field near the solar system. Nature 462:1036

Opitz A, Karrer R, Wurz P, Galvin AB et al (2009b) Temporal evolution of the solar wind bulk velocity at solar minimum by correlating the STEREO A and B PLASTIC measurements. Solar Phys 256:365

Papadopoulos KD, Freund HP (1978) Solitons and second harmonic radiation in type III bursts. Geophys Res Lett 5:881

Pogorelov NV, Heerikhuisen J, Mitchell JJ, Cairns IH, Zank GP (2009) Heliospheric asymmetries under strong interstellar magnetic field conditions and their impact on the 2–3 kHz radio emission. Astrophys J 695:L31

Prestage N (1994) IPS Radio and Space Services Technical Report TR-94-01, IPS, Haymarket

Pulupa M, Bale SD (2008) Structure on interplanetary shock fronts: type II radio burst source regions. Astrophys J 676:1330

Reiner MJ (2000) Interplanetary type II radio emissions associated with CMEs. In: Stone RG, Weiler KW et al (eds) Radio astronomy at long wavelengths, Geophysical Monograph Series 119. American Geophysical Union, Washington, DC, p 137

Reiner MJ, Kaiser ML (1999) High-frequency type II radio emissions associated with shocks driven by CMEs. J Geophys Res 104:16979

Reiner MJ, Kaiser ML, Bougeret J-L (2001) Radio signatures of the origin and propagation of coronal mass ejections from the solar corona and interplanetary medium. J Geophys Res 106:29989

Reiner MJ, Kasaba Y, Kaiser ML, Matsumoto H, Nagano I et al (1997) Terrestrial $2f_p$ radio source Location determined from wind/Geotail triangulation. Geophys Res Lett 24:919

Reiner MJ, Kaiser ML, Fainberg J, Stone RG (1998) A new method for studying remote type II radio emission from coronal mass ejection-driven shocks. J Geophys Res 103:29651

Reiner MJ, Kaiser ML, Fainberg J, Bougeret J-L, Stone RG (1997) On the origin of radio emission associated with the January 6–11, 1997, CME. Geophys Res Lett 25:2493

Riddle AC (1974) On the observation of scattered radio emission from sources in the solar corona. Solar Phys 35:153

Roberts JA (1959) Solar radio bursts of spectral type II. Aust J Phys 11:215

Robinson PA (1995) Stochastic wave growth. Phys Plasmas 2:1466

Robinson PA (1997) Nonlinear wave collapse and strong turbulence. Rev Mod Phys 69:507

Robinson PA, Cairns IH (1995) Maximum Langmuir fields in planetary foreshocks determined from the electrostatic decay threshold. Geophys Res Lett 22:2657

Robinson PA, Cairns IH (1998) Fundamental and harmonic emission in type III solar radio bursts: 1. Emission at a single location or frequency. Solar Phys 181:363

Robinson PA, Cairns IH (2000) Theory of type III and type II solar radio emissions. In: Stone RG, Weiler KW et al (eds) Radio astronomy at longwavelengths, Geophysical Monograph Series 119. American Geophysical Union, Washington, DC, p 37

Robinson PA, Cairns IH, Gurnett DA (1993) Clumpy Langmuir waves in type III radio sources: Comparison of stochastic-growth theory with observations. Astrophys J 407:790

Robinson PA, Cairns IH, Willes AJ (1994) Dynamics and efficiency of type III radio emission. Astrophys J 422:870

Robinson PA, Li B, Cairns IH (2006) New regimes of stochastic wave growth: theory, simulation, and comparison with data. Phys Plasmas 13:112103

Rodriguez P, Gurnett DA (1975) Electrostatic and electromagnetic turbulence associated with the Earth's bow shock. J Geophys Res 80:19

Russell CT (1985) Planetary bow shocks. Collisionless shocks in the heliosphere: reviews of current research. American Geophysical Union, Washington, DC, p 109

Scherer K, Fahr HJ (2003) Solar cycle induced variations of the outer heliospheric structures. Geophys Res Lett 30(2):17–1

Schmidt JM, Gopalswamy N (2008) Synthetic radio maps of CME-driven shocks below 4 solar radii. J Geophys Res 113:A08104

Scholer M, Matsukiyo S (2004) Nonstationarity of quasi-perpendicular shocks: a comparison of full particle simulations with different ion to electron mass ratios. Ann Geophys 22:2345

Scudder JD, Mangeney A, Lacombe C, Harvey CC, Wu CS, Anderson RR (1986) The resolved layer of a collisionless, high β, supercritical, quasiperpendicular shock wave 3. Vlasov electrodynamics. J Geophys Res 91:11075

Shapiro VD, Bingham R, Dawson JM, Dobe Z et al (1998) Electron energization by lower-hybridwaves as a possible source for X-ray emission from comets. Phys Scr T75:39

Sigsbee K, Kletzing CA, Gurnett DA, Pickett JS et al (2004) The dependence of Langmuir wave amplitudes on position in Earth's foreshock. Geophys Res Lett 31:L07805

Smerd SF, Sheridan KV, Stewart RT (1974) On split-band structure in type II radio bursts from the Sun. In: Newkirk G (ed) Coronal disturbances: proceedings of the IAU symposium. Springer, New York, NY, p 389

Steinberg J-L, Hoang S, Dulk GA (1985) Evidence of scattering effects on the sizes of interplanetary type III radio bursts. Astron Astrophys 150:205

Stewart RT, Magun A (1980) Radio evidence for electron acceleration by transverse shock waves in herringbone type II solar radio bursts. Proc Astron Soc Aust 4(1):53

Stone EC, Cummings AC, McDonald FB, Heikkila BC et al (2005) Voyager 1 explores the termination shock region and the heliosheath beyond. Science 309:2017

Stone EC, Cummings AC, McDonald FB, Heikkila BC et al (2008) An asymmetric solar wind termination shock. Nature 454:71

Suzuki S, Dulk GA (1985) Bursts of type III and V. In: McLean DJ, Labrum NR (eds) Solar radiophysics. Cambridge University, Cambridge, New York, p 289

Thejappa G, Wentzel DG, MacDowall RJ, Stone RG (1995) Unusual wave phenomena near interplanetary shocks at high latitudes. Geophys Res Lett 22:3421

Thejappa G, MacDowall RJ, Kaiser ML (2007) Monte Carlo simulation of the directivity of interplanetary radio bursts. Astrophys J 671:894

Thompson BJ, Plunkett SP, Gurman JB, Newmark JS, St Cyr OC et al (1998) SOHO/EIT observations of an Earth-directed coronal mass ejection on 12 May 1997. Geophys Res Lett 25:2465

Tidman DA, Krall NA (1971) Shock waves in collisionless plasmas. Wiley-Interscience, New York, NY

Toptyghin IN (1980) Acceleration of particles by shocks in a cosmic plasma. Space Sci Rev 26:157

Uchida Y (1974) Behavior of the flare produced coronal MHD wavefront and the occurrence of type II radio bursts. Solar Phys 39:431

Vrsnak B, Cliver E (2008) Origin of coronal shock waves. Solar Phys 253:215

Vrsnak B, Magdalenic J, Aurass H, Mann G (2002) Band-splitting of coronal and interplanetary type II bursts: coronal magnetic field and Alfven velocity. Astron Astrophys 396:673

Wang SJ, Yan Y, Zhao RZ, Fu Q, Tan CM et al (2001) Broadband radio bursts and fine structures during the great solar event on 14 July 2000. Solar Phys 204:155

Warmuth A, Mann G (2005) The application of radio diagnostics to the study of the solar drivers of space weather. In: Scherer K, Fichtner H, Heber B, Mall U (eds) Space weather, Lecture Notes Phys. 656. Springer, p 51

Warmuth A, Vrsnak B, Magdalenic J, Hanslmeier A, Otruba W (2004a) A multiwavelength study of solar flare waves. I Observations and basic properties. Astron Astrophys 418:1101

Warmuth A, Vrsnak B, Magdalenic J, Hanslmeier A, Otruba W (2004b) A multiwavelength study of solar flare waves. II Perturbation characteristics and physical interpretation. Astron Astrophys 418:1117

Wild JP (1950) Observations of the spectrum of high-intensity solar radiation at meter wavelengths. Aust J Sci Res Ser A 3:399

Wild JP, McCready LL (1950) Observations of the spectrum of high-intensity solar radiation at metre wavelengths. I. The apparatus and spectral types of solar burst observed. Aust J Sci Res A 3:387

Wild JP, Smerd SF (1972) Radio bursts from the corona. Annu Rev Astron Astrophys 10:159

Wild JP, Smerd SF, Weiss AA (1963) Solar bursts. Annu Rev Astron Astrophys 1:291

Wu CS (1984) A fast fermi process: energetic electrons accelerated by a nearly perpendicular bow shock. J Geophys Res 89:8857

Wu CS, Lee LC (1979) A theory of terrestrial kilometric radiation. Astrophys J 230:621

Wu CS, Gaffey JD, Zhou GC (1985) Induced emission of radiation near $2\omega e$ by a synchrotron-maser instability. Phys Fluids 28:846

Yan Y, Deng Y, Karlicky M, Fu Q, Wang S, and Liu, Y (2001) The magnetic rope structure and associated energetic processes in the 2000 July 14 solar flare. Astrophys J 551:L115

Yin L, Ashour-Abdalla M, El-Alaoui M, Bosqued JM, and Bougeret J-L (1998) Generation of electromagnetic f_{pe} and $2f_{pe}$ waves in the Earth's electron foreshock via linear mode conversion. Geophys. Res Lett 25:2609

Yuan X, Cairns IH, Robinson PA (2007) Simulation of electron bursts upstream of reforming shocks. Astrophys J 671:439

Yuan X, Cairns IH, Robinson PA (2008a) Numerical simulations of electron distributions upstream and downstream of high Mach number quasi-perpendicular collisionless shocks. J Geophys Res 113:A08109

Yuan X, Cairns IH, Robinson PA, Kuncic Z (2007) Effects of overshoots on electron distributions upstream and downstream of quasiperpendicular collisionless Shocks. J Geophys Res 112:A05108

Yuan X, Cairns IH, Trichtchenko L, Rankin R (2008b) Effects of shock parameters on upstream energetic electron burst events. J Geophys Res 113:A09106

Yuan X, Cairns IH, Trichtchenko L, Rankin R, Danskin DW (2009) Confirmation of shock reformation in two-dimensional hybrid simulations. Geophys Res Lett 36:L05103

Zakharov VE (1972) Collapse of Langmuir waves. Sov Phys JETP, Engl Trans 35:908

Zank GP (1999) Interaction of the solar wind with the local interstellar medium: a theoretical perspective. Space Sci Rev 89(3/4):1

Zank GP, Muller H-R (2003) The dynamical heliosphere. J Geophys Res 108(A6):SSH7

Zank GP, Pauls HL,Williams LL, Hall DT (1996) Interaction of the solar wind with the local interstellar medium: a multifluid approach. J Geophys Res 101:21639

Part VII
Coordinated Science in the Sun-Earth System

Chapter 24

Advances in Coordinated Sun-Earth System Science Through Interdisciplinary Initiatives and International Programs

Horst Fichtner and W. William Liu

Abstract Recent observational advances have allowed us to study the changing Sun with unprecedented detail. Many dominant research topics have become broader as we seek to connect changes at the Sun and of the Heliosphere to changes in the Earth's magnetosphere and atmosphere through a deeper understanding of the physics obtained from interdisciplinary approaches to problems. The challenge of conducting interdisciplinary research in a plasma system as large and complex as the one between the Sun and Earth within the heliosphere, which is itself embedded in the interstellar medium, demands long-term, carefully coordinated international cooperation. In response to this need, the "Climate and Weather of the Sun-Earth System" (CAWSES) and "International Living With a Star" (ILWS) programs were created. Complementary in their nature and goals, the two programs combine resources from the national and international communities to both advance the state of our knowledge and coordinate and plan Sun-Earth System research missions of the present and future. Part of the corresponding research that is related to space weather and space climate, including external forcing of the terrestrial climate and the atmospheric response, and that is put into an overarching context in the present paper is highlighted in the articles of this section.

H. Fichtner (✉)
Institut für Theoretische Physik IV, Ruhr-Universität Bochum, 44780 Bochum, Germany
e-mail: hf@tp4.rub.de

24.1 An Integrating View

The increasing complexity and demands of industrial civilisations at the beginning of the third millenium not only have an impact on the local, terrestrial environment and, possibly, climate, but are also affected by the geospace and interplanetary environment as well as the space climate conditions. On one hand, the impact of industrial civilisations on environment is no longer limited to the ground, the oceans, or the atmosphere but, in view of the space debris problem, already concerns near-Earth space. On the other hand, physical processes in geospace as well as in the interplanetary medium influence the terrestrial environment, and a connection between solar activity and terrestrial climate is evident. Consequently, the Earth cannot be understood as an isolated physical system: It must be considered as a part of the Solar System and the heliosphere that is influenced by the surrounding interplanetary medium and the Sun.

Due to such insight but also as a consequence of their increasing level of sophistication the research fields of atmospheric physics, geophysics, space physics, solar physics, and to some extent even astrophysics not only touch but significantly overlap each other resulting in increasing degrees of interdisciplinarity. This led to the development of newly defined research fields like "geospace physics" combining the physics of the atmosphere and the magnetosphere, or "heliophysics" comprising the physics of the Sun and the heliosphere. This change is also reflected in the approach by which future science will be performed. Although the idea of systems science has been with us for a long time, it was not until recently that our experimental and computational capabilities have become

M.P. Miralles, J. Sánchez Almeida (eds.), *The Sun, the Solar Wind, and the Heliosphere*, IAGA Special Sopron Book Series 4, DOI 10.1007/978-90-481-9787-3_24,

powerful enough for us to tackle a number of big questions systematically.

This partial fusion of research fields is reflected in the installment of various international programs that intend to foster joint activities of previously separate disciplines with partly converging research goals. The closer international cooperation is driven by a number of factors. Financial constraints confronted by many countries have created significant incentives for scientists to collaborate. More importantly, each country, by virtue of its history, geography and even culture, has something unique to bring to the international milieu. Besides a series of "International Years" in Geophysics (http://www.ipy.org/), Heliophysics (http://ihy2007.org/), and Astronomy (http://www.astronomy2009.org/), two of such programs are "Climate and Weather of the Sun-Earth System" (CAWSES, http://www.bu.edu/cawses), which is sponsored by the Scientific Committee on Solar-Terrestrial Physics (SCOSTEP), and "International Living With a Star" (ILWS, http://ilws.gsfc.nasa.gov/) which is supported by numerous national space agencies. The articles of this section are concentrating on activities within the latter two initiatives.

24.2 CAWSES and ILWS

Recent observational advances have allowed us to study the changing Sun with unprecedented detail. Many dominant research topics have become broader as we seek to connect changes at the Sun and of the Heliosphere to changes in the Earth's magnetosphere and atmosphere through a deeper understanding of the physics obtained from interdisciplinary approaches to various problems. The challenge of conducting interdisciplinary research in a plasma system as large and complex as the one between the Sun and Earth within the heliosphere, which is itself embedded in the interstellar medium, demands long-term, carefully coordinated international cooperation. In response to this need, the Climate and Weather of the Sun-Earth System (CAWSES) and International Living With a Star (ILWS) programs were created. Complementary in their nature and goals, the two programs combine resources from the national and international communities to both advance the state of our knowledge and coordinate and plan Sun-Earth System research missions of the present and future.

24.2.1 CAWSES

CAWSES is structured into four overarching themes, namely, (1) Solar Influence on Climate, (2) Space Weather: Science and Applications, (3) Atmospheric Coupling Processes, and (4) Space Climatology. Within the framework of various corresponding working groups the goal of CAWSES are intended to be the answers to the following questions:

- Can we link the end-to-end processes that produce geoeffective coronal mass ejections, facilitate their transfer through the heliosphere, their interaction with the magnetosphere, and the production of geomagnetic storms that affect the atmosphere?
- Can we identify evidence for long-term variations of solar luminosity related to solar activity and resultant impacts on global change, compared with other climate change mechanisms?
- To what extent are the magnetosphere and ionosphere-thermosphere systems modulated by solar activity on long time scales, including the solar cycle, and how do variations driven by different processes interact with dynamical and radiative forcing processes from below?
- Can we reconcile apparent responses of the middle and lower atmosphere to solar activity, identify the physical mechanisms, in comparison with anthropogenic influences, and estimate future ozone changes?

The program, that started officially in 2004, has been established with the aim of significantly enhancing our understanding of the space environment and its impacts on life and society. The main functions of CAWSES are not only to help coordinate international activities in observations, modeling, and theory crucial to achieving this understanding, but also to involve scientists in both developed and developing countries as well as to provide educational opportunities for students at all levels. This international initiative has triggered the financing of several national CAWSES programs in France, Germany (http://www.iap-kborn.de/DFG-Schwerpunktprogramm.63.0.html), India, Japan, and Taiwan.

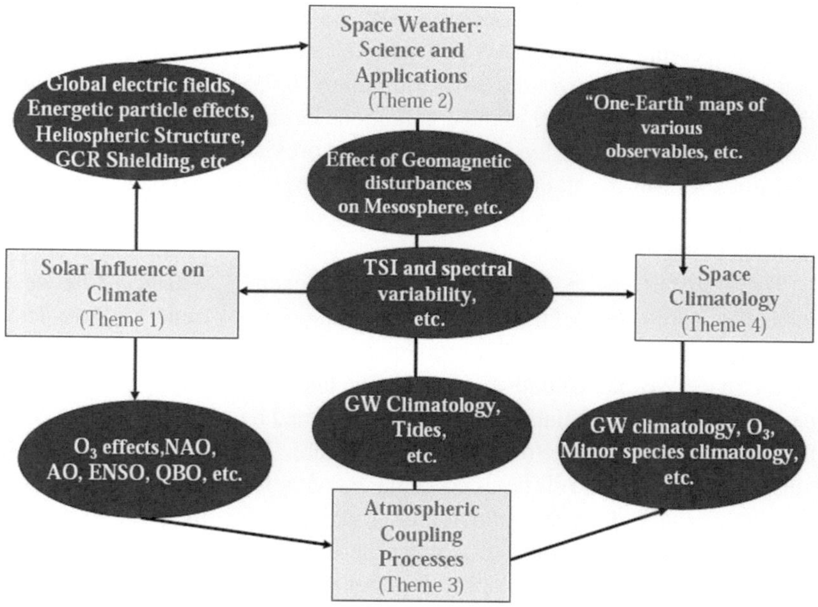

Fig. 24.1 The principle structure of the CAWSES program; from Basu and Pallamraju (2006)

24.2.2 ILWS

ILWS has the goal stimulate, strengthen, and coordinate space research to understand the governing processes of the connected Sun-Earth System as an integrated entity. Its specific objectives are to stimulate and facilitate (1) Studies of the Sun-Earth connected system and the effects which influence life and society, (2) Collaborations among potential partners in solar-terrestrial space missions, (3) Synergistic coordinations of international research in solar-terrestrial studies, including all relevant data sources as well as theory and modeling, and (4) effective and user-driven access to all data, results, and value-added products. Specific questions to be addressed are:

- How and why does the Sun vary?
- How does the Earth respond?
- What are the impacts on humanity?

The program started officially in 2002. Its emphasis on space missions is reflected by some of the following articles that highlight approved as well as planned missions to further explore geospace, the interplanetary medium, and the Sun. Aside from existing missions such as SOHO, Cluster, STEREO, IBEX, Hinode, and THEMIS that have graced newspaper headlines with their discoveries, ILWS is planning the next wave of missions to unlock the secrets of solar variability (PICARD and Solar Dynamic Observatory), propagation of solar energy outbursts from the solar wind (Solar Orbiter and Kuafu), energy release through magnetic reconnection (MMS, Cross-Scale, and SCOPE), magnetosphere-ionosphere coupling (Kuafu and ePOP), and ion-neutral interactions (SWARM) that can potentially impact the evolution of Earth's atmosphere.

24.3 Further Developments and Trends

There are various activities that indicate possible or even likely next steps. On the one hand there exist preparations for new programs providing frameworks to continue and intensify the interdisciplinary efforts. On the other hand there are recent scientific developments making a further joining-of-forces of different fields not only desirable but mandatory. In the following we briefly give examples for such activities.

24.3.1 Future Programs

While the ILWS is a long-term initiative, CAWSES has a defined duration. In view of the great success of the latter program a plan for CAWSES-II is well on its way (http://www.cawses.org/CAWSES/Home.html). Following the first CAWSES period, this second phase is named "Towards Solar Maximum" referring to the changing space climate conditions defined mainly by the Sun's activity. It will continue to provide structures for Sun-Earth community to make progress on science issues that cannot be achieved without international collaboration.

There are, of course, further activities. One example is the continuation of the "International Heliophysical Year" (IHY, http://ihy2007.org/) with the "International Space Weather Initiative" (ISWI). The latter is a new United Nations initiative that has been adopted by the United Nations Office of Outer Space Affairs. The purpose of this effort will be to continue encouraging existing and new scientific collaborations, and, particularly will begin the process of making the data from these new arrays part of the space weather alert system. Obviously, the ISWI also strongly complements the continuing ILWS.

24.3.2 Growing Interdisciplinarity

Examples for the increasing need of interdisciplinarity are recent activities in space physics and in astrophysics. In general, there exists a transfer of ideas and concepts like, e.g. magnetic field structures, reconnection, particle acceleration, or plasmadynamics, from heliophysics to astrophysics. The underlying processes taking place on micro-, meso- and macro-scales have recently been termed "universal physical processes". While this term is a bit unfortunate in unneccessarily stating the "universality" of physical processes, it intends to emphasise the ongoing transfer as well as the growing interdisciplinarity. In particular, there exist fields where specific knowledge about the Sun and the heliosphere is needed in order to understand astrophysical systems in the required detail. This concerns, first, the field of stellar activity. The improving observational basis calls for a more detailed modelling and this can be fed by what is known from solar activity. Second, the ultimate quest of detecting life outside the solar system is driving the search and study of extrasolar planets. Recently, it has clearly been recognized that the targets for the search of suitable candidates cannot be identified on the basis of planetary physics alone: The definition of habitable zones requires knowledge about the activity and wind of the host star as well as of the interstellar medium surrounding the corresponding astrosphere, so that, e.g., the cosmic ray flux at the location of the planet can be estimated. In order to obtain meaningful, i.e. quantitive estimates, one will tremendously benefit from the use of sophisticated heliophysical models. As mentioned in the introduction, one of the next interdisciplinarities will occur between helio- and astrophysics. Another trend we have detected is a decisive move to integrate inquiries of space plasma physics and atmospheric physics. Recent controversies concerning IPCC's conclusion on the Sun's role in climate change has highlighted the importance of research on how the Sun impacts Earth's atmosphere, both through variations in its radiative output and geospace activities driven by the solar wind. An interesting example can be found on the cover of recently published NASA heliophysics roadmap. Rather than the traditional cartoon featuring the Sun and Earth's magnetosphere, the new icon gives a much greater emphasis on Earth's atmosphere and its potential impacts on humanity. And this change is not merely cosmetic. The NASA roadmap contains four candidate missions that deal with geospace-atmosphere or sun-atmosphere relationships, out of seven targeted missions that were identified. The US example is indicative of an international trend to focus on interdisciplinary contact points in the Earth's atmosphere. In this sense, we see that planning processes in space agencies are becoming more aligned with the scientific priorities manifested by the community, for example, in the science themes of CAWSES-II.

24.4 Summary

We have outlined how recent developments in the research fields of atmospheric physics, geophysics, space physics, solar physics, and astrophysics have led to various interdisciplinary efforts and, eventually, to new fields like "geospace physics" combining the physics of the atmosphere and the magnetosphere, or "heliophysics" comprising the physics of the Sun and the heliosphere. We have briefly discussed that this partial convergence of research interests and requirements is well-recognized and resulted in various international programs with the goals (1) to facilitate the build-up of scientific interfaces between research fields, (2) to improve, enlarge and facilitate access to comprehensive databases, as well as to (3) financially support specific research projects. How to take advantage of the goodwill and plethora of opportunities to perform systems science at a level to truly gain new knowledge about our Planet and Solar System is a challenge both to the scientific community and funding agencies. With a glance at the following articles in this section, the emphasis is here on the CAWSES and the ILWS program.

References

Basu S, Pallamraju D (2006) Adv Space Res 38:1781

The *Climate and Weather of the Sun-Earth System* Program, http://www.bu.edu/cawses

The *Climate and Weather of the Sun-Earth System* Program, Phase 2: Towards Solar Maximum, http://www.cawses.org/CAWSES/Home.html

The German Priority Program *Climate and Weather of the Sun-Earth System*, http://www.iap-kborn.de/DFG-Schwerpunkt programm.63.0.html

The *International Living with a Star* Program, http://ilws.gsfc.nasa.gov/

The *International Heliophysical Year* Program, http://ihy2007.org/

The *International Polar Year* Program, http://www.ipy.org/

The *International Astronomical Year* Program, http://www.astronomy2009.org/

Chapter 25

Solar Orbiter: Linking the Sun and Inner Heliosphere

Richard G. Marsden and Daniel Müller

Abstract Solar Orbiter, a candidate mission in ESA's Cosmic Vision programme, is designed to study the Sun and inner heliosphere in greater detail than ever before. At the closest point on its heliocentric orbit, the Solar Orbiter spacecraft will be about 0.23 AU from the Sun, closer than any other satellite to date. In addition to providing high-resolution images of the solar surface, perihelion passes at these distances occur in near co-rotation with the Sun, allowing the instruments to track features on the surface for several days. The mission profile also includes a latitude cranking phase that will allow observations from up to 34° above the solar equator. Multiple Venus gravity assist manoeuvres will be employed to increase the inclination of the orbital plane. The combination of near-Sun, quasi-heliosynchronous and out-of-ecliptic observations by remote-sensing and in-situ instruments makes Solar Orbiter a unique platform for the study of the links between the Sun and the inner heliosphere. These aspects can be further enhanced by exploiting the joint capabilities of Solar Orbiter and NASAs Solar Probe Plus mission, which is planned to be launched in the same time-frame as Solar Orbiter. In this chapter, we review the science goals of Solar Orbiter and present the mission design.

25.1 Introduction

We live in the extended atmosphere of the Sun, a region of space known as the heliosphere. Understanding the connections and the coupling between the Sun and the heliosphere is of fundamental importance to addressing one of the major scientific questions of ESA's Cosmic Vision 2015–2025 programme: "How does the Solar System work?" The heliosphere also represents a uniquely accessible domain of space, where fundamental physical processes common to solar, astrophysical and laboratory plasmas can be studied under conditions impossible to reproduce on Earth, or to study from astronomical distances. The results from missions such as Helios, Ulysses, Yohkoh, SOHO, TRACE and RHESSI, as well as the recently launched Hinode and STEREO missions, have formed the foundation of our understanding of the solar corona, the solar wind, and the three-dimensional heliosphere. Each of these missions had a specific focus, being part of an overall strategy of coordinated solar and heliospheric research. However, an important element of this strategy has yet to be implemented. None of these missions have been able to fully explore the interface region where the solar wind is born and heliospheric structures are formed with sufficient instrumentation to link solar wind structures back to their source regions at the Sun. With previously unavailable observational capabilities provided by the powerful combination of in-situ and remote-sensing instruments on Solar Orbiter, and the unique inner-heliospheric mission design specifically tailored for the task, Solar Orbiter will address the central question of heliophysics: "How does the Sun create and control the heliosphere?" This primary, overarching scientific objective can be broken down into four interrelated scientific questions, which are discussed briefly below. We then present an overview of the mission implementation, including the mission profile, the scientific payload and spacecraft characteristics.

R.G. Marsden (✉)
ESA-ESTEC, 2200AG Noordwijk, The Netherlands
e-mail: Richard.Marsden@esa.int

M.P. Miralles, J. Sánchez Almeida (eds.), *The Sun, the Solar Wind, and the Heliosphere*, IAGA Special Sopron Book Series 4, DOI 10.1007/978-90-481-9787-3_25, © Springer Science+Business Media B.V. 2011

25.2 Scientific Goals

The four top-level scientific questions being addressed by Solar Orbiter are:

How and where do the solar wind plasma and magnetic field originate in the corona?

The solar corona continuously expands and develops into a supersonic wind that extends outward, interacting with itself and with the Earth and other planets, to the heliopause boundary with interstellar space, far beyond Pluto's orbit. The solar wind has profound effects on planetary environments and on the planets themselves. For example, it is responsible for many of the phenomena in Earth's magnetosphere and is thought to have played a role in the evolution of Venus and Mars through the erosion of their upper atmospheres. Two classes of solar wind – "fast" and "slow" – fill the heliosphere, and the balance between them is modulated by the 11-year solar cycle. The fast solar wind ($\sim$700 km/s and comparatively steady) is known to arise from coronal holes. The slow solar wind ($\sim$400–500 km/s) permeates the plane of the ecliptic during most of the solar cycle so it is important to Earth's space environment. The slow solar wind shows different mass flux and composition than the fast wind, consistent with confined plasma in the solar corona (Zurbuchen 2007). The specific escape mechanism through the largely closed magnetic field is not known since candidate sites and mechanisms cannot be resolved from 1 AU. Fast and slow wind carry embedded turbulent fluctuations, and these also display different properties compatible with different solar origins (Tu and Marsch 1995). It is thought that such fluctuations may be responsible for the difference in heating and acceleration between different solar wind streams. Understanding the physics relating the plasma at the solar surface and the heating and acceleration of the escaping solar wind is crucial to understanding both the effects of the Sun on the heliosphere and how stars in general lose mass and angular momentum to stellar winds.

How do solar transients drive heliospheric variability?

The largest transient events from the Sun are coronal mass ejections (CME), large structures of magnetic field and material that are ejected from the Sun at speeds up to 3,000 km/s (see Crooker et al. 1997, and references therein). CMEs are also of astrophysical interest since they appear to be the dominant way that stars shed both magnetic flux and magnetic helicity that build up as a result of the stellar dynamo. Interplanetary CMEs (ICME) are the major cause of interplanetary shocks, but the locations and mechanisms by which shocks form around them is not known since this occurs in the inner solar system. Similarly, the longitudinal structure of ICMEs is not observable from the ecliptic, while its extent has a large impact on the acceleration of energetic particles. ICMEs are a major cause of geomagnetic storms but their effectiveness at disrupting the magnetosphere is only loosely related to the parent CME, because the evolution of the propagating cloud with the surrounding heliosphere is complex and has not been well studied. These unknowns have direct impact on our ability to predict transient ("space weather") events that affect Earth.

How do solar eruptions produce energetic particle radiation that fills the heliosphere?

Like many astrophysical systems, the Sun is an effective particle accelerator. Large solar energetic particle (SEP) events produce highly energetic particles that fill the solar system with ionizing radiation (Reames 1999). CME driven shocks can produce relativistic particles on time scales of minutes, and many CMEs convert $\sim$10% of their kinetic energy into energetic particles. Other processes produce high energy particles on magnetic loops without involving shocks. The multiple processes operating in SEP events are not well understood or distinguishable from observations at 1 AU. In particular, particles accelerated in the corona and inner heliosphere are scattered by inhomogeneities in the interplanetary magnetic field (IMF) before they arrive at Earth, destroying much of the information they carry about the processes that accelerated them. Particle transport and scattering in the inner solar system are poorly understood since the turbulence properties cannot be determined from 1 AU. The actual seed population of particles energized by CME-driven shocks in the inner solar system is unexplored, and needs to be understood to construct a complete picture of particle acceleration in shock-related events.

How does the solar dynamo work and drive connections between the Sun and the heliosphere?

The Sun's magnetic field connects the interior of the star to interplanetary space and is dominated by a quasi-periodic 11-year sunspot cycle that modulates the form of the heliosphere and strongly affects the space environment throughout the solar system (Solanki et al. 2006). The large-scale solar field is generated in the Sun's interior, within the convection zone, by a dynamo driven by complex three-dimensional mass flows that transport and process magnetic flux. Despite notable advances in our knowledge and understanding of solar magnetism made possible by Ulysses, SOHO, and Hinode observations, as well as by recent theoretical models and numerical simulations, fundamental questions remain about the operation of the solar dynamo and the cyclic nature of solar magnetic activity. Of paramount importance to answering these questions is detailed knowledge of the transport of flux at high latitudes and the properties of the polar magnetic field. To date, however, the solar high latitudes remain poorly known owing to our dependence on observations made from the ecliptic. In addition to questions about the global dynamo and the generation of the large-scale field, there are unanswered questions about the origin of the small-scale internetwork field observed in the quiet photosphere. Is this weak field produced by turbulent local dynamo action near the solar surface?

These are outstanding fundamental questions in solar and heliophysics today. By addressing them, we will make major breakthroughs in our understanding of how the inner solar system works and is driven by solar activity. To answer these questions, it is essential to make in-situ measurements of the solar wind plasma, fields, waves, and energetic particles close enough to the Sun that they are still relatively pristine and have not had their properties modified by subsequent transport and propagation processes. This is one of the fundamental drivers for the Solar Orbiter mission, which will approach the Sun to ~0.23 AU. Relating these in-situ measurements back to their source regions and structures on the Sun requires simultaneous, high-resolution imaging and spectroscopic observations of the Sun in and out of the ecliptic plane. The resulting combination of in-situ and remote-sensing instruments on the same spacecraft, together with the new, inner-heliospheric perspective, distinguishes Solar Orbiter from all previous and current missions, enabling breakthrough science which can be achieved in no other way.

25.3 Mission Design

A mission profile for Solar Orbiter has been developed that will, for the first time, make it possible to study the Sun with a full suite of in-situ and remote-sensing instruments from inside 0.25 AU and provide imaging and spectral observations of the Sun's polar regions from out of the ecliptic. This proximity to the Sun will also have the significant advantage that the spacecraft will fly in near synchronization with the Sun's rotation, allowing observations of the solar surface and heliosphere to be studied from a near co-rotating vantage point for almost a complete solar rotation. The baseline mission is planned to start on 4 January 2017 with a launch on a NASA-provided Evolved Expendable Launch Vehicle (EELV) from Cape Canaveral, placing the spacecraft on a ballistic trajectory that will be combined with planetary Gravity Assist Manoeuvres (GAM) at Earth and Venus. The initial resonance with Venus is 4:3, switching to 3:2 after the third Venus GAM. The resultant operational orbit has an orbital period of 150 days, a perihelion radius of 0.23 AU and a solar inclination of 7.7°. A series of Venus gravity assists (every 450 days) will then increase the orbit inclination. The end of the nominal mission occurs 7.5 years after launch, when the orbit inclination relative to the solar equator exceeds 25°. The inclination may be further increased during an extended mission phase using additional Venus GAMs, to reach a maximum of 34°. The mission profile is shown in Fig. 25.1.

25.4 Scientific Payload

The scientific payload of Solar Orbiter will be provided by the ESA Member States and NASA, and has already been selected and funded for the definition phase through a competitive AO selection process. The 10 Principal Investigator-led hardware investigations are:

The in-situ instruments

- The Solar Wind Analyser (SWA) instrument suite (C. Owen, PI, UK) will fully characterize the major constituents of the solar wind plasma (protons, alpha particles, electrons, heavy ions) between 0.23 and 1.4 AU.
- The Energetic Particle Detector (EPD) experiment (J. R. Pacheco, PI, Spain) will measure the properties of suprathermal ions and energetic particles in the energy range of a few keV/n to relativistic electrons and high-energy ions (100 MeV/n protons, 200 MeV/n heavy ions).

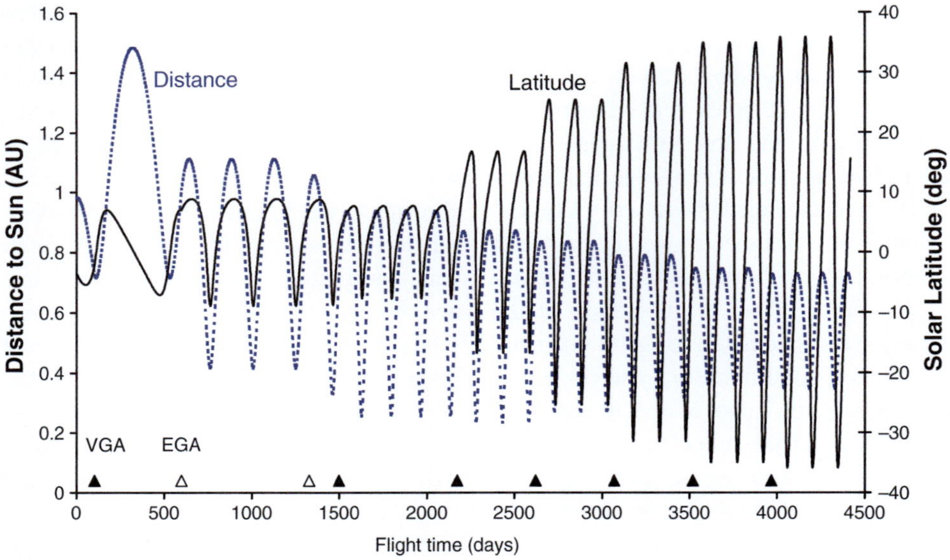

Fig. 25.1 The Solar Orbiter mission profile, showing heliocentric distance and solar latitude as a function of time since launch for a launch in 2017. Also indicated are the times at which gravity-assist manoeuvres at Venus (VGA, *solid triangles*) and Earth (EGA, *open triangles*) occur

- The Magnetometer (MAG) experiment (T. Horbury, PI, UK) will provide detailed in-situ measurements of the heliospheric magnetic field.
- The Radio and Plasma Wave (RPW) experiment (M. Maksimovic, PI, France) will measure magnetic and electric fields at high time resolution and determine the characteristics of electromagnetic and electrostatic waves in the solar wind from almost DC to 20 MHz.

The remote-sensing instruments

- The Polarimetric and Helioseismic Imager (PHI) (S. Solanki, PI, Germany) will provide high-resolution and full-disk measurements of the photospheric vector magnetic field and line-of-sight velocity as well as the continuum intensity in the visible wavelength range.
- The Extreme Ultraviolet Imager (EUI) (P. Rochus, PI, Belgium) will provide image sequences of the solar atmospheric layers from the photosphere into the corona.
- The Spectral Imaging of the Coronal Environment (SPICE) EUV Spectrograph (D. Hassler, PI, US) will provide spectral imaging of both the solar disk and in the corona to remotely characterize plasma properties of regions at and near the Sun.
- The Spectrometer/Telescope for Imaging X-rays (STIX) (A. Benz, PI, Switzerland) provides imaging spectroscopy of solar thermal and non-thermal X-ray emission from 4 to 150 keV.
- The Multi Element Telescope for Imaging and Spectroscopy (METIS/COR) Coronagraph (E. Antonucci, PI, Italy) will perform broad-band and polarized imaging of the visible K-corona and narrow-band imaging of the UV and EUV corona.
- The Solar Orbiter Heliospheric Imager (SoloHI) (R. Howard, PI, US) will image both the quasi-steady flow and transient disturbances in the solar wind over a wide field of view by observing visible sunlight scattered by solar wind electrons.

25.5 Spacecraft

As illustrated in Fig. 25.2, the Solar Orbiter spacecraft is a Sun-pointed, 3-axis stabilized platform, with a dedicated heat shield to provide protection from the high levels of solar flux near perihelion. Feed-throughs in the heat shield (with individual doors) provide the remote-sensing instruments with their required fields-of-view to the Sun. Two-sided solar arrays provide the capability to produce the required power through-

Fig. 25.2 Artist's impression of the Solar Orbiter spacecraft in operational mode (Courtesy of Astrium UK)

out the mission over the wide range of distances from the Sun using rotation about their longitudinal axis to allow switching between faces, as well as control of the Solar Aspect Angle to allow management of the array temperature throughout the mission, particularly during closest approach to the Sun. An articulated High Temperature High Gain Antenna provides nominal communication with the ground station, and a Medium Gain Antenna and two Low Gain antennas are included for use as backup. The design drivers for the Solar Orbiter spacecraft come not only from the need to satisfy the missions technical and performance requirements, but also from the need to minimize the total cost of the mission. The adopted philosophy is therefore to avoid technology development as far as possible, in order to maintain the cost-cap of the mission in keeping with its M-class ("Medium"-cost Cosmic Vision mission) status. The design of Solar Orbiter has therefore incorporated technology items from ESA's BepiColombo mission where appropriate. Furthermore, design heritage from ESA's Express series of missions, with their goal of rapid and streamlined development, has also featured heavily in the Solar Orbiter spacecraft design. The key features of the spacecraft, in particular the thermal control subsystem, are shown in Fig. 25.3.

25.6 Mission Operations

As noted above, one of the strengths of the Solar Orbiter mission is the synergy between in-situ and remote-sensing observations, and each science objective requires coordinated observations between several in-situ and remote sensing instruments. Another unique aspect of Solar Orbiter, in contrast to near-Earth observatory missions like SOHO, is that Solar Orbiter will operate much like a planetary encounter mission, with the main scientific activity and planning taking place during the near-Sun encounter part of each orbit. Specifically, observations with the remote-sensing instruments will be organized into three 10-day intervals centered around perihelion and either maximum latitude or maximum co-rotation passages. This is illustrated in Fig. 25.4. As a baseline, the in-situ instruments will operate continuously during normal operations. Another important aspect of this mission, from a science operations standpoint, is that every science orbit is different, with different orbital characteristics (Sun-spacecraft distance, Earth-spacecraft distance, etc.). Science and operations planning for each orbit is therefore critical, with specific orbits expected to be dedicated to specific science problems. This will

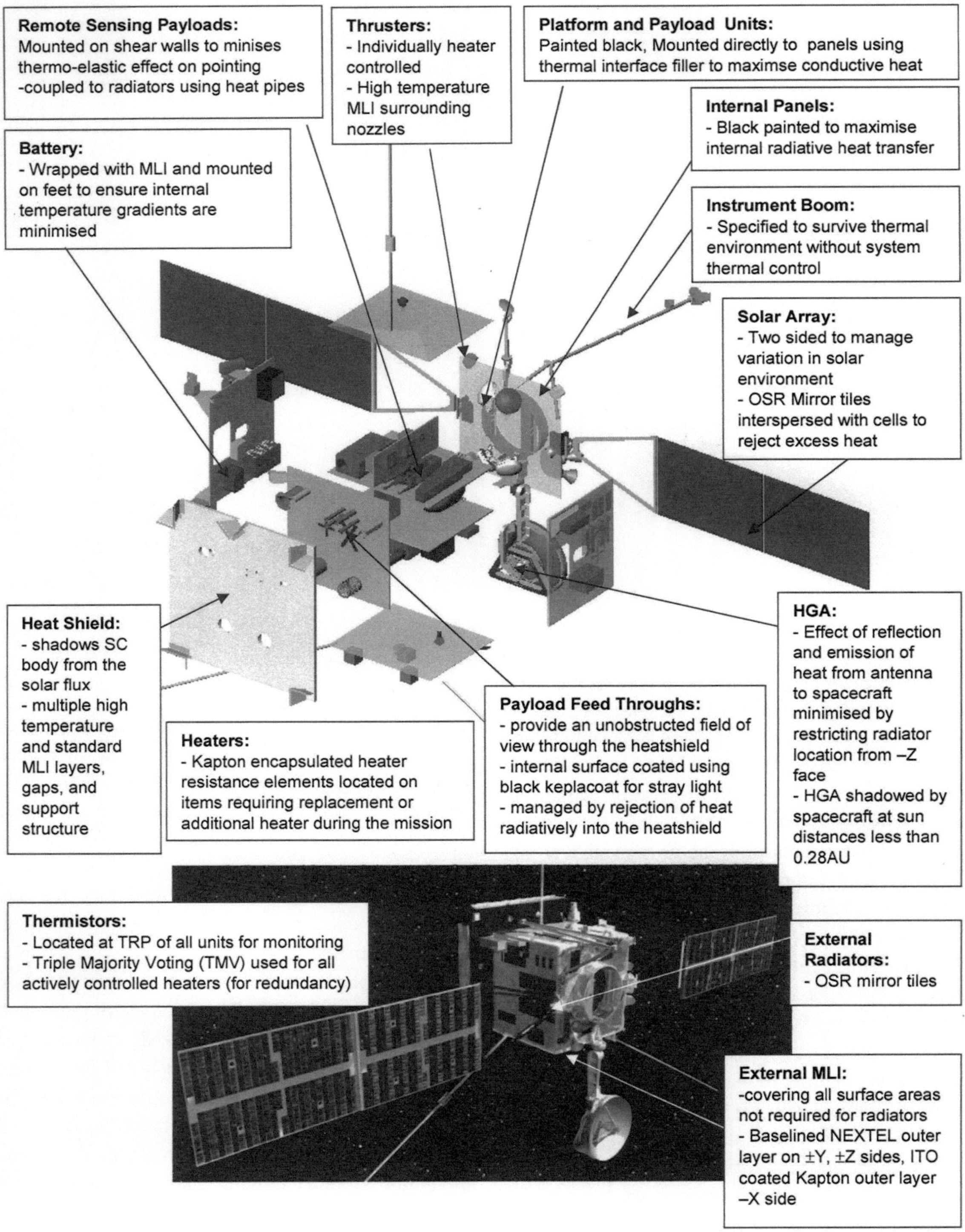

Fig. 25.3 Exploded view of the Solar Orbiter spacecraft, showing details of the thermal control system

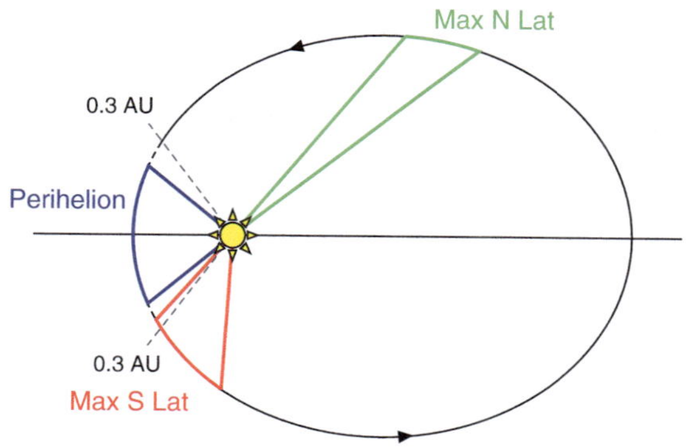

Fig. 25.4 Schematic representation of the science operation phases of the mission, showing a typical orbit with nominal 10-day windows centred on maximum northern and southern heliolatitude, and perihelion during which operation of the full payload is planned. As a baseline, only the in-situ instruments will be operated outside these windows. Also shown (*dashed line*) is the segment of the orbit below which the High Gain Antenna must be stowed in the shadow of the spacecraft body for thermal reasons (nominally below 0.3 AU heliocentric distance)

be similar to what has been used successfully in ESA's SOHO missions Joint Observation Programs (JOP).

25.7 Science Management and Data Archiving

Planning for Solar Orbiter is already quite mature, with science planning in particular already under way. Science teams have been formed for each science problem that include representatives from each instrument team, as well as theorists and modelers from the broader international scientific community. Data archiving will follow the same model as previous ESA PI-led solar and heliospheric missions, such as SOHO, with data made available to the scientific community through the ESA science data archive.

25.8 International Cooperation

Solar Orbiter is an ESA-led mission, but has strong NASA participation and substantial funded commitment. Specifically, NASA will provide the launch on an EELV, and significant parts (2 complete instruments, and portions of 2 instrument suites) of the scientific payload. The mission also has important synergies with NASA's Solar Probe Plus mission (McComas et al. 2007), and coordinated observations are expected to enhance greatly the scientific return of both missions. In the overall international context, Solar Orbiter has long been considered to be ESA's primary contribution to the International Living With a Star (ILWS) initiative, and joint studies incorporating data from all missions operating in the inner heliosphere (or providing remote-sensing observations of the near-Sun environment) will contribute greatly to our understanding of the Sun and its environment.

25.9 Status Update April 2010

At its meeting in February 2010, ESA's Science Programme Committee (SPC) selected Solar Orbiter as one of the three missions that will compete for the first two medium-class mission launch slots in its Cosmic Vision programme. The final decision as to which missions to implement will be taken after the completion of definition phase activities, foreseen for mid-2011, with the first launch no earlier than 2017. In order to reduce the technological risk, SPC stipulated that Solar Orbiter re-use solar generator technology that will be qualified for the BepiColombo mission to Mercury. This in turn required that the minimum perihelion distance for Solar Orbiter be increased from 0.23 to 0.28–0.29 AU. A revised mission profile that satisfies this new constraint while at the same time allowing all the key science goals to be accomplished is currently being designed.

25.10 Conclusion

Solar Orbiter is an exciting and mature mission with focused and timely scientific objectives directly relevant and important to the Cosmic Vision science programme. Its powerful combination of in-situ and remote-sensing instruments and unique mission design make Solar Orbiter ideally suited to answer several of the outstanding, fundamental questions in solar and heliophysics today. By addressing them, Solar Orbiter achieve make major breakthroughs in our understanding of how the inner solar system works and how it is driven by solar activity, as well as improve our understanding of fundamental physical processes common to all solar, astrophysical, and laboratory plasmas.

Acknowledgements This chapter uses material prepared for the Solar Orbiter Assessment Study Report provided by the members of the Solar Orbiter Science Working Team, together with the ESA Solar Orbiter Project Team, the industrial study team led by Astrium UK, and members of the science community.

References

Crooker N, Joselyn JA, Feynman J (eds) (1997) Coronal mass ejections. Geophysical monograph 99, American Geophysical Union

McComas DJ et al (2007) Understanding coronal heating and solar wind acceleration: case for in situ near-Sun measurements. Rev Geophys 45:RG1004. doi:10.1029/2006 RG000195

Reames DV (1999) Particle acceleration at the Sun and in the heliosphere. Space Sci Rev 90:413–491.doi:10.1023/A: 1005105831781

Solanki SK, Inhester B, Schuessler M (2006) The solar magnetic field. Rep Prog Phys 69:563 doi:10.1088/0034-4885/ 69/3/R02

Tu C-Y, Marsch E (1995) MHD structures, waves and turbulence in the solar wind: Observations and theories. Space Sci Rev 73(1):210. doi: 10.1007/BF00748891

Zurbuchen TH (2007) A new view of the coupling of the sun and the heliosphere. Annu Rev Astron Astrophys 45(297):338. doi: 10.1146/annurev.astro.45.010807.154030

Chapter 26

Scientific Objectives of the Canadian CASSIOPE Enhanced Polar Outflow Probe (e-POP) Small Satellite Mission

Andrew W. Yau and H. Gordon James

Abstract CASSIOPE is a Canadian small satellite scheduled for launch in 2011 into a polar orbit (325 × 1500 km, 80° inclination). The mission scientific objective of its Enhanced Polar Outflow Probe (e-POP) payload is to make observations of mesoscale and microscale plasma processes in the topside polar ionosphere at the highest-possible resolution, specifically to study the microscale characteristics of plasma outflow and related acceleration processes, the occurrence morphology of neutral escape, and the effects of auroral currents on plasma outflow and those of plasma microstructures on radio propagation. The e-POP payload will carry a suite of 8 scientific instruments, including imaging plasma and neutral particle sensors, magnetometers, dual-frequency Global Positioning System (GPS) receivers, charge-coupled-device (CCD) cameras, a radio wave receiver and a beacon transmitter. It will utilize a large (terabyte) data storage and downlink capacity onboard to support the planned high-resolution observations. The imaging plasma sensors will measure particle distributions and the magnetometers will measure field-aligned currents on a time scale of 10 ms and spatial scale of ~100 m. The CCD cameras will capture auroral images on a time scale of 100 ms. The GPS and radio-wave receivers will perform imaging studies of the ionosphere in conjunction with ground-based transmitters, as will the beacon transmitter in conjunction with ground receiving stations. In this chapter, we discuss the planned investigations of plasma outflow, wave propagation, and related plasma processes in the e-POP mission in the context of its mission scientific objective.

26.1 Introduction

The Enhanced Polar Outflow Probe (e-POP) is a part of the multi-purpose Canadian CASSIOPE small satellite mission.[1] One of its primary scientific targets is the in-situ observation of micro-scale characteristics of plasma outflow and related micro- and meso-scale plasma processes in the polar ionosphere. Its other two primary targets are the occurrence morphology of neutral escape in the upper atmosphere, and the effects of auroral currents on plasma outflow and those of plasma microstructures on radio propagation.

The escape of plasma from the polar ionosphere – its acceleration and subsequent transport towards the magnetosphere – is one of the most important processes in the ionosphere-thermosphere-magnetosphere system. This is because ion outflow from the auroral and polar ionosphere plays a very important role in magnetosphere-ionosphere coupling: it provides a significant source of plasma for the inner magnetosphere and the plasma sheet (Chappell et al. 2000), and is believed to influence the onset of magnetic reconnection on both the dayside and the nightside (Winglee 2004).

A.W. Yau (✉)
Department of Physics and Astronomy, University of Calgary, Calgary, AB, Canada T2N1N4
e-mail: yau@phys.ucalgary.ca

[1] A list of acronyms used in this paper is provided in the Appendix.

M.P. Miralles, J. Sánchez Almeida (eds.), *The Sun, the Solar Wind, and the Heliosphere*, IAGA Special Sopron Book Series 4, DOI 10.1007/978-90-481-9787-3_26, © Springer Science+Business Media B.V. 2011

A number of recent studies of ion acceleration and outflow, for example Peterson et al. (2001) and Abe et al. (2004), point to the importance of polar wind and auroral bulk upflow as a source of cold plasma for energetic ions at higher altitudes. These studies also underscore the scarcity of low-energy ion outflow observations below 3,000-km altitude relative to those at higher altitudes, and the need for such observations at high resolution to complement the considerable body of observations from Freja (Andre et al. 1998) and FAST (Strangeway et al. 2000), which were made at higher ion energies.

The observations on FAST demonstrate the important roles of the Poynting flux and precipitating soft electrons in controlling auroral ionospheric ion acceleration, and the connection between electromagnetic (EM) energy transmission from the magnetosphere along auroral field lines, via the Poynting flux, and energy dissipation in the auroral acceleration region. In the topside ionosphere, in particular, the Poynting flux is converted to heat through Joule dissipation, and ion-neutral collisions result in both the neutrals and ions being heated, and lifted to transverse heating altitudes because of increased scale height.

An important question is the micro-scale relationship between auroral ion bulk upflow, heating, and acceleration in the topside ionosphere on the one hand, and the associated auroral emissions, field-aligned currents, and plasma waves on the other. There are a number of latitudinal spatial scales associated with auroral arcs, from the 10–100 km wide band system that often appears as one single broad arc from space, to the 0.1–1 km thin curtain that can be observed using ground based high-resolution imaging (Borovsky 1993). A large variety of highly dynamic small-scale structures exist within the visible aurora, including auroral filaments, curls, and spirals, and features associated with extremely rapid motions (>10 km/s) and temporal variations (time constants of 1–60 s) (Trondsen and Cogger 1998) that have apparent widths on the order of 10–100 m at the magnetic zenith. Such features suggest the presence of auroral acceleration processes of electron inertial or ion gyroradius scale size.

As the polar wind and other low-energy ions flow upward along the geomagnetic field lines in the topside ionosphere, they undergo charge exchange reactions with background thermal neutrals and hot geo-coronal atoms. Such reactions involve the transfer of an electron from the neutral to the ion, and produce a streaming neutral atom, which will escape the Earth's gravitation if the original ion had sufficient energy before the reaction. The aforementioned microscale structure is embedded in characteristic mesoscale topology. Significant horizontal density gradients exist in the cusp, auroral-oval and adjoining regions, in structures such as troughs, polar patches and traveling ionospheric disturbances. It will be important to understand the ionospheric dynamics that couple the comparatively large scales to smaller ones associated with aurora. The radio propagation experiments in e-POP will be used to study ionospheric dynamics in the large-scale domain.

Depending on the local state of the ionosphere, the polar ionosphere can refract, scatter, amplify or damp EM waves passing through it, as well as decompose the waves through non-linearity. Through the collaborative use of a ground-based radio source in conjunction with a spacecraft over-flight, it is possible to conduct a 2-point high-frequency radio propagation experiment to study wave propagation through the ionosphere between the spacecraft and various ground transmitters, using various measured wave parameters to reconstruct the shapes of irregularities in the ionosphere.

The observation of spontaneous radio emissions of the ionosphere-magnetosphere system is of continuing interest, from ultra low frequencies characteristic of ions to electron processes at medium frequencies. The relation of these emissions to e-POP particle observations will be investigated. It will also be important to apply the detection of very low frequency – high frequency (VLF-HF) ground transmissions on e-POP to the understanding of the structure of the magnetosphere.

26.2 Instrument Complement

To achieve the science objectives of the mission, the e-POP mission will focus on in-situ measurements of small-scale plasma, waves, and fields, at the highest possible spatial-temporal resolution, and on imaging and tomographic measurements of the meso- and large-scale auroral morphology and ionospheric topology.

The CASSIOPE spacecraft will be placed in an elliptical polar orbit, and will have a perigee of 325 km, an apogee of 1,500 km, and an inclination of 80°, to achieve optimum altitude, local time and seasonal sampling for the respective planned science investigations. It will be 3-axis stabilized, to facilitate the high-resolution in-situ measurements and auroral imaging onboard. In addition, it will utilize the large (terabyte) onboard data storage and (350-megabits-per-second) telemetry downlink capacity of its communications payload to transmit the large volume of e-POP science data to ground.

To meet these measurement objectives, the e-POP science instrument payload will have a complement of 8 in-situ, imaging, and tomographic science instruments, including imaging plasma and neutral particle sensors, magnetometers, radio wave receivers, dual-frequency Global Positioning System (GPS) receivers, charge-coupled-device (CCD) cameras, and a beacon transmitter. Table 26.1 lists the respective instruments and their measurement characteristics, and Fig. 26.1 depicts their layout on the CASSIOPE spacecraft.

The two imaging plasma sensors – the imaging and rapid-scanning ion mass spectrometer (IRM) and the suprathermal electron imager (SEI) – will measure ion and electron distributions, respectively, on the time scale of 10-ms, which corresponds to an orbital distance of ~70 m. The measured ion distributions will cover the thermal and suprathermal energy ranges (0.5–100 eV) and will be used to derive the composition (density) of major and minor ions and the drift velocity and temperature of major ions. The measured electron energy and angular (pitch-angle) distributions will include thermal electrons, atmospheric photoelectrons, and soft electrons in the 1–200 eV range.

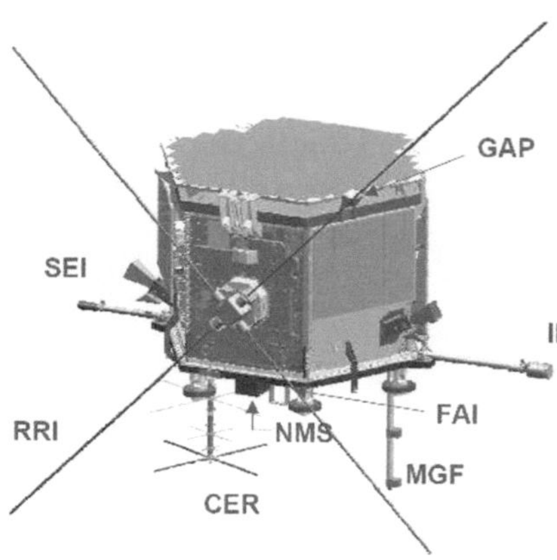

Fig. 26.1 e-POP instrument layout on CASSIOPE spacecraft

The magnetic field instrument (MGF), which consists of two spaced fluxgate magnetometers, will measure the vector magnetic field at 160 samples s^{-1}; from the measured field, field-aligned current structures will be inferred at ~90-m spatial resolution.

The neutral mass and velocity spectrometer (NMS) will measure the density and velocity distributions of major neutral atmospheric species, particularly O and N_2, and attempt to detect the possible presence of fast or hot atmospheric atoms or molecules resulting from charge exchange with outflowing ions or from other non-thermal processes.

The fast auroral imager (FAI) consists of two cameras, which will perform auroral imaging and measure the temporal and spatial distributions of auroral emission intensity at 630 nm and in the near-infrared (NIR) up to 850 nm, respectively. The NIR images will have an exposure time of 100-ms and a maximum pixel resolution of 400 m at spacecraft perigee.

The radio receiver instrument (RRI) consists of four 3-m monopole antennas, which will be mounted on the ram-facing side of the spacecraft, and will measure the electric field amplitude and polarization of naturally occurring VLF and HF waves as well as man-made EM radio waves from ground transmitters such as the SuperDARN and the Canadian Advanced Digital Ionosondes (CADI) ground facilities. The GPS-receiver based position, attitude and profiling instrument (GAP) consists of an array of 5 dual-frequency GPS receivers connected to a network of 4 patch antennas and an occultation antenna. The GPS receivers will measure the pseudo-range and carrier phase of the incoming GPS signals in the L1 and L2 bands, from which the spacecraft position, velocity and attitude and total electron contents will be derived. The spacecraft velocity data will have a precision on the order of 1 cm/s and will be used for accurate spacecraft ram velocity correction in the IRM ion drift velocity and corresponding convection electric field data. The radio occultation measurements in GAP will have a comparable time scale (50 ms) to that of the auroral images. The coherent electromagnetic radiation tomography instrument (CER) beacon transmitter will transmit at 3 VHF frequencies to perform total electron content measurements in conjunction with ground receiving stations.

Table 26.1 e-POP Instrument Complement

ID	Instrument	Investigator	Institute[a]	Measurements[b]
IRM	Imaging and rapid-scanning mass spectrometer	P.V. Amerl	Calgary	0.5–100 eV, 1–60 AMU ions
SEI	Suprathermal electron imager	D.J. Knudsen	Calgary	1–200 eV electrons
NMS	Neutral mass and velocity spectrometer	H. Hayakawa	JAXA/ISAS	0.1–2 km/s, 1–40 AMU neutrals
FAI	Fast auroral imager	L.L. Cogger	Calgary	630 nm, NIR
RRI	Radio receiver instrument	H.G. James	CRC	ELF-HF $\mathbf{E}(\omega)$ and $\mathbf{k}(\omega)$
MGF	Magnetic field instrument	D.D. Wallis	Calgary	$\Delta\mathbf{B}$, $j_{\parallel}$
GAP	GPS-receiver based attitude, position, and profiling instrument	R.B. Langley	UNB	spacecraft position, velocity, attitude; TEC
CER	Coherent EM radiation tomography instrument	P.A. Bernhardt	NRL	TEC and scintillation

[a] CRC = Communications Research Centre; JAXA/ISAS = Japan Aerospace Exploration Agency, Institute of Space and Astronautical Science; NRL = Naval Research Lab; UNB = University of New Brunswick.
[b] NIR = near infrared; $\mathbf{E}$ = electric field; $\mathbf{k}$ = wave vector; ω = frequency; $\Delta\mathbf{B}$ = magnetic field perturbation; $j_{\parallel}$ = field-aligned current; TEC = total electron content, AMU = atomic mass unit.

26.3 Planned Investigations

An initial operational phase of 12 months is planned for the mission. Science measurements will be made primarily during polar passes and less frequently over specific ground locations at lower latitudes, at various spacecraft altitudes between the perigee and apogee. The instrument complement will operate in a number of science operation modes targeted at specific scientific investigations.

In each investigation, the spacecraft will be placed in a specific attitude mode. In nadir-pointing mode, the FAI cameras will be viewing in the nadir direction. In ram-pointing, the RRI antennas will be normal to the spacecraft ram direction, and the entrance aperture planes of both IRM and SEI will be in the ram direction, to facilitate low-energy ion measurements in the ram direction. In inertial-pointing, the spacecraft will maintain its orientation so that a specific instrument will be viewing a certain inertial direction, e.g., in limb-pointing, FAI will view the limb to obtain aurora or airglow emission altitude profiles. In slew-pointing, the spacecraft will slew its attitude slowly while traversing a specific target on the ground or in space, so that one of the instruments will point continuously at the target; this will enable, for example, the FAI cameras to image the auroral emission in a localized region continuously or the RRI antenna plane to remain normal to the direction to a ground-radar or an ionospheric-heater transmitter.

Each of the instruments on e-POP is capable of operating in a number of instrument operation modes that are tailored to support specific investigations. In each investigation, selected instruments will operate in specific instrument modes and acquire measurement data at maximum data rate and resolution while others will acquire data at a reduced rate, depending on the scientific objective of the investigation. A central data handling unit will be used to coordinate the collection of data from the respective instruments during a pass and the subsequent data processing and storage.

Typically, up to 15 GB of data will be produced and transmitted to ground each day. The e-POP payload will use the terabyte data storage and Ka-band downlink telemetry system of the companion CASCADE payload onboard, to transmit downlink telemetry data to ground in a single 10-min pass at data rates exceeding 300 megabits-per-second (Mbps). Thereafter, the data will be processed at the e-POP Science Operations Center (eSOC) at the University of Calgary, and made available to the scientific community for access via the Canadian Space Science Data Portal (www.cssdp.ca).

The planned investigations will include the study of: sub-decameter ionospheric structures in the topside

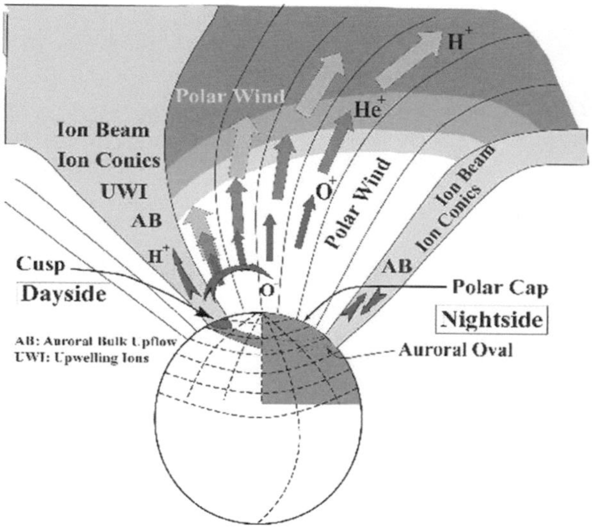

Fig. 26.2 Schematic depiction of planned investigation of important ion outflow populations in the polar ionosphere

ionosphere; small-scale aurora structures using auroral tomography; 3-dimensional polar wind velocity distributions; neutral upwelling; plasmasphere mass loading; ion outflow tomography; coordinated radio propagation experiments; and coordinated active heating experiment, to name a few. Figure 26.2 depicts the important ion outflow populations in the polar ionosphere that will be the focus of our investigation on e-POP. Figure 26.3 depicts the coordinated observation of radio propagation using the radio receiver instrument on e-POP in conjunction with ground HF and VHF transmitters.

Sub-decameter structures are frequently observed in the topside auroral ionosphere on sounding rockets. For example, LaBelle et al. (1986) reported highly localized packets of large-amplitude electric field waves in the 500–600 km altitude region during the activation phase of a large auroral substorm. These structures were termed spikelets and were subsequently interpreted as localized lower hybrid waves and in some cases lower hybrid solitary structures (LHSS), and they coincided with localized regions of transversely accelerated ions – the so-called perpendicular ion conics.

On sounding rockets, the spikelets were typically observed for only 1–2 ms each. This suggests that they have a lifetime or temporal scale on the order of 1 ms, or a vertical or horizontal spatial scale on the order of 1 m. This may explain why they have not been observed on satellites in the past, where the spatial resolution of measurements is typically no better than tens or hundreds of meters.

Burchill et al. (2004) recently reported the distribution of heating and density depletion widths of LHSS observed on the Geodesic sounding rocket. The heat-

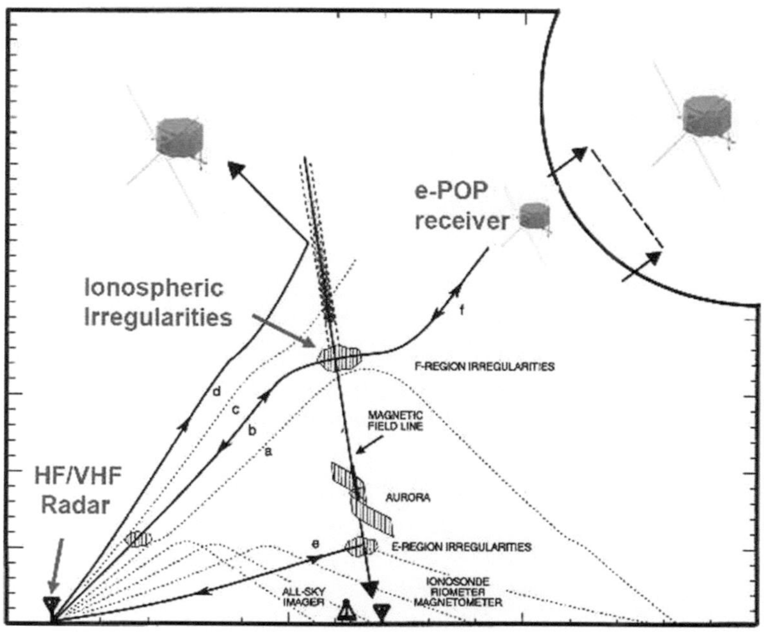

Fig. 26.3 Schematic illustration of coordinated observation of radio propagation using e-POP in conjunction with ground transmitters

ing width for each individual LHSS was calculated from the time difference between the first and the last perpendicularly heated ion distributions observed in each structure and the perpendicular rocket velocity. The width ranged from 13 m, which corresponds to the spatial resolution of the ion velocity measurements, to 190 m, which corresponds to the gyro-radius of a ~180 eV O^+ ion. The average width was 63 m, which is about 3 times larger than the average density depletion width of LHSS. The reason for the factor-of-3 difference between the heating and density depletion width is not clear.

On e-POP, high-resolution measurements of ions, electrons, field-aligned currents and electric field waves using IRM, SEI, MGF, and RRI, respectively, will be used to study the occurrence morphology of LHSS at various magnetic local times in the auroral ionosphere and the possible connection between the plasma and the electric field and currents inside LHSS.

Small, km- or sub-km scale structures in the aurora are presumably linked to sub-decameter structures in the plasma (density) and electric field observed in-situ. It is known that auroral arcs are associated with several latitudinal spatial scales, ranging from 10-100 km wide auroral band systems that often appear as a single arc, to 0.1–1 km wide (thin) curtains. Of the different types of dynamic small-scale structures in the visual aurora, auroral curl is probably one of the most interesting in terms of the physics of the instabilities involved. Trondsen and Cogger (1998) reported high-resolution ground-based television observations of auroral curl systems within breakup aurora and westward traveling surges, in which multiple curls of 1–2 km in scale size were evident within the ~10 × 14 km field-of-view of the television camera and would evolve at a time scale of ~1 s. The rotational shape and motion of the curl were counter-clockwise as viewed anti-parallel to the magnetic field.

On e-POP, auroral and airglow imaging will be performed using FAI at a number of spacecraft attitudes. As noted above, when FAI will be in nadir-viewing mode, the NIR images will have a pixel resolution of 400 m at perigee, which will be sufficient to resolve small, km- or sub-km scale spatial structures in the aurora. At apogee, FAI will be capable of viewing a scene element repeatedly in up to 120 consecutive frames. As illustrated in Fig. 26.4, this makes it possible to resolve dynamic auroral structures on time scales of 1 s or greater. Similarly, when FAI will be

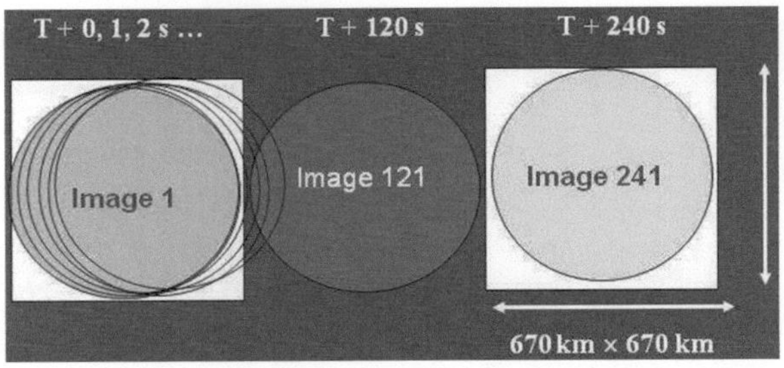

Fig. 26.4 Schematic illustration of FAI nadir-viewing imaging at apogee at a frame rate of 1 frame per second, resulting in up to 120 frames over a scene element

in slew-viewing mode, it will be able to image fast small-scale auroral structures to sub-km spatial and 1-s time resolution. In addition, in limb-viewing mode, FAI will be used to obtain auroral emission altitude profiles, which will provide information on the average energy characteristics of the precipitating auroral electrons.

Convection is believed to play an important role in the dynamics of the polar wind. The anti-sunward convection of plasma into the central polar cap brings the polar wind ions into regions of large magnetic field curvature as the polar wind reaches high altitude, where the process of centrifugal acceleration can result in accelerating the ions up to keV energies (Cladis and Francis 1985; Horwitz et al. 1994). On e-POP, measurements of 3D polar wind ion velocity will be made using IRM, and used to study the distribution of polar wind velocities in conjunction with measurements using the PolarDARN radars at Rankin Inlet and Inuvik, Canada (which cover the full polar cap) and the SuperDARN radars at auroral latitudes.

Using accelerometer measurements on the CHAMP satellite, Liu (2005) inferred neutral mass density enhancements near the cusp or the cleft at 410 km altitude. During times of large geomagnetic storms, such enhancements are often as large as 200%, and they strongly suggest the occurrence of neutral upwelling. The possible drivers of the upwelling include large-scale Joule heating in the E- and F-region, which results in enhancement in the neutral temperature and expansion in thermospheric density; soft electron precipitation, which results in auroral bulk ion up-flow and subsequent charge exchange between the upflowing ions and the ambient neutrals; and small-scale field-aligned current (FAC) structures. On e-POP, high-spatial-resolution measurements of ions, electrons, and neutrals using IRM, SEI and NMS, respectively, will be combined with corresponding FAC measurements using MGF to study the drivers for neutral upwelling and their possible auroral connections.

Ion composition measurements from DE-1 and other satellites have shown that helium is usually the second most abundant ion in the plasmasphere after H^+ at about 20% concentration, but just inside the plasmapause heavy ions such as O^+, O^{++} and N^+ sometimes increase by a factor of 10 or more in density when there is no corresponding variation in H^+ or He^+ ions. Fraser et al. (2005) showed mass loaded density profiles for the two component (H^+ and He^+) and three component (H^+, He^+ and O^+) plasmas, and found that the inclusion of He^+ doubles the mass density over L = 3–4 while the further addition of O^+ increases it by over one order of magnitude and essentially eliminates the plasmapause at L = 2.5. The increase in mass loading resulting from the presence of significant populations of He^+ and O^+ ions has a profound effect on the ULF wave field line resonance (FLR) harmonic structure. Normally, the decrease in H^+ and He^+ densities across the plasmapause would create a region of increasing FLR frequencies. However, the addition of O^+ can in some cases completely or significantly suppress the increase in resonance frequency across the plasmapause. On e-POP, ion composition measurements on outer plasmaspheric field lines using IRM will be used in conjunction with ground magnetic field and FLR frequency measurements to study mass loading in plasmaspheric refilling.

Using its side-viewing occultation antenna, the GAP instrument will be used to measure the total electron content (TEC) between e-POP and the GPS satellites occulted by the Earth's ionosphere or atmosphere below the e-POP horizon. The ionosphere refracts the L1 and L2 waves from the GPS satellites and affects their phase and amplitude in a manner that varies with the plasma density distribution. The relative motion between the GPS satellites and e-POP results in a tomographic sweeping of the ionosphere, and the TEC data may be de-convolved into an altitude distribution of plasma density below e-POP using tomographic techniques.

Using its four zenith-viewing patch antennas, the GAP instrument will also be used to measure the TEC between the e-POP spacecraft and GPS satellites above the e-POP spacecraft horizon, effectively turning the TEC mapping technique on its head. The TEC data will then be used to reconstruct an electron density map above the spacecraft and the electron density distribution along magnetic field lines can be used to infer the corresponding ion outflow distribution. This technique was successfully applied to the GPS data onboard FedSat (Yizengaw et al. 2006). The ion outflow tomography data on e-POP will be correlated with in-situ ion outflow measurements from IRM and with measurements using GPS receivers in the Canadian High Arctic Ionospheric Network (CHAIN). In addition, the TEC mapping of electron density above the spacecraft will be complemented by the CER TEC tomography below the spacecraft.

Many of the planned e-POP investigations will entail coordinated observations using Canadian and foreign ground facilities, including magnetic and optical observatories, radars, receivers and heaters. The RRI on e-POP will conduct trans-ionospheric propagation studies, in conjunction with SuperDARN and CADI transmitters on the ground. It will receive HF signals transmitted from these transmitters and refracted or diffracted by the underlying plasma irregularities. The inversion of the measured RRI data will be used to study the distribution of electron density irregularities on the order of 100 km or less in spatial scale, by comparing SuperDARN and CADI measurements on the ground with simultaneous measurements on e-POP.

The RRI will measure the electric field strength, direction of arrival, Doppler shift and signal delay of refracted waves as a function of wave frequency and the e-POP orbital position. The measurements will then be inverted mathematically to infer the 2-D shape of the refracting structure. The independent measure of refracting structure by other means such as incoherent scatter radar (e.g., Nicolls and Heinselman 2007) or CER tomography is attractive for checking the HF imaging. The RRI will also measure coherent backscatter from small-scale structure, in coordination with SuperDARN. The measurements will allow us to search for instances of oblique scatter that can be compared with the ground measurements. Also, the measured angular spectrum of scatter will allow us to test our concept of the coherent backscatter mechanism.

The RRI will also make measurements in coordination with HF ionospheric heaters such as the High

Altitude Auroral Research Program (HAARP) facility in Alaska during flights in or near the heater beams, to provide new understanding about the nonlinear interaction of HF waves with E or F-region plasma. Thanks to the slew-pointing attitude mode, RRI and FAI are ideal remote sensing tools of fixed targets. In one heating experiment, a high-powered EM wave is modulated at ELF or VLF frequency and launches a VLF wave that produces energetic electron flux near the equator, and subsequent ohmic heating of the ambient electrons in the ionosphere. This energization has already been clearly identified in observations on the DEMETER satellite of energetic fluxes linked to a high-power ground VLF transmitter (Gamble et al. 2008). Electron acceleration is also possible at HF via the parametric decay of Langmuir waves or the collapse of cavitons that may be formed by strong Langmuir turbulence. On e-POP, the upgoing EM waves may be observed using RRI and the artificial airglow (Kagan et al. 2005) or auroral emissions may be imaged using FAI. In addition, suprathermal electron fluxes resulting from the modulated EM beam may be detected using SEI.

In addition to these radio studies, other coordinated studies include investigations involving the EISCAT radar and the NSF Antarctic Polar Experiment Network for Geospace Upper Atmosphere Investigations (PENGUIN) facility. Investigations are also being planned involving conjunction opportunities with other operating spacecraft, for example Reimei, COSMIC (FORMOSAT 3), DSX and THEMIS.

26.4 Conclusion and Discussions

In conclusion, e-POP is a part of the multi-purpose CASSIOPE small satellite mission. Its goal is to study plasma outflow and the associated radio wave propagation and neutral escape. The e-POP mission is focused on the micro-scale physics of ion outflow and acceleration, and therefore requires in-situ plasma and field observations at the highest possible resolution, as well as detailed studies of 3D wave propagation and fast auroral imaging.

The e-POP payload has a complement of 8 plasma, magnetic field, optical, and radio instruments. The planned investigations will include the study of sub-decameter ionospheric structures in the topside ionosphere, small-scale aurora structures using auroral tomography; 3-dimensional polar wind velocity distributions; neutral upwelling; plasmaspheric mass loading; ion outflow tomography; and coordinated radio propagation and active heating experiments.

Understanding the multi-scale coupling arising from the competing (gravitational versus EM) forces and multiple species in the topside ionosphere is an important prerequisite to solving the problem of magnetosphere-ionosphere coupling. The energy generated through the large-scale dynamo in the magnetosphere is funneled in meso-scales through the topside ionosphere and ultimately dissipated in small scales through wave-particle interactions or classical collisions with the atmosphere. Some of the most important meso-scale processes take place at spatial scales of approximately 1–100 km, between the kinematic scale of particles ($\sim$100 m) and the scale of the transition region ($\sim$1,000 km). A notable example is field-line resonance (FLR), which results from the concentration of large-scale waves into a narrow shell of resonant oscillation in the magnetosphere and the nonlinear kinetic modification of incident Alfven waves on the electron inertial scale, and has latitudinal width between 10 and 100 km.

The question of how the topside ionosphere processes the large-scale magnetospheric energy inputs and how the resulting nonlinear wave-particle interaction may produce the small-scale structures and particle acceleration processes is central to the planned small-scale investigations in the e-POP mission. To understand the physics of the Sun-Earth system and the cause-and-effect relationships between its multi-scale processes requires a concerted approach involving investigations over a wide spectrum of multiple temporal and spatial scales: the e-POP mission is focused at the smallest end of the scale spectrum.

Acknowledgements We gratefully acknowledge the funding support for the e-POP project from the Canadian Space Agency (CSA) and the Natural Science and Engineering Research Council of Canada (NSERC). We also want to thank all e-POP Science Team members, for their instrument and other scientific contributions.

Appendix

Table 26.2 lists the acronyms used in this chapter.

Table 26.2 List of acronyms

2D	2-dimensional
3D	3-dimensional
AMU	Atomic mass unit
CADI	Canadian Advanced Digital Ionosonde
CASSIOPE	Cascade, Small Satellite, Ionospheric Polar Explorer
CCD	Charge coupled device
CER	Coherent EM radiation tomography instrument
CHAIN	Canadian High Arctic Ionospheric Network
CHAMP	Challenging Mini-Satellite Payload (German satellite)
COSMIC	Constellation Observing System for Meteorology, Ionosphere and Climate
CRC	Communications Research Centre
CSA	Canadian Space Agency
CSSDP	Canadian Space Science Data Portal
DE-1	Dynamic Explorer-1 (NASA satellite)
DEMETER	Detection of Electro-Magnetic Emissions Transmitted from Earthquake Regions
DSX	Defense Space Experiment
ELF	Extremely low frequency
EISCAT	European Incoherent Scatter Radar
EM	Electromagnetic
e-POP	Enhanced Polar Outflow Probe
eSOC	e-POP Science Operations Center
FAI	Fast auroral imager
FAST	Fast Auroral Snapshot (NASA satellite)
FLR	Field line resonance
GAP	GPS-receiver based attitude, position, and profiling instrument
GPS	Global Positioning System
HAARP	High Altitude Auroral Research Program
HF	High frequency
IRM	Imaging and rapid-scanning mass spectrometer
ISAS	Institute of Space and Astronautical Science
JAXA	Japan Aerospace Exploration Agency
LHSS	Lower hybrid solitary structure
Mbps	Megabits per second
MGF	Magnetic field instrument
NIR	Near infrared
NMS	Neutral mass and velocity spectrometer
NRL	Naval Research Lab

Table 26.2 (continued)

NSERC	Natural Science and Engineering Research Council
NSF	National Science Foundation
PENGUIN	Polar Experiment Network for Geospace Upper Atmosphere Investigation
PolarDARN	Polar Dual Auroral Radar Network
RRI	Radio receiver instrument
SEI	Suprathermal electron imager
SuperDARN	Super Dual Auroral Radar Network
TEC	Total electron content
THEMIS	Time History of Events and Macroscale Interactions during Substorms
ULF	Ultra low frequency
UNB	University of New Brunswick
VLF	Very low frequency

References

Abe T, Yau AW, Watanabe S, Yamada M, Sagawa E (2004) Long-term variation of the polar wind velocity and its implication for the ion acceleration process: Akebono suprathermal ion mass spectrometer observations. J Geophys Res 109:A09305. doi:2003JA010223

Andre M, Norqvis, P, Andersen L, Eliasson L, Eriksson AI, Blomberg L, Erlandson RE, Waldemark J (1998) Ion energization mechanisms at 1700 kilometer in the auroral region. J Geophys Res 103:419–422

Borovsky JE (1993) Auroral arc thicknesses as predicted by various theories. J Geophys Res 98:6101–6138

Burchill JK, Knudsen DJ, Bock BJJ, Pfaff RF Jr, Wallis DD, Clemmons JH, Bounds SR, Stenbaek-Nielsen H (2004) Core ion interactions with BB ELF, lower hybrid, and Alfven waves in the high-latitude topside ionosphere. J Geophys Res 109:A01219. doi:10.1029/2003JA010073

Chappell CR, Giles BL, Moore TE, Delcourt DC, Craven PD, Chandler MO (2000) The adequacy of the ionospheric source in supplying magnetospheric plasma. J Atmos Solar Terr Phys 62:421–436

Cladis JB, Francis WE (1985) The polar ionosphere as a source of the storm time ring current. J Geophys Res 90:3465–3473

Fraser BJ, Horwitz JL, Slavin JA, Dent ZC, Mann IR (2005) Heavy ion mass loading of the geomagnetic field near the plasmapause and ULF wave implications. Geophys Res Lett 32:L04102. doi:10.1029/2004GL021315

Gamble RJ, Rodger CJ, Clilverd MA, Sauvaud J-A, Thomson NR, Stewart SL, McCormick RJ, Parrot M, Berthelier J-J (2008) Radiation belt electron precipitation by man-made VLF transmissions. J Geophys Res 113:A10211. doi:10.1029/2008 JA013369

Horwitz JL, Ho CW, Scarbo HD, Wilson GR, Moore TE (1994) Centrifugal acceleration of the polar wind. J Geophys Res 99(A8):15051–15064

Kagan LM, Nicolls MJ, Kelley MC, Carlson HC, Belikovich VV, Bakhmeteva NV, Komrakov GP, Trondsen TS, Donovan E (2005) Observation of radiowave induced red hydroxyl emission at low altitude in the ionosphere. Phys Rev Lett 94:095004

LaBelle J, Kintner PM, Yau AW, Whalen BA (1986) Large amplitude wave packets observed in the ionosphere in association with transverse ion acceleration. J Geophys Res 91:7113–7118

Liu H, Luhr H (2005) Strong disturbance of the upper thermospheric density due to magnetic storms: CHAMP observations. J Geophys Res 110:A09S29. doi:10.1029/2004JA010908

Nicolls MJ, Heinselman CJ (2007) Three-dimensional-measurements of traveling ionospheric disturbances with the Poker Flat Incoherent Scatter Radar. Geophys Res Lett 34(21):L21104. doi:10.1029/2007GL031506

Peterson WK, Collin HL, Yau AW, Lennartsson OW (2001) Polar/Toroidal Imaging Mass-Angle Spectrograph observations of suprathermal ion outflow during solar minimum conditions. J Geophys Res 106:6059–6066

Strangeway RJ, Russell CT, Carlson CW, McFadden JP, Ergun RE, Temerin M, Klumpar DM, Peterson WK, Moore TE (2000) Cusp field-aligned currents and ion outflows. J Geophys Res 105:21129–21142

Trondsen TS,Cogger LL (1998) A survey of small-scale spatially periodic distortions of auroral forms. J Geophys Res 103:9405–9415

Winglee RM (2004) Ion cyclotron and heavy ion effects on reconnection in a global magnetosphere. J Geophys Res 109:A09206. doi:2004JA010385

Yizengaw E, Moldwin MB, Dyson PL, Fraser BJ, Morley S (2006) First tomographic image of ionospheric outflows. Geophys Res Lett 33:L20102. doi:10.1029/2006GL027698

Chapter 27

The Sun-Climate Connection Through Measurements and Modeling: The Picard Investigation

Gérard Thuillier, Steven Dewitte, Werner Schmutz, and the PICARD team

Abstract In order to understand and model the mechanisms that drive the changes observed in the Sun on several time scales, the PICARD mission will carry out several simultaneous measurements that include the total and spectral solar irradiance, solar diameter, limb shape and solar oscillations. The instruments consist of two radiometers, three four-channels sunphotometers, a bolometer, and a metrological imaging telescope. This set of instruments will be placed on board a microsatellite developed under the responsibility of the Centre National d'Etudes Spatiales. Picard is now in orbit: PICARD was launched on 15 June 2010. Since the instrumentation has already been described elsewhere, in this chapter we focus on related measurements, and solar and climate models developed within this mission.

27.1 Introduction

27.1.1 Global Solar Data

Global stellar measurements are high level information, which characterize the functioning and the properties of a star as whole. For the Sun, an important global property is the total solar irradiance (TSI), which allows computing the solar luminosity, and consequently the temperature of the corresponding emissive layer. Another important global measurement is the solar spectrum, which allows us to retrieve the solar atmosphere composition, and temperature of regions from which photons are emitted thus permitting to identify the properties of the photosphere, the chromosphere and the corona.

Helioseismology allows us to investigate the solar interior properties, in particular its temperature and rotation speed as a function of depth. Most of the pressure modes (p-modes) have been detected with the SoHO instruments and GONG and other ground-based networks. Gravity modes (g-modes) are suggested by the helioseismologic measurements on board SoHO. However, a definitive proof of their existence remains to be obtained. Their observations would permit the study of the dynamics of deep solar interior down to the core. The solar diameter is a fundamental quantity, which results from temperature, composition (through opacity of the solar atmosphere), magnetic field, dynamics and turbulence of the entire star, but primarily of the convective zone. Any change affecting these quantities will result in a change of diameter and limb shape (Thuillier et al. 2010), as well as the solar asphericity. The particular role of the turbulence within the convective zone has been pointed out by Sofia et al. (2005). Despite the fact that the solar diameter has been measured since the pioneering work of Jean Picard (1620–1682), no conclusion can be drawn concerning the its variation on the long term, or its relationship with solar activity from the entire data set essentially gathered from ground. Perhaps the most reliable information was obtained with the Solar Disk Sextant flown on board stratospheric balloons during the decreasing phase of cycle 22, which showed an increase of the diameter of about 200 milliarcseconds (Egidi et al. 2006; Djafer et al. 2008). Solar modeling includes all physical processes, which have been

G. Thuillier (✉)
LATMOS-CNRS, Bp3, 91371 Verrières-le Buisson, France
e-mail: gerard.thuillier@latmos.ipsl.fr

M.P. Miralles, J. Sánchez Almeida (eds.), *The Sun, the Solar Wind, and the Heliosphere*, IAGA Special Sopron Book Series 4, DOI 10.1007/978-90-481-9787-3_27, © Springer Science+Business Media B.V. 2011

invoked to explain the observations. Using laws of Physics and Chemistry, a model attempts to mimic the Suns properties, and in particular its variability. This is why, information as a function of solar activity is of prime interest. More information allows for a better validation of the model by comparison between observations and predictions made at different level of solar activity.

27.1.2 Global Available Measurements

Table 27.1 shows some of the most important global solar data available. Given the effect of the atmosphere on the optical measurements (absorption, scattering, ...), the most reliable solar data are obtained from space. TSI measurements started in 1978, and thanks to overlapping missions, continuity of observations was achieved. However, the TIM instrument (Kopp et al. 2005) on board SORCE (Rottman 2005), has measured a TSI having a difference of 5 W/m^2 (Kopp et al. 2005) with respect to the value provided by the radiometers on board SoHO as well as by those on board the International Space Station (SOLAR). This difference is presently not understood, and more measurements are needed to help in solving this problem, which is of importance for solar physics as well as for climate physics.

Spectral irradiance measurements are regularly performed since the launch of UARS, followed by SORCE, the SOLAR payload (Schmidtke et al. 2005) on board the International Space Station, and the Extreme ultraviolet Variability Experiment (EVE, Eparvier et al. 2004) on board Solar Dynamics Observatory (SDO). A similar situation exists with the helioseismic measurements thanks to SoHO, several ground based networks, with an ensured continuity with the Helioseismic and Magnetic Imager (HMI) instrument (Scherrer 2002) on board SDO, and PICARD.

It appears that the least reliable available data concerns the solar diameter. Thuillier et al. (2005) have pointed out several causes explaining the existing discrepancies. The main reason is the absence of metrological measurements due to a lack of geometrical references included in the instrument. The only reliable information has been produced by the SDS (Sofia et al. 1992; Egidi et al. 2006; Djafer et al. 2008), but is only based on four flights. For this reason, the principal scientific objective of PICARD is to carry out accurate diameter measurements.

27.1.3 PICARD Scientific Objectives

The PICARD mission includes:

- a spacecraft carrying several instruments,
- several investigations from the ground and from balloons,
- the development of solar models to interpret the measurements,
- the development of climate models.

PICARD will perform measurements of the following quantities and their variability (Table 27.2).

From space:

1. the solar diameter, asphericity, and limb shape in the photospheric continuum and in presence of Fraunhofer lines,
2. differential rotation,
3. solar oscillations to study the Sun internal structure,
4. total solar irradiance,
5. radiance/irradiance in UV and visible domains,

Table 27.1 Available solar data in relation with the PICARD mission

Total solar irradiance photosphere temperature	Solar luminosity	UARS, SoHO, SORCE, SOLAR, PSPT
Solar oscillations asphericity, dif. rotation	Internal structure	SoHO and ground network, SDO
Solar spectrum Solar limb	Composition temperature	ATLAS, UARS, SORCE, SOLAR
Solar diameter	Internal structure	SDS, ground-based instruments

Table 27.2 Measurements carried out during the PICARD mission in orbit and on the ground. SODISM I: diameter measurement in orbit. SODISM II, diameter measurement on the ground. SOVAP and PREMOS: measurement of the absolute Total Solar Irradiance (TSI), PREMOS: measurement in 5 spectral domains. MISOLFA: measurement of the local atmospheric turbulence from the ground

Themes	Measurements	Instruments
Solar physics	Variability diameter/luminosity	SODISM I, SOVAP, PREMOS
	Asphericity, limb shape	SODISM I
	Differential rotation	SODISM I
	Diameter/stellar reference	SODISM I
	Helioseismology	SODISM I, PREMOS
Climate	Diameter/luminosity	SODISM I, SOVAP, PREMOS
	Luminosity	SOVAP, PREMOS
	UV Variability	PREMOS, SODISM I
Atmospheric physics	Diameter and limb shape at ground Ozone photochemistry	Ground-based instruments: SODISM II, MISOLFA, PREMOS
Space weather	Images @ 215 and 393 nm (Ca II)	SODISM I

6. solar activity with specific measurements in the chromosphere,
7. solar diameter, asphericity, and limb shape in the photospheric continuum.

Our objectives are:

(1) Modeling of the solar machine using simultaneous measurements of several fundamental solar parameters and their variability. This study is aimed to understand the role of the magnetic field and the origin of the solar variability.
(2) For the climate modeling, the luminosity has to be reconstructed especially for the Maunder and Dalton minima. For that, it is expected that the solar model being well validated, will allow us to reconstruct the luminosity at certain periods. One way consists in using the relationship between diameter and luminosity determined by PICARD. Assuming that this relationship is not time dependent, it will be applied to the solar diameter determined from past solar eclipses starting in 1,715. This is an interesting time, where the solar activity resumes following the Maunder minimum, and where current luminosity reconstructions diverge most.
(3) The PICARD mission has a nominal duration of three years, and cannot obviously provide diameter long term trend. However, since the measurements are referred to angular distances between pairs of stars, by repeating the same measurements after 10 or 20 years, and taking into account the stars proper motions, the long term diameter trend will be derivable. The long term proper motions of the stars will be derived from the Hipparcos and GAIA missions.
(4) PICARD also aims at understanding the ground based measurements by using solar images gathered in space and on the ground.
(5) Since the solar activity has a stronger effect in the chromosphere than in the photosphere, the images at 215 and 393 may contribute to the study of Space Weather depending on the capability of retrieving these data with sufficient speed.

27.1.4 PICARD Measurements

- Solar diameter, limb shape and asphericity in the continuum at 535.7, 607, 782 nm,
- Solar diameter at 215 and 393 nm (Ca II),
- Solar activity at 215 and 393 nm,
- Solar spectral irradiance in several bands, with redundancy. All the above wavelengths are observed, plus some others including two channels dedicated to ozone photochemistry,
- Total Solar Irradiance measured by two independent radiometers as on SoHO, and one bolometric channel. This will be an important contribution to the discrepancy SORCE-TIM/SoHO,
- Solar oscillations on limb, macropixels at 535.7 nm and spectral channels.

27.2 Modeling

27.2.1 Convective Zone Modeling

From a stellar viewpoint, any variation of the global properties of the Sun on timescales shorter than evolutionary periods (which are measured in hundreds of million years) must be located within the outer convection zone. The principal reason for this is that within the convection zone there is a driving force (the dynamo magnetic field) with a short timescale (11 or 22 years), and it can itself readjust quickly. For example, hydrostatic equilibrium is established with the speed of sound (at most minutes), and significant energy can be transferred within a convective turnover timescale that is of the order of months. Because those effects are very small for the Sun (parts in 10^6), to adequately represent them requires models far more sophisticated than those used in stellar studies. In particular, the models must be formulated with a precision of better than a part in a million, and include processes that are neglected in routine modeling of stellar interiors (e.g. rotation, magnetic fields and turbulence). These upgraded models usually require multi-dimensional codes. Finally, they must take into account the fact that thermal equilibrium is not necessarily established during each time step, thus requiring an accurate accounting of energy inflow and outflow for each zone in the model. These studies were started by the Sofia group at The NASA/Goddard Space Flight Center in the late 1970s (cf. Endal et al. 1985), and are continuing at the present time (cf. Li et al. 2009, and references therein). The most significant results obtained to date include a significant role for turbulence, a feature not treated by the standard mixing length theory of convection, and the fact that changes of the solar diameter are highly inhomogeneous, being concentrated in the shallowest solar layers (Sofia et al. 2005). Moreover, exploratory studies conducted with these models indicate that in order to obtain an accurate model for the mechanism that drives solar variability at the years to millennia timescales requires simultaneous observations of all the global solar properties, plus oscillations. All those observations will be carried out by the instruments onboard the PICARD satellite.

27.2.2 Climate Modeling

One of the most critical aspects of the climate system, is that it naturally varies on a large range of time scales. Among the sources of natural variability in the Earths climate, is the change in the solar energy output. While the atmosphere is the fastest responding component of the climate system, it is also a challenge in itself in terms of modeling: the complex coupling between different levels of the atmosphere and the variety of feedback mechanisms operating make the problem highly non-linear. Recently the inclusion of the stratosphere in climate models has led to a new generation of models (chemistry-climate models, referred as CCMs). In fact, newer CCMs versions including some representation of solar variability forcing (for example Schmidt et al. 2006; Nissen et al. 2007; Marsh et al. 2007; Tsutsui et al. 2009) show better agreement with observations, thus encouraging further developments. Within the PICARD project, we work with four existing chemistry-climate models: the Canadian Middle Atmospheric Model (CMAM) in its various versions, (de Grandpré et al. 1997; Fomichev et al. 2002) the IGCM-FASTOC model (Taylor and Bourqui 2005) developed at McGil University, the LMDz-REPROBUS, a three-dimensional chemistry-climate model being developed by the Laboratoire de Météorologie Dynamique and LATMOS (Jourdain et al. 2008), and the SOCOL model being developed at PMOD/WRC (Egorova et al. 2005; Schraner et al. 2008). Those models represent different approaches to mimic the atmosphere. The CMAM contains a detailed representation of the middle atmosphere chemistry and dynamics from the ground to about 120 km altitude. With such a model we can study couplings among different altitude layers in the atmosphere. The IGCM-FASTOC is a model focusing on the stratosphere with a detailed representation of the dynamics and that incorporate a fast chemistry scheme allowing for non-expensive long term runs. The LMDZ-REPROBUS is a model with a realistic representation of the tropospheric chemistry and a detailed representation of the chemistry and dynamics of the middle atmosphere. The SOCOL model has detailed ion chemistry in the D region of the atmosphere allowing for studies of the effect of ionization particles to be done from the first principles and minimizing the need for parameterizations. While distinct in their approach to represent the atmosphere, the four models are very complementary in their capabilities. For all of them, we are implementing solar variability to study their capability of response, and in particular the dynamical coupling between stratosphere and troposphere. PICARD data will allow us to reconstruct the sun luminosity at certain specific periods (eighteenth to nineteenth). These reconstructions will be used to determine the climate of that time and compare with observations.

27.3 Instruments and Measurements

27.3.1 *Instruments on Board PICARD Spacecraft*

There are four PICARD instruments on board the spacecraft (Fig. 27.1):

SOVAP is a two-head radiometer measuring the total solar irradiance. This instrument has been developed by the Royal Institute of Meteorology of Belgium. PREMOS is composed of two radiometers of type PMO6 to measure the total solar irradiance and three four-channel sunphotometers observing the solar spectrum irradiance at several wavelengths. PREMOS has been built and characterized at the Physikalisch-Meteorologisches Observatorium Davos and World Radiation Center, Switzerland. Absolute spectral UV irradiance sensitivities have been calibrated at PTB Berlin with an expanded uncertainty of 3%, and the visual and IR channels have been calibrated at PMOD/WRC against a PTB calibrated lamp with an expanded uncertainty of 4% (Schmutz et al. 2009). The two types of radiometers (DIARAD and PMO6) ensure the continuity with SoHO radiometers. The DIARAD radiometers have a stability and accuracy of 0.04 and 0.8 Wm^{-2}, respectively. The PREMOS experiment will carry the first absolute radiometers into space that were calibrated in vacuum against the SI radiant power scale. The PREMOS radiometers are absolutely characterized to 300 ppm expanded uncertainty or 0.4 Wm^{-2} for one solar constant, respectively, and the SI traceable power

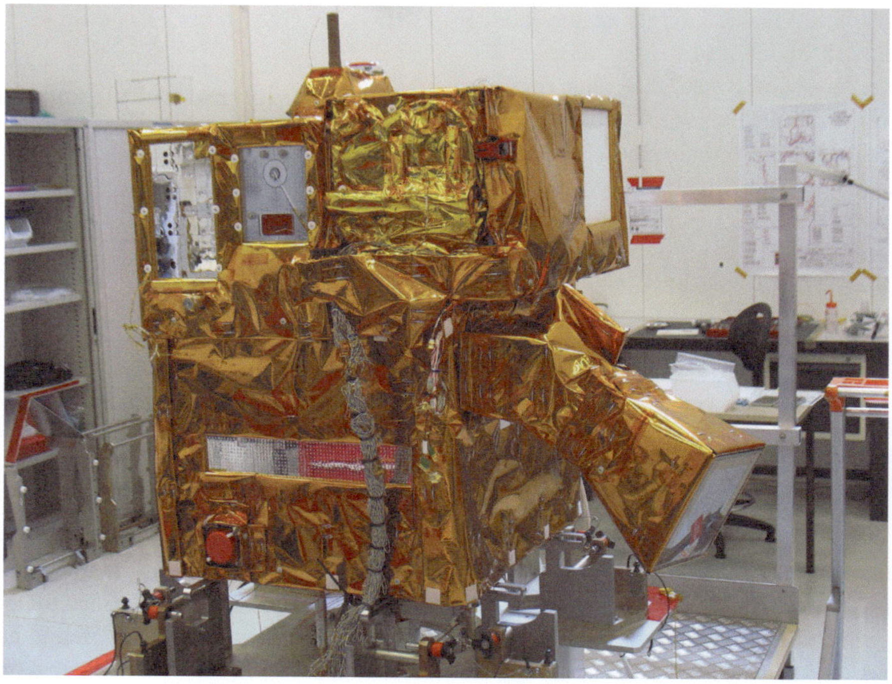

Fig. 27.1 This picture shows the PICARD satellite with the instruments mounted on the *top*. The PREMOS instrument is visible at the *upper left*. The SODISM instrument is placed at the *upper right* with its main door in the closed position. Between these two instruments, there are the SOVAP and BOS radiometers. At the *lower right*, we can see one of the two stars sensors

Table 27.3 Wavelength, nature of the data (full image, ring, one channel measurement), sampling. The measurements are synchronized by a one minute signal

Objectives	Measurements	λ(nm)	Sampling	Instruments
Diameter, asphericity, limb shape	Ring (40 pixels)	215, 393, 535.7, 607, 782	2/orbit/ λ	SODISM I
Helioseismology	Ring (22 pixels)	535.7	2 min	SODISM I
Helioseismology	Macropixels	535.7	1 min	SODISM I
TSI	1 channel	Full spectrum	2 min	PREMOS
TSI	1 channel	Full spectrum	3 min	SOVAP
Bolometric measurements	1 channel	Full spectrum	10 s	BOS
Spectral irradiance (SSI)	11 channels	215, 268, 535.7, 607, 782	10 s	PREMOS
SSI	1 channel	215	0.1 s	PREMOS
Activity	Full images	215, 393	1/orbit/ λ	SODISM I
Calibration	Full images	All λ	1/day	All instruments

calibration has an expanded uncertainty of 600 ppm (Schmutz et al. 2009).

A bolometric detector of high sensitivity will allow us to increase the sampling of the TSI measurements. This instrument has been built and characterized by the Royal Observatory of Belgium.

The fourth instrument, named SODISM, is a metrological imaging telescope especially designed to measure the solar diameter, limb shape and asphericity. This instrument is the first space instrument including an angular reference calibrated on stars angular distances. Its precision is 3 mas per single image, and the solar diameters, will be referred to stars angular distances with an accuracy of 2 mas (that value depends of the dark current value and platform stability).

The four instruments have their measurement synchronized by a one minute signal. The details of the measurements and sampling, as well as the role of each, are shown in Table 27.3.

These instruments are described in detail by Thuillier et al. (2006).

27.3.2 Associated Measurements and Missions

27.3.2.1 Measurement of the Diameter and Limb Shape from SDS

The Solar Disk Sextant (SDS) is an instrument designed to measure diameter variation, which includes an angular reference made by a wedge. Placed on stratospheric balloons, the SDS (Sofia et al. 1984) flew four times from 1992 to 1996 taking advantage of the solar activity decrease during that period. A diameter variation of 200 mas was measured in antiphase with the solar activity (Egidi et al. 2006; Djafer et al. 2008).

The instrument was refurbished, and had another flight on October 17, 2009. Plans exist for an additional flight per year starting in 2010. These flights will ensure an appropriate correlation and validation of the space measurements, and will allow to extend the PICARD measurements after the completion of the mission.

27.3.2.2 Measurement of the Solar Diameter and Limb Shape from the Ground

Comparison of a limb shape measured on the ground with one measured outside the atmosphere shows two significant differences (i) the ground based solar edge measurement is generally noisier than the one in space, (ii) the slope of the limb measured on the ground is weaker than the one measured in space. These effects are attributed to the scattering of the photons and to the turbulence of the Earth's atmosphere. Additionally, the ground based measurements differ significantly from each other. Among several reasons contributing to explain those discrepancies are differences in the instrument characteristics (psf, sampling, ...). This is why the PICARD mission was originally proposed to include ground based measurements made with the same instrument as in space. On the ground, the Qualification Model of the space instrument will be installed to run simultaneously with another instrument, named MISOLFA, to measure the local turbulence. Comparing the data gathered by these two instruments with observations at the same wavelength from space will allow us to quantitatively evaluate the role the atmosphere, to validate the numerical corrections, and explore the origin of the existing residual (if any) after correction.

27.3.2.3 Measurement of the Diameter and Limb by Eclipse Method

The very precise knowledge of the Moons shape and motion allows us to determine the solar diameter between the first and the second contact, as well as the limb shape. This method uses spectrometric recordings at high speed. The limitation of the method remains the atmosphere and the knowledge of the Moons mountains and valleys. Preliminary results were obtained at the 22 July 2008 eclipse.

27.3.2.4 Measurement of the Diameter by Solar Transit

The measurement is made at noon culmination using a high speed image recording. The weakness of the method is the presence of the atmosphere. Its advantage is the possibility of long term observations. The instrumentation will be first validated at Observatoire de Paris, and then moved to the Observatoire de Haute Provence.

27.3.2.5 Other Sources of Images

The PSPT imager records images in the Ca II (393 nm) and 607 nm, which will be provided to the PICARD data processing center. These images will be used for validation and to develop algorithms for corrections. Similarly, Observatoire de Meudon will provide Ca II images, however with 0.1 nm resolution.

27.3.2.6 SDO and SOLAR

The Solar Dynamics Observatory (SDO) was launched in February 2010, it will allow us to compare results of the oscillation modes measured by two different techniques: velocity variations, and photometric variations.

SOLAR will still be running on board the International Space Station. Two spectrometers measure the solar spectrum from 17 to 3000 nm. These measurements will be used to verify the stability of the PICARD instrumentation, and to provide data to the Climate Modeling Group.

27.4 Mission Preparation

The mission preparation is mainly concerned by the commissioning phase and the data processing center. Its duration is about three months, with one month dedicated to the outgassing of the spacecraft and the payload. During this period, the door will remain closed to protect the instruments entrance. However, several operations will be conducted:

- Unlocking the covers of each instrument,
- check commanding of each instrument,
- verify the functioning of all mechanisms (shutter, filters wheels),
- cleaning of the CCD by heating,
- establish the particles precipitation map to determine the best condition for measuring the solar diameter,
- adjust the coefficients for the thermal regulation of each instrument,
- check the data compression code using dark current images.

When the door will be open, we will adjust the offset angle to place the solar image at the imaging telescope detector center. Then, we will run the following sequence:

- record of SODISM images of reference for each wavelength,
- record of signals of reference for each filters and each channel,
- record reference dark current for each instrument
- run the helioseismologic and diameter measurements sequence,
- run specific sequence of measurements for radiometers and photometers,
- check the compression coefficients on real images,
- measure the instruments flatfield,
- measure the SODISM optics distortion,
- calibrate the SODISM detector using the star field.

The data processing center, located in Brussels, will receive the raw data, send commands and run the software up to level 2A, allowing us to evaluate the performance of the instruments. Higher levels will be generated by data processing centers located at each of the institutes that have developed the instrumentation.

27.5 Conclusion

The PICARD investigation is now ready for operations, as is the data processing software. This mission will benefit from information obtained with the Solar Dynamics Observatory, the SOLAR platform on the International Space Station, and several other investigations either on the ground or on stratospheric balloons.

Acknowledgements The PICARD mission is under the responsibility of CNES, which supported the design and construction of the spacecraft. Design and construction of the three instruments are carried out by the laboratories, LATMOS (CNRS) for SODISM, the Royal Meteorological Institute of Belgium for SOVAP and BOS, and Physikalisch-Meteorologisches Observatorium Davos and World Radiation Center (CH) for PREMOS. SODISM is funded by CNES, while the development is supported by CNRS. The PREMOS instrument is funded by the Swiss PRODEX program of ESA, and the SOVAP instrument development as well as the PICARD operation center located in Brussels is supported by the Belgian Science Policy Office through the European Space Agency PRODEX program. We are very grateful to Dr. S. Melo and Pr. S. Sofia, both Co-Investigators on the PICARD project, for their contribution in the climate and solar modeling part of this chapter.

References

deGrandpré J, Sandilands JW, McConnell JC, Beagley SR, Croteau PC, Danilin MY (1997) Canadian middle atmosphere model: preliminary results from the chemical transport module. Atmos Ocean 35:385–431

Djafer D, Thuillier G, Sofia S, Egidi A (2008) Processing method effects on solar diameter measurements: use of data gathered by the Solar Disk Sextant. Solar Phys 247:225–248. doi:10.1007/s11207-007-9079-2

Egidi R, Caccin B, Sofia S, Heaps W, Hoegy W, Twigg L (2006) High-precision measurements of the solar diameter and oblateness by the Solar Disk Sextant (SDS) experiment. Sol Phys 235:407–418

Egorova T, Rozanov EV, Zubov VA, Manzini E, Schmutz W, Peter T (2005) Chemistry-climate model SOCOL: a validation of the present day climatology. Atmos Chem Phys Discus 5:1557–1576

Endal AS, Sofia S, Twigg LW (1985) Changes of solar luminosity and radius following secular perturbations in the convective zone. Astrophys J 290(2):748–757

Eparvier FG, Woods TN, Crotser DA, Ucker GJ, Kohnert RA, Jones A, Judge DL, McMullin D, Berthiaume GD (2004) The EUV Variability Experiment (EVE) aboard the NASA Solar Dynamics Observatory (SDO), SPIE Proceedings, 5660, pp 48–55

Fomichev VI, Ward WE, Beagley SR, McLandress C, McConnell JC, Mc-Farlane NA, Shepherd TG (2002) Extended Canadian middle atmosphere model: zonal-mean climatology and physical parameterizations. J Geophys Res 107(D10) 9-1, cite ID 4087. doi:10.1029/2001JD000479

Jourdain L, Bekki S, Lott F, Lefebvre S (2008) The coupled chemistry-climate model LMDz-REPROBUS: description and evaluation of a transient simulation of the period 1980–1999. Ann Geophys 26:1391–1413

Kopp G, Lawrence G, Rottmann G (2005) The total irradiance monitor (TIM): science results. Solar Phys 230(1):129–140

Li, LH, Sofia S, Ventura P et al (2009) Two-dimensional stellar evolution code including arbitrary magnetic fields II. Precision improvement and inclusion of turbulence and rotation. Astrophys J Suppl 182(2):584–607

Marsh DR, Garcia RR, Kinnison DE, Boville BA, Sassi F, Solomon SC, Matthes K (2007) Modeling the whole atmosphere response to solar cycle changes in radiative and geomagnetic forcing. J Geophys Res 112:D23306

Nissen KM, Matthes K, Langematz U, Mayer B (2007) Towards a better representation of the solar cycle in general circulation models. Atmos Chem Phys 7:5391–5400

Rottman GJ (2005) The SORCE mission. Solar Phys 230(1):7–25

Scherrer Ph (2002) The helioseismic and magnetic imager for the solar dynamics observatory, American Geophysical Union, Fall Meeting 2002, SH52A-0494

Schmidt H, Brasseur GP, Charron M, Manzini E, Giorgetta MA, Diehl T, Fomichev VI, Kinnison D, Marsh D, Walters S (2006) The HAMMONIA chemistry climate model: sensitivity of the mesopause region to the 11-year solar cycle and CO2 doubling. J Climate 19:3903–3930

Schmidtke G, Fröhlich C, Thuillier G (2005) ISS-SOLAR: Total (TSI) and spectral (SSI) irradiance measurements. Adv Space Res 37:255–264

Schmutz W, Fehlmann A, Hülsen G et al (2009) The PREMOS/PICARD instrument calibration. Metrologia 46:S202–S206

Schraner M, Rozanov E, Schnadt Poberaj C, Kenzelmann P, Fischer AM, Zubov V, Luo BP, Hoyle CR, Egorova T, Fueglistaler S, Brönnimann S, Schmutz W, Peter T (2008) Chemistry-climate model SOCOL: version 2.0 with improved transport and chemistry/microphysics schemes 8. Atmos Chem Phys 8:5957–5974

Sofia S, Heaps W, Twigg LW (1994) The solar diameter and oblateness measured by the Solar Disk Sextant on the 1992 September 30 balloon flight. Astrophys J 427: 1048–1052

Sofia S, Basu S, Demarque P, Li L, Thuillier G (2005) The non-homologous nature of solar diameter variations. Astrophys J 632:L147–L150

Taylor CP, Bourqui MS (2005) A new fast stratospheric ozone chemistry scheme in an intermediate general-circulation model. I: Description and evaluation. Q J Roy Meteor Soc 131(610):2225–2242

Thuillier G, Sofia S, Haberreiter M (2005) Past, present and future measurements of the solar diameter. Adv Space Res 35(3):399–340

Thuillier G, Dewitte S, Schmutz W (2006) Simultaneousmeasurement of the total solar irradiance and solar diameter by the PICARD mission. Adv Space Res 38:1792–1806

Thuillier G, Claudel J, Djafer D, Haberreiter M, Mein N, Melo S, Schmutz W, Shapiro A, Short CI, Sofia S, (2010) The shape of the solar limb: theory and observations. Solar Phys, 10.1007/s11207-010-9664-7

Tsutsui J, Nishizawa K, Sassi F (2009) Response of the middle atmosphere to the 11-year solar cycle simulated with the Whole Atmosphere Community Climate Model. J Geophys Res 114:D0111

Chapter 28

The International Space Weather Initiative (ISWI)

Joseph M. Davila, Nat Gopalswamy, Barbara J. Thompson, Tom Bogdan, and Mike Hapgood

Abstract The International Heliophysical Year (IHY) provided a successful model for the deployment of arrays of small scientific instruments in new and scientifically interesting geographic locations, and outreach. The new International Space Weather Initiative (ISWI) is designed to build on this momentum to promote the observation, understanding, and prediction space weather phenomena, and to communicate the scientific results to the public.

28.1 Introduction

The International Heliophysical Year (IHY) was an international program of scientific collaboration involving thousands of scientists from more than 70 countries, which was conducted from February 2007 to February 2009. Along with programs of research, outreach, and IGY history preservation, activities included the deployment of new instrumentation arrays especially in developing countries. A detailed account of all IHY activities is reported by Thompson et al. (2009).

It was recognized early in the planning of the IHY that the understanding of the global ionosphere and its linkage to the near-Earth space environment was limited by the lack of observations in key geographical areas, e.g. near the magnetic dipole equator. To address this need, a series of workshops were held to facilitate collaborations between research scientists in scientifically interesting geographic locations, and researchers with the expertise to build scientific instrumentation. From these meetings scientific teams emerged. Each team consisted of a lead scientist who provided the instruments or fabrication plans for instruments in the array. Support for local scientists, facilities and data acquisition was provided by the host nation. All scientists participate in the analysis of the data from the instrument array. As a result of the this program, scientists from many countries now participate in the instrument operation, data collection, analysis, and publication of scientific results, working at the forefront of science research.

The instrument deployment program was one of the major successes of the IHY. Arrays of small instruments such as magnetometers to measure Earth's magnetic field, radio antennas to observe solar coronal mass ejections, GPS receivers, VLF radio receivers, and all-sky cameras to observe the ionosphere, and muon particle detectors to observe energetic particles were installed around the world. These arrays continue to provide global measurements of heliospheric phenomena.

An interesting side benefit of the instrument program was the seeding of heliophysics research groups in universities where there had been none before, and the strengthening of existing heliophysics research groups where new instruments were installed.

Building on this concept, in February 2009 the International Space Weather Initiative (ISWI)was proposed to the Science and Technology Subcommittee (STSC) of the United Nations. The program will continue the study of universal processes in the solar system that affect the interplanetary and terrestrial environments, and to continue to coordinate the deployment and operation of new and existing

J. M. Davila (✉)
Goddard Space Flight Center, Greenbelt, MD 20771, USA
e-mail: Joseph.M.Davila@nasa.gov

M.P. Miralles, J. Sánchez Almeida (eds.), *The Sun, the Solar Wind, and the Heliosphere*, IAGA Special Sopron Book Series 4, DOI 10.1007/978-90-481-9787-3_28,

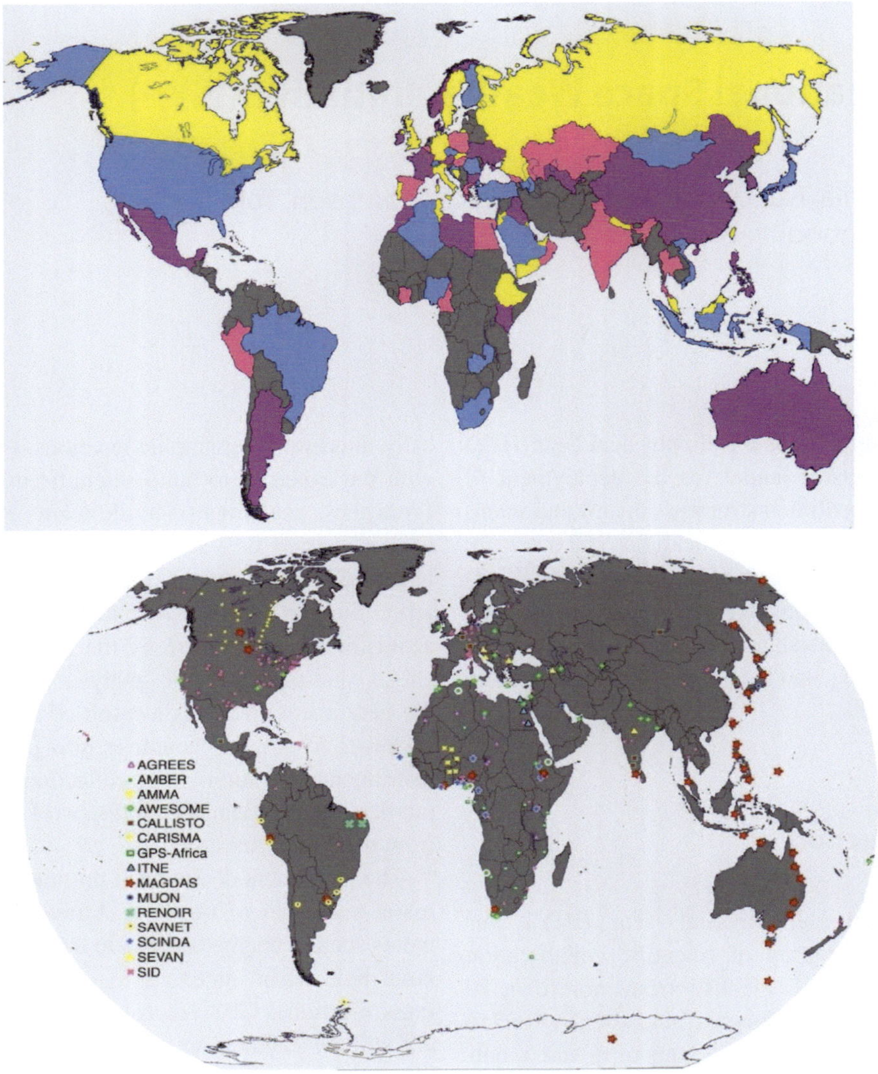

Fig. 28.1 The *top* panel shows countries where IHY planning committees were established (*darker grays*), the *bottom* panel shows the location of instruments participating in the instrument deployment program

instrument arrays aimed at understanding the impacts of Space Weather on Earth and the near-Earth environment. The ISWI was adopted by the Committee for the Peaceful Uses of Outer Space (COPUOS) in June 2009, and approved the UN General Assembly in the Fall of 2009. In addition to the United Nations, ISWI is supported by the National Aeronautics and Space Administration (NASA), the European Space Agency (ESA), the Japan Aerospace Exploration Agency (JAXA), and the International Committee on Global Navigation Satellite Systems (ICG). Additional information on the ISWI is available at http://iswi-secretariat.org.

In this chapter, we describe the goals and objectives of the ISWI program as it is currently envisioned.

28.2 Goals and Objectives

The ISWI will help develop the scientific insight necessary to understand the physical relationships inherent in space weather, to reconstruct and forecast near-Earth space weather, and to communicate this knowledge to scientists and to the general public. This will be accomplished by (1) continuing to deploy new instrumentation, (2) developing data analysis processes, (3) developing predictive models using data from the instrument arrays, and (4) continuing to promote knowledge of heliophysics through education and public outreach.

28.2.1 Instrument Array Development

The ISWI will continue to expand and deploy new and existing instrument arrays following the successful model demonstrated during the IHY. The basic principles of this model are simple. Each instrument team is led by a single scientist. The lead scientist or principle investigator, funded by his/her country, provides instrumentation (or fabrication plans) and data distribution. In a few cases, where resources allow, the hosting country will pay for the instrument. The host country provides the workforce, facilities, and operational support necessary to operate the instrument. This is typically at a local university or government laboratory. Host scientists become part of science team. All data and data analysis activity is shared within the science team, and all scientists participate in publications and scientific meetings where possible.

The current list of instrument providers is shown in Table 28.1. This list is not expected to remain static. Through workshops and other means, the ISWI will

Table 28.1 Current list of instrument arrays active within the ISWI

Instrument	Lead scientist	Country	Science objective
African Dual Frequency GPS Network	C. Amory-Mazaudier (CETP/CNRS)	France	To increase the number of real-time dual-frequency GPS stations worldwide for the study of ionospheric variability, response of the ionospheric total electron content (TEC) during geomagnetic storms over the African sector.
African GPS Receivers for Equatorial Electrodynamics Studies (AGREES)	M. Moldwin (U. Mich.) and E. Yizengaw (Bos College)	USA	Understand unique structures in equatorial ionosphere, low/mid latitude plasma production, effect of ionospheric and plasmaspheric irregularities on communications
African Meridian B-field Education and Research (AMBER)	M. Moldwin (U. Mich.) and E. Yizengaw (Bos College)	USA	Understand low latitude electrodynamics, ULF pulsations, effect of Pc5 ULF on MeV electron population in inner radiation belts
Atmospheric Weather Education System for Observation and Modeling of Effects (AWESOME) and SID (Sudden Ionospheric Disturbance Monitor)	U. Inan and D. Scherrer (Stanford)	USA	Lightning, sprites, Elves, relation to terrestrial Gamma Ray flashes , whistler induced electron precipitation, conjugate studies,
Coherent Ionospheric Doppler Radar	T. Garner (UT, Arlington)	USA	To tomographically reconstruct the ionosphere and to provide input to Data Assimilation models
Compound Astronomical Low-cost Low-frequency Instrument for Spectroscopy and Transportable Observatory (CALLISTO)	A. Benz and C. Monstein (ETH-Zentrum)	Switz	Study the magnetic activity of a wide range of astrophysical objects with emphasis on the Sun and cool stars
H-alpha Telescope	K. Shibata (Kyoto)	Japan	Solar activity, flares, filaments, filament eruptions

Table 28.1 (continued)

Instrument	Lead scientist	Country	Science objective
Magnetic Data Acquisition System (MAGDAS)	K. Yumoto (Kyushu)	Japan	Study of dynamics of geospace plasma changes during magnetic storms and auroral substorms, the electro-magnetic response of iono-magnetosphere to various solar wind changes, and the penetration and propagation mechanisms of DP2-ULF range disturbances
Muon Detector Network	K. Munakata (Shinsu U)	Japan	To identify the precursory decrease of cosmic ray intensity that takes place more than one day prior to the Earth-arrival of shock driven by an interplanetary coronal mass ejection
Remote Equatorial Nighttime Observatory for Ionospheric Regions (RENOIR)	J. Makela (U. Illinois)	USA	Study the equatorial/low-latitude ionosphere/thermosphere system, its response to storms, and the irregularities that can be present on a daily basis.
Scientillation Network Decision Aid (SCINDA)	K. Groves (ARFL)	USA	Study equatorial ionospheric disturbances to aid in the specification and prediction of communications degradation due to ionospheric scintillation in the earth's equatorial region
South Atlantic Very Low frequency Network (SAVNET)	J.-P. Raulin (U Presbiteriana)	Brazil	Study of the South American magnetic anomaly region at low ionospheric altitudes and its structure and dynamics during geomagnetic perturbations
Space Environment Viewing and Analysis Network (SEVAN)	A. Chillingarian (Aragats)	Armenia	To improve short and long-term forecasts of dangerous consequences of space storms

actively seek to identify additional instruments, and instrument providers that could benefit from the ISWI process, as well as new instrument hosts.

28.2.2 Data Analysis

The ISWI program will promote the coordination of data products in a form useful for input into physical models of heliospheric processes. These data will be used for both retrospective analysis aimed at physical understanding of space weather, and for predictive models to predict future space weather conditions.

To be useful for space weather prediction, data must be available in near real-time. However, today internet connections are intermittent or slow in many locations in the developing world, making near real-time data return impossible. Eventually, as internet connectivity improves, these data will be made available in near real-time in a form where they can be ingested into predictive models. In the near term, other strategies like data transfer during selected time periods, or on recorded media like DVDs and tapes are adequate for the retrospective scientific studies of space weather events, and the development of physical models.

Data from the instrument arrays will be deposited in publicly available archives. For the most part, these will be existing data archives, like the virtual observatory systems which are currently under development, similar to the Virtual Heliospheric Observatory (http://vho.nasa.gov/) or the Virtual Solar Observatory (http://sdac.virtualsolar.org/cgi/search). This will make data from ISWI instruments available to the broader community of researchers.

28.2.3 Training, Eduction, and Outreach

During the IHY space science schools in US, China, India, Brazil, and Nigeria provided training to hundreds of graduate students and new researchers. The ISWI will continue to provide support for space science schools. The ISWI will continue to promote space science and the inclusion of space science curricula in universities and graduate schools. This has been most effective when combined with the installation of instrumentation at the university.

The ISWI will continue to support public outreach projects. It is essential to communicate the excitement, the beauty, and the relevance of our science to scientists from other disciplines, and to the public at large. We will continue to develop public outreach materials unique to the ISWI, and coordinate the distribution these materials through individual contacts and outreach workshops.

28.3 Collaboration with Other Programs

The ISWI will continue to collaborate with other programs like SCOSTEP/CAWSES, ILWS, ICTP schools, scientific organizations, and funding agencies. Through these collaborations the ISWI will maximize the return from its programs and avoid duplication of effort.

28.4 Summary and Conclusions

The ISWI will continue a portion of the IHY program, providing a forum for the formation of scientific collaborations between instrument providers and instrument hosts. Initially data will be used primarily for understanding the physical processes important for space weather phenomena. Later, ISWI will move toward near real-time data availability as internet connectivity improves, allowing data ingest predictive modeling. A robust program of outreach is envisioned, with a continuation of the space science schools, support for university space science curricula, and a public outreach program.

Reference

Thompson BT, Davila JM, Gopalswamy N (2009) Putting the 'I' in IHY. Springer, Heidelberg

Index

M.P. Miralles, J. Sánchez Almeida (eds.), *The Sun, the Solar Wind, and the Heliosphere*, IAGA Special Sopron Book Series 4, DOI 10.1007/978-90-481-9787-3, © Springer Science+Business Media B.V. 2011

MIX
Papier aus verantwortungsvollen Quellen
Paper from responsible sources
FSC® C105338

If you have any concerns about our products,
you can contact us on
ProductSafety@springernature.com

In case Publisher is established outside the EU,
the EU authorized representative is:
Springer Nature Customer Service Center GmbH
Europaplatz 3, 69115 Heidelberg, Germany

Printed by Libri Plureos GmbH
in Hamburg, Germany